## Animal Teeth and Human Tools

A Taphonomic Odyssey in Ice Age Siberia

The culmination of more than a decade of fieldwork and related study, this unique book uses analyses of perimortem taphonomy in Ice Age Siberia to propose a new hypothesis for the peopling of the New World.

The authors present evidence based on examinations of more than 9000 pieces of human- and carnivore-damaged bone from 28 late Pleistocene and two special Holocene archaeological and paleontological sites, including cave and open locations, which span more than 2000 miles, from the Ob River in the West to the Sea of Japan in the East. The observed bone damage signatures suggest that the conventional prehistory of Siberia needs revision and, in particular, that cave hyenas had a significant influence on the lives of Ice Age Siberians.

The findings are supported by more than 250 photographs, which illustrate the bone damage described and provide a valuable insight into the context and landscape of the fieldwork for those unfamiliar with Siberia.

**Christy G. Turner II** is Regents' Professor Emeritus of the School of Human Evolution and Social Change, Arizona State University. He is internationally recognized for his work on human dentition and, more recently, for his taphonomic studies of cannibalism in the American Southwest.

**Nicolai D. Ovodov** is Chief Research Collaborator at the Institute of Archaeology and Ethnography, Novosibirsk, Russia. He is well known in Russia for his important contributions to Siberian paleontology and paleoanthropology.

**Olga V. Pavlova** was a translator with the Russian Academy of Sciences for more than 30 years in both the Institute of Geology and Geophysics and the Institute of Archaeology and Ethnography.

# Animal Teeth and Human Tools

## A Taphonomic Odyssey in Ice Age Siberia

CHRISTY G. TURNER II
*Arizona State University*

NICOLAI D. OVODOV
*Institute of Archaeology and Ethnography, Novosibirsk*

OLGA V. PAVLOVA

CAMBRIDGE
UNIVERSITY PRESS

CAMBRIDGE UNIVERSITY PRESS
Cambridge, New York, Melbourne, Madrid, Cape Town,
Singapore, São Paulo, Delhi, Mexico City

Cambridge University Press
The Edinburgh Building, Cambridge CB2 8RU, UK

Published in the United States of America by Cambridge University Press, New York

www.cambridge.org
Information on this title: www.cambridge.org/9781107030299



First published 2013

Printed and bound in the United Kingdom by Bell and Bain Ltd

*A catalogue record for this publication is available from the British Library*

*Library of Congress Cataloguing in Publication data*
Turner, Christy G.
Animal teeth and human tools: a taphonomic odyssey in ice age Siberia / Christy G. Turner II,
Nicolai D. Ovodov, Olga V. Pavlova.
pages cm
Includes bibliographical references and index.
ISBN 978-1-107-03029-9
1. Tools, Prehistoric – Russia (Federation) – Siberia 2. Teeth, Fossil – Russia
(Federation) – Siberia 3. Excavations (Archaeology) – Russia
(Federation) – Siberia 4. Paleontology – Russia (Federation) – Siberia 5. Siberia
(Russia) – Antiquities. I. Ovodov, Nicolai D. II. Pavlova, Olga V., Turner II, Christy G. III. Title.
GN855.R9T87 2013
947′.01–dc23

2012040134

ISBN 978-1-107-03029-9 Hardback

# Contents

# Acknowledgments

This project eventually came about after meeting Sergey Arutyunov in 1979 at the Pacific Science Conference held in Khabarovsk. It was he who explained to the senior author how one goes about getting permission to carry out research in Russia. Financial aid came from the National Geographic Society (grant #6454-99); the Wenner Gren Foundation for Anthropological Research (grant #6588); the senior author's Arizona State University Regents' Professorship research account; the Institute of Archaeology and Ethnography, Academgorodok; the Archaeology Laboratory of Krasnoyarsk Pedagogical University, and ASU Emeritus College grant. Personal help came from the following individuals:

## Irkutsk

German I. Medvedev (Director, Laboratory of Archaeology, Irkutsk State University): collections access and visits to Mal'ta, other nearby sites and to Lake Baikal and the Shamenka site. Ekaterina A. Lipnina: discussions about Mal'ta. Y. M. Ineshin: access to Bolshoi Yakor collection. P. E. Shmygun: information about Kurla I. Yuri A. Mochanov (Department of Archaeology and Human Paleoecology, Academy of Sciences of the Sakha Republic, Yakutsk): discussions and examination of the Diring-Yuryakh collections in Irkutsk.

## Kiev (Ukraine)

Yuri Kukharchuk (Institute of Archaeology): collections access. Vadim Stepanchak: collections access. Victor Karchenko (Institute of Archaeology at the Museum of Natural Science): review of exhibits and examination of hyena remains.

## Krasnoyarsk

Nicolai I. Drozdov (President, Krasnoyarsk Pedagogical University; Director, Laboratory of Archaeology): trips to the Yenisei River Afontova Gora site, and sites at Kurtak field camp, collections access, lodging, explanations about the Pleistocene Kurtak complex. Eugene V. Artemiev: collections access. Nicolai I. Martynovich

(Krasnoyarsk Regional Museum): assistance in excavation at Dvuglaska Cave and visit to Razboinich'ya Cave paleontological site; helped sort collections from Bolshoi Yakor, excavated by Yevgeny M. Ineshin. Lena Popkova: compilation of part of Appendix A.

## Moscow

Alexander K. Agadjanyan (Institute of Paleontology): collections access, logistics. Natalia B. Leonova (Moscow State University): collections access.

## Novosibirsk

Anatoly P. Derevianko (Director of the Institute of Archaeology and Ethnography): formal invitations to carry out research in Siberia; all Altai field trips, meeting invitations, collections access, permission to conduct excavation at Dvuglaska Cave. Anatoly I. Kurbatov (Deputy Director, Business Affairs). Vyacheslav I. Molodin: Denisova Cave and Kaminnaya Cave site visits. Alexander V. Postnov: Ust-Kan Cave site visits and collections access. Sergei V. Markin: Okladninov Cave and Kaminnaya Cave site visits. Sergei K. Vasiliev: species identifications and Ob River Krasny Yar paleontological site visit. Elena Yu. Pankeyeva: references and editorial advice. Olga Volkova-Kozintseva: references. Nonna M. Shakhmatova: references. Tatiana A. Chickisheva: collections access. Nelly A. Kuzema and Valery V. Pazelsky: Russian official government affairs. Irina I. Kedrova: museum access.

## St. Petersburg

Alexander G. Kozintsev (Senior Researcher, Museum of Anthropology and Ethnography): references and discussions about human evolution in Siberia.

## Tomsk

Sergei V. Lechshinskiy (Tomsk State University): collections access and site visit to Tomsk mammoth locality.

## Ulan-Ude

Ludmila V. Lbova (Institute of Mongolian, Buddist and Tibetan Studies): visit to Kamenka, Varvarina Gora, other sites, collections access and lodging arrangements.

## United Kingdom

We would like to express our sincere appreciation to the several individuals at Cambridge University Press who helped to see this volume through to completion: Martin Griffiths, Edward Bailey, Lynette Talbot, Gary Smith, Chris Miller, and Beata Mako.

## United States

Korri Dee Turner (Bioarch, LLC, Arizona): copy editing, word-processing assistance. Gary Haynes (University of Nevada, Reno): references and access to African taphonomy collection. John Hoffecker (Alpine and Arctic Research Center, University of Colorado, Boulder): references. R. Dale Guthrie (University of Alaska, Fairbans): references. Kali T. Holtschlag (Adams Ranch-Texas Canyon, Arizona): dead bull photographs. Mathew J. Betz (Arizona State University, Emeritus College): living hyena photographs. Gerald A. Bair (retired, Bureau of Indian Affairs): editorial advice. Carole A. Travis-Henikoff (author): professional contacts. Alan Mann (University of Pennsylvania): French hyenas. Susan Anton (Rutgers University): hyena advice. Gary J. Galbreath (Northwestern University): hyena information. Esther Jacobson-Tepher (University of Oregon): Mongolian–Siberian rock art information.

## Vladivostok

Alexander N. Popov (Director, Anthropology Museum, Far Eastern State University): collections access. Sergei Trofimov: logistics. Lena Trofimova: logistics. Yaroslav V. Kuzmin (Pacific Institute of Geography [now with Institute of Geophysics, Novosibirsk]): carbon-14 dating information.

There are dozens of other Russians, whose names are unknown or not remembered, but without their help we could never have carried out this study. They range from the charming female conductors of the Trans-Siberian Railroad, shop keepers, maids, students, waitresses, dacha vegetable vendors, minibus and taxi drivers, and on and on. *Spasibo!*

# Note on photograph identifications

All images are cataloged using a coded system; for example: CGT neg. IHPP 2-15-84:21. The first letters in capitals are the initials of the photographer. The second term – neg. or color – indicates a black-on-white negative or an original Kodachrome slide. The next letters, either in capitals or a geographic name, indicate where the photograph was taken. The next three numbers indicate date as month–day–year. The last number is the frame exposed. Thus, the above example means: Christy G. Turner, black-on-white, Institute of History, Philosophy and Philology, February 15, 1984: frame 21.

Initials of institutions, mostly units of the Russian Academy of Sciences, refer to:

FESU, Far East State University, Vladivostok
IAE, Institute of Archaeology and Ethnography (formerly IHPP)
IEL, Institute of Ethnography, Leningrad
IHPP, Institute of History, Philosophy and Philology, Academgorodok, Novosibirsk
KSPU, Krasnoyarsk State Pedagogical University, Krasnoyarsk
LPR, Laboratory of Plastic Reconstruction, Moscow
MBSC, Mongolian and Buriat Scientific Center, Ulan-Ude
PIPM, Paleontology Institute, Paleontology Museum, Moscow
TU, Tomsk University, Tomsk
"Odyssey" refers to illustrations that help establish context, locality, people, travel, and related considerations.

# 1 What is perimortem taphonomy, and why study it in Siberia?

Landscapes have the power to teach, if you query them carefully. And remote landscapes teach the rarest, quietest lessons.

(David Quammen 2003:9)

## Introduction

### Preliminary remarks

Our landscape, in Quammen's sense, is the scene of Ice Age Siberia. Remote in both the realms of time and space, Siberia is an area larger than the United States or Australia. Our curiosity takes us to open and cave sites east and south of Lake Baikal, along the Yenisei and Ob rivers, and into limestone karst caves located in the Altai Mountains near Mongolia. In this territory, marvelously well-preserved faunal remains have been found by Russian archaeologists, paleontologists, and speleologists. We focus on the human and non-human inhabitants of 30 sites. In particular, we are interested in the perimortem taphonomy of late Pleistocene humans and cave hyenas (*Crocuta speiaea*), apparently a sub-species of the African spotted hyena.

### Analogy and taphonomy

"The taphonomic approach is most creative when ancient archaeological patterns are analyzed and interpreted in relation to modern analogs in which relevant cause and effect relationships are known from direct observations" (Bunn 1991:435).

The archaeological record of Ice Age northern hunters, fishermen, and gatherers is almost invisible when compared with, say, the Siberian ethnographic items illustrated by Fitzhugh and Crowell (1988). Our sites contain no perishable items. Artifacts are almost all made of stone. Nevertheless, we believe that we have learned some rare and quiet lessons during our study of ancient bone damage.

This book contains two linked stories. The main story deals with *perimortem taphonomy*, the study of damage done to bones at or near the time of death of the creatures in question. Illustrations of damage types and variations on this story are in Chapters 2–4.

Most of the assemblages we examined are of late Pleistocene age. The main exception is a Pacific coastal Holocene site that was added so that our study could include an approximation of the Siberian Far East oceanic coastal habitat that was flooded due to rising sea levels at the end of the Pleistocene.

The second – and much shorter – story is illustrated in various figures and told in their captions, which include the activities that led up to our collaborative *personal odyssey* of travel and study from 1998 to 2006, the years we collected our taphonomic observations (Figs. 1.1–1.25 and elsewhere). This story hopefully will help readers unfamiliar with Siberia to better understand the context and landscape within which our work was carried out.

## How this study began

The senior author first met the other team members of this project, Nicolai D. Ovodov and Olga V. Pavlova, on a very cold and gray Siberian day in January 1984. Accompanied by his youngest and anthropologically trained daughter, Korri Dee, we had traveled to Novosibirsk to examine prehistoric human dental remains in the collections of the Institute of Archaeology and Ethnography (formerly the IHPP), Academgorodok (which means "Academy City"). It is called so because this unique Russian government enterprise has thousands of scientists studying the natural and human cultural world of Siberia in numerous specialized institutes. The dental studies, mentioned above, were permitted by Institute Director and paleoarchaeologist, Academician Anatoly P. Derevianko, seeking to add information for a long-term study on the origin and dispersal of anatomically modern humans, particularly Native Americans (Turner 1992a, 1998, and elsewhere). Olga Pavlova had been assigned to help us with translations. She was a soft-spoken, petite woman with marvelous skills in English and Russian languages.

Our first meeting with Nicolai was in his cold little attic-level fourth-floor lab, which was down a long, dark, unheated hallway from the lab space provided for our three-week study of Siberian Holocene human teeth. Nicolai was a friendly, hospitable person, then 45 years old. He had no pretensions, and was highly knowledgeable about mammalian osteology and the Pleistocene fauna whose remains filled up all the shelves in his cramped lab. On the highest shelf sat a silver samovar and an Ice Age wooly rhinoceros skull – symbols of where much of our future team's remaining careers would be focused. Most of the faunal remains collected during 40 years of archaeological and related paleontological research were kept in a large attic storeroom, collections that began to accumulate with the many excavations led by the Institute's first director, Academician A. P. Okladnikov.

## The landscape

Most mammalian life plays out on the ground surface of the planet, and its non-uniform landscape selects for organic variation best suited for specific conditions. The landscape of Siberia can be roughly divided into three great geographic provinces: a vast western plain, possibly the largest plain in the world; a rumpled mineral-rich hilly and

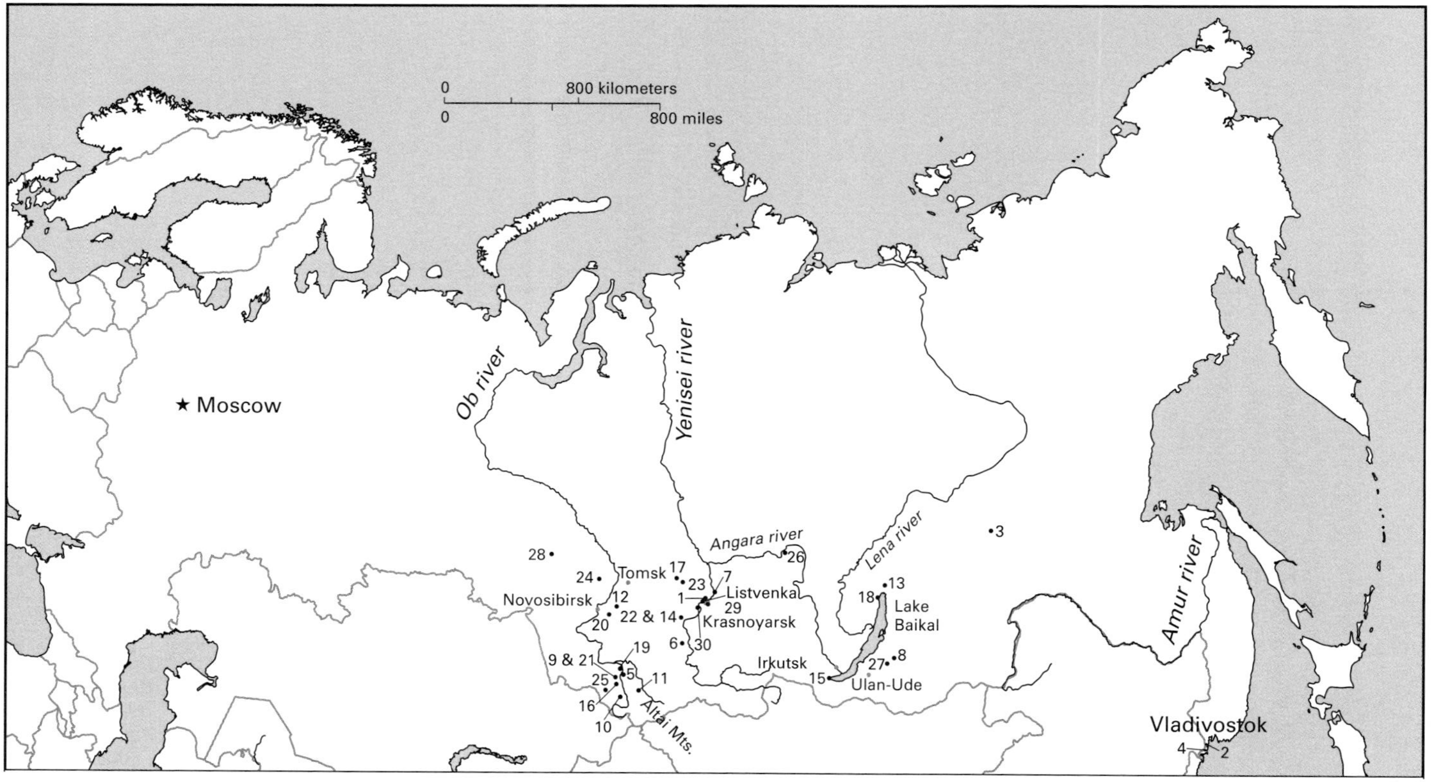


1. Afontova Gora
2. Boisman II
3. Bolshoi Yakor I
4. Borabashevskaya
5. Denisova Cave
6. Dvuglaska Cave
7. Gosudarev Log I
8. Kamenka
9. Kaminnaya Cave
10. Kara-Bom
11. Kyrkalinskaya Cave
12. Krasny Yar
13. Kurla I
14. Malaya Seeya
15. Mal'ta
16. Maly Yaloman Cave
17. Mokhovo Mine
18. Nizhneudinskaya Cave
19. Okladnikov Cave
20. Proskuryakova Cave
21. Razboinich'ya Cave
22. Sarala Cave
23. Shestakovo
24. Straschnaya Cave
25. Ust-Kan Cave
26. Ust-Kova
27. Varvarina Gora
28. Volchiya Griva
29. Yelenev Cave
30. Zhemchuzhnaya Cave

**Fig. 1.1** Map of Siberian site locations.

**Fig. 1.2** Odyssey. Khabarovsk mammoth. A life-size replica of a Pleistocene wooly mammoth, specially erected outside the Sport and Science Museum, Amur Blvd., Khabarovsk, symbolized the unique Siberian venue for the 1979 Pacific Science Conference. It was at this conference that the senior author met Russian researchers who would in time come to be close friends. Although he and the junior author, Olga Pavlova, did not actually meet face-to-face, years later, when looking back at photos he had taken at the conference, he discovered that he had taken available-light color photos of Ruslan Vasilievsky and later, Robert Ackerman, with Pavlova translating for both presenters in a large lecture hall. Pavlova recalls Turner's presentation and the laudatory remarks made afterwards by Russia's most brilliant anthropologist and a speaker of nine languages, Sergey Arutyunov, chairman of the Caucasus section, Institute of Ethnography, Moscow.

This gathering of Pacific Rim scientists was at a tense time between the Soviet Union and the West. Nevertheless, the conference venue, far from Moscow, made it possible for Soviet and Western scholars to exchange information and ideas more-or-less openly, particularly about future research collaboration. Hence, the senior author made plans for future research in the Soviet Union with the help of Sergey Arutyunov, Ruslan Vasilievsky, and apparently with the approval of Academician A. P. Okladnikov, who attended the conference aided by his archaeologist-translator, Alexander Konopatsky. Okladnikov would months later send his famous book on the prehistoric art of the Amur to the senior author (CGT color Khabarovsk 8-23-79:3).

mountainous region in the eastern half; and a high mountainous southern region bordering on Mongolia, China, and Kazakhstan. Siberia possesses some of the world's largest rivers, notably the Ob in the west, and the Yenisei, Lena, and Amur in the east. Into these rivers, and others like them, drain hundreds of smaller rivers. Most of the

**Fig. 1.3** Odyssey. February in Novosibirsk. The two women who helped get the work started for this book. The senior author began data collection in Siberia in February 1984, aided by his daughter Korri Dee, shown on the left with Olga Pavlova. The temperature was −29°C (−20.2°F) at the time the photograph was taken. The pair are standing at the entrance to the Institute of Archaeology and Ethnology (then called the Institute of History, Philosophy and Philology). The flags barely visible in the upper right corner were commemorating the death of Yuri Andropov. Korri and the senior author learned of his death 48 hours before it was announced to the Soviet public because they had heard the news in a BBC broadcast they picked up on the small shortwave radio they carried with them to Siberia. It was during this visit that we were shown faunal specimens in Ovodov's lab and that the seed was planted for research that would culminate in this book (CGT color IHPP 2-14-84:19).

**Fig. 1.4** Odyssey. A private dinner. A Friday evening dinner party at the Academgorodok apartment of Academician Anatoly P. Derevianko (right) and his then wife, physical anthropologist Tatiana Chickisheva (second from left). Archaeologist Alexander Konopatsky (center) served as translator for the gathering. The senior author's late wife, Jacqueline, is at the left. It was this gathering that led to the Turners' trip to the Altai and, subsequently, the clandestine Moscow examination of Mousterian teeth found in two of the Altai cave sites (at that time it was illegal for foreigners to study and publish on Soviet materials before Russian scholars) (CGT color Academgorodok 5-29-87:7).

greatest rivers originate in the general region of the third and mountainous area, which we will refer to hereafter as simply the Altai. With the exception of the Amur and other smaller eastern rivers, the great rivers of Siberia flow northward, providing natural but long looping routes to the Arctic seas. In the spring, summer, and fall these waterways can be followed by boats, and in the winter by foot, sled, or pack animals on the thickly frozen surfaces. Lakes abound in Siberia, the greatest of all being Lake Baikal, whose single outlet, the Angara River, runs through the old Siberian city of Irkutsk founded by Russian *promyshleniki* (fur hunters) in 1651.

Today, the vegetative zones in the area encompassed by our study in southern Siberia, extending from the Ob River in the west to Primorsky Territory along the Sea of Japan in the east, are characterized as forest, steppe, meadow, alpine tundra, or varying combinations of these plant communities. In mixed zones, the first word in a hyphenated combination indicates the dominant vegetation. For example, if there is more steppe than forest, then the region would be termed steppe-forest. Steppe includes drought-resistant

**Fig. 1.5** Odyssey. Flight into the Altai. Jacqueline Turner (right), Olga Pavlova, the senior author, and about 20 other passengers flew from the old mining city of Barnaul to the Altai farming village and "county administrative center" of Soloneshnaya in the single-engine biplane visible in the background. These aircraft were known to Russians as "corn" planes because they could land almost anywhere. From there, a bottomless muddy mountain road was cautiously navigated by the driver, Volodya Tekunov (left), all afternoon to the archaeological field camp at Denisova Cave (CGT color Soloneshnaya 6-9-87:21).

plants and small animals, even lizards, and vast tracts of tough, drought-resistant grass. The steppe is a key plant zone for the initial colonization of the New World. Pollen, soil, and faunal studies indicate that, other than temperature and seasonal duration, the late Pleistocene environmental influence on humans in southern Siberia was not vastly different from what it is today. Forested regions in the past, as now, tended to be higher, colder, better watered, and more mountainous than the steppe country. Drainage affected the type of forest vegetation. Over the course of many millennia in the late Pleistocene, climate fluctuated so that today's plant zones shifted slightly in latitude and elevation in accompaniment with changes in temperature, albedo (ground surface reflectivity), and precipitation. Vast steppe grasslands and steppe-forest existed and supported herds of bison, horses, and other hoofed animals. Skeletal remains of various deer and other browsers, along with habitat-specific rodents and insects, indicate the existence of the forest condition, as do the occasional bones of beaver and other species dependent on a forested habitat. Small animal bones in owl pellets provide a remarkable source of information about the habitat surrounding ancient nesting sites. The relatively large number of remains of top-feeders such as cave hyenas, cave lions, wolves, and other

**Fig. 1.6** Odyssey. On the way to Denisova Cave. A photo stop in the forest-covered steep limestone Altai Mountains. We traveled along a winding route from Soloneshnaya to Denisova camp. The route paralleled the rushing Annui River, and passed through two small, poor villages settled in small relatively open mountain valleys. Livelihood was based mainly on milk production. Not far from this location, in later years a very deep lower Middle Paleolithic open site called Kamara was discovered. Left to right: Jacqueline Turner, Vyacheslav Ivanovich Molodin, Olga Pavlova, and driver (CGT color Altai 6-9-87:22).

large carnivorous predators speak to the large number of prey upon which they fed. Exactly why many of the megafauna, including mammoth and rhinoceros, went extinct at the end of the Pleistocene remains an unresolved question in Ice Age biological history. During the course of our multi-year study, we thought possibly that we could help answer the megafauna extinction question, and to some extent we may have. We have assembled bits and pieces of diverse information that hint at a significant human contribution, even though climate and habitat changes patently occurred in Siberia as elsewhere in the world. English-speaking readers wanting more details about the late Pleistocene northland landscape will find R. D. Guthrie (1990), J. F. Hoffecker (2005), N. Ukralutseva, L. Agenbroad, and J. Mead (1996), H. K. Vereschagin and G. F. Baryshinkov (1984), and F. H. West (1996) helpful.

## Studying perimortem taphonomy

The concept of, and term, *taphonomy* were coined by the Russian paleontologist, Ivan Antonovich Yefremov (1940), whose base of operations was the Institute of

**Fig. 1.7** Odyssey. Native-style residence. Shown is the entrance to the Karakol village property of a retired teacher, Galina Erosiva (right). At left is Olga Pavlova, and in the background is a developmentally disabled villager who helps Erosiva and others with simple tasks such as hauling water. This home is like the strictly Russian village homes and out-buildings, but differs in also having a round, cribbed, hogan-like native Altai structure. Out of sight in the distant mountains is the Paleolithic Kaminnaya Cave site (CGT color Karakol 7-10-99:34).

Paleontology in outer Moscow (Figs. 1.22–1.24). He defined taphonomy to mean the study of the "burial" of plants and animals, their geological placement, and all agencies, associations, changes, and damage to the organisms that were involved in the "processes of embedding" into the lithosphere. Although Yefremov did not specifically mention humans as an agency of embedding (he was clearly concerned with very ancient life), it is clear that he meant *all* possible agencies. One of the near-universal characteristics of human culture is the burial of our dead, at least family and friends. This alone makes us an element in his definition of taphonomy.

While some workers look upon taphonomy as little more than a destructive or information-destroying process, we believe that there is often much useful information about how a bone or shell accumulation and its damage occurred, as well as a number of related considerations. We are far from alone in this view. For example, see Brain (1981), Lyman (1994), Grupe (2007), and many others we will cite in the course of this book.

**Fig. 1.8** Odyssey. A Karakol village round house. Galina Erosiva's windowless log round house (*aiyel*) is 40 years old. It stands in the 200-person village (500 before World War II) whose cash economy is based mainly on raising wild deer for antler export to China. Both Altai and Navajo cribbed round houses of the American Southwest face eastward. An Altai native man must build an aiyel for his bride-to-be before the marriage can take place. He will also construct a rectangular Russian-style windowed log house (CGT neg. Karakol 7-10-99:00).

The boundary between destructive taphonomy and damage-producing information such as tooth marks is not easy to define, especially in deep time (Plummer 2004). It is our explicit theoretical view that all damage is information. The problem is determining what the damage means. This is especially true when working with museum collections that generally have little contextual information.

We chose Siberia to carry out this research for both scientific and personal reasons. One of the scientific reasons was the hope that, in Quammen's sense, we might help understand why the colonization of the New World from Siberia was so late in relation to the much earlier colonization of most of the rest of the world. Another was curiosity about bone damage that was ignited by the senior author's research on prehistoric cannibalism in the American Southwest and Mesoamerica. The personal reasons were many, not the least of which was the feeling of true adventure wherever we traveled, with every assemblage we studied, and with everyone whom we met in the field, in museums and institutes, and at conferences (Fig. 1.25). The following allegorical story abbreviates the scientific evidence and inferences in this book.

Fig. 1.9 Odyssey. In an Altai village yard. After a thunderous afternoon rain and hail storm, Nicolai Ovodov (left), Sergei V. Markin (middle), Olga Pavlova (right), a sow, chickens, ducks, and dogs emerge from cover (CGT neg. Karakol 7-10-99:24).

## An imaginary tale of winter death

Falling snow flakes and steamy breath from frost-rimmed nostrils blurred the view of Old Long Clitoris. She was so named because of the unusual size of what is normal anatomy of female hyenas, often and falsely believed to be hermaphroditic. She was trying to see into the short twisting canyon wherein a small group of humans huddled in a gray limestone cave whose wide opening was level with the snow-covered canyon floor. The little canyon stream was frozen solid. Providing some protection from the icy afternoon wind, the cave entrance faced out upon the frozen treeless steppe that stretched far beyond the distant horizon. In thousands of years to come, this vast land would be called Siberia. Now, it had no general name. Only specific places had names, like the cave, which was known as "two-eyed" cave, because through the darkened ceiling light entered from two round openings, giving the feel of a monstrous, watching cave lion. Aside from a few miniature and deformed willow bushes rooted to the canyon walls, the only other fuel for a warming fire was a copse of dwarf birch trees rooted in the canyon's thinly soiled floor. The humans knew the canyon was a dangerous place since there was no way to escape should any predators enter the canyon mouth. But what choice did they have in their storm-stalled southward trek – stay in this

**Fig. 1.10** Odyssey. Afternoon dinner. We were guests of Galina Erosiva who, helped by her daughters (right), put on a table full of traditional dishes and drink, including tea and lots of home-made vodka (*arachka*) distilled from fermented cow's milk (the five-gallon stainless steel milk can in the lower right was full of freshly made vodka).The senior author noted: "We had [a] lavish meal of potato and meat soup, sour cream, fresh bread (Altaic and Russian styles), two kinds of home-made jam, butter, tea with milk, milk vodka, egg salad, a cottage cheese and onion salad, buckwheat, and candy." The aiyel had a central fire hearth on the earth floor directly below an open smoke hole in the center of the roof. Smoke from the smoldering wood fire, the smell of roasting meat, the yeasty aroma of fresh-baked bread, and other smells reminded the senior author of his earlier anthropological days on the Navajo and Hopi Indian reservations of northern Arizona. Substituting animal hides for the colorful rugs hanging on the walls, and removing the modern utensils and furniture, one can easily imagine aiyel-like structures as having been the winter residences of the Paleolithic Altai people. Wind-proof, easily heated, and erected near firewood sources, aiyel-like structures would have been much better winter residences than the cold drafty archaeological cave sites. While the cave sites are not hard to find, open habitation and camp sites like Mal'ta and Afontova Gora depend on considerable luck to locate. Left to right: Nicolai Ovodov, Olga Pavlova, Sergei Markin, Galina Erosiva, and her family (CGT neg. Karakol 7-10-99:24).

bone-littered shelter that had been occupied by carnivores, or freeze to death like the stiff gazelle carcass that the people had stumbled across on the snow-covered steppe not far from the mouth of the canyon. Ancient habits had taught them to eat meat raw, or briefly cooked as soup warmed with heated stones dropped into a skin bag containing meat, crushed bone, melted snow, and pieces of aromatic plants. Earth ovens worked only when the ground was not frozen. Either way, the food was more nutritious than when roasted, despite the mouth-watering smell of wood smoke and fatty, roasting

**Fig. 1.11** Odyssey. A camping tent. Shown is a lower Ob River Sel'kup bark tent. Light and easily transported, this type of structure was favored by fishermen, hunters, and trappers. Constructions like this might have been erected in the Altai cave sites, but unquestionable tent rings and post holes are unknown. Korri Dee Turner (right) (CGT color IHPP 2-13-84:21).

meat. Besides, the humans dared not gather branches from the birch thicket this late in the day because they feared that some carnivore might be hiding there.

As night came on, a shift in the wind brought whiffs of animal smell from the little forest. Hyena stink! Many! The humans knew that 50 (counted by the fingers of five people) or more adult animals were common in a hyena pack, many more than the number of the human band. Hyenas were not only ferocious killers, they ate anything – including humans – live or long dead. When hungry they were the most dangerous of the big carnivores because they hunted in packs of many individuals, unlike the more solitary bears and lions. Even wolf packs avoided confrontations with hyenas because the hyena's powerful jaws could crush a leg or head in an instant. Hyenas were cruel killers and never revered as were the cave lion and other Ice Age predators. The group's powerful shaman, an old deformed man given to fits, fondling the group's young boys, and unerringly able to predict future events, said hyenas were evil. Even his guardian spirit, a tiny but frightful vole, had no power to combat them.

Despite her age and many battle scars, Old Long Clitoris had perfect unbroken and unworn teeth, her reward as matriarch to command the best soft meat and internal organs from a kill. Low-ranking family members had chipped, cracked, and worn

**Fig. 1.12** Odyssey. A modern hunter. Professional hunter Demitry Medvedev, son of German Medvedev (Fig. 1.17), shows Olga Pavlova (center) and Nicolai Ovodov (right) two of his shotguns. Medvedev's mother is at far right. The photograph was taken in Irkutsk after a day traveling by fishing boat along the southwest coast of Lake Baikal. In his youth Ovodov did a lot of hunting. On one outing he killed a large bear with his single-shot rifle – the only kind of firearm a person could possess legally in the former Soviet Union (CGT color Irkutsk 10-5-00:12).

teeth because they had to often eat the less desirable hard bones, horns, and hooves. Frost also covered the sloping, spotted back of Old Long Clitoris, whose shaggy fur rose and fell in the gusty, cold, snow-flecked wind. Nevertheless, her aged hearing relayed to her small but highly intelligent brain the sounds of bones being smashed open by the people, and her starving sense of smell told her that meat and marrow were being eaten, meat and bone fat that she wanted for her own starving body. Old Long Clitoris was not alone in watching the humans. Others of her clan were also crouching in the birch stand and, like her, had not eaten for the many days that the winter storm had made finding food nearly impossible. They were now full-time scavengers, not having seen a living animal in the past 15 days, except for occasional watching ravens. What made matters even worse, the humans were huddled in the sheltering cave that she and her ancestors had often used in stormy emergencies for tens of thousands of winters.

She had learned through many seasons of experience that there were two kinds of humans, just as there were different kinds of look-alike horned creatures she killed and devoured. The older kind of humans were dangerous, but not as much as the humans who arrived recently. But cold and scarcity of food were now worse than any time in her many

**Fig. 1.13** Odyssey. Snow leopard pelts. Demitry Medvedev shows Olga Pavlova and Nicolai Ovodov two pelts of the very rare snow leopard. Winter-killed pelts like these would have been valuable to the Paleolithic Siberians for warmth as well as symbolic and status purposes (CGT color Irkutsk 10-5-00:7).

**Fig. 1.14** Odyssey. Archaeological site tour. Travel during our project included the use of various forms of water-craft. In this image, Nicolai Drozdov (left) and Olga Pavlova board a patrol boat near Kurtak, joining others for a day-long tour of Paleolithic archaeological sites discovered and excavated by Drozdov and his team along the upper lake level of the Yenisei River Reservoir. While thousands of Pleistocene animal bones are washing out of the loess shoreline, few have been found in the Yenisei Reservoir archaeological sites. This lack of preservation seemingly adds to the view that some of these sites are very ancient; either that or many of the artifacts have been redeposited. The regional vegetation is steppe and forest parks. CGT notes for the day included: "Day was clear, warm, and very enjoyable. Dinner was beef shishkabob, cucumbers, bread, cabbage, peas, onions and lots of vodka" (CGT color Kurtak 8-7-98:11).

years as the born-to-be matriarch of the family. Sooner or later, as has happened before, a young or elderly human will wander away alone, separated from the safety of its group. Then, she and her fellow hunters could attack and carry the torn body to another shelter elsewhere in the thinly pine-forested limestone foothills. Hunting solitary humans in the winter, when a child wandered away from the protection of its terrifying fires and barking little wolves, became ever more common as the winter cold deepened and lasted longer, season after season. Summers barely began when the first frosts arrived. These earlier, colder, and longer-lasting winters started the southward migrations of the great herds of hoofed creatures, drove underground the many kinds of small, tasty, chisel-toothed species, and brought life to a cold desperate standstill. The world had changed to a terrible, never-ending, lifeless cold that far in the future human scientists would call the Late Glacial Maximum.

**Fig. 1.15** Odyssey. Wild strawberries. Along the tour mentioned in Fig. 1.14, we stopped at a summer camp for young people. Here, Nicolai Drozdov was shown a pail full of small wild strawberries that had been gathered that day. Made into a jam, they are delicious and have an over-powering fruity fragrance. Drozdov is not only a renowned Siberian Paleolithic archaeologist, he is also President of the Krasnoyarsk Pedagogical University (CGT color Kurtak 8-4-98:20).

**Fig. 1.16** Odyssey. Irkutsk station. Ovodov and Pavlova at the Irkutsk train station as snow was falling in early October. We had just completed a lengthy study of some of the Mal'ta faunal remains at the Archaeological Laboratory of Irkutsk State University, and were returning to Novosibirsk (CGT color Irkutsk 10-8-00:29).

**Fig. 1.17** Odyssey. Examining graves. German Medvedev and his wife, Ekaterina A. Lipnina. Steppe grasses are abundant here. Out of sight in the background is a land-locked saline lake 376 m above sea level, only a few kilometers from the Yenisei River, and 272 km south of Krasnoyarsk (CGT color Kurtak 7-28-00:3).

**Fig. 1.18** Odyssey. Ulan-Ude celebration. East of Lake Baikal, in even more severe steppe country, referred to as the Trans-Baikal or Buryatia, the capital of which is Ulan-Ude. The image shows a national Buryat holiday that occurred during our visit. The ceremony was taking place on a Saturday in the city's central square. Costumes have a strong Chinese influence. The Buryat Republic was formed 80 years ago under Soviet direction. The city itself is, however, 337 years old. Here we studied the large faunal collection from Kamenka excavated by Ludmila Lbova (CGT color Ulan-Ude 7-5-03:19).

Hunting the lone victim in overpowering numbers was an ancient strategy of many predators, but her kind excelled by the fact that they hunted as well at night as during the day; they were a well-coordinated hunting team; they had evolved digestive and dental systems that extracted nutrients even from the bones and teeth of their victims. No other predators were this efficient. In this way they could extract nutrients from the bones, hooves, even fur left at kills by lions, wolves, and all other large predators. Unlike the more solitary hunters of the taiga and steppe – the cave lions, cave bears, and tigers, or the easily intimidated wolves, foxes, and raptors – only the new humans who also hunted in packs were near equals. Evolution had crafted both human and hyena species to be highly intelligent, socially organized, and gifted with great endurance. Humans were able to wear down horses, for example, by forcing an animal to walk all day until it collapsed in exhaustion. The shaggy matriarch and her kind had three other advantages. First, their kind had lived in their lands for thousands of generations and knew every detail about the vast landscape and its inhabitants. The new humans knew almost nothing about the region, often making dangerous mistakes like camping in this box canyon.

Fig. 1.19 Odyssey. Attic collection storage area. Olga Pavlova and Nicolai Ovodov search for specific Paleolithic faunal assemblages recovered from excavations by members of the Institute of Archaeology and Ethnography, Novosibirsk. This anthropological component separated from the original IHPP, first directed by Academician A. P. Okladnikov, with scientific assistance from Ovodov and others, has assembled the largest Siberian archaeological, ethnographic, and physical anthropological collection in Siberia, rivaling all the anthropological institutes combined in Moscow and St. Petersburg. Here, the faunal collections are curated in a very large attic space of a huge U-shaped four-story building that houses the IAE and other divisions of the Siberian Branch of the Russian Academy of Sciences. Hundreds of thousands of pieces and whole bones are in this storage area. There is a severe shortage of storage space all over Siberia, as expected where even human housing is at a premium. Most of the collections we studied were curated in hot, freezing, or damp attics or basements and had to be moved to a study area (CGT IAE neg. 7-30-99:13A).

Second, she and her kind would eat humans whenever they needed, alive or long rotted. The new humans had the odd behavior of burying their dead, rarely eating them as was common in hyena packs. Despite burial, the hyena family easily found and dug up the cold, rancid bodies. There has never been news of humans eating their kind. Far in the future, archaeologists would find hundreds of rock and mobile art images of predators and other creatures. Her kind were very rarely depicted. Humans did not admire hyenas. Last, when all food resources were gone, she had less reservation about eating her own kind than did the humans. Both species were cannibalistic when desperate. While sharing many similarities with humans, her kind had superior immune resistance to disease and toxins found in rotted carcasses. While they relished freshly killed fatty game, they could live off almost anything along the scale of organic diagenesis.

**Fig. 1.20** Odyssey. In Ovodov's little IAE laboratory, the senior author looks for bone damage in wooly rhinoceros mandibles from the Ob River paleontological site of Krasny Yar (CGT neg. IAE 6-21-00:32).

**Fig. 1.21** Odyssey. Also in Ovodov's lab is Olga Pavlova, who translated most of the labels and notes associated with each of hundreds of boxes of bones. When time permitted she translated articles describing various sites from which the faunal remains were recovered and other related publications (CGT neg. IAE 6-21-00:35).

**Fig. 1.22** Odyssey. At the Institute of Paleontology, Moscow. The remaining three photographs in this chapter are to commemorate Ivan Antonovich Yefremov, the Russian paleontologist who conceived the idea of taphonomy. It is a Russian tradition to take a photograph at the plaque of a visited institute, here the "Russian Academy of Sciences Institute of Paleontology, Paleontology Museum, after Academician Yefremov." Left to right: large-mammal paleontologist Irene V. Krilova, the senior author, and Alexander Karenovich Agadjanian, a small-mammal paleontologist who arranged for our visit to the museum. Small mammals are very sensitive to temperature, and Agadjanian has developed a marvelous late Pleistocene climate sequence for Denisova Cave based on these tiny creatures (OVP color PIPM 6-8-01:36).

Thus, with these advantages, the cave hyenas were able to keep the new human predators from over-running their territory, ruled for thousands and thousands of generations. But the terrible unending cold now was murderous. Food had become much scarcer, neutering her weaker male companions so that few of her sisters became pregnant. Her kind was vanishing.

The more recent humans thrived in the cold. They were clothed in tightly sown skins from head to foot. The older humans also wore animal skins, but their seams were sown with less precision, allowing the cold winds to suck away precious body heat that could best be replenished by eating fat. Hunting and killing large animals like the wooly mammoth produced the greatest amounts of fat. And watching the recent humans revealed to Old Long Clitoris that these kinds of humans were much more skilled hunters than the old humans. Moreover, the new humans used their small wolves (*volchata*) to help hunt as well as using them to help carry the many kinds of things that the moderns used for their daily living and travel.

**Fig. 1.23** Odyssey. Siberian mammoths. One of the many halls in the Paleontology Museum that range from the earliest fossil forms of life found in ancient Russia to Pleistocene megafauna as illustrated here. Mammoth fossils have been found all over Siberia in both paleontological and archaeological sites. Direct evidence of their having actually been hunted is very scarce (CGT color PIPM 6-8-01:11).

Old Long Clitoris' shriveled and long-empty stomach pained her. Uncontrollably, she began to inch her way toward the cave entrance, getting so close that she could smell individual humans. They smelled like no other creatures because when they were very active their bodies would glisten with unpleasant-smelling moisture. She learned long ago that sweat was the hallmark of human odor, and humans tried to avoid sweating in the winter so that they would not freeze inside their protective clothing. Sweating was something that animals did who migrated from the south, so she thought that these humans originated from somewhere to the south. In fact, there were no humans known in the lands to the north where the sun disappeared for many weeks in the unbearable winter cold. Nor did her kind go to the northern edge of the land where all plant life was dwarfed and grew no higher than her foot.

Her cramping hunger drove her ever closer to the cave. Her movements were watched closely by her family, and they crept out stealthily from the birch stand toward the cave when they sensed from her upright ears that she was going to attack. When they were near, she rose to her full height; with her almost 100 kg weight she charged into the dim cave, followed only seconds later by the others. Instinct and experience told them to attack the smallest and youngest, or the crippled and elderly. Because they

**Fig. 1.24** Odyssey. Exhibit of Yefremov's work. Strangely minimal and inconsistent with his worldwide fame, the inconspicuous exhibit is placed in a poorly lit first-floor corner. The captions read (Left): "Basic zones of hydrodynamic section. [sampling] of the taphocenoses of terrestrial organisms. In the primary sorting, the movement of large creatures will remain in situ. They will not go into the deposit and will be destroyed, small bones of smaller forms will be brought out of the sedimentation area; they will not get into the deposit and will be destroyed as well. The remains of middle-sized forms will predominantly stay. I. A. Yefremov 1950." Center: "The sequence of stages of burial." Right: "All the evidence about life in the past centuries we obtain through the study of fossil remains of animals and plants. Only a small amount of organisms, which lived at one or another time are preserved in the fossilized condition. The regularities (laws) of burying fossil remains in deposits and types of their location are the subject of taphonomic science. The founder of taphonomy is I. A. Yefremov, well known, not only as a paleontologist, but fiction writer, as well." The small caption beneath Yefremov's photograph reads: "Ivan Antonovich Yefremov, Professor, Awarded the Laureate for the Laboratory of Quadrupeds, Director" (CGT color PIPM 6-8-01:31).

outnumbered the humans five to one, they easily grabbed the gazelle torso and pulled down two children by their heads. Amid the chaotic screams, cackling, barking, slashing teeth, and growls, an old woman was knocked to the ground, dropping the small bone tailoring needle she was using to repair a rip in a child's boot. Two male hyenas tore her face off. In seconds, the hyenas dragged two young ones, the old woman, and the frozen gazelle to the darkening birch thicket where the pack tore the children and woman to pieces after they first eviscerated them alive. No humans or little wolves followed.

The hyenas had won another victory. Their feeding was total. No archaeologist in the future would ever learn of this event, nor of many other such events that left no traces of

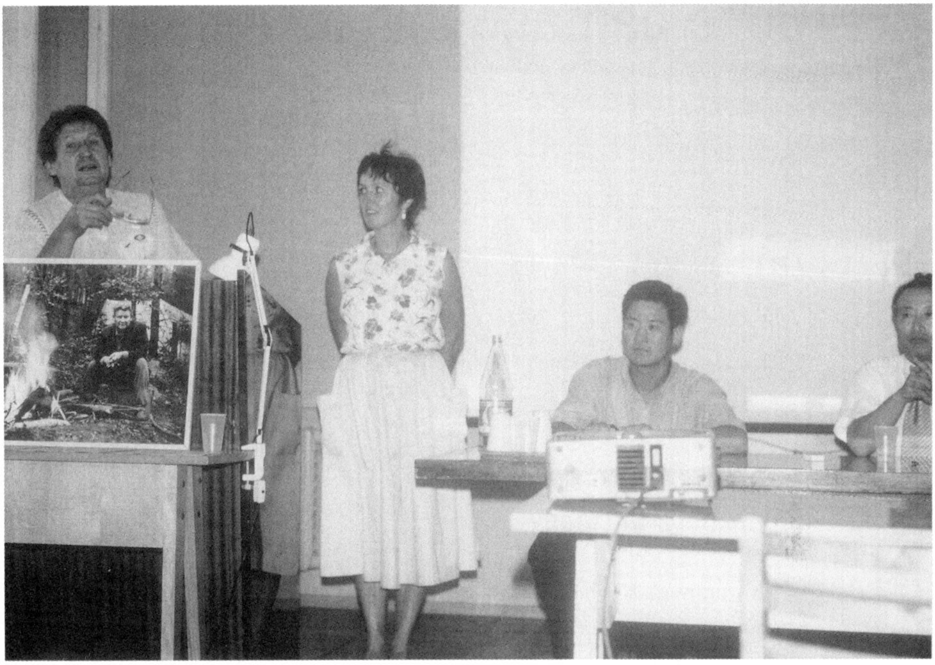

**Fig. 1.25** Odyssey. Reporting scientific findings. IAE Director, Academician Anatoly P. Derevianko (left), presents the wrap-up of a paleoanthropological conference on "Pleistocene Paleoecology at the Kaminnaya Locality, Siberia, and Adjacent Territories," held at the Institute in memory of Academician A. P. Okladnikov, first Director of the Institute (photo on podium). Translating for Derevianko is Elena Y. Pankeyeva (center). Present also are the final discussants, Heon-Jong Lee from Korea (second from right), and Gi-Kill Lee, also from Korea (right). The Siberian archaeologists and related earth scientists are almost compulsive in reporting their findings at these sorts of conferences, which unlike elsewhere are not based on formal membership in a particular scientific organization. As far as the senior author could learn, there are no formal organizations like the American Association of Physical Anthropologists or the Society of American Archaeology anywhere in Russia. Reporting of scientific accomplishments is driven by topical specialists such as Derevianko. We have reported our taphonomic research at a few conferences in Siberia (the senior author speaking, Olga Pavlova translating) (CGT color IAE 7-24-98:10).

the deadly competition between Siberian cave hyenas and humans. But vague clues to these events remained, and it is this book that attempts to tie these clues together into a larger story about Ice Age Siberia.

Next, we discuss the strange life of the author of taphonomy, and define and illustrate the criteria for our taphonomic observations.

# 2 Bone damage and its meaning

Trash and cemeteries are the best source of information.

Gordon Child, *Progress and Archaeology*;
cited by Gennady Markovich Prashkevich, 2002

## Taphonomy: the man

World-renowned Siberian writer, Gennady Markovich Prashkevich, whom we met several times in Academgorodok, the huge "science city" upriver from Novosibirsk, told us about his personal relationships with Ivan Antonovich Yefremov, the Russian paleontologist and writer who coined the term *taphonomy* and developed the theory behind the term.

For us, working on an anthropologically oriented taphonomic project of prehistoric animal remains, it was of special interest to learn about the life, work, and personality of Yefremov, who was a kind of mentor to Prashkevich. Traditional obituary summaries of Yefremov's life lack many of the details that follow. The senior author of this book could not help but think about strange points of intersection between Yefremov and our back-in-time-traveling odyssey into Ice Age Siberia.

One of these intersections was meeting Prashkevich, a famous Russian writer and poet, whose scientific interests include paleontology and archaeology. His relationship with Yefremov began in the 1950s when he wrote to the renowned paleontologist on behalf of himself, then 16 years old, and some Taiga Village school children interested in fossils. Yefremov replied from Moscow on April 23, 1957, explaining in much detail how fossils and the study of paleontology were related, and encouraged the students to seek out the remains of a small parrot-beaked dinosaur known from a fossil locality 150 km from Taiga (about a two-hour drive). A small expedition was formed and travelled to the fossil locality, but no dinosaurs were found because the locality was inundated by the adjacent Kiya River, then in flood stage. Despite this disappointment, Prashkevich continued his correspondence with Yefremov, and in the course of time visited him in Moscow.

In our dinner conversations with Prashkevich and his geophysicist wife, Dr. Lydia Kiseleva, there emerged undertones of a possible duplicitous side to Yefremov's life, possibly even serving as an English or Russian government spy. Moreover, there seems to be some mystery surrounding his death.

On this subject, Prashkevich (2002:n.p.) wrote:

> "What could be in common between the author of the famous 'Robinson Crusoe,' Daniel Defoe, and the great science fiction writer, Ivan Yefremov?" asked the popular [Russian] newspaper *Agrumenty i Fakty*.

And its rhetorical response was: "the former created the English intelligence service, and the latter, probably has been its agent . . . the sudden death of Ivan Yefremov that had happened in an hour after receiving a strange letter from abroad gave some grounds for this version. The letter could have been treated with some special chemicals, inhaling of which could cause death." This element of intrigue should be weighed against the fact that Yefremov was in poor health at the time. Prashkevich included the above in his book, *Novel on Numerous Perfect Things*. Prashkevich also wrote a piece in 2000 simply entitled "Ivan Antonovich Yefremov." In this biographical review, Prashkevich notes that Yefremov was born on April 22, 1908, in the village Vyritsa, near St. Petersburg. He served and engaged in military action with the Red Army from 1919 to 1921. In 1923 he met Professor and Academician P. P. Sushkin, Director of the Reptile Gallery in the Geological Museum (then Leningrad). In 1925, after taking courses in biology, Yefremov became an assistant to Sushkin. His duties included extracting fossils from rock and carrying out fieldwork in search of fossils. He moved up in rank and carried out extensive fieldwork for the rest of his professional career. In 1935 he defended his candidate dissertation and in 1941 defended his doctoral dissertation. He spent several years working on Mongolian fossil reptile deposits, making taphonomic interpretations in the overlap area between biology and geology. His 1935 paper, which spelled out his basic ideas of taphonomy, was entitled "Falling out of transitional forms in the conditions of burial of the oldest quadruped." In this regard, Prashkevich (2000) wrote:

> Taphonomy, as Yefremov called a new branch of science, is the knowledge about regularities of burying organic remains, in other words, the knowledge about those regularities (processes) that promote the transition of organic remains from the biosphere to the lithosphere. The main objective of taphonomy is to work out sufficiently exact concepts, corresponding to reality, a part of the organic world of past geological epoches, that fell out of the geological record, and to define the limits of precision of theoretical constructions in paleontology.

Of special importance to Yefremov were "transitional life forms." He believed that these forms existed only briefly and in limited numbers, explaining their absence or rarity in the geological record. In addition to the fossils themselves, Yefremov felt that the fossil localities were of equal importance as "windows into the past." He offered numerous contextual inferences regarding preservation, with the following being necessary: large numbers of living individuals; total collapse of a rich fauna under unfavorable circumstances; occurrence of specific conditions that concentrated skeletal remains; total fossilization; rapid burial; preservation of bone-bearing deposits in the lithosphere; and their subsequent exposure.

Due to heart disease that began with an intense fever in the early 1940s, Yefremov retired in 1957 from professional paleontological work, for which he received the Russian Academy of Sciences Presidium award for his "Taphonomy and the Geological Record," and the 1952 State Prize. His retirement years were spent writing novels and stories of adventure and science fiction. He died in Moscow on October 5, 1972.

## A few historical accounts of perimortem taphonomy

An early – if not the first – instance of human taphonomy (zoo-archaeology) has been attributed by Simon J. M. Davis (1987:20) to an Englishman named John Frere, who found stone tools in association with bones of enormous size in Suffolk. Frere reported his discovery in the journal *Archaeologia* in 1800, and as expected in the strict biblical interpretative climate of the times, was coolly received. Frere's find was followed by others elsewhere in Europe, so that by the middle of the nineteenth century the antiquity of humans was reasonably inferred to be greater than the 4004 BCE date for the world's beginning, as calculated by Archbishop James Usher. It is these sorts of associations that Yefremov would later be concerned with. How had the human artifacts and extinct animal remains come to be deposited together?

The earliest *classification* of perimortem damage that we have come across in the eighteenth- to twentieth-century literature is a scheme presented by Gabriel and Adrian de Mortilje (1903). They distinguished between the use of metal and stone tools based upon the depth and clarity of cuts on broken animal bones found in Roman period sites vs. those from prehistoric cave- and lake-dwelling sites. Their typology included (1) sawing; (2) chopping (dints or impressions, and deep scratches or eraillures); (3) cuts (dints made by pressing, not percussion); and (4) cuts, usually shallow and not clearly made by any pointed stone tool or blade. We came across the Mortiljes' work several years after we had developed our own damage classification, which includes pseudo-cuts. Much to our surprise, the Mortiljes also had a class they called pseudo human marks, marks left by animal teeth (we independently have a class called pseudo-cuts). They recognized the type of marks left by rodents and those by carnivores. They proposed that only hyena and dog genera do significant bone damage, whereas other carnivores only rarely leave tooth marks on bone.

An early Siberian example of perimortem taphonomy comes from a mammoth bone and human artifact assemblage found in 1896 in the city of Tomsk. N. F. Kaschenko excavated from the upper right bank of the Tom River the remains of a mammoth skeleton that had cut marks and a few associated Upper Paleolithic stone tools. He promptly reported this work in the publications of the Imperial Academy of Sciences of St. Petersburg (Kaschenko 1901). Figs. 2.1–2.4 replicate illustrations in that work. Kaschenko, a professor of the newly opened Tomsk University, considered the assemblage to represent an opportunistic event for the humans, who came across the mammoth, perhaps already dead, and minimally utilized the great carcass. Judging from the illustrations in his report, we find his interpretation to be quite reasonable. Kaschenko did something else of a pioneering nature for Siberia, if not the entire world of prehistoric research. He sealed in glass vials samples of charcoal that he found with the mammoth. Recently, these samples were discovered in Kaschenko's personal effects. Radiocarbon dating was attempted. Remarkably, the assay indicated an age of 18 300 BP +/− 1000 years, almost exactly what was to be expected given artifact types and contextual considerations (S. Lechshinskiy, personal communication, June 27, 2002). Other, even earlier, research in the Tomsk region was reported by Fischer (1834) regarding bones found in the sites called Tscharych and Khankara.

Fig. 2.1 Location of the Tomsk mammoth site. The senior author and Sergei V. Leshchinskiy stand near the location where N. F. Kaschenko excavated a mammoth and associated artifacts from the right bank of the Ob River where it ran through the old Siberian trading center of Tomsk. The river has shifted and the excavation site is now covered with forest. They are holding the monograph prepared by Kaschenko wherein he describes his 1896 excavation and perimortem taphonomic study, the first in Siberia (OVP neg. Tomsk 6-29-02:21).

Researchers outside Russia have made perimortem bone taphonomy an integral part of archaeological and paleoanthropological analysis. A pioneering and classic study of perimortem taphonomy are the analyses that followed the massive excavation at Star Carr (Clark 1954). A listing of prominent analysts would include almost every one in forensic anthropology and forensic taphonomy (Haglund and Sorg 1997), and minimally, workers such as Noel Boaz and Russell Ciochon (2004), A. K. Behrensmeyer and A. P. Hill (1980), C. K. Brain (1981), Lewis R. Binford (1981), Robson Bonnichsen and Marcella Sorg (1989), R. E. Chaplin (1971), Simon Davis (1987), B. M. Gilbert (1990), Curtis Marean (1991), Gary Haynes (2000), Liora Kolska Horwitz and Julian Kerbis (1991), R. L. Lyman (1994, 2002), Travis Pickering (2002), Mary Russell (1987), Pat Shipman (1981), C. G. Turner (2007), D. West (1990), Tim White (1992), and David Yesner (2001) would be included. Attempts to reconstruct ancient events have both scientific and humanistic appeal.

There are some other early Russian studies involving perimortem taphonomy – e.g., G. A. Bonch-Osmolovsky (1931). This analyst produced a brief study of notching and cutting of Crimean archaeological faunal remains. Today, N. K. Vereschagin (1981 and elsewhere) is probably Russia's best-known pioneer researcher concerned with

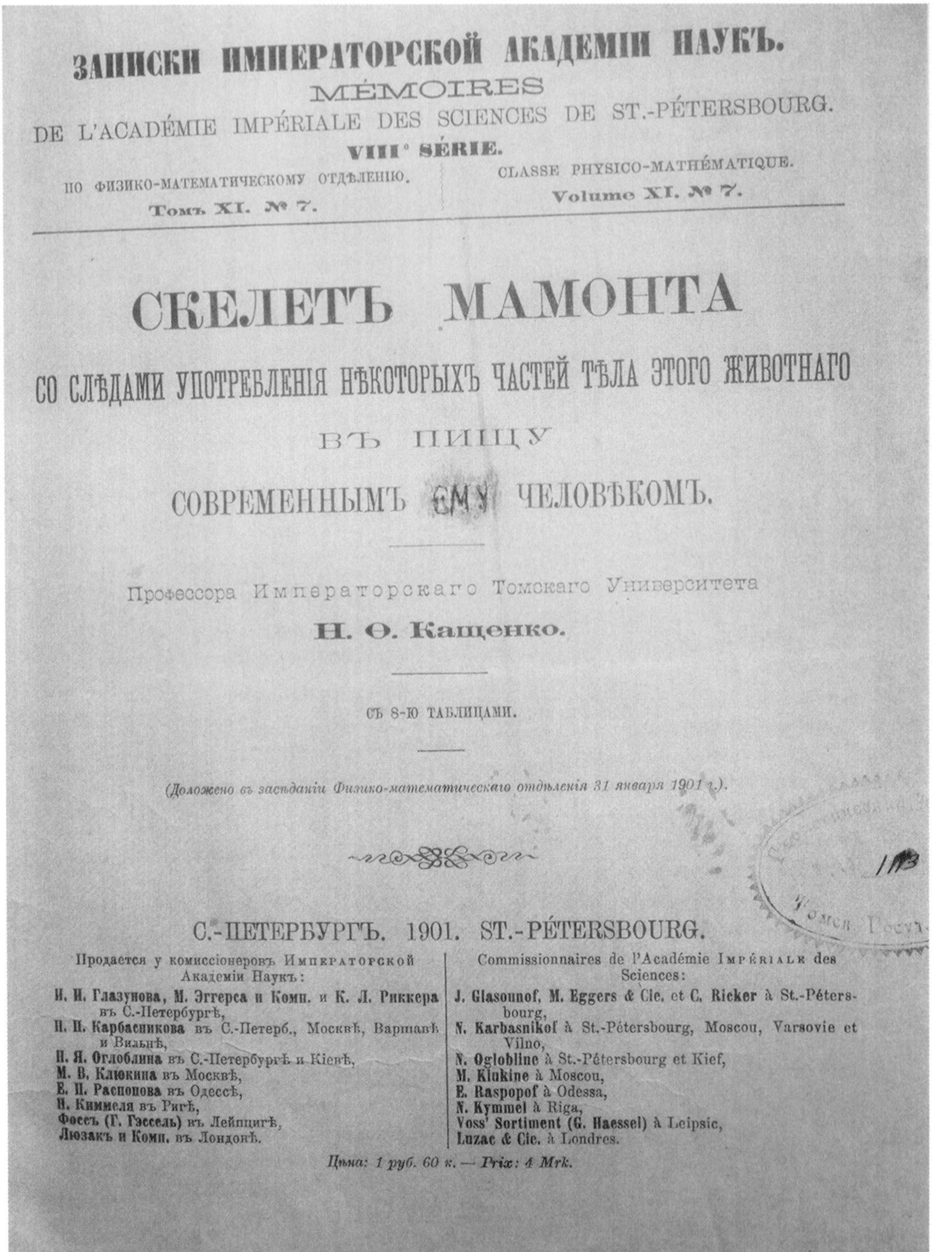

ЗАПИСКИ ИМПЕРАТОРСКОЙ АКАДЕМІИ НАУКЪ.

MÉMOIRES

DE L'ACADÉMIE IMPÉRIALE DES SCIENCES DE ST.-PÉTERSBOURG.

VIIIᵉ SÉRIE.

ПО ФИЗИКО-МАТЕМАТИЧЕСКОМУ ОТДѢЛЕНІЮ. Томъ XI. № 7.

CLASSE PHYSICO-MATHÉMATIQUE. Volume XI. № 7.

СКЕЛЕТЪ МАМОНТА

СО СЛѢДАМИ УПОТРЕБЛЕНІЯ НѢКОТОРЫХЪ ЧАСТЕЙ ТѢЛА ЭТОГО ЖИВОТНАГО

ВЪ ПИЩУ

СОВРЕМЕННЫМЪ ЕМУ ЧЕЛОВѢКОМЪ.

Профессора Императорскаго Томскаго Университета

Н. Ѳ. Кащенко.

СЪ 8-Ю ТАБЛИЦАМИ.

*(Доложено въ засѣданіи Физико-математическаго отдѣленія 31 января 1901 г.).*

С.-ПЕТЕРБУРГЪ. 1901. ST.-PÉTERSBOURG.

Продается у комиссіонеровъ Императорской Академіи Наукъ:
И. И. Глазунова, М. Эггерса и Комп. и К. Л. Риккера въ С.-Петербургѣ,
Н. П. Карбасникова въ С.-Петерб., Москвѣ, Варшавѣ и Вильнѣ,
Н. Я. Оглоблина въ С.-Петербургѣ и Кіевѣ,
М. В. Клюкина въ Москвѣ,
Е. П. Распопова въ Одессѣ,
Н. Киммеля въ Ригѣ,
Фоссъ (Г. Гэссель) въ Лейпцигѣ,
Люзакъ и Комп. въ Лондонѣ.

Commissionnaires de l'Académie Impériale des Sciences:
J. Glasounof, M. Eggers & Cie. et C. Ricker à St.-Pétersbourg,
N. Karbasnikof à St.-Pétersbourg, Moscou, Varsovie et Vilno,
N. Oglobline à St.-Pétersbourg et Kief,
M. Klukine à Moscou,
E. Raspopof à Odessa,
N. Kymmel à Riga,
Voss' Sortiment (G. Haessel) à Leipsic,
Luzac & Cie. à Londres.

*Цѣна: 1 руб. 60 к. — Prix: 4 Mrk.*

**Fig. 2.2** The Kaschenko title page. The Tomsk mammoth excavation by Kaschenko was published as a memoir of the Imperial Academy of Sciences, St. Petersburg, in 1901. The title reads: "Mammoth's Skeleton, with the marks of using some parts of this animal body for food by contemporaneous humans" (CGT color TU 6-27-02:19).

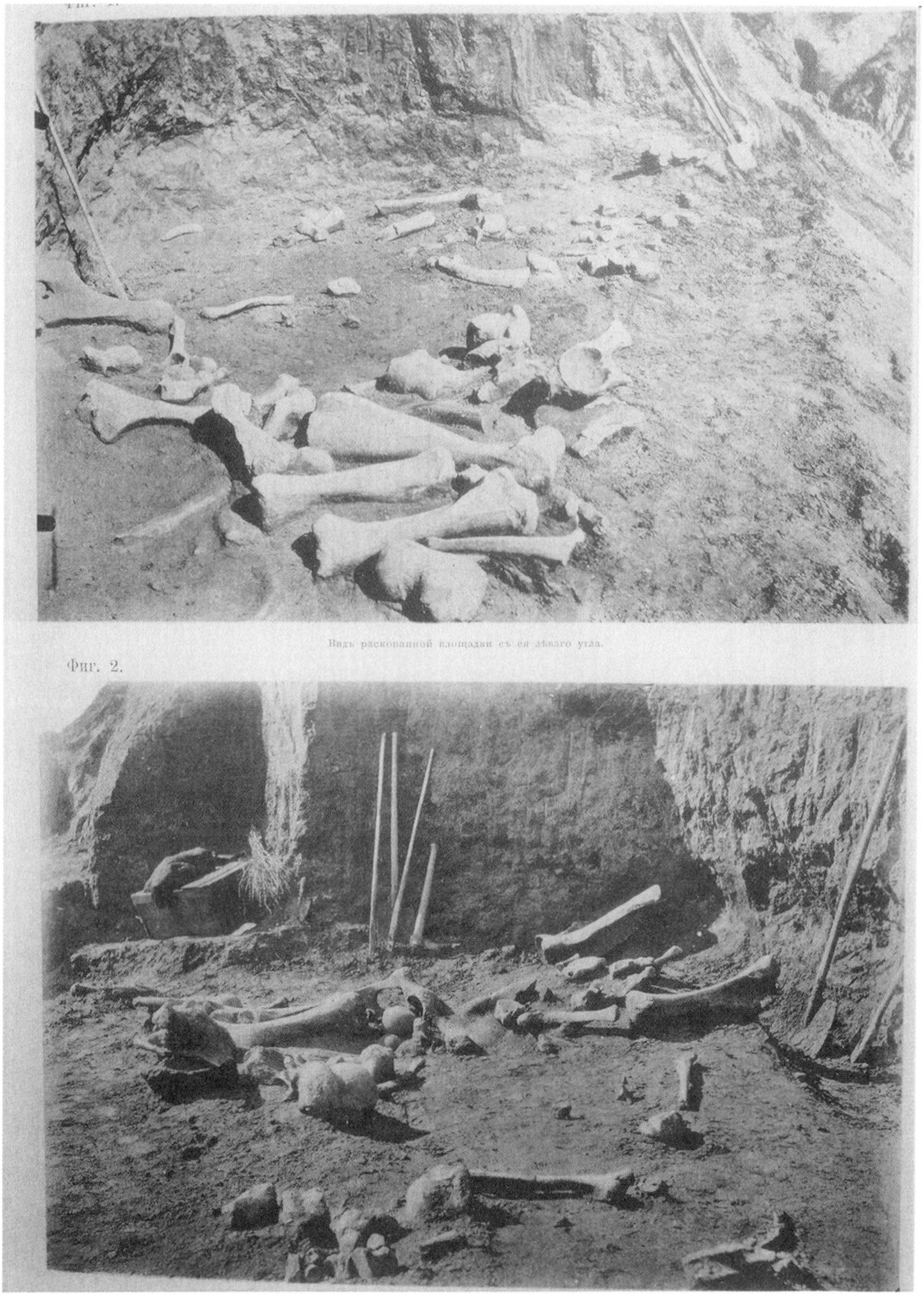


**Fig. 2.3** Photographs taken in 1896 showing the Tomsk mammoth remains that were covered by subsequent flooding of the Ob River. Caption reads: "Fig. 1. View of the excavated area from the left corner" (CGT color TU 6-27-02:22).

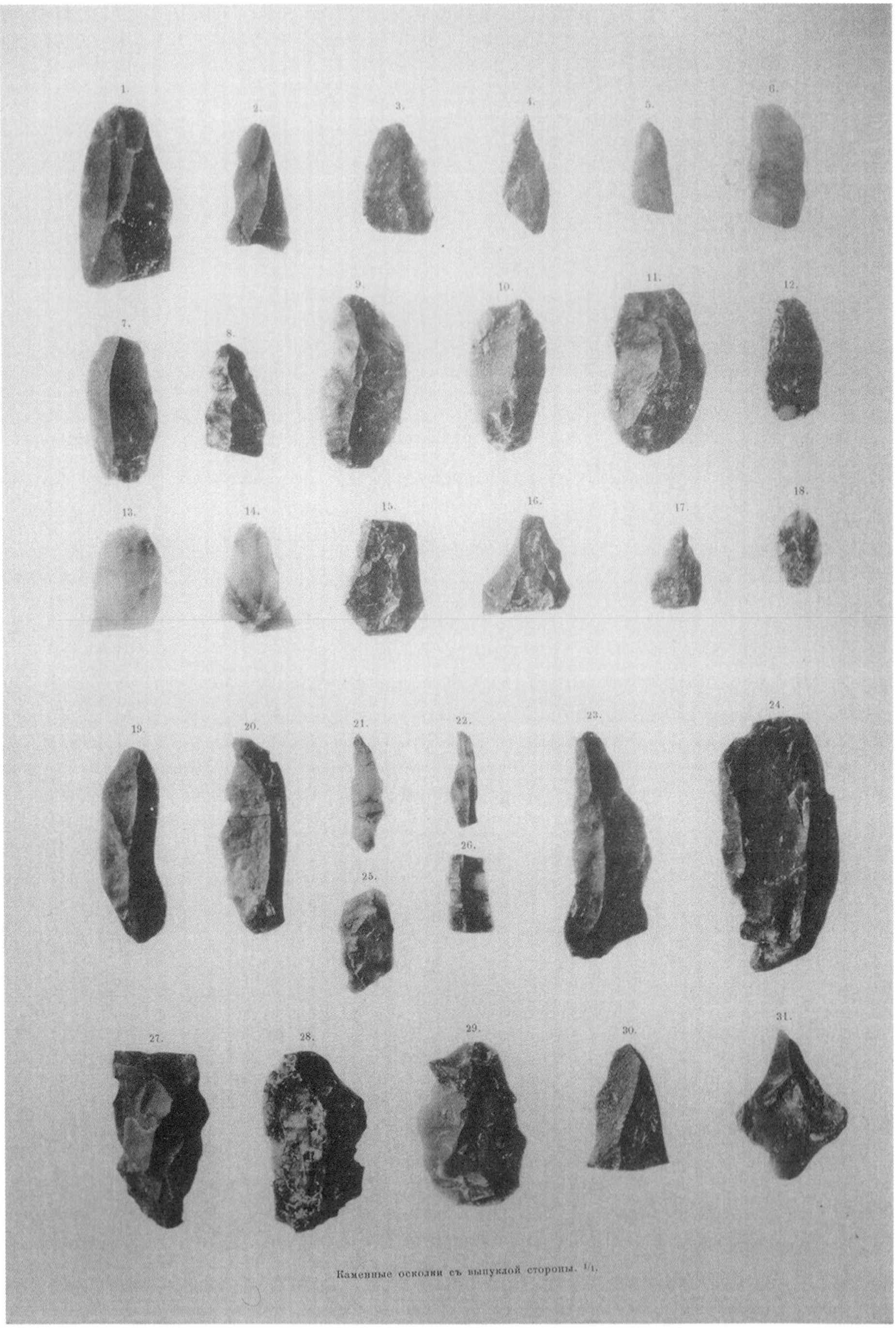


**Fig. 2.4** Artifacts found with the Tomsk mammoth. Small expedient stone flakes and blades that Kaschenko found among the Tomsk mammoth bones. Kaschenko believed that the mammoth was minimally utilized and was probably found dead, not hunted by the Upper Paleolithic folk who left these artifacts. The caption reads: "Stone fragments on their convex side" (CGT color TU 6-27-02:24).

perimortem taphonomy of paleontological and archaeological faunal remains. He identified earlier almost all of our independently recognized damage criteria. His discussion of mineral flowers ("dendrites" of magnesium and iron hydrates) is the best we have come across in our literature survey. Like Yefremov, Vereschagin developed a temporal and contextual sequence for bone damage and potential total destruction. He also recognized natural pathological conditions in fossil bones and teeth, such as those holes in skulls of ancient lions that some Russian journalists proposed were due to cosmic alien electric bullets. After Vereshchagin unquestionably comes Gannady F. Baryshnikov (2003 and elsewhere), today Russia's best-known researcher concerned with perimortem taphonomy of archaeological faunal remains. He has examined some of the assemblages we describe herein. It is only a small leap from Baryshniknov to the rules we used to classify the variation of our study. We first presented our perimortem damage definitions at an international conference held in the Kurtak archaeological field camp, directed by N. I. Drozdof in July 2000 (Turner *et al.* 2000).

## Definitions of 26 perimortem taphonomic variables

There is an abundance of research published on fossil bone taphonomy, one focus of which is the determination of whether fossil bone accumulations were made by Plio-Pleistocene hominids or carnivores. This literature has been succinctly summarized by Travis R. Pickering (2002) in his review of criteria that have been proposed to accomplish this differentiating purpose. Pickering discusses seven criteria that have been proposed by various workers (Cruz-Uribe 1991, in particular), and suggests that only three are really up to the task: (1) carnivore–ungulate MNI ratios (carnivore assemblages have more carnivore remains); (2) intactness of limb bones (relatively high frequency of long bone mid-shaft survival in carnivore assemblages); and (3) bone surface modifications (chewing marks in carnivore assemblages). In the assemblages we studied, species identification was very difficult due mainly to small fragment size, so we are hesitant to employ our crude ungulate–carnivore minimum number of individuals (MNI) ratios for any sort of even semi-rigorous analysis. However, our assemblages were well suited for the second and third criteria. In light of our forensic orientation, we could identify many more perimortem taphonomic features, as the reader will see in the following listing of what we looked for. Unfortunately, our examinations could rarely identify scavenged bones where teeth and/or tool marks were evident (Selvaggio 1994). As will be discussed in various assemblage descriptions, we assume that scavenged bone was abundant when a hyena presence could be recognized. Taphonomic research is also aided by actualistic study, as illustrated in Figs. 2.5–2.8.

**1** *Provenience.* Most of the assemblages were excavated by archaeologists and paleontologists using horizontal and stratigraphic units, such as squares, levels, layers, and distance from some monument or datum point. However, much of that information was not kept with most of the faunal collections; that is, point provenience was lacking for nearly all small pieces. While we have tried wherever possible to obtain this precise point of origin information, we have been far from successful. But this deficency is

**Fig. 2.5** A large dead animal, time sequence 1. First in a series to illustrate natural surface diagenesis of a large beef bull. Day 2 after death. Bloating has begun. Flies are abundant. No scavengers yet. November 7, 1988. Relatively isolated Upper Sonoran habitat on southeastern Arizona cattle ranch (Kali T. Holtschlag color Adams Ranch 11-7-88:0).

probably not important because wherever we have been able to check for significant stratigraphic and temporal differences in our perimortem damage variables – the ones we are mainly concerned about – none was found. We have generally been unable to test for significant stratigraphic differences in species content due to the limited number of bone pieces with specific identifications. Moreover, where we do have stratigraphically controlled sub-samples, there are rarely enough pieces in each unit to give much statistical confidence, even in simple two by two chi-square comparisons. Hence, all things considered, both good and bad, we present our observations on a pooled intra-site basis. In a sense, then, we are presenting our assemblages as representative of a generalized late Pleistocene time except for the Primorsky Territorial site of Boisman II and the Yenisei River site of Yelenev Cave, both of Holocene time. We follow Kuzmin and associates for late Siberian Pleistocene carbon-14 dates (Kuzmin and Orlova 1998, Vasili'ev *et al.* 2002).

**2** *Species*. Whole or nearly whole skulls, and many types of teeth are, with experience, easy to identify (examples shown in Figs. 2.9–2.10). Many isolated mammalian whole bones and even fragmentary pieces with only one anatomical end, such as the distal or proximal end of a femur, can also usually be identified to at least the genus level, and often to the species level. Most of the genus or species identifications in this book were made or confirmed by Ovodov. Ovodov is very conservative in his identification of

**Fig. 2.6** A large dead animal, time sequence 2. Bloating pronounced. Limbs raised because of internal gas. Flies abundant. Undated but film frame number near to Fig. 2.5. Odor pronounced. Tomsk mammoth was probably exploited for food before this stage of decomposition, or afterwards if only bone and oil were sought (Kali T. Holtschlag color Adams Ranch undated:4).

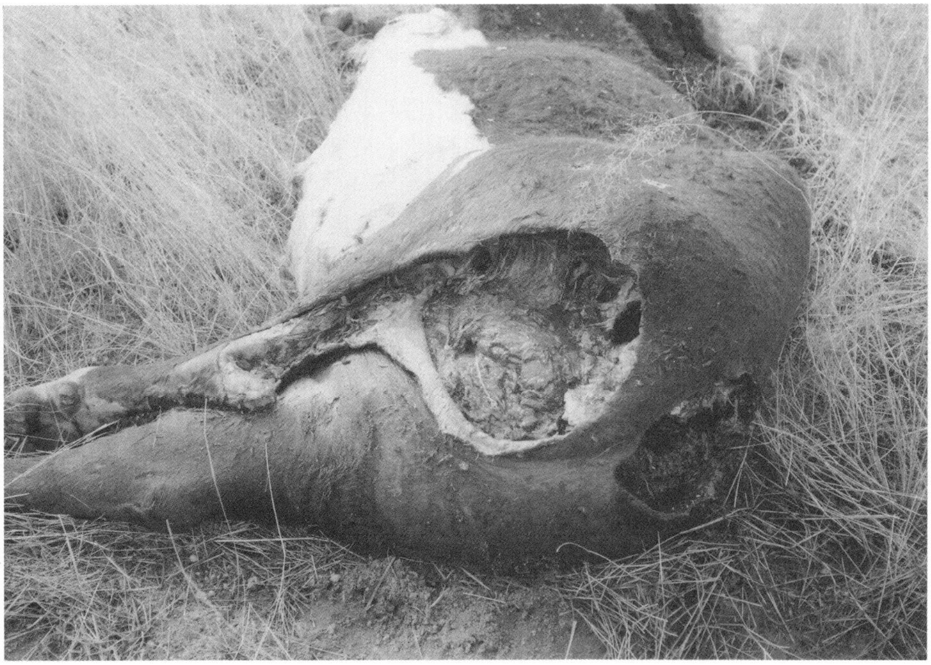

**Fig. 2.7** A large dead animal, time sequence 4. Most gas has escaped. Scavengers have chewed into carcass from anal region. Tail is gone. Flies abundant (Kali T. Holtschlag color Adams Ranch undated:28).

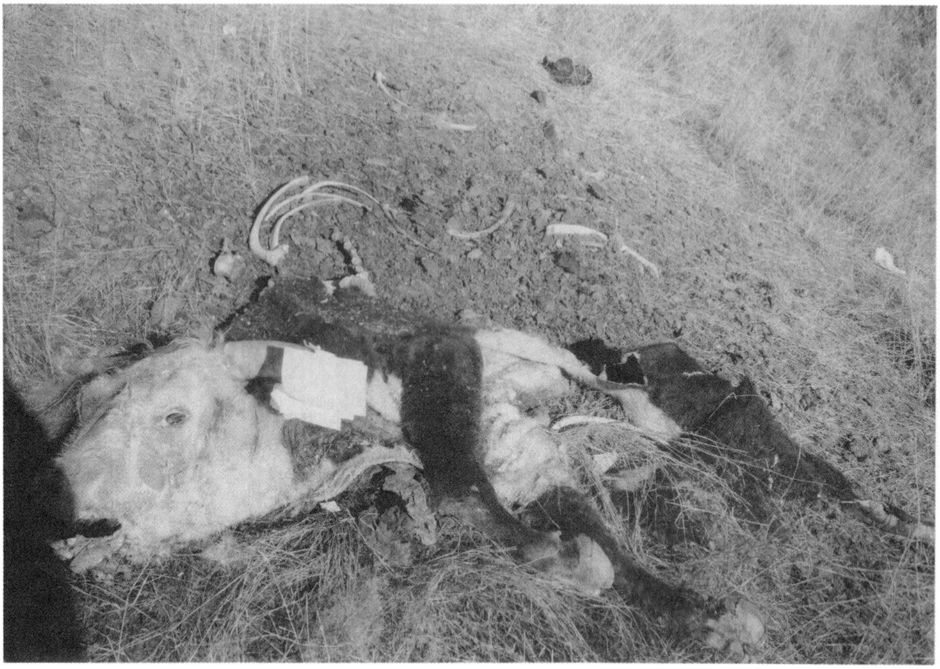

**Fig. 2.8** A large dead animal, time sequence 7. About two months after death, scavengers have scattered ribs, other elements, and intestinal contents. Head is still attached to cervical vertebrae. About half of the animal is missing. Large scavengers in this southeastern Arizona high desert area include turkey vultures, crows, coyotes, and wild dogs. Many smaller animals, including birds, are known to scavenge also (Kali T. Holtschlag color Adams Ranch 1-19-89:32).

species or genus, and even many relatively large pieces he has declined to classify. However, not all the unknowns are attributable to Ovodov, since he did not participate in the study of all assemblages. Many of the unknowns were scored as such by Turner. Our study largely excludes insectivores, bats, rodents, and birds. The rare exceptions are pieces with human processing marks.

If too much anatomical information is missing from a broken piece of bone to make genus identification possible, the piece may still possess enough morphology and size information so that it can be classified as a herbivore or carnivore, or even less definite, as a small-, medium-, or large-sized animal. We consider terrestrial animals such as fox, marten, marmot, and various hares to be small; deer, wolf, and boar to be middle-sized; and horse, bison, rhinoceros, and mammoth to be large animals. Elk and hyenas would be in the upper end of the middle-sized range, or in the lower end of the large-sized range. M. G. L. Mills (1990:11) gives average sizes for Kalahari brown and spotted hyenas. The latter are larger, males averaging 59.0 kg (130 lb) while females average 70.0 kg (156 lb). The Siberian cave hyenas may have been even larger than their African cousins if their average mass corresponded with Bergmann's rule (mammalian species that live in colder environments are frequently larger than similar species living in warmer habitats). Bunn (1991:442) envisions small animals to be less than 50 lb (Thompson's gazelle), medium-sized to be

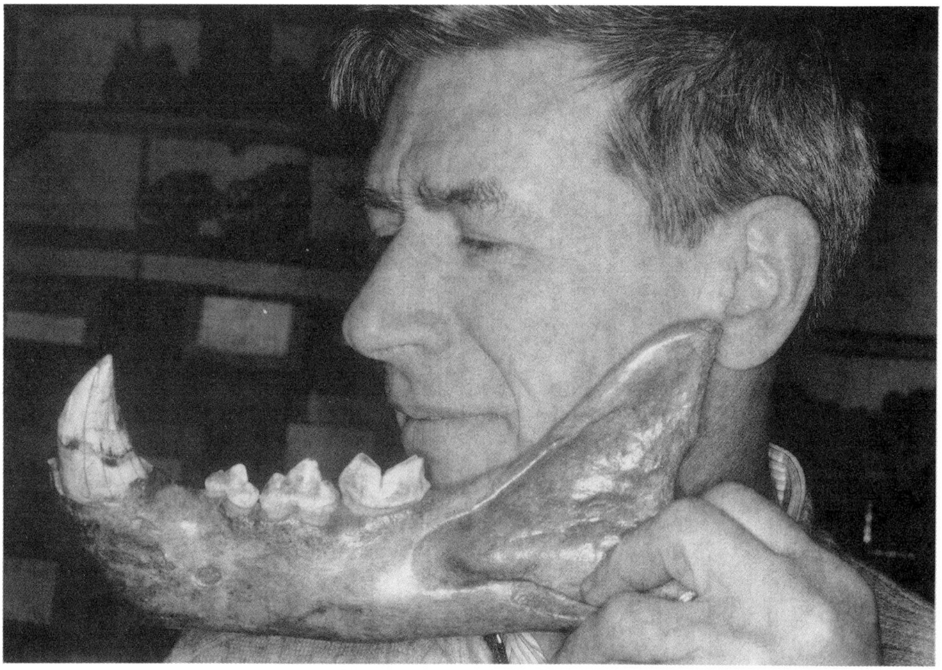

Fig. 2.9 Cave lion mandible. Nicolai Ovodov holds a cave lion jaw found in a cave near Irkutsk to show its relative size with that of a human. Dental morphology of lions is distinctive and easily identified (CGT color IHPP 6-4-87:27).

50–250 lb (impala), large to be 250–750 lb (zebra), very large to be 750–2000 lb (buffalo), larger animals to be >2000 lb (giraffe), and the largest to be more than 6000 lb (elephant).

In the maritime context of our Primorsky Territorial sites, sea otters would be small, various seals would be middle-sized, and walrus and whales would be large. With the exception of many teeth, broken pieces of long bone smaller than 10 cm in length are rarely identifiable, even to our size categories. The majority of broken vertebrae and ribs cannot be identified generically, but do fit our threefold animal-size categories reasonably well.

In this study we have paid relatively little attention to species identification for two reasons: (1) An overwhelmingly large number of recovered bone pieces could not be identified. In establishing our 2.5 cm minimal size criterion we found early on that almost no bone fragments smaller than 2.5 cm could be identified by Ovodov. Those that could be provisionally identified were almost always anatomically distinctive skull pieces. (2) We expected size biasing caused by preservation differentials (big bones preserve better), and human hunting practices. As David R. Yesner (2001:1) has made clear in his studies of ancient Alaskans: "Northern hunters generally focus on large game, whenever they can, especially those (like caribou) that aggregate in large numbers, both for reasons of availability and energetic efficiency."

**3** *Skeletal element.* There are a number of excellent guides to the skeletal anatomy of animals commonly found in archaeological contexts (Davis 1987, O'Connor 2000, and several others). Experience with one mammalian species is usually enough to make accurate

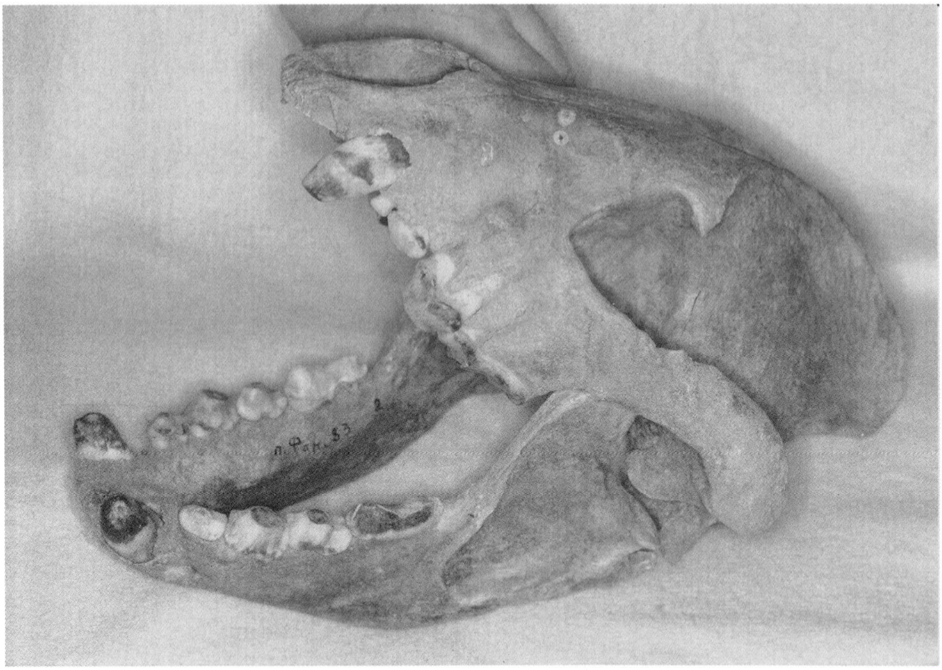

**Fig. 2.10** Cave hyena skull belonging to a nearly complete old adult hyena skeleton found in Fanatic's Cave, Khakasia, 1977. It was part of a paleontological assemblage found on the floor at the bottom of a vertical limestone shaft. The remains are estimated to be 30 000–40 000 years old. The animal was large, some 90 cm in height at the shoulder (Ageeva *et al.* 1978). As with the cave lion, cave hyenas are easily identified by teeth and cranial elements. This and other middle Yenisei river basin specimens means that there was likely no cis- or trans-Baikal gap in the longitudinal spread of these creatures (CGT color IHPP 6-4-87:28).

whole-bone skeletal element identifications for many other species. For example, femurs of terrestrial mammals all possess similar biomechanical anatomy – ball and socket proximal joint, heavy slightly curved shaft, and rounded double surface distal joint. Whole vertebrae, crania, ribs, metapodials, and many other elements are too distinctive to ever be confused with one another. Teeth and bones can always be distinguished from one another. However, with breakage and fragmentation, reliable identification rapidly goes downhill. Reliability is usually high if a broken bone has an articular end or a distinctive cross-section, such as those of oval-shaped ribs, metapodial fusion vestige, and complex surfaces of cranial bones. Crania possess some of the most distinctive and easy to identify elements of the entire skeleton. For this reason, when the frequency of cranial elements is low in an assemblage, the odds are very good that some sort of segregation took place in the past. Hunters may have removed the heads and did not carry them back to their camp; the heads might have received special ritual placement away from the camp; or the heads or parts of heads were subsequently carried off by scavengers. Bunn (1991:444) remarked that skeletal elements are potentially helpful in identifying whether an archaeological site is a camp or kill site, or had some other function. As most of our assemblages came from caves, concern over site function is of less interest to us because all cave bone had to be carried in. Cave sites are not kill sites. Of more concern to us is who or what carried the bone into the caves.

**4** *Age*. Because of the high degree of fragmentation in our assemblages, we limited age identification to adult, sub-adult, adult?, and unknown classes. While we very much wanted to have a highly refined set of age categories for considerations of seasonality, prey preference, MNI estimates, and other topics, such information was simply not contained in the remains we had to work with. In fact, simple age identification of adult and sub-adult was possible in only 70% of our total. Adult age was assigned when an epiphysis was fully fused to its diaphysis, when teeth were worn well into the dentine, when cortex bone was thick and dense, and when there was no surface indication of bone growth remodeling. Sub-adults would show the opposite of these mature conditions, i.e., incomplete epiphysial union, little or no crown wear, thin cortex bone relative to the amount of spongy bone, and the presence of porous growth remodeling areas. As with most of our perimortem taphonomy variables, there is a substantial inverse relationship with identification and piece size.

**5** *Completeness*. There are several ways to assess and describe element reduction. Our simple three-category classification of whole, one anatomical end, and no anatomical ends, was created to help identify carnivore activity. These are illustrated in several of the sites discussed in Chapter 3. Many other workers have also observed that carnivores more often chew off the relatively weak ends of bones in their effort to extract marrow grease, rather than attack the harder mid-shaft regions. Humans, with the same objective in mind, more often smash the shafts of long bones with stone or bone/antler hammers and anvils, many times leaving the ends intact. However, the greater the degree of fragmentation, the more remote becomes the relationship between carnivores and completeness. Hence, the ease and reliability of observing this variable is outweighed to some extent by its containing causal information that differs from the initial and end phases of skeletal element destruction. Nevertheless, this variable is especially useful when combined with other indicators of carnivore or human processing activity.

**6** *Maximum size*. This variable deals with the maximum dimension of a bone or piece, be it width, length or a combination of both. It is not a measure of anatomical length such as zygomatic breadth, cranial length, or crown–root height. Among its many uses are parametric tests for significant differences in bone reduction by carnivores and humans, and differences in archaeological assemblages from different localities. It is especially valuable in the analysis of pieces that have been digested by hyenas. We measured maximum size to the nearest millimeter using a 15 cm ruler or dial calipers for small pieces, and a measuring tape for pieces larger than 15 cm. Note that the size ranges in Table A1.6 are rarely lower than 2.5 cm. This is because the minimum size used in this study was set at 2.5 cm. There were thousands of pieces smaller than 2.5 cm in our various assemblages. Most had very little perimortem damage information other than color and breakage. Occasionally a smaller piece was scored for all 26 variables because it had one or more special qualities that we wanted to keep track of. Note also that there are no pieces larger than 82.5 cm. This is also partly our doing, since we did not consider whole bones for this study if they lacked perimortem damage. Keep in mind the objective of this work. It is not a study of anatomy or species, or traditional taphonomy. It is a more experimental study of perimortem damage.

Fig. 2.11 Damage types. The form of bone damage corresponds with skeletal element. Left to right: Top row: two maxillary fragments, tooth-bearing maxilla with teeth, two cranial pieces. Middle row: 11 splinters (horizontal), five flakes (vertical), one butt. Bottom row: one vertebra spine, one vertebra body, four ribs, one segment. Not shown are cracked open phalanx, and irregular shaped fragments (pelvis, scapula). Specimens mainly from Razboinich'ya Cave (CGT neg. IAE 7-6-00:1).

7 *Damage type/shape* (Fig. 2.11). In the initial stage of this investigation we wanted to create an original descriptive typology for the shape or form of a broken piece, something along the lines of classifications used in stone tool manufacturing, i.e., cores, flakes, blades, etc. Terms like "spiral fracture" or "blown-out tooth socket" are valuable for identifying conditions, but they do not identify shapes. Try as we might, we were unable to develop a set of shapes that was not so generalized as to be useless. Binford's (1981) breakage classification did not fit well with the damage we were observing. Moreover, no classification that we dreamed up applied to all skeletal elements because the form of a broken piece is very much dependent on which bone it comes from. Cranial pieces break up into irregular shapes very different from those of ribs. The long bones, including metapodials, can splinter and flake, but rarely wind up in the chunky forms common to vertebrae. Phalanges when broken open look for all the world like cigarette butts, a shape rarely seen in other broken skeletal elements, etc. Thus, the terminology used in Table A1.7 is basically anatomical and element dependent.

Four damage categories include more than one skeletal or anatomical unit: (1) Long bone fragments, flakes, and splinters. We include many metapodial pieces here as well as any other skeletal elements broken into these shapes. (2) Undamaged. Any skeletal element can be placed in this class. Undamaged is basically elements that have only a few tooth dints or tooth scratches, or one or two cut marks. Minor damage of minimal extent, if any, defines this class. (3) Irregular. Odd-shaped pieces of pelvis, scapula, cranium, and unidentifiable are the main components of this category. (4) Mostly whole. Any element that is approximately 90% complete is placed in this group. Although we are far from satisfied with our damage shape classification, it turns out to be a useful adjunctive descriptive tool for characterizing overall perimortem damage. We would have been even more dissatisfied had we not developed some manner of shape and form damage classification.

**8** *Color.* We initially tried to score color by using various color standards such as soil and paint reference scales. These standards soon proved to be needlessly precise. A single piece of bone might have six or even seven slightly different shades of color, the recording of which was time consuming, much too exact for our needs, and decidedly prone to intra-observer judgment variation. Judgment was plainly influenced by differences in institutional lighting, quality of bone cleaning, and end-of-the-day observer eye fatigue. Super precision was senseless because all bone is the same color before it acquires stain or is burned. We eventually settled on four categories that can be easily identified: ivory (unstained, creamy pale yellow to light brown due to washing and archaeological processing); white (like chalk); brown (moderately to heavily stained or charred); and black (burned or manganese-stained). Unstained ivory color characterizes cave bones. White is found in bone left in the open to become sun-bleached and heavily weathered. White can also occur with calcining, that is, severe burning. Calcined bone can be scratched very easily with one's fingernail. Brown characterizes bone color from most open sites where staining is caused by mineral-bearing groundwater contamination. Rarely is the staining much more than surficial, as shown in scratches resulting from excavation. When a piece is colored brown throughout its cross-section, we have usually found the piece to have been scorched by fire. Black bone is considered stained when the color is surficial. This is usually in the form of "manganese flowers," which are surprisingly common on bones from cave sites (see examples in various figures). Black bone is considered burned when the color is present throughout the cross-section, that is, the piece is carbonized or charcoaled. Burned black bone rarely holds a polish, whereas bone stained black usually has end-polishing. Only in one site, Okladnikov Cave, did we have difficulty deciding whether a few pieces were brown or black. These we scored as black brown. Burning is very important taphonomic information for identifying human presence in a site. But since burning is almost always recognized, at least initially, by color, we delegated burning to our color classification, rather than setting it up as another variable classification. Said another way, all black pieces are black in color, but not all black pieces are necessarily burned.

**9** *Preservation* (see examples in various site assemblages). Bone preservation can be easily classified as good or poor, that is, ivory or chalky in hardness. Ivory-quality bone cannot be scratched with one's fingernail, whereas chalky bone is readily scratched. Bone preservation is generally good in cave sites and poor in open sites. Special conditions such as permafrost or high calcium carbonate content of midden in open sites can enhance the quality of preservation. Chalky bone indicates that one or several destructive processes have acted on the bone. These processes range from seasonal dehydration and wetting, sunlight, weathering, thermal stress, denaturing, and so forth. We found that in a few sites some pieces of bone were both ivory and chalky. These were classified as intermediate.

**10** *Perimortem breakage* (see examples in various site assemblages). Breakage may occur at any time in the life and death history of a bone. A fracture that exhibits bone reaction in the form of healing or infection is a life history event, an antemortem break. A fracture that has no sign of repair or inflammation, but does have a clean sharp surface that is smooth like broken glass and sometimes spiraled around the long axis of the broken bone, is a perimortem break. That is, a break that occurred at or around the time of death (or later in some circumstances). There are various forms of perimortem breakage, including embedded fragments (caused by a tooth puncture or hammer stone impact), circular conchoidal breakage (caused by bullet- or bat-like trauma), radiating skull fractures that originate at the point of weapon impact, and many other forms. All share the same sort of smooth splitting type of break comparable to the shattered trunk of a lightening-struck tree.

A fracture that has a step-like surface, and shows a marked color difference from the remainder of the bone, is a postmortem break, one that occurred long after death. Postmortem breakage is commonly associated with small crumbs of bone that detach due to the dissolution and loss of collagen, the protein binder in bone. While perimortem bone breakage can occur in living and recently deceased individuals, it also can occur in very old bone that has been well preserved in dry desert caves and sands, in Arctic and Alpine permafrost, and in other such settings of very low biomechanical energy. Under these conditions bacterial and other agencies of diagenesis may be slowed down or even stopped altogether. Frozen mammoths and accidental mummification are well-known examples of the remarkable preservation that can occur, and which would allow perimortem-like breakage long after death.

Determining who the bone breaker was is more difficult than telling when the breakage occurred. Gary Haynes (2000) has cleverly argued that bones and tusks broken and polished, which appear to be products of human activity, were in fact done by elephant trampling at African water holes. Manifestly, context is critical but even so may be misleading. Bone breakage in an archaeological setting, where there is physical evidence of hyenas having also occupied the site, does not mean that all or even most of the breakage was done by humans. Observations of hyena chewing clearly reveal that they are capable of cracking open very large and heavy bones, just as well as can humans using bone or stone hammers. One such observation by Owens and Owens (1984:73) is representative. The event they observed was when a brown hyena, after cautiously ensuring herself that the killers, lions, had left the scene, approached and ate at the leftover carcass of a gemsbok:

> After nibbling at morsels of stringy meat, tendons, and sinew, she opened her jaws wide and began to crush leg bones as thick as baseball bats and to swallow splinters at least three inches long. (We measured these later, by fecal analysis.) A brown hyena's teeth are veritable hammers specialized for processing bone. The premolars are flattened and enlarged, unlike the sharp, scissorlike cutting blades of other predators. Tilting her head to one side, she wedged her teeth between the ball and socket of a hind leg until it tore free.

**11** *Postmortem breakage* (see examples in various sites; see also Binford 1981). Bone breakage that occurs months or years after death often exhibits a step-and-ledge fracture surface that may be granular instead of smooth, and sometimes even crumbly or flaky due to the lack of collagen. Postmortem breakage occurs often in bones left on the ground surface, which even when mechanically undisturbed can exfoliate small flakes or large chucks. Old trampled bone usually shows postmortem breakage, and archaeological or paleontological excavation procedure sometimes causes postmortem breakage. Postmortem breakage should be expected in open sites because of the greater opportunity for weathering and collagen disintegration. Bone left on the ground or slightly buried is subject to exfoliation due to mineral crystals forming as a result of drying out after being wet in water containing dissolved minerals.

**12** *End-hollowing* (see examples in various sites). This variable is a hallmark of carnivore chewing activity at the ends of long bones, ribs, vertebral spines, pelvic edges, and other parts of a skeleton where the cortex is thin. This chewing may remove most of an anatomical end or border, leaving the remainder with a cupped appearance, sometimes ragged, but often smooth and polished. Tooth marks are commonly nearby. End-hollowing is the result of carnivores attempting to remove marrow-rich spongy bone. We know of only one other agency that produces end-hollowing – the tumbling and abrasive action that bone can receive from wave action along a lakeshore, river, or in an energetic oceanic inter-tidal zone. Under these abrasive conditions, spongy bone is less resistant to erosive damage than compact bone, leading to differential loss and the formation of end-hollowing.

**13** *Notching* (see examples in various sites; see also Binford 1981). We first observed this form of perimortem damage on the edges of fracture surfaces of bones from Razboinich'ya Cave. Because this cave's larger occupants were mainly carnivores, we inferred that these notches were due to jaw and tooth pressure-flaking of one or more small pieces of bone at the tips of the carnassial molars and premolars responsible also for the bone cracking or fracture. These notches are usually large (>3 mm diameter), and common in big animal bones, making hyenas good candidates for their production. Since then, a few notches have been found suggesting human impact blows must have produced them with a bone billet or stone hammer. When a notch appears to have been caused by humans, that is, if the notched area shows the crushed appearance of hammer stone breakage, and if there are no nearby tooth dints, scratches, polishing, or end-hollowing, we assign the notch to human causality. Marean and Bertino (1994) suggested that spatial analysis of bone assemblages would be aided by focusing on midshaft fragments with hammer stone damage. In this way the effect of carnivore disturbance of the bone deposit could be subtracted and the hominid discard placement better defined. We do not disagree, but emphasize that distinguishing human and carnivore damage by a single criterion is not always possible. Cruz-Uribe (1991) would seem to agree.

**14** *Tooth scratches* (see examples in various sites; see also Binford 1981). Slippage of a bone in the tooth-studded clamped jaws of a carnivore can produce scratches comparable to slippage scars left by a bench vice, the jaws of a pipe wrench, common pliers, and other gripping devices. These scratches usually have one deeper and straighter end, with the other being shallower and ending in faint drifting-curved or curved-and-recurved track. Nearly all have a U-shaped cross-section. We attribute fine and delicate scratches to small carnivores, whereas larger carnivores are thought to be responsible for broader and deeper scratches. We have observed such differences on bones chewed by modern small and large dogs.

**15** *Tooth dinting* (see examples in various sites). Indentation pits caused by the surface penetration of pointed cusps of carnassial and other teeth are commonly seen with tooth scratches. These tooth dints are sometimes found in clusters of three or four, which vaguely record the shape of the responsible tooth. Occasional bone pieces, notably whole segments of a carnivore-ravaged long bone, will show dinting on both sides where the upper and lower jaws had gripped the bone. Like tooth scratches, we regard (1) small, shallow and (2) large, deep dints as having been left by small and large carnivores, respectively. We have recorded a few dints that could well have been produced by some unknown form of human bone processing. These rare examples are suspected of human causality because they occur near a hammered impact injury, and the pieces have no other signs of carnivore damage.

**16** *Pseudo-cuts* (see examples in various sites). As the name we have given this variable suggests, these are cut-like perimortem features. They were originally observed in the Razboinich'ya hyena cave, the context of which indicates that they were very unlikely human-caused cut marks. Instead, they are mimics of stone tool cut marks left by carnivores, presumably the cave hyenas that utilized Razboinich'ya for thousands of years. Nearly all the pieces we scored as having pseudo-cuts also had one or more indications of carnivore processing. However, as Bunn (1991:445) notes, hyenas in particular are attracted to bone refuse left by African hunter-gatherers. They visit these kill or camp sites within hours of abandonment. Hence, both human and carnivore damage can occur on a piece of bone, which means that pseudo-cuts on bones in carnivore dens might have been on bone cut by humans elsewhere and later transported to the den.

Pseudo-cuts are straight, narrow and deep V-shaped gouges just like stone tool cut marks. They are not U-shaped and they do not wander about randomly like tooth scratches. They are not only like stone tool cut marks; they are also reminiscent of the very fine grooves that can be produced on bone with a freshly struck stone flake or burin. Conceivably, pseudo-cuts are the result of gouging by very sharp unworn teeth of a young animal, or by a cusp that had been chipped, leaving in a sense a kind of burinized edge. As of this writing, we have found no qualitative or quantitative microscopic difference between stone tool cut marks and pseudo-cut marks. Internally, both may even contain the hinge fracture remnants where minute bone shavings broke off within a cut mark.

**17** *Abrasions* (see examples in various sites). Abrasions are an any-sized set of very fine closely spaced parallel striations that resulted from a bone slipping on an abrasive surface, such as having been placed on an anvil stone, or receiving a glancing blow from an implement with an abrasive surface, again stone, and again abrasive. Most sets of abrasions are associated with archaeological finds, although a few are not. The latter must then represent accidental trampling, slipping, or grinding on some manner of a finely abrasive surface, as limestone cave floors and mixed sand and icy beaches can be. While usually straight, a set of abrasion grooves may make a sharp turn due to additional slippage in another direction. Abrasions are scored by the number of parallel grooves that can be counted with a ×20 hand lens on a given piece, regardless of the number of sets.

**18** *Polishing* (see examples in various sites). Polishing occurs when a bone fragment is subjected to some manner of fine abrasive force that physically removes a very small amount of surficial bone, presumably the result of periostial gnawing. In so doing, the minutely rough surface is smoothed to a mirror-like condition that reflects light in a less scattered fashion than in the original state. We have observed this sort of polishing resulting from family dogs chewing on leftover BBQ steak bones. Polishing can occur any place where abrasive contact is possible. This most often is on the ends of fragments. We prefer to score polishing only on perimortem fracture surfaces. If polishing has occurred, the usually sharp fresh border is altered to a more rounded and reflective edge. We score end-polishing when it is present on one or both ends. Polishing on the middle of a piece can only be scored if there is a perimortem fracture present on one or both sides of a more-or-less rectangular fragment. When polishing occurs on both the end and middle, we refer to the condition as "both."

The polished area can be as little as 0.5 mm in diameter (usually at a pointed end), and up to 10 cm or more in length (almost always along the middle portion). Polishing can be seen with a ×10 hand lens, but we prefer ×20. An assemblage may contain many pieces that are also extensively polished, as in the river-deposited Krasny Yar collection. This polishing is readily attributed to tumbling and scouring in an abrasive context. Very extensive polishing in a few pieces and less in others, as occurs in the Razboinich'ya hyena cave assemblage, suggests more-or-less random carnivore chewing. Low amounts of polishing are expected in low-energy settings such as swampy soil, permafrost, and rarity of humans. In archaeological sites with minimal carnivore presence but much polishing, some thought has to be given to trampling and the possibility of bag-boiling with the use of heated stones. The stirring of the bone fragment and adhering soft tissue porridge could create the polishing on any surface, whereas stirring the same mixture in a pottery cooking vessel can produce polishing only on the ends of fragments due to the geometry of the pot (White 1992). Finally, we think of polishing as an indicator of taphonomic activity, both perimortem and postmortem.

**19** *Embedded fragments* (see examples in various sites). Small adhering pieces of bone situated in or at a notch, dint, tooth puncture, or impact point are referred to as embedded fragments. The frequency of embedded fragments in an assemblage is the combined result of the production agent (carnivores, humans, or both), and the predominant species and skeletal elements. We would in general expect more embedding by carnivore bone puncturing, especially in the ribs, pelvis, and ends of long bones of small- to medium-sized animals, than we would for human processing. Carnivore chewing is a relatively low-energy "squeezing" action, whereas human hammering is a relatively high-energy "smashing" action. Breaking an egg open by squeezing produces many times more incomplete hinge fractures than smashing it. Smashing produces many more completely separated pieces. For large animals we suspect the opposite effect, namely carnivore puncturing would be less due to bone density, so fewer embedded pieces should occur. Human smashing of big bones may lack the shattering power to produce unhinged fractures, so embedding would be more frequent. These expectations are based on an idealized homogeneous material, which of course is not the case for content or form of a carcass or skeleton. Nevertheless, we

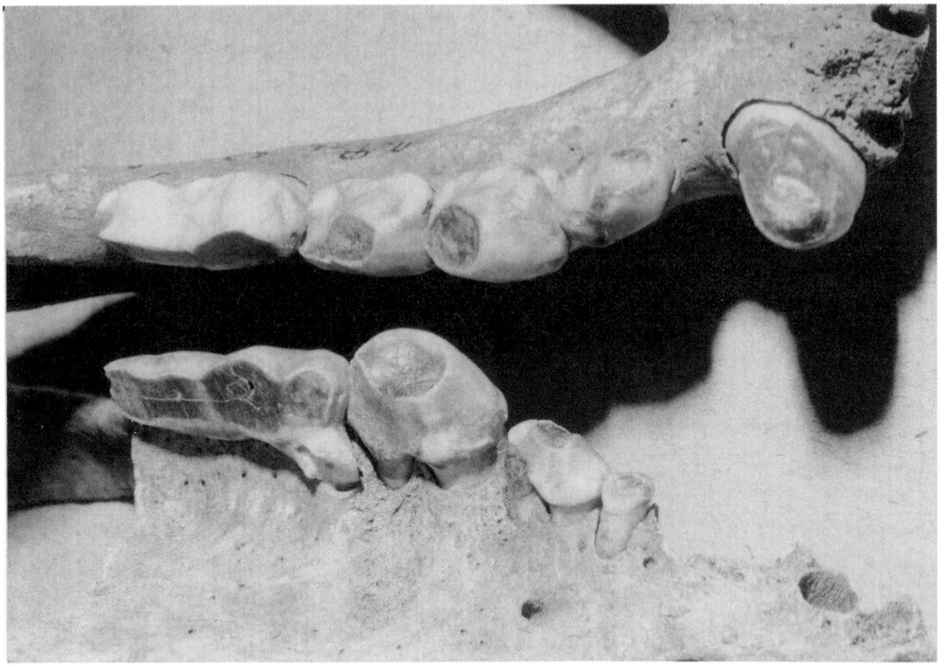

Fig. 2.12 Tooth wear. These hyena teeth of an animal estimated to be 15 years old illustrate the wear grade at which dentine is exposed but some degree of cusp anatomy remains. This photograph was taken to show that enamel thickness alone is not necessarily a strong indicator of diet. Hyena teeth actually have a rather thin layer of enamel, despite their bone-crushing dietary habit. Fanatic's Cave (CGT color Krasnoyarsk Regional Museum 8-7-98:5).

have scored embedding as a variable that we feel helps to identify the perimortem bone damage signature of large carnivores.

**20** *Tooth wear* (Fig. 2.12). We added tooth wear to our variable list in the hope that it would provide useful information about prey age preference and seasonality. There is a great deal of natural history research on carnivore hunting behavior, especially wolves, but also hyenas and others. It was our hope that we could compare the ages of animals found in archaeological and paleontological sites to see if humans and carnivores hunted and scavenged in a similar fashion. Unfortunately, our selection procedures caused our samples of usable teeth to be too small for such an ambitious but secondary objective. In any event, we scored tooth wear using the same standard as used in the Arizona State University Dental Anthropology System (Turner *et al.* 1991): no wear = 0; dentine exposed = 1; cusps largely worn off = 2; pulp or secondary dentine exposed = 3; root stumps are all that remain = 4. Intermediate grades are such as slight wear but dentine not yet exposed = 0–1, etc. We considered an animal to be "young" when its tooth wear scored 0 or 0–1. Obviously, to achieve "real" ages we would have had to standardize by using only teeth of the same kind that erupted at the same time for most species, a precision far beyond the capability of our assemblages, even if we had assessed wear in every single tooth. Hence, this variable has proven to be disappointing except in the most general way, as shown in Table A1.20.

**21** *Acid erosion* (see examples in various sites). Pieces of bone, teeth, and even small whole bones that have been swallowed by carnivores will be exposed to the destructive processes of part or the entire digestive track if not regurgitated before passing as feces. These acid-eroded pieces are commonly referred to in Russia as "stomach bones." This term is more useful than, say, "gut bones." The former allows for regurgitation, whereas once a bone or tooth reaches the gut it is not likely to be regurgitated. Destruction involves the dissolution, rounding, and polishing of bone or piece surfaces, be they spongy, compact, or both. As we have noted elsewhere (Turner *et al.* 2001b:26), 92 stomach bones from Dvuglaska Cave averaged about 4.0 cm in diameter, with a range of 1.7 cm to 7.5 cm. Presumably some percentage of very small stomach bones are eroded completely and leave no physical trace of their having ever existed, although the distinctive white color of hyena coprolites represents bone calcium residue. Because bone contains nutritionally useful collagen, powerful bone-crushing hyenas not only can gain access to energy-rich marrow, but also by accident and evolutionary adaptation extract protein from bone itself. Acid-eroded pieces are remarkably distinctive, and as such are one of our best indicators of carnivore presence, especially cave hyena presence. The occurrence of stomach bones in an archaeological site signals discontinuous human occupation, as well as probable stratigraphic disturbance due to the digging behavior of hyenas and other carnivores. Observers of hyena behavior regularly note that these creatures regurgitate parts of their meals, both as a means to rid themselves of huge undigestible "hair balls," and as a way to feed pups. A description by van Lawick and van Lawick-Goodall (1970:248) is insightful:

> Hyenas are frequently sick [vomit], and always they roll in it. It was not for some while that I realized that, for the most part, the hyena is not being sick in the normal sense of the word, but is actually regurgitating a mass of undigestible hair. Often, before or after rolling on this hair mass, a hyena picks out fragments of partially dissolved bone – when the hyena chews on them it seems that they are soft for there is no sound.

**22** *Rodent gnawing* (see examples in various sites). Finely chiseled, flat-bottomed, parallel tooth grooves, particularly at a sharp, bony edge, are unmistakable evidence of rodent activity. Some of their bone gnawing is nutritionally motivated to extract grease, minerals, and trace elements, but some gnawing probably serves other purposes such as controlling the size of the continuously erupting incisors. Despite the ease of recognizing and scoring this variable, it has not proven to be an especially valuable perimortem taxonomic trait because of its generally infrequent occurrence in our Siberian assemblages. However, we do consider rodent gnawing to be an indicator of site vacancy or abandonment (Table A1.22).

An interesting experiment and excellent literature review by Meg Thornton and Jennifer Fee (2001) found that in 150 skeletal elements belonging to four species (deer, squirrel, raccoon, and chicken), parallel grooving left by domesticated mice occurred in only 17 pieces. This amount was only about 10% of the total damage done by the mice. The other and more frequent damage included missing pieces and roughened areas. In our study, roughening and missing pieces could just as well have been done by carnivores. Hence, while our criterion for rodent damage is reliable, the actual frequency was likely greater.

**23** *Insect damage.* Small holes and surface traces are sometimes left in bone by burrowing insects. These traces can be distinguished from root damage by the latter's branching dendritic pattern with older and wider branches thinning down to younger and narrower branches. As was noted in 2001 (Turner *et al.* 2001a:27), the identity of these insects, and possibly some other form of invertebrate, remains unknown to us.

**24** *Human bone and teeth.* This special variable considers whether humans have damaged human bone. The damage forms include cutting, perimortem breakage, abrasion, polishing, burning, and (many) missing vertebrae. If these six traits are identified in an assemblage of human bones, then cannibalism can be suggested.

**25** *Cut marks* (see examples in various sites). Inasmuch as we consider stomach bones to be a major taphonomic characteristic of hyenas, cut marks are similarly important as indicators of human carcass processing. Stone tool cut marks are generally V-shaped in cross-section due to their being the result of a slicing action. Within a cut itself there can be ledges, hinge fractures, and other irregularities. These can often be seen with a ×20 hand lens, although a ×75 dissecting microscope reveals these minutia much better. Cut marks result when stone knives, blades, or flakes used for butchering, dismemberment, and tissue slicing accidentally cut into underlying bone. As indicators of butchering, cut marks are often found near anatomical joints, indicating dismemberment of limbs, heads, jaws, and other body parts. Cut marks are variable in length, breadth, and number in a set. A set presumably represents a repeated effort to dismember a body part, to remove a muscle mass, or to separate hide from underlying muscle. We have found as many as 200 cut marks on a single piece, although most cut pieces have only two or three cut marks. Paul Volkov (2006) has written and illustrated a marvelous analysis of stone tool production technology; it seems from our reading that there are few differences in the resulting cut marks of stone tools. The goal of the various ways of making stone cutting tools is to produce a slicing function. In the early months of our study we tried to identify a given cut mark with a particular form of stone tool. We were unable to recognize any macroscopic variation that we could link to time period (tool type) or location (stone types). Perhaps we should have examined cut marks more closely, but our goal was more broadly focused. We did not distinguish deep cut marks from shallow cut marks. However, we attributed this difference to the act and actor of butchering, not to the form of the cutting tool.

**26** *Chop marks* (see examples in various sites). Whereas cut marks are produced by a slicing or back-and-forth sawing action, chop marks are produced by hammer or hatchet impact action. The result is a round and sometimes deep crater. If the hand axe, heavy flake, or worked core is dull, the chop mark may have rough and ragged internal surfaces. If the chopping edge is sharp, the internal surface can be smooth like a cut mark, but wider and non-linear, and can often show smashing at the bottom of the crater. Some small percentage of chop marks are likely due to a bone having been pounded on a rough rocky surface or anvil stone. The resulting craters are indistinguishable from those produced by a stone tool. We have seen a couple of chop marks that seem more likely to have been produced by hyena teeth than by stone tools.

## Piece selection

The bone that we actually had the opportunity to examine was earlier subjected to one or more mechanical, identification, size, and preservation treatments. Mechanical treatment began at the archaeological site, where midden was usually water-screened after identifiable pieces of bone, stone, and artifacts had been bagged by the person(s) digging any given unit. Pieces smaller than the screening mesh would have been lost. Selection of bone to be curated was influenced by size and completeness, with the exception of ribs and vertebrae that were not always saved before or after the excavated items were brought back to the archaeological and paleontological laboratories. We selected pieces to study if their maximum diameter was greater than 2.5 cm and if plant root damage was minimal (Figs. 2.13–2.14; see also Binford 1981). Most of the pieces of bone that were curated in various institutions were less than 2.5 cm in diameter. Very few of these unstudied pieces had perimortem taphonomic information other than size, breakage, color, or quality. At least one million of these small pieces went unstudied.

## Grand total

We pooled individually the 26 variables for all the assemblages in order to create our best possible approximation of a hypothetical taphonomy universe. By comparing an individual assemblage we could see how different it was when compared to the universe, and could apply a chi-square test to see if the difference was statistically significant at the $p < 0.05$ level of confidence. Given what we know about the complexity of our data sets, we regard the chi-square comparisons as only suggestive, not definitive.

## Summary

In addition to taphonomic variable definitions, this chapter discusses the nature of perimortem taphonomy and its use in paleoanthropological and related research. The definer of the term taphonomy, I. A. Yefremov, is discussed. Examples of early studies of perimortem taphonomy are provided. The context of where archaeological bones are found uses the common distinction of cave versus open sites. There is variation in both the types of cave sites and environmental conditions of the open sites. As elsewhere in the world, hard-tissue preservation is better in cave sites than in open sites. The taphonomic

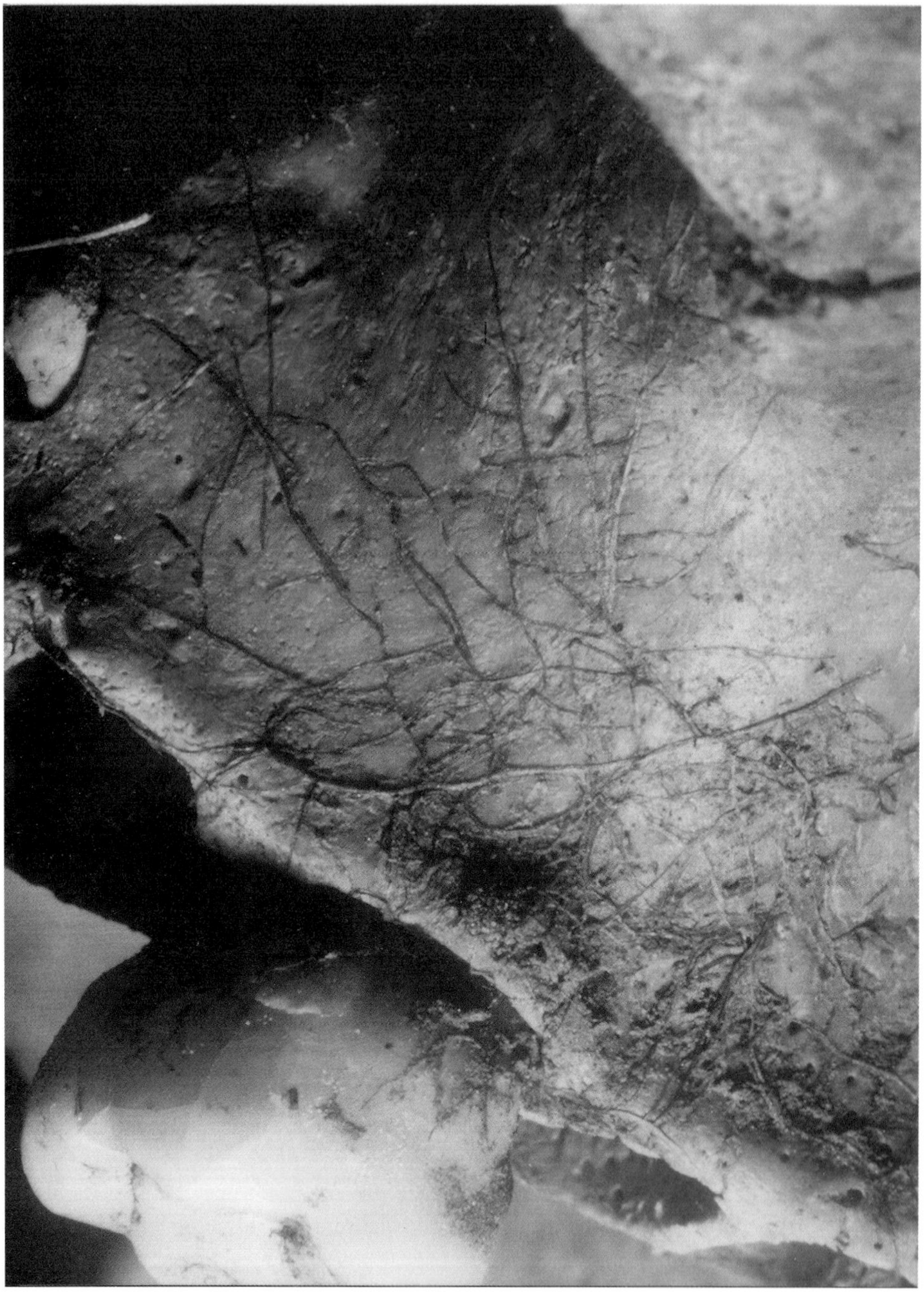

Fig. 2.13 Minimal plant root damage. Roots cover the bone surface and had been lying in faintly etched tracks at the time of excavation. Human female maxilla from Schobonova VIII (CGT color IAE 6-29-99:22).

**Fig. 2.14** Moderate plant root damage. Here, root damage on the interior rib surface of a human skeleton from Ozkerki I (1993) had the deeply etched characteristic dendritic pattern, but considerable bone surface remains undamaged. This degree of damage was acceptable but very close to eliminating the specimen for our study purposes (CGT color IAE 6-29-99:7).

features that were studied herein are listed, defined, and illustrated. A few comments about live hyenas are provided. Perimortem taphonomy supplements other lines of evidence about past human activity and events, just as it does in contemporary human forensic taphonomy (Haglund and Sorg 1997, Boddington *et al.* 1987) and disaster archaeology (Gould 2007).

Next, in Chapter 3, we describe and briefly discuss our taphonomic observations for each of our 30 assemblages. A more detailed discussion is provided in Chapter 4.

# 3 The 30 Siberian archaeological and paleontological sites, distributed from the Ob River to the Sea of Japan

I have to see a thing a thousand times before I see it once.

(Thomas Wolfe, *You Can't Go Home Again*, 1942)

The purpose of this chapter is to provide: (1) background and contextual information for each of the 30 sites from which our observations on their prehistoric faunal assemblages were derived. (2) To describe and tabulate the perimortem taphonomic conditions of these assemblages using the 26 variables defined in Chapter 2. The descriptions and comparisons are drawn largely from the quantitative information about each variable provided in Tables A1.1–A1.26. These are located in Appendix 1, with assemblages listed by site number. Our first conference presentation of this research was at an international field meeting held at the Kurtak archaeological station directed by N. I. Drozdov (Turner *et al.* 2000). A brief English version appeared in Turner *et al.* (2001a).

Some additional untabulated information about specific pieces of bone is drawn from our data-collecting forms. For instance, we think that in some cases it is useful to describe several variables for a given piece, whereas in most cases this additional description does not seem worthwhile.

We have chosen to arrange the order of presentation of our assemblages alphabetically for the following reasons: (1) We initially thought that we would easily develop a dichotomous classification containing two distinctive sets of sites – archaeological and paleontological. This soon proved not to be possible due to our discovery of major carnivore presence in several of our "archaeological" sites. Some sites that have been previously identified as archaeological have turned out to possess a highly significant carnivore presence, as inferred from the types and amount of perimortem damage to their faunal content. In fact, we doubt if any archaeological site described in this book has not been influenced to some extent by carnivores that also occupied these sites, regardless of whether they were cave or open-air types. (2) We considered other forms of presentation such as regional, ecological, temporal, etc., but abandoned each because of the information that we felt the reader would have to already possess in order to make the ordering useful instead of being a burden of additional background learning. Moreover, the senior author is well aware of how very little geographic, archaeological, and natural history knowledge non-Russians scholars possess about Siberia. (3) On the basis of many faunal studies of large Pleistocene animal remains, it is surprising how often the same set of animals turns up in northern Eurasian assemblages – mammoth, horse, bison, reindeer, wooly rhinoceros, cave hyena, cave lion, wolf, and several smaller species. While predator and human hunting selection, as well as preservability, may influence this community, nevertheless, these species were present in nearly all of the Pleistocene sites. This substantial faunal homogeneity suggests that there were no marked environmental differences except in the very far north and the very far south of Siberia. Some species are more common in one assemblage and less common in others. This may not reflect environmental differences as much as human and carnivore hunting preferences affected by seasonality, population size, sex ratio, mean age of human and carnivore groups, and various unfixed environmental considerations such as local forest fires, epidemics, severe winter storms, etc. (4) The names of several of these sites will be strange to most readers of this book, so the simplest possible ordering, namely alphabetical, was thought to be the most useful form of organization.

A few remarks need to be made about statistical analysis. At the beginning of our study we had assumed that multivariate statistical analyses would be ideal. We soon abandoned this viewpoint as it became clear that there were a host of data acquisition problems that would make such sophisticated analyses meaningless. These included much missing data; limited and even impossible identifications (such as species, skeletal element, etc); hard-to-standardize types of variation (damage form); subtle-to-significant differences in the ways the assemblages were originally collected, saved, partitioned, cataloged, and curated; and other problems of creating a database utilizing non-standardized, between-institution procedures and practices. In other words, we suspect that there is an undefinable but real amount of non-comparability between some of our assemblages, and even non-comparability within a specific assemblage, that makes a general plan of complex statistical analysis nonsensical and meaninglessly expensive. We have developed a very respectable database, but because we are so familiar with it, we are certain that it is not up to the task of supporting complex statistical analysis, and cannot meet many of the assumptions of multivariate analysis. Therefore, we limit ourselves to a few univariate statistical comparisons, and these only when we feel the data sets are adequately matched in quality, or where we need a statistical test to help decide if a probable real difference exists between two sets of not especially similar information.

# 1 Afontova Gora

## Background

Afontova Gora (Afon's Hill) has one near-river-level component that overlooks the wide, cold, gray Yenisei River, which flows past the old Siberian city of Krasnoyarsk. Other components are further away from the river and on higher ground (Fig. 3.1). Afontova Gora is located at 56°01′ N 96°50′ E. The site complex consists of a laminated series of stratified Upper Paleolithic camps in at least five locations near each other, all buried in gray-yellow silty loess deposits that front onto the Yenisei River. Afontova Gora is only a few long city blocks from the central district of Krasnoyarsk. The site was first dug into by a local teacher named I. T. Sevenkov, who in August 1884 found artifacts suggesting the site was of Paleolithic age (Astakhov 1999). For nine years Savenkov observed cross-sections dug by clay miners and railroad construction. In 1892 Savenkov presented the results of his archaeological studies at the International Anthropological Congress held in Moscow. By then he had collected 1500 faunistic pieces and 250 "ancient implements" (Ovodov notes). In addition to his and later local teachers' and professional archaeologists' excavations, there were discoveries of stone tools, stone tool manufacturing refuse, other cultural elements, semi-subterranean dwellings, and faunal remains including those of a possible dog. A small amount of human bone and teeth was found in a dwelling, suggesting to some the possibility of cannibalism (Chard 1974:30). M. P. Gryaznov (1932:144) noted that a human ulna fragment had been split open without any sign of animal damage. In 1937 participants of the 17th International Geological Congress visited Afontova Gora. During this visit, French archaeologist S. Fromazhe found a human skull fragment that would later serve as the basis for suggesting the Afontova people were racially Mongoloid (see Finding 24. Human bone).

Afontova Gora is an important source of information about the physical anthropological characteristics of the late Pleistocene inhabitants of southern Siberia. The human remains and artifacts, location, and dating of Afontova Gora make it a key site in considerations about the peopling of the New World by interior "Siberian Paleolithic" people (Chard 1974:20).

All three present authors have together visited two of the Afontova Gora localities, and have examined much of the more recently recovered stone artifacts curated in Krasnoyarsk. Due to a lack of time we were unable to arrange a study of the earlier excavated human remains housed in the Hermitage Museum, but we were invited to

Fig. 3.1 Afontova Gora V site. Located near the left shore of the Yenisei River where it flows through the city of Krasnoyarsk, Afontova Gora is the designation for five or more large Upper Paleolithic concentrations of human occupation debris and structures. These concentrations spread horizontally for many meters in the silty Pleistocene deposits. This excavation directed by Nicolai I. Drozdov, near hotel Solar House, revealed archaeological materials said to be 35 000 years old. It is being examined by Japanese Paleolithic specialist, Hideaki Kimura. The Yenisei River is out of sight in the background. The bone taphonomy reported in this monograph came from a nearby excavation (CGT neg. KSPU 8-3-98:9).

examine a large collection of faunal remains in Krasnoyarsk, within which some pieces were suspected of being human. We found none. Carbon-14 analyses indicate a late Upper Paleolithic terminal Pleistocene occupation. Uncalibrated dates cluster around 13 000 BP (Kuzmin and Orlova 1998).

S. N. Astakhov (1999) has reviewed and written a lengthy account of the history of archaeological investigation of Afontova Gora. For our purposes, he inventoried and briefly described the human remains (pp. 65–66), as well as identifying the faunal species (pp. 66–68). Afontova Gora has almost a century-long history of archaeological exploration, description, and discussion (Astakhov 1999, Auerbakh and Sosnovsky 1924, Chard 1974, Abramova *et al.* 1991, Derev'anko *et al.* 1998, Drozdov *et al.* 2000, Drozdov and Chekha 2003). These excavations have produced a wealth of Upper Paleolithic material culture, including drilled beads, eyed needles, blade and microblade stone tools, etc. Unlike cave sites where the purpose of habitation is easily understood, the aboriginal choice for open site locations is less clear. Afontova Gora is no exception, although the faunal remains suggest a good hunting locale and possibly a large stand of riverside trees.

Astakhov (1999:196) notes that:

> the majority of Upper Paleolithic span the period of the Sartan glacial . . . [the regions surrounding the Krasnoyarsk basin] . . . during the Sartan were cold periglacial steppes with scattered stands of trees along river valleys and ravines. The open steppes were favorable for abundant herds of ungulates, as well as for mammoth and wooly rhinos.

## Findings

**1 Provenience (Table A1.1, site 1).** The sample that we briefly examined on August 12, 2006, came from Afontova Gora II. As in previous excavations, it was obtained by Nicolai Drozdov and Eugene Artemmiev.

**2 Species.** Our sample was extremely broken up by human processing, making species identifications nearly impossible. More than 95% of some 5000 pieces were unidentifiable except for a few pieces of bone and teeth of mammoth, bison, reindeer, and other mammals of the late Pleistocene Arctic steppe (steppe-forest) community. One cave lion tooth was found, as have been others in previous excavations. Since hyenas are a major animal concern of our project, we were especially careful to look for even the tiniest fragment that would reveal the actual presence of this species in the sample. None was found. When other archaeological samples we've examined had cave lion, they usually also had hyenas. Auerbakh and Sosnovsky (1924) had better luck with species identification. They reported *Elaphus primgenius*, *Cerves tarandus*, *Lepus* sp., *Alopex lagopus*, *Bison pricus*, *Equus caballus*, *Canis lupus foss*., *Canis vulpes*, *Gulo cf. borealis*, *Antelope (colus) saiga*, and *Ovis* sp.

According to Astakhov (1999:198), "26 animal species have been identified: 30% belong to reindeer, 20% to Polar fox, 11% to hare, and 4% to mammoth." Ovodov *et al.* (1992b) found similar values.

**3 Skeletal elements.** Not recorded.

**4 Age.** Not recorded.

**5 Completeness.** Effectively none.

**6 Maximum size.** Fragment size was generally less than 2.5 cm maximum diameter, so no measurements were taken.

**7 Damage shape.** Not recorded, but all forms were present, especially flakes and splinters.

**8 Color.** Nearly all of the 5000 pieces were ivory colored. One piece was black and clearly had been burned.

**9 Preservation.** While preservation was good, there was much root damage to the surfaces of the fragments. The quality of the bone was universally ivory; that is, it could not be scratched with a fingernail.

**10 Perimortem breakage.** As for human processing, nearly every one of the 5000 pieces had perimortem impact breakage.

**11 Postmortem breakage.** None was recognized.

**12 End-hollowing.** While one piece had end-hollowing, the degree of damage indicated a carnivore smaller than a hyena, possibly a dog.

**13 Notching.** None found.

**14 Tooth scratches.** Very few.

**15 Tooth dints.** Very few.

**16 Pseudo-cuts.** None recognized.

**17 Abrasions.** None recognized.

**18 Polishing.** Similarly, there was very little end-polishing, and no identifiable midshaft-polishing. In fact, there was effectively no carnivore damage to speak of as no tooth scratches or dints could be identified with certainty.

**19 Embedded fragments.** None found.

**20 Tooth wear.** Not recorded.

**21 Acid erosion.** There was no sign of the typical severe hyena chewing damage nor of their distinctive digestive damage to swallowed bone and tooth fragments.

**22 Rodent gnawing.** None.

**23 Insect damage.** None recognized.

**24 Human bone.** Our examination of about 5000 pieces turned up no human bone fragments or human teeth. However, of the past finds, the most important is a fragment of a child's frontal bone with adhering nasal bones, the shape of which gives the child a flattened facial profile that is characteristic of modern northeast Asians. On this morphological basis, Russian physical anthropologists (V. Alexeev following G. Debets 1946) have suggested that the Afontova Gora people, if not all Upper Paleolithic Siberians, were Mongoloids (Alexeev 1998). Such an inference should be viewed with skepticism because most children lack nasal prominence until later in childhood. Moreover, unerupted permanent teeth of a child from the Upper Paleolithic site of Mal'ta, far to the east of Afontova Gora, have a distinctly European appearance (Turner 1990a, Haeussler and Turner 2000). Mentioning human skeletal remains is relevant to this brief account despite its title, because in addition to our examination of perimortem bone damage, our primary objective with this collection was to see if any human remains were in the 8000-piece faunal collection recovered in 2006 at Afontova Gora II. We had time to examine more than half. There were no identifiable fragments of human bone or teeth. This absence continues the mystery of the nearly total absence of human remains in Upper Paleolithic Siberian sites, when these sites have excellent preservation of non-human animal remains.

**25 Cut marks.** Only six pieces had one or more cut marks.

**26 Chop marks.** Thirteen pieces had one or more chop marks. Three pieces had both cut and chop marks.

## Discussion

The overall Afontova Gora perimortem damage decidedly excludes carnivores as having contributed much to the taphonomic condition. Perimortem breakage was extensive and seemingly all done by human smashing of most bones. This conclusion was reached from the minimal signs of carnivore activity, as well as with the good preservation, especially so for an open site. The lack of chalkiness to bone surfaces suggests very little weathering in the open, as well as absence of postmortem breakage and very little carnivore damage. Even rodents left no record of their presence, commonly seen in bones left on the surface for several days or weeks. Taken together, these conditions suggest rapid burial that would protect the bone assemblage from air and sun weathering and scavengers, possibly by rain puddling and mud covering, clouds of dust blowing off the river banks and terraces that would fall to cover the cultural material. Shallow burial would protect against most destructive processes except for plant growth and root penetration, which was abundant in the assemblage.

The near total absence of carnivore damage, and the associated rich material culture inventory makes Afontova Gora one of the best Siberian sites from which to infer diet and hunting cultural practices of these Paleolithic Siberian hunters and their families. In contrast, nearly all of our other archaeological sites, both Paleolithic and Neolithic, where the hyena/carnivore bone damage signature is markedly strong, raise the question as to how the bone refuse got into the archaeological sites. Were the bones and adhering flesh carried to the sites by hyenas, other carnivores, or humans? Who or what were the primary hunters or scavengers in the archaeological sites with a marked hyena/carnivore taphonomic presence?

We are exploring herein a concept of a northern Eurasian "hyena barrier" or "hyena wall" that, along with cold and other factors, might have inhibited human movement to Beringia before hyenas went extinct just before the close of the Pleistocene. The absence of hyena remains in the Afontova Gora refuse, and the general rarity of these creatures in the Yenisei basin (Ovodov, unpublished observations), allows us to speculate that the Yenisei might have been a major route out of Siberia, used by the ancestors of Paleo-Indians to reach Beringia if hyenas had been a migration-retarding factor. However, we do have traces of hyenas in one site we tested away from the river in the region of Abakan. This site is Dvuglaska Cave, which we will discuss later.

While we believe the hyena barrier idea to be original, we recognize that Valerius Geist (1989) earlier proposed that huge Pleistocene carnivores such as the short-faced bear of Alaska might have kept Paleo-Siberians out of eastern Beringia. Like the Siberian cave lion and cave bear, the Alaskan short-faced bear was a formidable creature, but none of these three species were social animals that hunted in packs. One of these larger carnivores could not have equaled the mass of even a small pack of hyenas, let alone the multiple ripping jaws that such a pack would have had. Hence, we consider the hyena to have been a larger threat to humans than the larger cave lions and cave bears. Fossil evidence for the magnificent Siberian tiger, also a lone predator, is sparse outside of the Russian Far East.

Afontova Gora perimortem taphonomy shows almost no carnivore contribution to the bone breakage, the reason for which we believe was extraction of all nutrients possible. Cooking of these bone fragments with their adhering bits of flesh would have helped extract even more nutrients. However, one burned bone out of 5000 can hardly be considered sufficient physical evidence to infer cooking. How food was prepared by late Pleistocene Siberians remains a puzzle.

# 2 Boisman II

## Background

Boisman II is a Neolithic shell mound located south of Vladivostok at 42°47′20″ N, 131° 16′30″ E. It sits at the junction of the Riazanovka River alluvial plain and a coastal hill 500 m back from the present-day shore of the Sea of Japan in the Gulf of Peter the Great. Archaeologist Alexander N. Popov has been the principle investigator (Fig. 3.2). Faunal remains recovered by Popov and associates suggest that at the time of site occupation, the environment was much like it is today, and as will be explained below, we assume both are not too different from conditions in the late Pleistocene, perhaps with exception of pack ice conditions and greater sea mammal densities.

Boisman II is one of only two assemblages in our study that date entirely in Holocene times. We include it because it provides unique comparative perimortem bone processing information from a coastal habitat in the Russian Far East. On various grounds discussed elsewhere (Turner 2002), we assume that there were northern coastal settlements in the late Pleistocene, and likely with economies similar to those of Boisman II. Such sites are today drowned from the terminal Pleistocene 100 m rise in sea level. We further assume that rich coastal marine resources were similar in both the Pleistocene and Holocene, at least with respect to most fish, bird, and sea mammal species. In addition, the sea cow may have once existed south of the north Pacific pack ice. All coastal fauna would have been largely displaced to the southern border of the ice, perhaps near the mouth of the Amur River and the island of Sakhalin. Inter-tidal invertebrate species edging the Pacific Ocean, the Sea of Okhotsk, and the Sea of Japan may have been displaced even farther south due to winter ice scouring their near-shore habitat in the colder phases of the Pleistocene. However, deeper-water invertebrates such as clam, crab, octopus, and other forms probably were unaffected by whatever ice conditions came and went through time. Pleistocene humans probably had nothing to do with these deeper-water species. In any event, we use Boisman II as a bioarchaeological model for the probable way of life on the Primorsky coast in late Pleistocene times, except for the coastal marine invertebrate community. We suspect that Pleistocene coastal sites differed to some extent from Pleistocene riverine sites hundreds of kilometers inland, if for no other reason than sea mammal exploitation would have been rare to non-existent in the interior.

According to Popov's excavations, oyster shells make up volumetrically more than 90% of the food refuse in the midden deposits. Fish, birds, sea mammals, and terrestrial

Fig. 3.2 Boisman II workers and Olga Pavlova. Director of the Far Eastern State University Natural History Museum, and chief archaeologist for the Boisman II excavations, Alexander N. Popov sits at far right. His museum technicians, Natasha and Olga, are on the left. The photograph was taken in the archaeology laboratory of the museum, the most modern and best organized of all the institutions we visited and worked in. Closer to Japan than to Moscow, the museum is obviously being upgraded for Far Eastern tourist expectations. Boisman II is a coastal shell mound located south of Vladivostok at 42°47′20″ N, 131°18′30″ E (CGT color FESU 6-14-00:3).

species have been identified (Popov and Yesner 2006). David R. Yesner (personal communication September 13, 2009), working with his graduate student Rhea Hood, suggests that sea mammals make up 20% of the Boisman II mammal remains, with 50% being cervids and 30% being suids. Artifacts found by Popov and associates include worked shell, bone points and spear heads, fish hooks, needles, awls, chipped and ground stone tools, pottery vessels, and other material culture items. Excellently preserved human burials were discovered, the first in Primorsky. The skeletal remains radiocarbon date at approximately 5000 BP (Popov *et al.* 1997:85). Tatiana Chiksheva's comparisons indicate that the Boisman II Neolithic crania are most similar to Deer Chuckchi who live far to the north.

## Findings

**1 Provenience.** In 1999 and 2000 Olga Pavlova and the senior author studied 840 pieces from collections made throughout most of the midden areas of Boisman II, excavated prior to those two years. These materials were curated in the Natural History Museum of the Far Eastern State University, Vladivostok. We found no meaningful stratigraphic or spatial differences except those related to pockets of shell, localized burials, and other minor cultural-depositional events well known for coastal shell middens in the north Pacific, if not worldwide. The pooled Boisman II assemblage total is 9.5% of our grand total of 8813 pieces (see Table A1.1, site 2). Numerically, Boisman II overwhelms some of our smaller Pleistocene assemblages. We try to compensate formally and informally for these dissimilar sample sizes in our various interpretations and inferences.

**2 Species.** The most common classes identified among 840 Boisman II pieces include indeterminable (59.9%), big mammal (10.4%), *Canis* (8.2%), and small mammal (4.4%) (see Table A1.2, site 2). Compared with the pooled assemblage averages, Boisman II has more big mammal and indeterminable, and fewer goat-sheep and all mammals that went extinct at the end of the Pleistocene. A faunal analysis of Boisman II that will include birds, fish, and more specific identifications is in progress by David Yesner and students.

**3 Skeletal elements.** Of 1242 Boisman II pieces classified, the most common are unknown (50.7%), long bone (8.9%), vertebrae (6.7%), foot (5.4%), and mandible (4.5%) (see Table A1.3, site 2). Compared with the pooled assemblage, Boisman II has fewer ribs, long bones, and toes, and more unknown pieces. All in all, there are many more similarities than significant differences between Boisman II and the pooled assemblage. Given the marked ecological difference between these two data sets, should we expect to see more or less general difference in skeletal elements? And if we do not see much difference, is this because of identification potential, preservation potential, or utilization?

**4 Age.** Boisman II has 7.6% sub-adults (39 / 513) (Table A1.4, site 2), which is the same as the pooled assemblage.

**5 Completeness.** Out of 501 Boisman II pieces, there are 13.8% whole bones, 37.5% with one anatomical end, and 48.7% with no anatomical ends (Table A1.5, site 2). The Boisman II pieces are slightly more complete than those of the pooled assemblage.

**6 Maximum size.** The mean maximum Boisman II piece size is 7.4 cm, and the range is 2.5–22.5 cm (Table A1.6, site 2). These values are slightly less than those of the pooled assemblage. Taken together with completeness, maximum size suggests that average animal size was smaller in Boisman II than in most of the other assemblages, even excluding the very big forms such as mammoth, rhinoceros, and others. Comparing the maximum size values with undamaged long bone lengths provided by Vera Gromova (1950: table 27) shows that the Boisman II upper range limit is similar to that of boars and wolves, but less than the long bone lengths of larger forms similarly expected in the Primorsky region (bear, elk, others).

**7 Damage shape.** This variable was not scored in Boisman II as it was still being formulated at the time of our studies in Vladivostok.

**8 Color.** There is only a small number of black pieces (burning) in the Boisman II assemblage (7 / 1247; 0.6%) (Table A1.8, site 2). Most pieces are ivory colored (99.1%). Compared with the pooled assemblage averages, burning is nearly the same, but there are fewer ivory colored pieces in the pooled assemblage. Elsewhere it will be shown that brown colored pieces are fairly common in open sites. Since Boisman II is an open site, the difference in the color ratio is attributable to antiquity, soil composition, or both.

**9 Preservation.** Out of the 513 Boisman II pieces scored for quality, 98.0% were ivory hard and only 1.9% were chalky (Table A1.9, site 2). For an open site, this high-quality condition may be due to a combination of modest site antiquity and the buffering of soil acid by the presence of calcium carbonate-rich invertebrate shells in the midden. Also, rapid burial aids preservation. Compared with the pooled assemblage average for ivory quality, Boisman II is better.

**10 Perimortem breakage.** Boisman II has 81.8% perimortem breakage in 1022 pieces (Table A1.10, site 2). This is only slightly less than the pooled assemblage average, suggesting substantial similarity in bone reduction behavior by humans and carnivores through time and under widely different environmental conditions.

**11 Postmortem breakage.** Out of 1099 Boisman II pieces, postmortem breakage occurs in 8.2% of the assemblage (Table A1.11, site 2). This is about half of the pooled assemblage average. We have no idea what lies behind this difference except that rapid burial may be involved.

**12 End-hollowing.** Inasmuch as domestic dogs were found at Boisman II, and they are ethnographically ubiquitous in Siberia, it is not surprising that end-hollowing occurs in Boisman II. It does so in 3.2% of the 1232 scorable pieces (Table A1.12, site 2). Compared with the pooled assemblage average, Boisman II has about three times less end-hollowing. Again, rapid burial comes to mind.

Boisman II matches the hyena-free open site of Ust-Kova (site 26, 3.5%) for end-hollowing frequency. Because of the dog–wolf size similarity, in considerable contrast to the larger body weights and jaw sizes of hyenas, bears, and lions, we are strongly tempted to suggest that the frequency of end-hollowing might help identify relative size, and thus the species of the carnivore(s) involved. We have no idea how to factor in considerations like hunger and play.

**13 Notching.** Notching was scored in 541 Boisman II pieces. It occurs in only 2.6% of the assemblage (Table A1.13, site 2). Most of these 14 pieces have only one notch. Compared with the pooled assemblage average, Boisman II has fewer notched pieces. Again, the dog–wolf size similarity comes to mind, especially when much more notching is associated with the more powerful hyenas, as identified on multiple grounds at Razboinich'ya (site 21, 23.6%) and Maly Yaloman (site 16, 23.0%) caves.

**14 Tooth scratches.** Out of 1235 Boisman II pieces, tooth scratching is present in only 1.6% (Table A1.14, site 2). The number of scratches ranges between one and six per piece. Compared with the pooled assemblage average, Boisman II has fewer scratched pieces. To turn again to the dog–wolf vs. hyena comparison, Razboinich'ya and Maly Yaloman have 37.1% and 28.6% of pieces with scratching, respectively.

**15 Tooth dints.** There are 8.2% of 512 Boisman II pieces with one to more than seven dints per dinted piece (Table A1.15, site 2). Compared with the pooled assemblage average, this shows three times less dinting in the Boisman II assemblage, as well as much less than in Razboinich'ya and Maly Yaloman.

**16 Pseudo-cuts.** There are only two Boisman II pieces with pseudo-cuts (0.4%) (Table A1.16, site 2). This is several times less than the pooled assemblage average, as well as less than what Razboinich'ya (3.1%) and Maly Yaloman possess (6.3%). These sites were hyena dens.

**17 Abrasions.** The Boisman II assemblage has 1.8% of pieces with abrasions (9 / 512 pieces) (Table A1.17, site 2). This value is slightly greater than the pooled assemblage average. Since abrasions are uncommon in our study, this difference is probably meaningless, although in all assemblages bone breaking seems to have been done far less often with abrasive stone than with bone, horn, or wood implements.

**18 Polishing.** Boisman II possesses 14.2% end, 0.8% middle, and 1.6% both end and middle polishing in 485 scored pieces (Table A1.18, site 2). Compared with the pooled assemblage averages, it is clear that there is significantly less polishing in Boisman II. Polishing is much more common in Razboinich'ya and Maly Yaloman caves.

**19 Embedded fragments.** Out of the 513 Boisman II pieces, 4.7% possess 1–6 embedded fragments (Table A1.19, site 2). This frequency is nearly the same as the pooled assemblage average. Razboinich'ya, but not Maly Yaloman, has more pieces with embedded fragments.

**20 Tooth wear.** Boisman II has the largest number (38) of teeth for assessing wear in our study (Table A1.20, site 2). Using grades 0 and 0–1 (no wear and wear present but dentine not exposed) as criteria for defining "young" individuals, the Boisman II dental wear group suggests that three-quarters (76.3%) were young. This is twice the number of young individuals in the pooled assemblage. We are generally disappointed with the tooth wear variable because of its overall small sample size (293 pieces), which means statistical inference is weak. Nevertheless, young animals suggest summer or fall kills. In Aleutian shell mounds, sea mammal tooth wear is rare except for sea otters.

**21 Acid erosion.** Boisman II has no examples of acid erosion in 553 scored pieces (Table A1.21, site 2). Since we have long suspected acid erosion to be associated primarily with hyenas, Boisman II offers a useful comparative sample to support this view due to the lack of hyenas and the presence of middle-sized carnivores (dogs) in the site. The amount of acid erosion that we could possibly attribute to wolf digestion at Bolshoi Yakor I, Kamenka, Varvarina Gora, and other sites, is very low to absent. Hence, Boisman II, despite its Holocene age, nicely provides comparative information for differentiating between Siberian wolf and hyena perimortem taphonomy where acid erosion is a critical variable.

**22 Rodent gnawing.** The largest amount of rodent gnawing in our study occurs in the 553 scored Boisman II pieces (Table A1.22, site 2). This value is much greater than the pooled assemblage average. The only other site with this magnitude of occurrence is found in Krasny Yar (4.9%), a naturally deposited riverside accumulation of large and small mammal remains. We have no satisfactory explanation why these two sites stand out so against a background of general absence of rodent gnawing in our study. Perhaps in both localities the bones remained upon the ground surface longer than at sites with no rodent gnawing, or possibly both site localities had larger rodent populations, thus increasing the chance for occasional gnawing. Porcupines are well-known gnawers of human bone in tree burials on the northwest coast of British Columbia. Ovodov has identified at least one porcupine in Pleistocene Siberia, so these creatures cannot be excluded from other rodents.

**23 Insect damage.** Like most of the assemblages we investigated, none of the 513 Boisman II pieces showed any sign of insect damage or unusual damage that might be attributed to insects (Table A1.23, site 2).

**24 Human bone.** The human skeletal remains from intentional burials were removed from the faunal collection and sent to physical anthropologist Tatiana Chicksheva at IAE, Novosibirsk. There, the senior author examined most of the remains. No perimortem damage was identified. Morphometric analysis of the human remains can be found in Popov *et al.* (1997).

**25 Cut marks (Figs. 3.3–3.6).** Cut marks were plentiful (21.3%) on 553 Boisman II pieces (Table A1.25, site 2). While the number of cut marks per piece varied from one to more than seven, most common was one cut per piece (6.1%). Compared with the pooled assemblage average, Boisman II has three times more pieces with cutting. There is nothing much to be made of this difference since some of the pooled assemblage sites are mainly paleontological, not archaeological.

**26 Chop marks.** Out of 513 Boisman II pieces, 44 have chop marks (8.6%) (Table A1.26, site 2). As with cutting, the number of chop marks per piece varies from one to more than seven, slightly skewed toward lower numbers of chops. The frequency of Boisman II chop marks differs only slightly from the pooled assemblage average.

## Discussion

Although of Neolithic age, Boisman II is included in our study because it offers a glimpse of human economic life in a coastal setting. For several reasons we view Boisman II as a model of what late Pleistocene coastal northeast Asian life might have been like. Despite the strong difference between interior Siberian and coastal Primosky ecology and environments, there are interesting similarities as well as obvious differences in perimortem taphonomy of the two regions. Despite being an open site, the Boisman II faunal remains are preserved in an excellent condition. This we presume to be due to time as

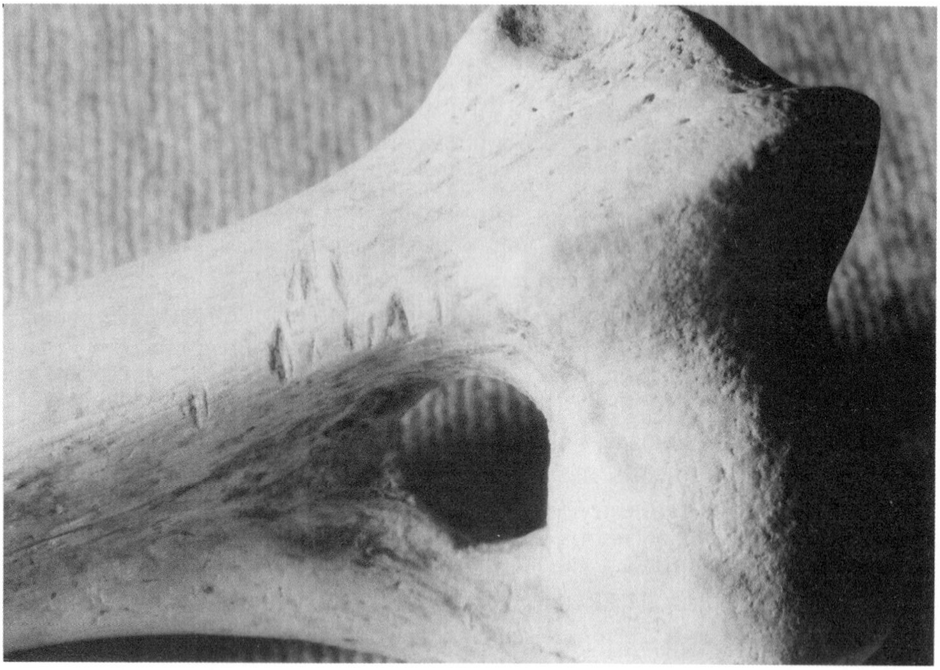

**Fig. 3.3** Boisman II dismemberment cut marks. Ten or more cut marks occur near the distal end of this humerus. Actual width of image in photograph is 3.3 cm (CGT neg. FESU 6-9-00:32).

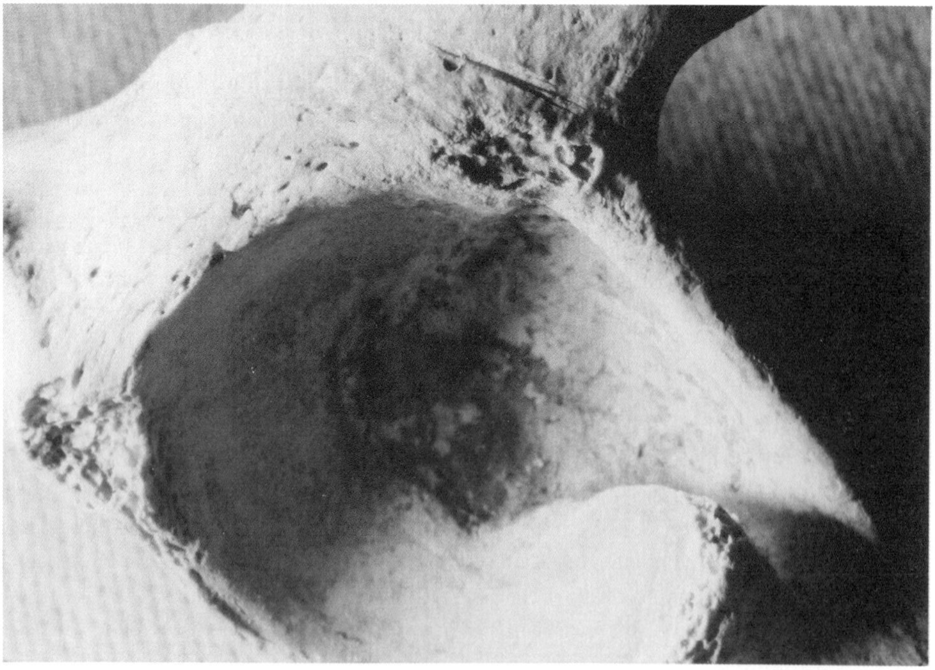

**Fig. 3.4** Boisman II dismemberment cut mark. One cut mark occurs near the acetabulum of this pelvic bone. Actual width of image in photograph is 3.3 cm (CGT neg. FESU 6-9-00:31).

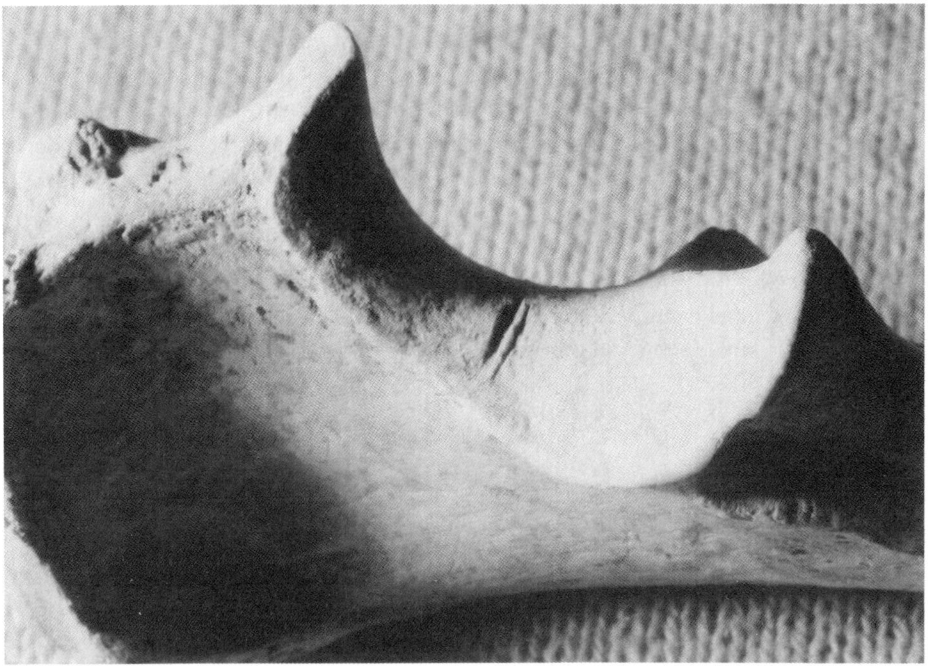

**Fig. 3.5** Boisman II dismemberment cut marks. Two cut marks occur in a proximal ulna joint. Actual width of image in photograph is 3.3 cm (CGT neg. FESU 6-9-00:28).

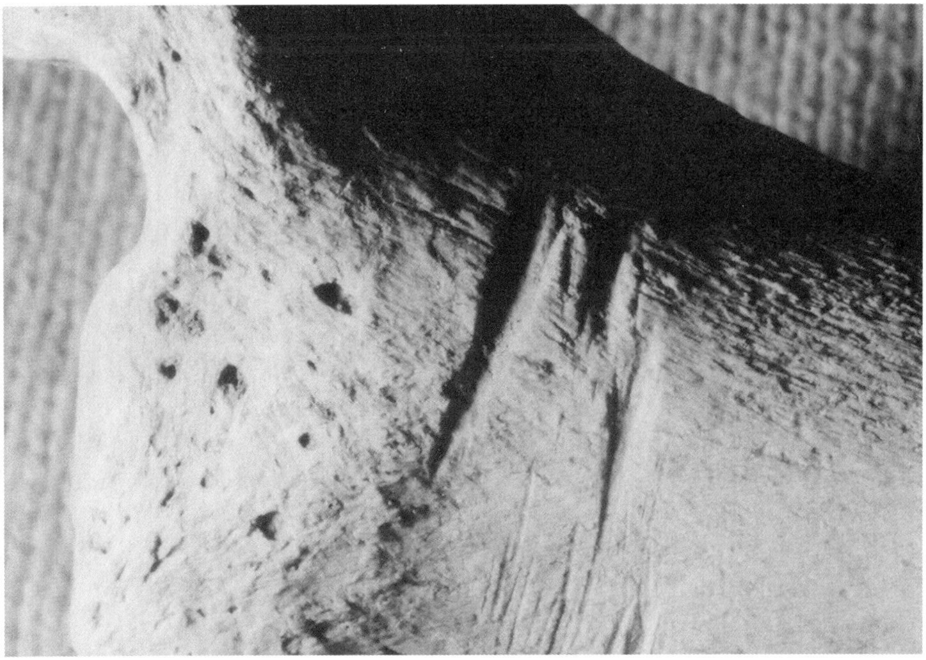

**Fig. 3.6** Boisman II dismemberment cut marks. At least 12 cut marks occur near the distal end of a tibia. Actual width of image is 3.3 cm (CGT neg. FESU 5-9-00:23).

well as soil conditions. Bone breakage patterning is much like that of our interior sites. This suggests similarities in human and carnivore processing behaviors in both regions. Several carnivore indicators, including end-hollowing, notching, tooth scratches, tooth dints, polishing, and acid erosion, are infrequent, but nevertheless demonstrate the presence of carnivores at Boisman II. Since dog remains have been found in Boisman II, the site's bone damage that we attribute to carnivores must have been produced by village dogs. Given that dogs and wolves are more similar in body weight and jaw size than either is to hyenas and larger carnivores, the Boisman II carnivore damage provides a useful perimortem taphonomic signature to help differentiate between damage caused by middle-sized carnivores like the wolf, and that done by the larger hyenas of our Pleistocene interior sites.

# 3 Bolshoi Yakor I

## Background

Bolshoi Yakor I (Big Anchor) is located in the mountainous gold-mining region northeast of Lake Baikal (Ineshin and Tetenkin 1995). Kuzmin and Orlova (1998:14), in reporting radiocarbon dates, place it at 56°22′ N and 115°73′ E. The site was discovered in 1985 by E. M. Ineshin on the 14 m right bank terrace of the northwesterly trending Vitim River, downstream from the left bank mouth of the Mamakan River, and 12 km from Bodaibo village. The Vitim River originates not far from Chita and joins the Lena that flows northward to the Arctic Ocean. The senior author (Turner 1998) once proposed that this drainage system (Vitim–Lena) would have been a natural corridor for reaching Beringia and the subsequent colonization of the New World, given that the pre-Siberian ancestors of Native Americans originated in Mongolia and north China. Because of the senior author's long-standing research into the peopling of the Americas, the location, content, and dating of Bolshoi Yakor I is of considerable interest.

Nine cultural levels have been identified by Ineshin and associates at Bolshoi Yakor I, ranging in depth from 0.6 m to 2.7 m below the present ground surface. Some 26 carbon-14 dates indicate that these cultural levels were deposited in the waning years of the Pleistocene, around 10 000–13 000 years ago (Ineshin *et al.* 2003). Ineshin *et al.* (1998) report that Bolshoi Yakor I yielded many finely splintered bones of reindeer, elk, Arctic fox, and hare, as well as remains of white partridge and fish. Dental annuli ("growth rings") of the continuously erupting reindeer molar teeth showed that the Bolshoi Yakor I people camped at this place during the fall migrations of reindeer traveling to their winter foraging grounds. The site may have been situated near a river crossing used by reindeer herds. Ineshin and Tetenkin (1995) illustrate some of the stone artifacts recovered from Bolshoi Yakor I. Included are end-scrapers made of macroblades, microblades, wedge-shaped cores, and waste associated with microblade production. These types of stone artifacts have been recovered in numerous sites in northeast Asia (Mochanov 2007, 2010, Mochanov and Fedoseeva 2010). While the latter do not illustrate any cylindrical cores or bifacial objects other than the preliminary blank that eventually is discarded as an exhausted wedge-shaped core, the sorts of artifacts that are illustrated relate better to those of proto-Aleut-Eskimo and Paleo-Arctic or Alaska and British Columbia than with Paleo-Indian. On the basis of rare stone types used in artifacts found at Bolshoi Yakor I, and knowledge of the stone's source location, Ineshin *et al.* (1998) have identified probable travel routes that extend from 60 km to 600 km from Bolshoi Yakor I. Various other sorts of information can be found in Ineshin and Tetenkin (2000) and Ineshin *et al.* (2004, 2005).

In August 2001, Ineshin helped Olga Pavlova and the senior author to study the site's faunal remains, which are curated in the Laboratory of Archaeology at Irkutsk State University. In addition to helping us access the collection, he explained the conditions and history of the multiple summers of excavation at Bolshoi Yakor I. Our meeting with Ineshin in the Laboratory of Archaeology was helpfully arranged by Ekaterina (Katya) Lipnina and German Medvedev. Nicolai D. Ovodov and Nicolai V. Martynovich made a preliminary sorting a week earlier for a small part of our studied sample. Examination time was limited due to the approaching end of our field season, so the collection was sampled instead of examining it in its entirety. The older deposits were favored, all of which are of terminal Pleistocene age. Kuzmin and Tankersley (1996:578) list the average carbon-14 ages for Bolshoi Yakor Levels 3 and 4 as being 10 237 ± 352 BP; Level 6, 11 733 ± 154 BP; and Level 7, 12 330 ± 250 BP. Additional dates are listed by Vasili'ev *et al.* (2002:527), and most of the 15 cluster tightly around 12 500 BP. We examined faunal remains from Levels 4 to 7 (Table A1.3). Other details about Bolshoi Yakor I, its location, and recent analyses can be found in Ineshin and Tetenkin (1995, 2000) and Ineshin *et al.* (2004, 2005).

## Findings

**1 Provenience.** From a much larger series, a total of 263 pieces of bone were selected for in-depth study (Table A1.1, site 3). These fragments make up 3.0% of our grand total from all sites. Assuming that the stratigraphic level designations for different excavation years each belong to the same stratum, then our sample breaks down to being 0.8% Level 4, 44.5% Level 5, 9.9% Level 6, and 44.9% Level 7.

**2 Species.** Working with small fragments of bone, species identification was as difficult as we have found in most other Siberian sites studied to date. Only one fragment out of 263 could be identified (reindeer), whereas nearly 90% were unidentifiable (Table A1.2, site 3). We might have done a little better had co-author Ovodov been with us. However, even with his many years of experience, Ovodov usually has strong reservations about making identifications when few anatomical landmarks are present. Identifying species from bone splinters and flakes, which make up 85% of the Bolshoi Yakor sample, has almost never been attempted by Ovodov. Ineshin and Tetenkin (1995) list the following forms as having been present: polar fox, fox, elk, reindeer, roe deer, white partridge, perch, and pike. They do not list any extinct species.

**3 Skeletal elements.** Cranial fragments, however small, are generally easy to identify, even if the species to which they belong are not. Hence, the very low frequency (4.9%) of cranial elements, including antler, listed in Table A1.3 (site 3) indicates that these elements are actually missing from our sample. Equally apparent is the low frequency of vertebrae (1.1%) and the absence of pelvic fragments. Pieces of long bones are abundant, and without exception in this sample of 263 pieces, all were broken open. Ribs are well represented, and to some extent this likely results from their easy recognition. Foot bones are also easy to identify, even more so since most are largely undamaged, containing as they do but little nutrient and even less usefulness for bone tool fabricational purposes. Given the overall severe damage, coupled with the missing cranial elements, no reliable MNI (minimal number of individuals), or other such characterizations can be provided. NISP values (number of identified specimens) are, however, possible to generate and are provided in Table A1.2 (site 3). Compared with our pooled assemblage averages, the main exceptional difference is that Bolshoi Yakor has more long bone pieces. It also has more unknown pieces.

**4 Age.** Sub-adulthood was identified on the basis of incomplete epiphysial closure or tooth eruption. Although the age of half of the 263 pieces could not be assessed (Table A1.4, site 3), it would appear that sub-adults were not common. Of the 127 pieces for which relative age could be determined, only two (1.6%) were sub-adult. There were no fragments suggesting fetal, neonatal, or very aged individuals. Compared with our pooled assemblage average for sub-adults, there is definitely a deficiency in the frequency of Bolshoi Yakor sub-adults. This suggests possible seasonality, perhaps late fall or winter hunting.

**5 Completeness.** Of 262 pieces, there were no whole bones in the Bolshoi Yakor sample (Table A1.5, site 3). More than 90% of all skeletal elements lacked both anatomical ends, indicating very intensive processing, coupled with the fact that no very large mammal such as mammoth, rhinoceros, or bison is represented in the assemblage. Very large mammal bones are much more difficult to break up than those of smaller mammals such as reindeer. Compared with our pooled assemblage average, Bolshoi Yakor has significantly more pieces lacking both ends.

**6 Maximum dimensions.** Since all bones in this sample had been broken into two or more fragments, the size statistics (Table A1.6, site 3) are indicative of the extent of reduction. In our sample selection procedure we used only specimens whose maximum dimension was greater than 2.5 cm. The mean of 6.0 cm is only slightly greater than two times the minimum selection size. Similarly, the standard deviation (4.0 cm) is small, and taken together with the mean clearly shows a considerable amount of bone fragmentation and reduction. Presumably this severe fragmentation indicates bone boiling to extract the calorie-rich fatty tissue (marrow) from the interior of skeletal elements, especially the long bones. Fragmentation of bones without much marrow probably represents the desire to boil off the small amounts of edible tissue still attached to bones after butchering. When Bolshoi Yakor is compared with the values for our pooled assemblage, the latter has a larger mean, standard deviation, range, and standard error. Looking at undamaged long bone lengths provided by Vera Gromova (1950, table 27), the Bolshoi Yakor I upper range limit of 50.5 cm matches the upper range limit of three out of five elk bones, and is larger than any of the long bones of reindeer or roe deer.

**7 Damage shape.** Despite the numerous descriptive categories of damage types possible for the 263 Bolshoi Yakor pieces (Table A1.7, site 3), the vast majority (91.0%) are flakes, fragments, and splinters. Combining these three damage forms in our pooled assemblage adds up to a considerably smaller value. Again, this clearly reflects the intensive butchering and perimortem processing this assemblage received.

**8 Color.** More than 75% of the 262 Bolshoi Yakor pieces are ivory colored, the condition that accompanies good preservation (Table A1.8, site 3). All of the black and white pieces represent burned bone, as does one ivory colored piece. Two of the brown pieces were probably also burned. Altogether, burning can be identified in 4.6% of this sample. This amount of burning is several times greater than our pooled assemblage average. There were additional burned and calcined Bolshoi Yakor fragments, but they were not scored because their size was smaller than the minimal study size of 2.5 cm.

**9 Preservation.** A hard ivory quality, in contrast with a soft chalky condition, occurs in 90.1% of the 263 Bolshoi Yakor pieces (Table A1.9, site 3). As with color, this hardness indicates good preservation conditions. Most of the pieces were not exposed to weathering (cracking, exfoliation, bleaching, and denaturing) that occurs during an extended period of time on or just beneath the ground surface. Most must have been rather quickly embedded into the cold protective soil of the site. This embedding could have been readily caused by human trampling due to heavy or repeated use of the site area. Some of the soft chalky pieces are calcined as a result of intense burning. The pooled assemblage has a generally poorer quality.

**10 Perimortem breakage.** One piece (0.4%) of the 263 in this Bolshoi Yakor sample lacked perimortem breakage (Table A1.10, site 3). This is an exceptionally large amount of processing, which correlates well with the assemblage's small average fragment size (mean: 6.0 cm). Compared with our pooled assemblage's perimortem breakage average, Bolshoi Yakor has significantly more (99.6%). The extreme amount of perimortem bone breakage, following carcass processing (slaughtering and dressing out the meat), documents the Bolshoi Yakorians' desire for the calorie-rich marrow and bone grease. Perhaps a small amount of the breakage might have been caused by scavengers after the departure(s) of the hunters and their families.

**11 Postmortem breakage.** This type of damage occurs at a much later time after death, a time when bone has become sufficiently denatured so that when broken, the fracture surfaces lack the curved and smooth appearance of perimortem breakage. Most postmortem breakage is the result of exposure to trampling, redeposition, decomposition, and archaeological recovery methods. Only seven of the 263 Bolshoi Yakor pieces (2.7%) exhibit postmortem breakage (Table A1.11, site 3). Compared with the pooled assemblage average, Bolshoi Yakor has much less postmortem breakage.

**12 End-hollowing.** One of the diagnostic perimortem taphonomic features of prolonged carnivore chewing is end-hollowing, which is rare in archaeological sites, but common in carnivore sites such as the Razboinich'ya hyena cave (Turner *et al.* 2001b). Only one (0.4%) of 263 Bolshoi Yakor fragments shows end-hollowing (Table A1.12, site 3). This suggests that scavengers did not stay long at Bolshoi Yakor after human departure, or at least there was but little activity by scavengers within the area of the site excavated so far. Compared with the pooled assemblage average, end-hollowing occurred less often.

**13 Notching.** The edge of a fracture may show one or more semicircular puncture-like notches caused by carnivore teeth or human bone smashing. Human-caused notching sometimes also shows some manner of indentation rarely seen in carnivore notching. One can further distinguish between the two agencies by accompanying tooth scratches or tooth dints. Out of 263 pieces, Bolshoi Yakor has 6.5% with notches (Table A1.13, site 3). Six of the 17 Bolshoi Yakor notched pieces also have tooth scratches or tooth dints (tooth cusp marks); 11 do not. Although some tooth damage may in fact have been caused by stone tools or bone mallets used by humans to break up whole bones, it is very doubtful that all six of the notched and scratch-dinted pieces resulted from human butchering. Hence, carnivores such as dogs must have been in the camp at the time the humans were present, or other carnivores scavenged the bone refuse after the humans left or even as they slept. (In the senior author's archaeological fieldwork in the Aleutian Islands of Alaska, he frequently had foxes come into camp in the evening to dig in the trash pit for anything edible.) As fragments with notching, tooth scratches, and/or dints were found in differing levels (Levels 5–7), joint human–carnivore activity would appear to have occurred repeatedly, perhaps during or following each time Bolshoi Yakor was occupied by its human inhabitants. Compared with our pooled assemblage average, Bolshoi Yakor has fewer examples of notching.

**14 Tooth scratches.** Carnivore tooth scratches occur most often on the outer cortex of a bone fragment. At Bolshoi Yakor tooth scratches are short and shallow, suggesting they were left by relatively small carnivores such as dogs or foxes. Out of 263 pieces, there are only five with tooth scratches, amounting to only 1.9% of the total sample (Table A1.14, site 3). One of the five pieces has 30 scratches. This possibly might have been caused by human mimicry of carnivore damage. However, it is the largest piece in the sample, being a 50.5 cm fragment of reindeer antler, and it does have the carnivore indicators of end-hollowing and tooth dinting. A large piece of chewable bone or antler would probably be the target of carnivore activity rather than most of the much smaller pieces that make up this sample. Compared with the pooled assemblage average, Bolshoi Yakor has fewer tooth scratches.

**15 Tooth dints.** Out of 263 Bolshoi Yakor pieces, 8.0% have two or more tooth dints (Table A1.15, site 3). The number of dints per piece ranges from two to more than seven. Of the 21 pieces with dinting, 19.0% have associated tooth scratches. Given that there are only five pieces with scratches, the dint–scratch association is substantial. Compared with the pooled assemblage, Bolshoi has fewer pieces with tooth dints, suggesting a limited carnivore presence during or after humans occupied the site.

**16 Pseudo-cuts.** There are three Bolshoi Yakor pieces out of 263 with damage considered to be pseudo-cuts (Table A1.16, site 3). Each of the three has a single pseudo-cut. This is a damage type that by itself cannot be readily distinguished from stone tool cut marks. However, each Bolshoi Yakor pseudo-cut is associated with carnivore damage done at the same time, hence the pseudo-cut designation. One of these fragments, a 12.6 cm splinter, also has five tooth dints. The second, a 10.0 cm splinter, has eight tooth dints and polishing in the middle of the fragment. The third, a 4.4 cm flake, has four tooth dints and polishing on both an end and the middle. Polishing of the middle portion of a fragment is more likely to have been done by carnivores than polishing on the ends, which could have resulted from some form of boiling and stirring of fragments to obtain bone grease. Cooking of this sort could have been done in a water-filled leather bag brought to a near-boiling temperature with cobbles heated in a campfire. Compared with the pooled assemblage average, Bolshoi Yakor has fewer pseudo-cuts, supportive of our view that the bone assemblage signals only a minor carnivore presence.

**17 Abrasions.** This form of multiple and parallel striation damage is experimentally known to occur when a struck bone slips on an anvil stone, or from a glancing impact from a rough hammer stone. There are only three (1.1%) Bolshoi Yakor examples with abrasion damage out of 263 (Table A1.17, site 3). The number of abrasion striations ranges from 10 to 45. It would appear that most of the bone breakage was done with a non-abrasive tool, such as a long bone billet or antler mallet, and whatever anvil was used to help crack bones open, it was smooth, like bone or wood, not abrasive and granular like most types of stone. Compared with the pooled assemblage average, Bolshoi Yakor abrasion frequency is nearly identical.

**18 Polishing.** As mentioned above, polishing at or near the middle of a fragment is thought to represent carnivore activity, whereas fragment end-polishing could be produced by human or carnivore activity. As Table A1.18, site 3, shows, 12.1% of 257 Bolshoi Yakor pieces have polishing. Two of the 31 pieces have only mid-piece polishing. This is assumed to be due to carnivore chewing. Allowing for the possibility that some of this mid-fragment polishing resulted from non-carnivore activity, such as trampling, there is, nevertheless, a fairly strong suggestion of carnivore activity that matches rather well the amount suggested by tooth dinting, scratching, and end-hollowing. However, the polishing is not heavy like that characteristic of bone fragments from the Razboinich'ya hyena cave, so, again, the carnivores are presumed to be relatively small dog- or fox-sized animals.

**19 Embedded fragments.** This damage feature was initially identified in the Razboinich'ya assemblage in large and deep tooth dints and in notched fracture surfaces. Subsequently, it has been observed in archaeological sites as well. Thus, it occurs in both carnivore- and human-derived bone assemblages. Of 263 Bolshoi Yakor pieces, 3.1% of the sample have one or more embedded fragments (Table A1.19, site 3). No embedded fragments are associated with tooth dinting, scratching, end-hollowing, and mid-piece polishing. Two are associated with notching and two are associated with cutting or abrasions. Four have no other association besides perimortem breakage. Thus, pieces with embedding at Bolshoi Yakor are more likely the result of bone breakage by humans than by carnivores. Compared with the pooled assemblage average, Bolshoi Yakor has nearly the same frequency of embedding.

**20 Tooth wear.** There are only two tooth-bearing pieces in the Bolshoi Yakor assemblage (Table A1.20, site 3). Both have fox-sized tooth crowns that are moderately worn into the dentine (grade 1); however, the crowns are not worn off completely (grade 2). This degree of wear likely represents middle age.

**21 Acid erosion.** Surface erosion is common in bone fragments from the Razboinich'ya hyena cave, and some are as large as 7.0 cm in diameter. Only one of the 263 Bolshoi

Yakor pieces greater than the minimum study size has stomach acid erosion (Table A1.21, site 3). It is a 3.6 cm, irregularly shaped foot bone fragment belonging to an unknown species. There were two other acid-eroded fragments, but both were smaller than the minimal 2.5 cm diameter and so were not counted. Each of these small pieces was presumably swallowed accidentally by a relatively small carnivore, such as a dog or fox. They were found in Level 5 (1989), in which were also found burned fragments and fragments with stone tool cut marks. Since we currently lack experimental evidence on the possibility that the human digestive system can also produce acid erosion, these eroded pieces should be regarded as having been swallowed by carnivores.

**22–24 Rodent gnawing, insect damage, human bone.** There are no pieces in this sample of 263 Bolshoi Yakor pieces that has either of these markings, nor is there any human bone.

**25 Cut marks.** Cut marks are generally distinguishable by their straightness, V-shaped cross-section, and one end of the cut usually being deeper than the other. Tooth marks are characteristically U-shaped in cross-section, and rarely straight. Plant root marks are semicircular in cross-section, if old, and almost always exhibit a random somewhat dendritic wandering pattern with one end commonly deeper than the other. A substantial 12.2% (32 / 263) of the Bolshoi Yakor assemblage has cut marks (Table A1.25, site 3). The number of cuts per cut piece ranges from 1 to 21, with the average being 5.1 incisions. Cut marks were present on pieces of long bones, vertebrae, metapodials, mandibles, ribs, one scapula, and on several unidentifiable pieces. Butchering dismemberment can be recognized in three pieces by cut marks near a joint of each one. Compared with the pooled assemblage average, Bolshoi Yakor has more cut marks and ranks in the upper third for the number of cut pieces. This amount of cutting goes along with our view that the Bolshoi Yakor site represents one having had intensive carcass processing.

**26 Chop marks.** Chop marks differ from cuts in being larger, deeper, often circular, and often with ragged edges. They are presumably produced by a large stone tool and delivered forcefully as impact blows rather than the slicing motion that produces cut marks. Only 2.2% (6 / 263) of the Bolshoi Yakor sample has chop marks (Table A1.26, site 3). Of the six chopped pieces there are 1–5 chops per piece. Altogether, there are 17 chop marks, making the average 2.8 chops per chopped piece. Chop marks occur on pieces of long bones, one rib, and one metapodial. Compared with the pooled assemblage average, Bolshoi Yakor has fewer chop marks.

## Discussion

Comparing several of the perimortem taphonomic findings at Bolshoi Yakor with another late Pleistocene open site, Ust-Kova, is interesting because of the substantial similarity between the two, except for preservation quality, which is much better at Bolshoi Yakor. Both differ markedly from cave assemblages such as Ust-Kan (Turner *et al.* 2001b). Ust-Kova is located at the junction of the Angara and Kova rivers. It is described by Derev'anko *et al.* (1998:122–123, 133–137).

The Ust-Kova sample contained mainly caribou (reindeer) and horse remains. The bone condition was mainly soft and chalky (65.5%), a poor condition six times more common than at Bolshoi Yakor (9.9%). Mean size was about the same in both sites, but Ust-Kova had more whole bones (11.2%). Tooth scratches and dints at Ust-Kova were 4.3% and 5.2%, respectively – values not much different from those at Bolshoi Yakor. Polishing was more frequent at Ust-Kova (33.3%), but pseudo-cuts (Ust-Kova, 1.7%) and stomach acid erosion (Ust-Kova, 0.0%) were very similar in frequency. Stone tool cut marks at Ust-Kova were 14.7%, nearly the same as Bolshoi Yakor (12.2%), as were chop marks. Most of the other taphonomic features were very similar; however, there was no sure sign of burning at Ust-Kova.

The sample of Bolshoi Yakor faunal refuse shows a pattern of perimortem taphonomy that resulted mainly from human butchering, cooking, and breaking the bones of small (hare-sized), medium (fox-sized), and large (deer-sized) animals. The fragmentation was so extensive that most fragments could not be identified as to skeletal element, species, or MNI represented in the sample. However, it is fairly certain that very large animals such as mammoth, rhinoceros, or bison are absent from our sample. The intensive processing by these ancient hunters and their families shows clearly that they were not wasting any of the nutrients available in each slain animal, regardless of its size, age, or species. In addition to the carcass processing by humans, there is good evidence of small carnivores having also chewed on some of the remains. The amount and degree of carnivore damage is felt to have been done by a few small- to medium-sized carnivores such as dogs or foxes. There is no sign of the more severe types of damage caused by hyenas, bears, or wolves in any of the fragments.

# 4 Borabashevskaya

## Background

Borabashevskaya is a small rock shelter whose faunal remains we examined very briefly in 1999. The site may have served as a temporary camp by humans, but its use was probably mainly by non-human animals. The site is located south of Vladivostok on a secondary river in the interior of southern Primorsky Territory, and is under investigation by Nina A. Kononenko.

## Findings

**1 Provenience.** Kononenko recovered 30 pieces of bone in 1998 from midden at the entrance to the rock shelter. The collection constitutes only 0.3% of our grand total of 8813 pieces (Table Al.1, site 4). Inasmuch as hyena remains and other late Pleistocene fauna were found in the Geographic Society Cave north of Vladivostok, near Nahodka (Ovodov 1977b), there was some thinking that Borabashevskaya might also date to the late Pleistocene.

**2–3 Species and skeletal elements.** None of the pieces could be identified for these variables (Table A1.2, site 4; Table A1.3, site 4).

**4 Age.** All of the 30 Borabashevskaya pieces are adult? (Table A1.4, site 4).

**5 Completeness.** All 30 of the Borabashevskaya pieces lack both anatomical ends (Table A1.5, site 4).

**6 Maximum size.** The Borabashevskaya range was estimated to be 2.5 cm to 5.0 cm (Table A1.6, site 4).

**7 Damage shape.** All of the 30 Borabashevskaya pieces are fragments (Table A1.7, site 4).

**8 Color.** Most of the 30 Borabashevskaya pieces are ivory colored, although some are darkened and superficially appear to have been burned (Table A1.8, site 4).

**9 Preservation.** Each of the 30 Borabashevskaya pieces is ivory hard, although there is some small amount of surficial exfoliation (Table A1.9, site 4).

**10 Perimortem breakage.** All 30 of the Borabashevskaya pieces have perimortem breakage (Table A1.10, site 4), none of which can with certainty be attributed to human processing.

**11 Postmortem breakage.** There is no postmortem breakage in the 30 Borabashevskaya pieces (Table A1.11, site 4).

**12 End-hollowing.** There is no end-hollowing in the 30 Borabashevskaya fragments.

**13 Notching.** Notching was not assessed in the Borabashevskaya collection.

**14 Tooth scratches.** One of the 30 Borabashevskaya fragments has one tooth scratch with the characteristic U-shaped cross-section (Table A1.14, site 4).

**15 Tooth dints.** Dinting was positively identified but not counted.

**16 Pseudo-cuts.** This variable was not identified, nor were there any cut marks that it might have been confused with in the Borabashevskaya collection.

**17 Abrasions.** No examples of abrasions occur in the 30 Borabashevskaya fragments (Table A1.17, site 4).

**18 Polishing** was common.

**19–20 Embedded fragments and tooth wear.** These variables were not looked for in the Borabashevskaya collection.

**21–26 Acid erosion, rodent gnawing, insect damage, human bone, cut marks, and chop marks.** No examples of these variables were seen in the 30 Borabashevskaya fragments (Tables A1.21, site 4–A1.26, site 4).

## Discussion

Our inspection of the Borabashevskaya collection was limited to a very brief period of study. Nevertheless, it was long enough to determine that we were not going to be able to help decide whether the rock shelter dated to the Pleistocene or Holocene, or whether it was used by humans or by carnivores. This is because the perimortem damage is non-distinctive, although there was more of a sense of carnivore damage than human damage in the small assemblage. Our plan was to examine additional material during our return to Vladivostok in 2000. However, Kononenko was out of the city during our second visit and we were unable to add to her 1999 collection.

# 5 Denisova Cave

There have been hints that Denisova was not a [continuously occupied] habitation site (the absence of charcoal is one indicator), but the large numbers of bones with heavy perimortem cracking and breakage, coupled with scores of specimens with tooth dints and scratches indicates a rather heavy usage of Denisova by hyenas.

(C. G. Turner field notes, Denisova Cave, July 13, 1999)

## Background

From the spacious entrance of Denisova Cave, one looks down 20 m to see the Anui River rushing through the narrow gorge on its long journey to the Ob River (Figs. 3.7–3.10). The cave is located in the northwestern Altai Mountains of southern Siberia at 51°22′50″ N 84° 41′20″ E, some 80 km southwest of Gorno-Altaisk city. It is situated near the base of a high and nearly vertical limestone cliff, like several other cliffs in the dramatic canyon locality. One reaches the cave from the canyon bottom by ascending a steep rocky talus slope. The cave is large and variable in shape. Denisova is named after a nineteenth-century monk, Denis, who used it for his *podvig* (ascetic hermitage deed), and charcoaled his name on its cold, stained gray limestone walls. In addition to the main entrance, there is an unreachable small opening in the high vaulted ceiling several meters in from the entrance. Water seeps down the walls in some places and drips from the ceiling in others. In contrast to our baseline hyena cave (Razboinich'ya) that is relatively narrow, totally dark, cold, and dry, Denisova Cave is a spacious, twilight-like interior, chilly and damp with water seeping down from its high, round ceiling. Even at the height of the summer, the student archaeologists wear several layers of clothing, knee-high rubber boots, gloves, and hats and/or scarves while on their shifts in the chilly, damp cave.

Denisova Cave, as an archaeological monument, was "born" in the summer of 1977 when Nicolai Ovodov, having obtained on his own initiative special permission to conduct exploratory field research in the Altai, made the first scientific test excavations just inside the entry drip line of the cave (Okladnikov and Ovodov 1977; Ovodov 1979). His two trenches of 4 m depth produced a wealth of Neolithic and Paleolithic artifacts, as well as numerous bones and teeth of large and small animals and other residue of human and animal use of the cave. The excavated loose sediments showed irregular structure, indicating the need for careful long-term study. This supreme prehistoric locality has attracted the attention of archaeologists and all sorts of earth scientists because of the excellent preservation and abundance of its anthropological

**Fig. 3.7** Denisova Cave. Alexander Soloviev photographing the cave (small black rectangle in white area in center of the image) from across the narrow Annui River canyon in the Altai Mountains. The cave entrance is 20 m above the level of the river (CGT color Denisova 6-10-87:23)

and paleontological content, its many millennia of deeply stratified deposits, and other elements of much scientific interest.

Because Denisova Cave was unquestionably a very important prehistoric resource site, the Institute of Archaeology and Ethnography (then the Institute of History, Philology, and Philosophy) established a permanent field camp in 1982 for long-term excavations here and at other Pleistocene sites in the area that Ovodov had tested. That year Denisova Cave became the subject of a multi-year research program of excavation and analysis under the directorship of A. P. Dereviyanko, and the field supervision in differing years by V. I. Molodin, S. V. Markin, and, currently, M. V. Shunkov. Denisova Cave probably has more published accounts than any other Siberian Paleolithic site, most in Russian, but increasingly more in English (Derevianko *et al.* 1998k). To mention only a few, and in chronological order: Derevianko *et al.* (1985, 1997d, 1998a, 1998b, 1999a, 1999f, 2002b, 2003a, 2003b), Ivleva (1990), Maloletko and Panychev (1989), Germonpré (1993), Derevianko and Molodin (1994), Agadjanyan (1999), Shunkov and Agadjanyan (2000), Panteleyev (2002), Baryshnikov (2003), Agadjanyan and Serdyuk (2005), Ovodov and Orlova (1977), Wrinn (n.d.). Additional information published in English can be found in Derev'anko *et al.* (1998).

Vasili'ev *et al.* (2002) have compiled an inventory of new and previously published carbon-14 dates for Paleolithic sites in Siberia, including Denisova Cave. For the

**Fig. 3.8** Denisova Cave. View from the cave mouth back toward the mountainside where Fig. 3.7 was taken. The student is running the looped cable bucket hoist that takes excavated cave dirt by gravity down to the river for water-screening, and at the same time returns the lighter empty bucket. We paid close attention to excavation and curation procedures to assess possible postmortem bone damage. Water screening and cleaning of soft, damp bone may have contributed somewhat to wear, polishing, and pseudo-cuts (CGT color Denisova 6-10-87:3).

Mousterian assemblages (Layer 21; Entrance Layer 9) their dates range from >34 700 BP to 46 000 BP. For the Upper Paleolithic (Layer 11) they list one date: >37 235 BP

There are some very much earlier thermoradiolucence (TRL) dates according to Michael Shunkov (excavation supervisor, personal communication, July 14, 1999): Layer 14, 69 000 BP TRL date; Layer 21, 110 000 BP TRL date; definite Mousterian; Layer 22, 224 000 BP TRL date. On various grounds these TRL dates are rejected by workers unaffiliated with the IAE.

The present authors have twice visited Denisova together, first in 1987, then in 1999. Turner and Pavlova, without Ovodov, visited this site and others in July 2006. The first visit was made possible by Derevianko for the purpose of helping the senior author and his late wife, Jacqueline A. Turner, gain familiarity with this important Siberian archaeological district. During the second visit, Turner, Ovodov, and Pavlova made the observations that follow. The third visit was made in 2006 by Turner and Pavlova for the purpose of checking a number of concerns about stratigraphy and verifying whether or not ongoing excavation in the rear of the main cave was producing additional hyena-damaged faunal remains. In fact, the 2006 bone refuse examination showed even more hyena damage than previous examinations. The damage was mainly in the form of many small acid-eroded bone fragments.

Fig. 3.9 Denisova Cave field crew. Shown are some of the 1987 Denisova field crew and visitors under the kitchen and dining tent following breakfast. Left to right: Irina Troitskaya, Marina Chepalova, Valery Trofimov, Vyacheslav Molodin (field supervisor), Anatoly Zenin, three students, Olga Pavlova, and Jacqueline Turner (CGT color Denisova 6-10-87:30).

## Findings

**1 Provenience.** The 116 pieces scored during 1999 make up 1.3% of our 8813-piece grand total (Table A1.1, site 5). The material was recovered from Layers 7–22, with most originating in Layers 10 and 22. Excavation years range from 1984 to 1997. Stratigraphic comparisons for species, age, maximal length, and human and carnivore processing produced no statistically significant differences. Hence, we have pooled the assemblage and treat it as a single temporal unit. In July 2006 we informally examined some faunal remains as they were being cleaned and labeled by student helpers. We noted an especially strong presence of hyena, perhaps because the material was being excavated from the very rear of the cave; however, we made no quantitative observations.

**2 Species.** For the 116 Denisova pieces, the most commonly identified groups are: indeterminable (74.1%), goat-sheep (7.8%), and horse (5.2%) (Table A1.2, site 5). Compared with the pooled assemblage averages, Denisova has fewer bear, big mammal, hyena, mammoth, and reindeer, and twice the number of indeterminable pieces (Table A1.2, site 4).

Our project did not study bat bones, but they have been found in Denisova Cave. Their use of caves is generally regarded as a sign of human absence (Ovodov 1974).

**Fig. 3.10** Deepest horizon in Denisova Cave. A young student stands in the cave's deepest level. The marked difference between Holocene and Pleistocene deposition is clearly evident in this view. Twenty-two levels have been identified (CGT color Denisova 7-14-99:6).

**Fig. 3.11** Denisova Cave Mousterian artifacts. Cores, Levallois and irregular-shaped pieces from Layer 14. Unifacial objects such as these would have been responsible for the cut marks found on various pieces of stratigraphically associated animal bone. Scale is 3.0 cm (CGT color IHPP 6-2-87:17).

**3 Skeletal elements.** The most commonly recognized groups in the 116-piece Denisova assemblage are long bones (37.9%), unknown (22.4%), metapodials (8.6%), ribs (7.8%), and foot bones (3.4%). Compared with the pooled assemblage averages, Denisova has fewer mandibles, vertebrae, humeri, toes, and twice the number of long bones (Table A1.3, site 5).

**4 Age.** Only one (0.9%) of the 116 Denisova pieces belongs to a sub-adult (Table A1.4, site 5). This low frequency of sub-adult pieces is several times less than the average for the pooled assemblage.

**5 Completeness.** Completeness (and size) influences to a large degree the ability to identify species, skeletal elements, and age. The 116 Denisova pieces are very incomplete: whole (6.0%), one anatomical end (7.8%), no anatomical ends (86.2%) (Table A1.5, site 5). The pooled assemblage averages show much more completeness.

**6 Maximum size.** The mean size of the 116 Denisova pieces is 5.2 cm; the range is 2.3 cm to 13.8 cm (Table A1.6, site 5). Compared to the pooled assemblage mean and range, it is clear that there has been marked bone processing and piece size reduction in Denisova. Looking at the whole long bone lengths of goat-sheep and horse provided by Vera I. Gromova (1950: table 27) shows that the Denisova upper range limit is much smaller than her values.

Fig. 3.12 Denisova Cave Mousterian artifacts. Stone artifacts from Layer 19. Scale is 3.0 cm (CGT color IHPP 6-2-87:10).

**7 Damage shape.** No observations were made on damage form in the Denisova sample, because at the time of our study we were still formulating the definitions.

**8 Color.** Out of 116 Denisova pieces, 94.8% are ivory colored, and 5.2% are brown. None are black (burned) (Table A1.8, site 5). Compared with the pooled assemblage, Denisova has more ivory colored, and fewer brown colored pieces.

**9 Preservation.** All 116 Denisova pieces are ivory hard (Table A1.9, site 5). The pooled assemblage has chalky pieces. Preservation is excellent at Denisova Cave.

**10 Perimortem breakage.** Most of the 116 Denisova pieces have perimortem breakage (96.5%) (Table A1.10, site 5). This frequency is greater than the pooled assemblage average.

**11 Postmortem breakage.** The frequency (18.3%) of postmortem breakage in the 116 Denisova pieces is nearly identical to the pooled assemblage average (Table A1.11, site 5).

**12 End-hollowing.** Out of 115 Denisova pieces, only 2.6% have end-hollowing (Table A1.12, site 5). This frequency is about one-third of the pooled assemblage average, and less than that seen in Razboinich'ya (8.1%) and Maly Yaloman (9.5%). We attribute the low frequency of end-hollowing to the simple fact that there are very few pieces with ends, as indicated by the mean maximum piece diameter.

**Fig. 3.13** Denisova Cave Mousterian artifacts. Specimens from Layer 22 show Levallois manufacturing technique. Scale is 3.0 cm (CGT color IHPP 6-1-87:9).

**13 Notching.** Out of 115 Denisova pieces, 22.6% have 1–6 notches (Table A1.13, site 5). One notch per notched piece is the most common condition. The Denisova frequency is greater than the pooled assemblage average (14.6%). It is very similar to the amount of notching found at Razboinich'ya (23.6%) and Maly Yaloman (23.0%). Because of this frequency of notching, as well as other damage characteristics, we feel that Denisova was used rather commonly by large carnivores.

**14 Tooth scratches.** Denisova's 115 pieces have from one to more than seven scratches in 20.9% of the assemblage (Table A1.14, site 5). Two scratches per piece is the most common number. The pooled assemblage average is almost exactly the same. Both Razboinich'ya (37.1%) and Maly Yaloman (28.6%) have more pieces with tooth scratches.

**15 Tooth dints.** Out of 115 Denisova pieces, 22.6% have dinting (Table A1.15, site 5). The most common number is one dint per piece. Compared to the pooled assemblage average, Denisova has slightly fewer dinted pieces. Razboinich'ya has more than twice the number of dinted pieces (52.8%), while Maly Yaloman has 32.6%.

**16 Pseudo-cuts.** Our 115 Denisova pieces have a very high frequency of pseudo-cuts (28.6%) (Table A1.16, site 5). This value is so out of line with the pooled assemblage average, and those of Razboinich'ya (3.1%) and Maly Yaloman (6.3%), that two

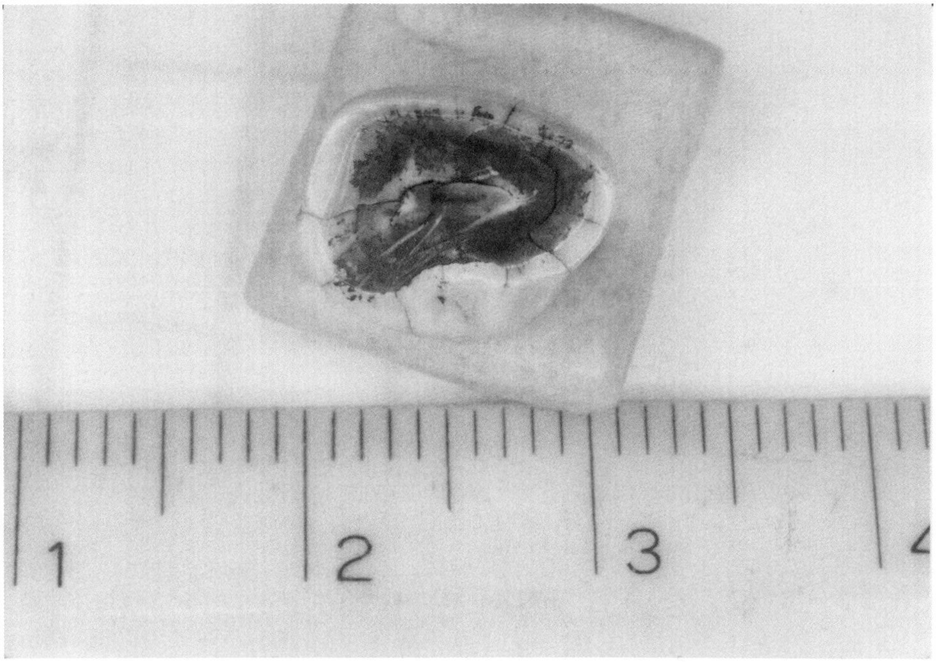


**Fig. 3.14** Denisova Cave human incisor (occlusal view). This view of an adult upper left central incisor shows that much of the crown had worn off before death. Secondary dentine fills the exposed pulp chamber. Rounding and polishing of the enamel suggests this tooth may have been partly digested by a hyena. There is no indication of labial surface flattening, lingual shoveling, or double-shoveling that might be expected had this person (sex indeterminable) been genetically affiliated with East Asian populations. Robustness favors a Neanderthal affiliation rather than anatomically modern humans. There is no pathology. Found in 1984 section D7, 119–220 cm deep (CGT color Moscow 6-25-87:20).

possibilities must be considered. First, it is possible that our scoring 33 pieces as having pseudo-cuts is grossly incorrect. Using the pooled assemblage average as a rough expectation value, there should be only five or six Denisova pieces with pseudo-cuts. Second, since we are unable to distinguish pseudo-cuts from stone tool cut marks, it could be that we were influenced by the presence of carnivore tooth scratches, dints, notching, and/or end-hollowing on the piece. A comparison of pieces with pseudo-cuts and carnivore damage vs. pieces with pseudo-cuts but no carnivore damage showed that there were 30 pieces with pseudo-cuts and carnivore damage vs. three without carnivore damage. Denisova pseudo-cuts are strongly related with carnivore damage to bone (dints, scratches, polishing, notching). Nevertheless, the high Denisova pseudo-cut frequency is disturbingly out of line with our other assemblages.

**17 Abrasions.** The Denisova sample of 115 pieces has 3.5% with abrasions (Table A1.17, site 5). Most have more than seven striations per piece. This is somewhat more than the average for the pooled assemblage. The abundance of limestone spall from the

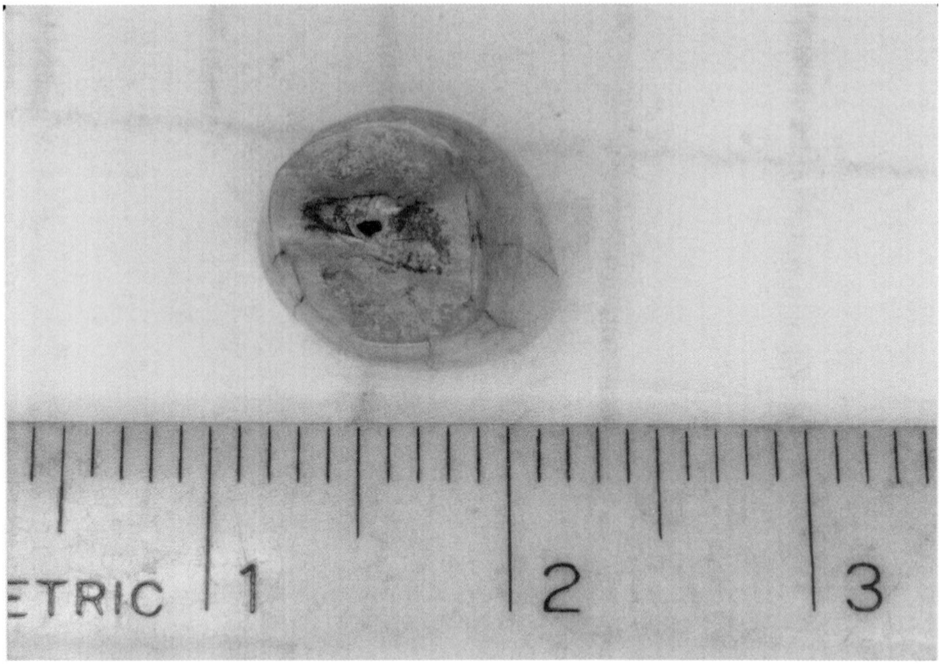


**Fig. 3.15** Denisova Cave Neanderthal incisor (root end). This view of the tooth from Fig. 3.14 is included to show the rounding and polishing of the surface where the root had broken off. This polishing of hard dentine where none should exist additionally suggests the tooth had been swallowed by a hyena and partly dissolved, like so many bone fragments and teeth from these Pleistocene Altai caves (CGT color Moscow 6-25-87:21).

cave ceiling suggests that at-hand stone was used a little more frequently for bone smashing than for our grand total.

**18 Polishing.** Denisova has a relatively high occurrence of polishing (83.0%; 93 / 112 pieces) (Table A1.18, site 5). Polishing is more frequent in Denisova than in the pooled assemblage, Razboinich'ya (60.0%), and only slightly less than in Maly Yaloman (89.3%). Here, as with notching, carnivore presence must have been extensive.

**19 Embedded fragments.** Only 83 Denisova pieces could be scored for embedding, and only 1.2% have the condition (Table A1.19, site 5). This frequency is less than the pooled assemblage average, much less than Razboinich'ya (8.8%), but similar to Maly Yaloman (2.3%).

**20 Tooth wear.** Only two Denisova teeth could be scored for wear. Both belonged to adults.

**21 Acid erosion.** Out of 115 Denisova pieces, 2.6% showed acid erosion (Table A1.21, site 5). This frequency is less than the pooled assemblage average, Razboinich'ya (7.3%), and much less than Maly Yaloman (48.0%). We consider the acid-eroded pieces in Denisova as documenting the use of the cave by hyenas, perhaps not as often or as long as in other Altai cave sites. The July 2006 visit turned up many pieces with acid erosion.

Most, at least 100 pieces, were less than 2.5 cm in length, so they would not meet our minimal size standard for formal study.

**22–23 Rodent gnawing, insect damage.** No cases of either variable were identified in the 115 Denisova pieces.

**24 Human bone.** A few human teeth and bone fragments have been found in Denisova Cave. They were stored separately from the fauna remains. In 1987 they were in Moscow on loan to Valery Alexeev. The remains will be discussed in Chapter 4. Here, we simply remark that tooth morphology and bone DNA indicate Neanderthals or Neanderthal-likes were the Paleolithic occupants of Denisova Cave.

**25 Cut marks.** One-fifth (20.0%) of the 115 Denisova pieces have from one to more than seven cut marks. The most common number is one (Table A1.25, site 5). Compared with the pooled assemblage average (7.6%), Denisova has more than twice the number of pieces with cut marks. In fact, Denisova is one of our five or so assemblages with more than 15% cut pieces.

**26 Chop marks.** As for cut marks, Denisova has a high frequency of chop marks (22.7%) (Table A1.26, site 5). This is five times greater than the pooled assemblage average (5.4%). The Denisova chopping frequency is exceeded only by one other assemblage – Kurla I (30.6%).

## Discussion

The Denisova Cave excavations have produced a wealth of late Pleistocene stone artifacts, stone tool manufacturing debitage, a small amount of human bone and teeth, pollen data (Malaeva 1999), and a great deal of faunal remains. Bone preservation in Denisova Cave is excellent. Bone pieces that were modified by humans (cutting, chopping, and breaking) and by non-human animals (tooth scratches, tooth dints, acid erosion, and so forth) are easily distinguishable. The number of acid-eroded pieces in our sample, which we attribute mainly to hyenas, is not as great as in other Altai caves. Nevertheless, there is enough acid erosion, as well as actual hyena remains, for us to feel confident that hyenas frequently if not intensively used Denisova Cave for shelter and denning. Thus, we propose that these scavenger-hunters carried some of the larger animal bones into the cave, rather than it all having been introduced by humans. Also, blurring (bioturbation) of stratigraphic horizons due to hyena digging should be considered when issues of cultural and human population continuity or replacement are being considered. The Denisova human teeth have been examined by three independent analysts, each of whom has concluded that they possess an archaic quality. They do not belong to anatomically modern humans as do the teeth from Siberian Upper Paleolithic sites such as Mal'ta and Afontova Gora.

# 6 Dvuglaska Cave

## Background

Dvuglaska Cave (Two Eyes) was brought to the attention of archaeologists by Krasnoyarsk speleologists in 1962. It is located in southern Siberia's Khakasian Autonomous Republic, about 400 km north of its border with the vast, high and arid Mongolian steppe country. The main city of Khakasia, Abakan, through which flows the Yenisei River, is south-southeast 46 km from Dvuglaska. This cave is located at 54°08′ N and 91°04′ E. It faces southward from the floor of a short steep-walled twisting box canyon about 5 km from a poor, dirt-streeted, two-water-well village named Tolcheya, which is dependent on a run-down remote dairy enterprise. A large, clear and permanent stream about 0.5 km from Dvuglaska Cave comes out of the nearby foothills, flowing to and through the village, eventually contributing to the Yenisei River. Situated at the relatively low elevation of 600 m, the regional steppe climate is considered as mild. Snow cover has usually melted by the end of March, and there is only about one half-month of frozen ground. The interior cave temperature in mid-July was 15°C (59°F) at 10:00 a.m., and 17°C (63°F) at 5:30 p.m.

Dvuglaska Cave is situated at the lower boundary of the craggy limestone Batenev foothills that look out over the vast treeless Bogradsky steppe plain, set in the center of the Minusinsk Depression (Figs. 3.16–3.23). The foothills have thin stands of tall old conifer trees mixed with sage, short-grasses, and low flowering plants. The bowl-shaped cave is about 10 m wide, 15 m deep, and from the present-day cave floor, about 4 m high. It is well-formed for protection against the heaviest summer rain storms and winter blizzards. However, being situated in a box canyon without a direct view of the steppe plain, it would have been a risky place for humans to live on a long-term basis. There would have been no natural protection from either predatory humans or large carnivores chancing into the box canyon from the steppe. Today, the cave's occupants are a few silent rodents, a small coming-and-going flock of pleasant-cooing rock-pigeons, and rare visits from nearby villagers partying around evening campfires.

At the mouth of the cave there is about 3 m of rocky fill, while at its rear there is almost no trash or roof spall. Judging from the very large number of sharply angular cobble- and pebble-sized rocks in the cave fill, there has been a slow and steady rain of rock from the highly fractured roof. The cave's name derives from two circular sky-lit holes in its roof that look like shining feline eyes in the darkened grotto – hence "Two-Eyes" cave.

**Fig. 3.16** Dvuglaska Cave. Z. A. Abramova and associates carried out the initial excavations of this cave in 1974. In various publications she referred to it as a "Mousterian Grotto" and a "Neandertal dwelling." Our brief re-testing, aimed at primarily securing charcoal for new carbon-14 dating, produced almost no cultural material but plenty of hyena and other large-carnivore-damaged bone. The cave is situated in a short box canyon whose entrance is out of sight to the left. (CGT color Dvuglaska 7-18-00:2).

In 1974, 1975, 1978, and 1979 excavations in Dvuglaska Cave were directed by St. Petersburg's Z. A. Abramova (Abramova 1980, 1981, 1985, Abramova *et al.* 1975, 1991, Abramova and Ermolova 1976, Vasili'ev 2000). Abramova, Ermolova, Eritsan, and Ovodov tested the floor of the box canyon immediately outside the cave and its opening earlier, in 1971–1973. In 1975 the cave and excavation were examined in detail by geologists M. Muratov and V. A. Panychev, archaeologist Yu. V. Grichan, and vertebrate paleontologist N. D. Ovodov. By the end of 1979 an area of 38 $m^3$ at the cave entrance had been excavated through the loose rocky sediments to the basal bedrock at 3.8 m depth. This multi-year expedition recovered bones and teeth of extinct Pleistocene animals and a few stone artifacts. The types of artifacts led these workers to view the occupants to have been *Mousterians*, the term used in publications about the cave's excavation (Abramova *et al.* 1975, Abramova and Ermolova 1976, Abramova 1981, 1985). Other publications that focus on or include Dvuglaska Cave are: Abramova *et al.* (1991); Vasili'ev (2000); and Ovodov and Martynovich (1992).

Abramova and the geologists identified eight stratigraphic levels at the entrance to Dvuglaska Cave. The oldest bone and artifact-bearing levels were 5–7 (Level 8 was sterile). Levels 5 and 6 had faunal remains of species that suggested a relatively warm and

**Fig. 3.17** Dvuglaska camp. A few pieces of lumber borrowed from a nearby village served for a low dining table and benches. Left to right: driver, Nicolai Ovodov, Olga Pavlova, Lena Popkova (behind Pavlova), and Nicolai Martynovich (CGT neg. Dvuglaska 7-20-00:36A).

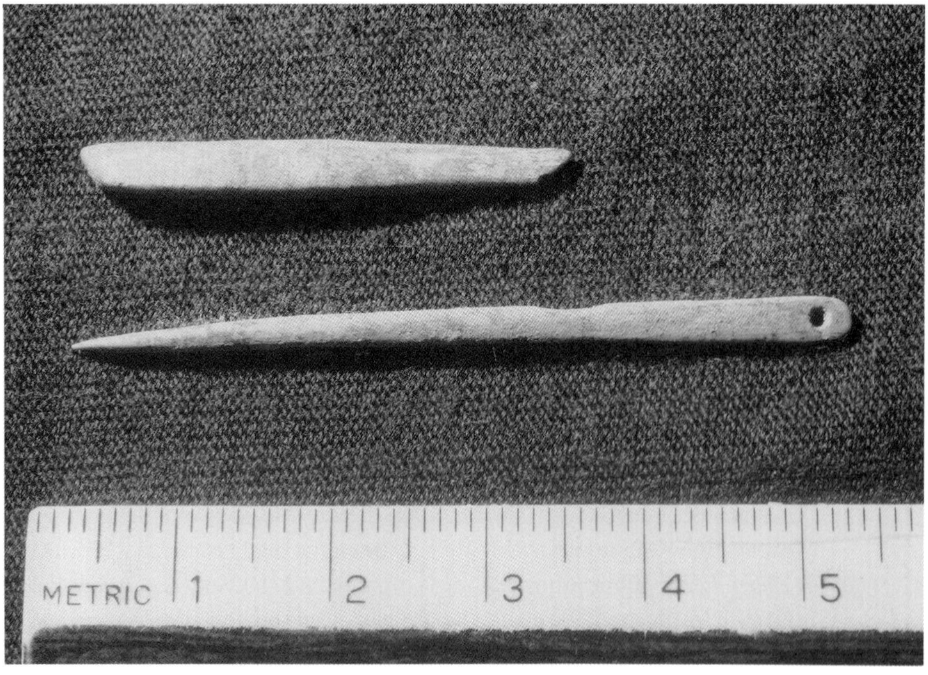


**Fig. 3.18** Dvuglaska artifacts. A 5.1 cm long, eyed, sewing needle and a sliver of polished bone were found in the 20 000 BP upper level of the cave deposit. These were the only artifacts we found, and well above the lower levels thought by Abramova to be Mousterian (CGT color Dvuglaska 7-20-00:26).

**Fig. 3.19** Dvuglaska bone and tooth fragments. Coarse-screened faunal remains. Much of this breakage appeared to have been perimortem. Pocket knife in lower right is ca. 10.0 cm long (CGT neg. Dvuglaska 7-20-00:36).

**Fig. 3.20** Final sorting. Nicolai Martynovich (left), bird paleontologist, and Irina Foronova, mammoth specialist, do the final sorting of tiny teeth and bones of Dvuglaska rodents, bats, and birds. None of these pieces met our size selection criterion (>2.5 cm) for this perimortem taphonomic study. Nevertheless, an informal search for perimortem damage turned up nothing (CGT color Kurtak 7-23-00:30).

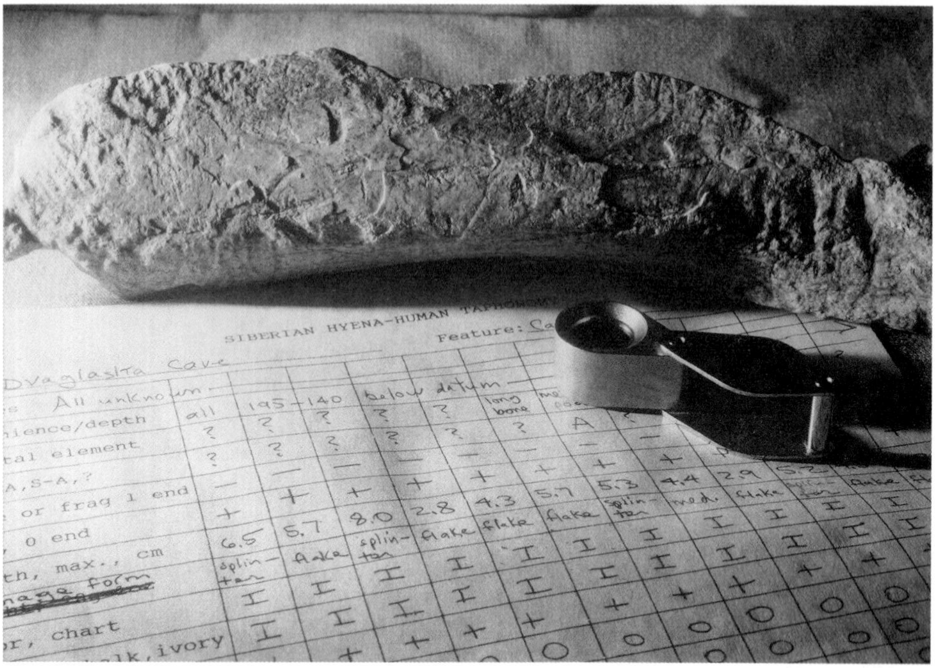


Fig. 3.21 Large Dvuglaska bone fragment. Most of the perimortem damage found in the Dvuglaska bone assemblage is attributable to hyenas and other large carnivores. For example, this bison radius, 25.5 cm in length, shows chewing scratches and dints, and end-hollowing, but no signs of human damage. Found at 140–195 cm depth (CGT neg. IAE 8-15-00:16).

dry climate. Ermolova (in Abramova *et al.* 1975) identified bones of Onager (*Equus hemionus*) and horse (*Equus cf. caballus*) as being the most frequent species. She also identified rhinoceros (*Coelodonta antiquitatis*), bison (*Bison pricus*), wild sheep (*Ovis ammon*), red deer (*Cervus elaphus*), saiga (*Saiga tatarica*), cave hyena (*Crocuta spelaea*), cave lion (*Panthera spelaea*), bear (*Ursus* sp.), gray wolf (*Canis lupus*), fox (*Vulpes vulpes*), glutton [wolverine] (*Gulo gulo L.*), and pika (*Ochotona* sp.). Remains of mammut and caribou were very rare.

Despite Dvuglaska Cave providing a relatively comfortable camp site, especially in the summer, the evidence for human use is limited. Abramova identified artifacts mainly from Level 4. These included two cores, two point fragments, a retouched blade, two scrapers, three screblos, and a modified piece of antler. Levels 5 and 6 yielded a few Levallois points, large disc-shaped cores, and some very large flakes (Abramova *et al.* 1991). On typological grounds these were viewed as being culturally Mousterian. Relative to the abundant evidence of Middle and early Upper Paleolithic occupation in the Altai, it would appear that the Minusinsk Depression during those time periods had very little human presence.

While our testing produced no stone artifacts, those that Abramova and associates found were later examined in 1986 by John Hoffecker (personal communication, April 19, 2003). He felt that they were Levallois flakes and points. He expressed surprise that

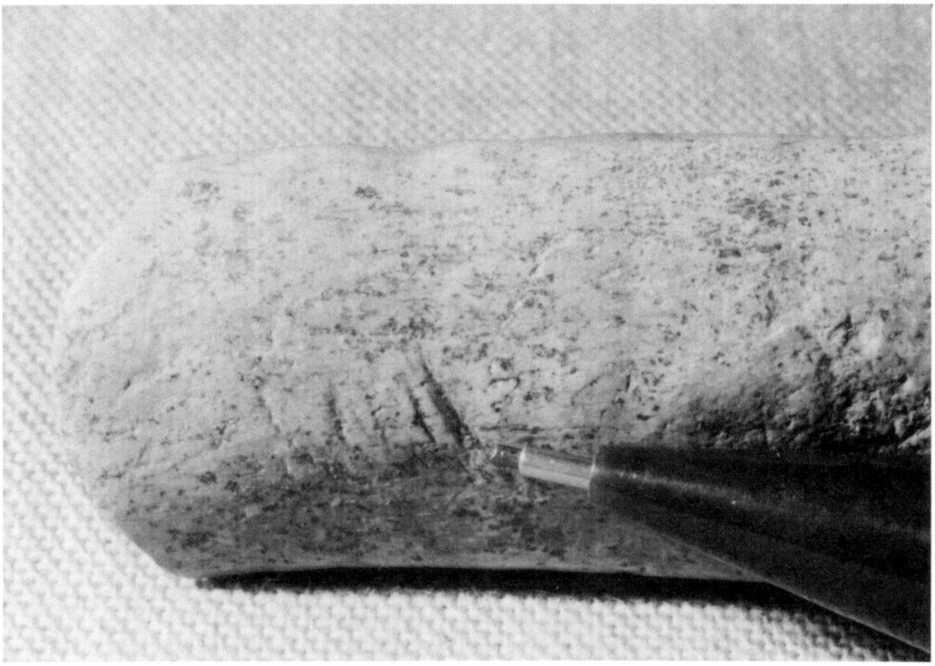

**Fig. 3.22** Dvuglaska pseudo-cuts. Five marks that look very much like cut marks, but each one has a rounded surface edge rather than a sharp border. The longest of these is <0.6 cm. Actual width of image is 3.3 cm (CGT neg. IAE 8-15-00:30).

**Fig. 3.23** Dvuglaska stomach bone. This piece of pelvic acetabulum has a maximum diameter of 7.2 cm, making it one of the largest ingested pieces seen during this project (CGT neg. IAE 8-15-00:27).

Neanderthals were living so far north and east, possibly because conditions were much warmer than today.

The present authors, assisted by Nicolai Martynovich, Elana Popkova, and a hired driver and his personal four-wheel-drive microbus, visited, set up camp, and carried out a small-scale broadside excavation on July 17–21, 2000. This work was proposed to, and approved by, IAE director, Academician Derevianko. The objectives were to: (1) enlarge the faunal sample for our taphonomy project; (2) re-collect charcoal samples for renewed dating of the cave's human occupation; and (3) for the senior author to see how paleontological excavation was done in Siberia and what sort of tools were used that might accidentally leave false cut and chop marks. Our work was a two-step procedure. First, we cleared away the rocky debris that had slid down over the front face of the cave deposit for some 25 years. This provided a fresh vertical face to the cave fill at its opening. The fresh face extended from the present cave floor surface with its modern trash of cans, bottles, and remnants of camp fires, to a vertical depth of about 3.0 m. The width of the broadside was about 2 m. Second, we excavated the fresh face inwards to a horizontal distance of about 25 cm, producing a volume of some 1.5 $m^3$. We recovered several thousand pieces of small rodent, bird, and bat bones and teeth, along with bones and teeth of bison, horse, rhinoceros, sheep-goat, and pieces of other large but unidentifiable mammals. We screened the entire excavated sample three times, first with half-inch hardware cloth, then with one-sixteenth-inch mesh all the material that had passed through the first screen. The second screening was then water-washed and dried to later extract the tiny bones and teeth of rodents, birds, and bats. All the larger bones were either recognized in actual excavation or caught in the first screening.

The two carbon-14 dates that we obtained from the upper and lower charcoal samples are 17 000 BP and 29 000 BP, respectively. Our dates may have little to do with the human use of Dvuglaska, inasmuch as the charcoal lenses seemed not to be hearths or even the scattered ashes of camp fires. The charcoal flecks could well have been blown into the cave from a nearby lightning-caused summer wildfire. However, the dates do give a good basis for assuming that the cave was available to humans and non-human animals alike for at least 30 000 years. Vasili'ev *et al.* (2002:521) list one carbon-14 date based on bone for Dvuglaska – namely 27 200 BP. Near the upper charcoal layer, a small-eyed bone tailoring needle was found. If the charcoal and needle are in true stratigraphic association, then the needle is one of the earliest known in Siberia. We also found a polished bone splinter, much like a toothpick. It is quite evident the cave was only rarely used by humans. Its main large occupants were hyenas, whose chewing damage is abundantly evident in our faunal sample, as are the numerous stomach bones.

## Findings

**1 Provenience.** Bone was initially saved as coming from upper, lower, and mixed levels. Analysis showed no meaningful differences between upper and lower levels, so all three are presented here as a single assemblage of 205 pieces . This number is 2.3% of our study total (Table A1.1, site 6). We ignored the bat, bird, and rodent pieces, but measured the maximal lengths of an additional 220 pieces whose only perimortem information was size.

**2 Species.** Out of the 205 Dvuglaska pieces, only big mammals (4.9%), bison (2.4%), horse (0.5%), and rhinoceros (0.1%) could be identified (Table A1.2, site 6). While we are confident that Dvuglaska Cave was mainly occupied by hyenas (chewing damage, stomach bones), we recovered no identifiable bones or teeth of this species. Compared with the pooled assemblage averages, Dvuglaska is similar except for the absence of bear, goat-sheep, hyena (bones/teeth), mammoth, and roe deer, and it has a lower frequency of horse.

**3 Skeletal elements.** Our Dvuglaska sample of 205 pieces contained mainly long bones (47.8%), unknown elements (25.4%), ribs (7.3%), and metapodials (5.9%) (Table A1.3, site 6). Compared with the pooled assemblage averages, Dvuglaska is similar except that it has fewer mandibles, vertebrae, humeri, toes, and more long bones and unknown elements.

**4 Age.** Of the 205 Dvuglaska pieces, there are almost no identifiable sub-adults (0.5%) (Table A1.4, site 6). Slightly more than half (56.1%) were assigned adult status. Compared with the pooled assemblage averages, both sub-adult and adult pieces are fewer in the Dvuglaska sample.

**5 Completeness.** Our Dvuglaska sample of 205 pieces has only 3.4% whole bones, and a very large number of pieces with no anatomical ends (91.7%) (Table A1.5, site 6). Compared with the pooled assemblage, Dvuglaska has three times fewer whole bones, and 32.5% more pieces with no anatomical ends. The large amount of incompleteness underlies the difficulty in species, skeletal element, and age identification. Dvuglaska even exceeds Razboinich'ya Cave in its amount and degree of incompleteness.

**6 Maximum size.** Overall, 429 Dvuglaska pieces were measured (Table A1.6, site 6). Incompleteness also influences size. The average maximum dimension is only 5.2 cm, whereas the few whole bones and large pieces extend the range from 1.7 cm to 28.2 cm. All the Dvuglaska statistics are lower than the pooled assemblage averages. Comparing the upper range limit with the lengths of whole long bones of bison and horse provided by Vera Gromova (1950) shows that Dvuglaska pieces are considerably smaller.

**7 Damage shape.** Out of 205 Dvuglaska pieces (Table A1.7, site 6), the most common damage forms are flakes (62.9%), splinters (15.1%), fragments (6.4%), and medial rib pieces (5.9%). Viewed against the pooled assemblage averages, Dvuglaska has more flakes and splinters, and fewer fragments, undamaged, and mostly whole pieces.

**8 Color.** Dvuglaska has 205 pieces, their color being almost entirely ivory (98.0%) (Table A1.8, site 6). There is only one black piece (0.5%), which represents burning, and one brown piece that may have been burned. Compared with the pooled assemblage averages, Dvuglaska has considerably more ivory colored pieces, and many fewer brown pieces.

**9 Preservation.** Table A1.9 (site 6) shows that Dvuglaska's 205 pieces are mostly ivory hard (94.1%), with only 5.9% being chalky (Table A1.9, site 6). Compared with the pooled assemblage averages, Dvuglaska has more ivory, and fewer chalky pieces. The quality of the Dvuglaska assemblage is almost exactly the same as that of Razboinich'ya Cave, suggesting similar preservational conditions shared by both caves, even though they are geomorphically very different, as well as having very different temperatures.

**10 Perimortem breakage.** Almost every one (99.5%) of the 204 scorable Dvuglaska pieces has perimortem breakage (Table A1.10, site 6). This amount of perimortem breakage is more than that found in the pooled assemblage average, and even more than what was found in Razboinich'ya (92.4%).

**11 Postmortem breakage.** Out of 204 Dvuglaska pieces, 10.3% have postmortem breakage (Table A1.11, site 6). Compared with the pooled assemblage average (17.8%), Dvuglaska has somewhat less postmortem breakage. Dvuglaska and Razboinich'ya (10.9%) are almost exactly the same.

**12 End-hollowing.** Dvuglaska has 4.4% (9/205) end-hollowing (Table A1.12, site 6). This value is less than the pooled assemblage average and the Razboinich'ya frequency (8.1%).

**13 Notching.** Out of 205 Dvuglaska pieces, 15.1% have notching (Table A1.13, site 6). Of these, most have one (8.8%) or two (3.4%) notches. The maximum number of notches is five. Compared with the pooled assemblage, Dvuglaska is almost exactly the same in the occurrence of notching, as well as being very similar in terms of the number of notches. Razboinich'ya has more notched pieces (23.6%).

**14 Tooth scratches.** More than one-third (38.5%) of the 205 Dvuglaska pieces have tooth scratching (Table A1.14, site 6). The largest percentage (10.7%) of these have only one scratch, but 8.0% have more than seven scratches. Compared with the pooled assemblage, Dvuglaska has almost twice as many pieces with tooth scratches. Dvuglaska is extremely similar to Razboinich'ya (37.1%) in its frequency of tooth scratches.

**15 Tooth dints.** Almost half (43.4%) of the 205 Dvuglaska pieces have one or more tooth dints (Table A1.15, site 6). Of these, they range from having one dint (9.8%) to having more than seven dints (9.9%). There are considerably fewer occurrences of tooth dinting in the pooled assemblage. Razboinich'ya Cave has about 10% more cases (52.8%).

**16 Pseudo-cuts.** There are only a few examples (2.9%) of pseudo-cuts in the 205-piece Dvuglaska assemblage (Table A1.16, site 6). Most of these (2.0%) have only one pseudo-cut. The maximum number of cuts per piece is five. The occurrence of pseudo-cuts in the pooled assemblage is slightly greater. Razboinich'ya is very similar (3.1%) to Dvuglaska.

**17 Abrasions.** There are no examples of abrasions on any of the 205 Dvuglaska pieces (Table A1.17, site 6). This finding adds to our belief that abrasions are produced mainly by human bone processing.

**18 Polishing.** Most of the 203 scorable Dvuglaska pieces have polishing; only 5.4% do not (Table A1.18, site 6). Polishing occurs on one or both ends (9.4%), the middle (0.5%), and most often on both an end and the middle (84.7%). The absence of polishing in the pooled assemblage average is much greater. Razboinich'ya has far more unpolished pieces (40.6%) than Dvuglaska.

**19 Embedded fragments.** There are very few (1.0%) of the 205 Dvuglaska pieces with an embedded fragment (Table A1.19, site 6). This frequency is less than the pooled assemblage average, and very much lower than the Razboinich'ya frequency (8.8%).

**20 There are no teeth in the Dvuglaska assemblage (Table A1.20, site 6).**

**21 Acid erosion.** Pieces with acid erosion are common (22.1%; 45 / 204) in the Dvuglaska Cave assemblage (Table A1.21, site 6). In addition to other variables indicating intensive carnivore bone processing, the many acid-eroded pieces point to Dvuglaska having been used by hyenas much more often than by humans, as assessed by the very few artifacts found by Abramova (we found two, a polished splinter and an eyed needle). Compared with the pooled assemblage average, Dvuglaska has more than three times the amount of acid erosion. The same can be said for Razboinich'ya (7.3%).

**22–26 Rodent gnawing, insect damage, human bone, cut marks, and chop marks.** There are no examples of these variables in the 205 Dvuglaska pieces (Tables A1.22–A1.26, site 6).

## Discussion

The prehistory of Dvuglaska Cave seems straightforward on the basis of the work by Abramova and associates, and on our own small effort in perimortem taphonomy. There is practically no evidence for human presence in any of our 205 pieces of bone. Only the one black-colored piece of bone may be linked to human activity, i.e., accidental burning in a campfire. We did find an eyed bone needle and a polished bone splinter in the 20 000-year-old horizon (the same horizon from which we obtained charcoal that carbon-14 dated at 17 000 BP. Our other charcoal sample came from deeper in the broadside and provided a date of 29 000 BP). At the older depth we found no sign of human presence.

On the other hand, the severe bone damage and the evidence of unequivocal carnivore processing, including the critically important acid erosion of bone, all point to the cave as having been used much more by hyenas and other carnivores than by humans. Moreover, the site's defensively poor location for human occupancy, as well as the very limited amount of stone and bone artifacts, stone tool manufacturing debris, and lack of hearths can best be interpreted to mean that Dvuglaska was used only on a rare and temporary basis by humans. While humans had unquestionably left a few stone and bone artifacts in the cave, requiring us to classify Dvuglaska as an archaeological site, in our view the cave's deposits are overwhelmingly more paleontological than archaeological. While we do not doubt that Abramova found actual Mousterian artifacts in Dvuglaska Cave, our failure to do so raised some question in our mind about those artifacts. However, John F. Hoffecker (personal communication, April 19, 2003), who has examined them, assured the senior author that they are man-made.

The combination of hyenas, a tailoring needle, and other evidence of human presence in the middle Yenisei basin at ca. 20 000 BP is potentially important for a better understanding of why the colonization of the New World was so late in comparison with all other parts of the Old World except Polynesia. We will go into more detail later about this combination. We only mention this subject now to help the reader appreciate that we feel that the lives of hyenas, other carnivores, and the northern dispersal of humans seemingly are competitively linked.

# 7 Gosudarev Log I

## Background

Gosudarev Log I (State Ravine I) is a late Pleistocene open-air site on the eastern outskirts of Krasnoyarsk, 4 km east of the suburb community called Berezovka, and 6 km east of the right bank of the Yenisei River, at approximately 56°4′ N, and 93°17′ E. It is situated in a broad 3 km long gully (ravine) that cuts almost to the edge of the highest Yenisei River terrace (120–135 m, up to 150 m), as well as cutting through and exiting at a middle Pleistocene terrace 60–70 m high. The right side of the ravine is steep and about 50 m high, the left side is 10 m high, and the ravine floor is 20–30 m wide. There is flowing water at Gosudarev Log I only in the spring. The archaeological site is situated on the right slope of the ravine. An area of 354 $m^2$ was excavated in 1999–2000 by members of the Krasnoyarsk City Center for Protection and Use of Historical and Cultural Monuments, and Research Fellows from Krasnoyarsk Pedagogical Institute's Laboratory of Archaeology and Paleogeography of Central Siberia (Drozdov *et al.* 2000).

Archaeological materials and faunal remains were recovered in the 6 m thick layer of buried soil characterized by traces of solifluction, pseudomorphoses, and cryogenic deformations. The limited amount of archaeological materials came from the upper and middle portion of the buried soil, whose geomorphological characteristics match those elsewhere dated to the middle and second half of the Karginsk period. While the very few artifacts (flake, scraper, core) lack diagnostic temporal or cultural attributes, their position in the Kargin soil layer corresponds with those found at the up-river Afontova Gora site complex (56°00′ N, 92°75′ E; Kuzmin and Orlova 1998:9) in Krasnoyarsk city. Ovodov identified the faunal remains as Asiatic wild ass, onager (*Equus hemionus*), bison (*Bison priscus*), and mammut (*Mammonteus primigenius*).

## Findings

**1 Provenience.** The Gosudarev Log I site is described above. Our sample of 14 pieces is only 0.2% of our 8813-piece total (Table A1.1, site 7).

**2 Species.** Of the 14 Gosudarev Log I pieces, 92.9% are big mammal, and 7.1% are horse (Table A1.2, site 7). Given the identifications made by Ovodov at an earlier time, some of the unknown pieces probably belong to one or another of the non-horse species.

**3 Skeletal elements.** Scapula (21.4%), pelvis (21.4%), mandible (14.3%), vertebra (14.3%), femur (7.1%), long bone (7.1%), foot (7.1%), and unknown (7.1%) make up the elements in the 14-piece Gosudarev Log I assemblage (Table A1.3, site 7). Given the small sample size, there is no value to making comparisons with the pooled assemblage.

**4 Age.** There are no discernible sub-adult pieces in the Gosudarev Log I assemblage (Table A1.4, site 7). Of the 14 pieces, 92.9% are positively identified as adult.

**5 Completeness.** The relatively high amount of skeletal element identification is made possible by the degree of completeness. Whole bones (7.1%) and bones with one anatomical end (42.9%) constitute half of the 14-piece Gosudarev Log I collection (Table A1.5, site 7).

**6 Maximum size.** Gosudarev Log I contains some relatively large pieces. These range in size from 4.7 cm to 30.0 cm, with a mean of 11.7 cm (Table A1.6, site 7). Completeness and size go hand in hand here. Compared with the pooled assemblage values, in the same order, 9.0 cm, 3.2 cm to 31.0 cm, suggests a certain similarity. Comparing the whole long-bone lengths of horse published by Vera Gromova (1950: table 27), shows that the upper range limit for Gosudarev Log I is somewhat smaller than her values.

**7 Damage shape.** Pieces of long-bone fragments (35.7%), mandibular symphysis (14.3%), vertebral spine (14.3%), pelvis center (14.3%), undamaged (7.1%), irregular (7.1%), and mostly whole (7.1%) make up the Gosudarev Log I damage inventory (Table A1.7, site 7). As above, maximum size and completeness relate to damage-form identification.

**8 Color.** There is one brown (7.1%), and 13 ivory colored pieces in the Gosudarev Log I collection (Table A1.8, site 7). The brown piece was almost certainly burned, although it is remotely possible that it was stained by the overlying sandy soil that was reddish in color.

**9 Preservation.** Like other open sites where preservation is generally poor, Gosudarev Log I has mostly chalky pieces (85.7%; 12 / 14) (Table A1.9, site 7).

**10 Perimortem breakage.** Out of the 14 Gosudarev Log I pieces, there is a relatively low occurrence (28.6%) of perimortem breakage (Table A1.10, site 7).

**11 Postmortem breakage.** This condition is common (85.7%) in the 14-piece Gosudarev Log I collection (Table A1.11, site 7). It correlates strongly with preservation quality.

**12 End-hollowing.** Some carnivore activity is represented by the Gosudarev Log I frequency of 15.4% (2 / 13 pieces) end-hollowing (Table A1.12, site 7).

**13 Notching.** Further evidence of carnivore activity comes from one Gosudarev Log I notched piece (1 / 14; 7.1%). It has one notch (Table A1.13, site 7).

**14–17 Tooth scratches, tooth dints, pseudo-cuts, and abrasions.** None of the 14 Gosudarev Log I pieces exhibits these perimortem damage features (Tables A1.14, site 7 to A1.17, site 7).

**18 Polishing.** Only end-polishing is present in the Gosudarev Log I collection (28.6%; 4 / 14) (Table A1.18, site 7).

**19 Embedded fragments.** None of the 14 Gosudarev Log I pieces has embedded fragments (Table A1.19, site 7).

**20 Tooth wear.** There are no Gosudarev Log I teeth from which to assess wear.

**21–23 Acid erosion, rodent gnawing, insect damage.** None of the 14 Gosudarev Log I pieces show these features (Tables A1.21, site 7 to A1.23, site 7).

**24 Human bone.** No human bone was found at Gosudarev Log I.

**25 Cut marks.** In our judgment there are no prehistoric cut marks on any of the 14 Gosudarev Log I pieces (Table A1.25, site 7). However, there are three pieces with cut marks that are very clean and postmortem in appearance. We believe these were caused by excavation procedures.

**26 Chop marks.** Two of the 14 Gosudarev Log I pieces have chop marks (14.3%). One piece has one chop mark; the other has three chop marks (Table A1.26, site 7).

## Discussion

Despite the small sample size of Gosudarev Log I, it is an informative assemblage. First, it has perimortem damage consistent with its being an open-air site – the bone is chalky and has considerable postmortem breakage. Second, it appears to be a kill site of one or more horses by humans, because perimortem chopping and probable burning are present. Third, after leaving the scene of their successful hunt or scavenging, the bone refuse left by the humans was chewed on by one or more carnivores. What they did not chew up or drag away was left on the ground surface to be eventually buried by natural processes that prevented further scattering and bone loss.

# 8 Kamenka

## Background

The west Trans-Baikal region of Siberia, wherein is found Kamenka (Bath House Stove) and other Paleolithic open sites (Figs. 3.24–3.26), has a substantial history of scientific investigation despite its distance from Moscow and St. Petersburg. The archival research by Ludmila V. Lbova and E. A. Khamzima (1999) found that the first Trans-Baikal scientific expeditions began in 1724, led by D. G. Messershmidt, who noted numerous stone graves that looked like a "fossil army" (Fig. 3.27). The members of the second Kamchatka expedition (G. Miller, I. Gmelin, others) undertook the first excavation in 1743 of burials in the area of today's Yeravninski reservoir. More than two centuries had to pass before the deeply buried Kamenka site would be discovered.

The Kamenka site is located near a settlement named Novaya Bryan, about 2 km west of the Bryanka River that flows through a sandy steppe valley flanked by hills darkened by small stands of conifer trees (51°46′ N, 108°20′ E). It consists of an accumulation of cultural and faunal remains of a repeatedly occupied large Upper Paleolithic open camping area. Kamenka is named after a low mountain on whose southeastern lower slope it appears, at about 600 m elevation. Artifacts and faunal remains were first discovered at Kamenka in 1989 by local school children and their history teacher due to exposure caused by workers removing sand and gravel from a large barrow pit for use in construction elsewhere (Abdulina and Tashak 1990). The Kamenka locality was saved from further destruction by the actions of Lbova, who had Kamenka legally designated as a cultural monument. She then directed the scientific sampling of the site from 1989 to 1995 for the Buryat Institute of Social Sciences of the Siberian Division of the Russian Academy of Sciences (now the Institute of Mongolian, Buddhist and Tibetan Studies) (Lbova 1991a, 1991b, 1996, 2000; Lbova and Volkov 1993; Lbova and Germonpré 1994; Lbova and Shteinokova 1996; Germonpré and Lbova 1996; Lbova *et al.* 2003). Judging from published site maps (Lbova 2000:162–165; and elsewhere), approximately 300 $m^2$ was excavated. Artifacts included knives, linescrapers, perforators, points, and burins. Wear analysis shows most were used for butchering animals, treating hides, and producing bone tools (Lbova 2002, Lbova *et al.* 2003). Hearths and a possible tent ring were also found deeply buried under sandy soil with various amounts and units of fine gravel, clays, and some calcification (Lbova 1991a, 1991b, 2000, Germonpré and Lbova 1996). Paleopedological features suggest that the main human occupation occurred during the warmer Karginsk interstadial, which was followed by the cold Sartan stadial.

**Fig. 3.24** The Kamenka site. Excavated by Ludmila Lbova, the Upper Paleolithic cultural horizons are near the level where the people are standing (CGT neg. Kamenka 7-8-03:16).

**Fig. 3.25** The Kamenka site. Visitors are examining the Upper Paleolithic cultural horizon on which they are standing. Left to right: Ludmila Lbova, Nicolai Ovodov, driver, Olga Pavlova (CGT neg. Kamenka 7-8-03:11).

Fig. 3.26 Odyssey. A small herd of horses. On the trip to Kamenka and Varvarina Gora, Ludmila Lbova also showed the authors her ongoing excavations at Khotyk. This locality, like much of the region around the Trans-Baikal population center of Ulan-Ude, is steppe. One evening a herd of horses trotted by our tents, provoking a mental image of what Upper Paleolithic hunters might have similarly seen in this locality (CGT neg. Khotyk 7-8-03:7).

Vasili'ev *et al.* (2002) have compiled many of the published carbon-14 dates for Paleolithic Siberia, including ten obtained from Kamenka (they did not include for technical reasons a date of 28 000 BP, included in the 11 dates published by Lbova [2002:63]). Bone and charcoal carbon-14 dates range from 24 625 BP to 40 500 BP, with an average of 30 200 BP. Oxford University radiocarbon dating laboratory dates released in June 2003 are 41 350 BP and 39 290 BP (Lbova, personal communication, June 30, 2003). Thus, Complex A (C) (middle Karginsk time) ranges from 26 760 to 41 350 BP, and Complex B from 24 625 to 28 000 BP (late Karginsk time). This range of dates conforms with established dates for the Karginsk interstadial. Site components A and C have typological stone tool similarities in their blade forms, whereas Component B is more varied, although dates for the three components are much the same. On the whole, the existence of this hunting camp (dates, bone-processing technology, and stone tool types) is viewed as transitional from Mousterian to the Upper Paleolithic. We restricted our taphonomic study to bones recovered only from Component A, generally the deepest of the three, and with many times more specimens.

As stated in the methods section of Chapter 1, we limited our formal observations to pieces whose maximum diameter was greater than 2.5 cm, and whose surfaces had little or no root damage. Rarely, we scored pieces with heavy root damage when it was limited

**Fig. 3.27** Odyssey. Forest fire smoke. The sun was almost completely blotted out by the acrid and eye-watering smoke from lightning-ignited forest fires far to the north of Ulan-Ude. Unreachable, these huge fires burn for months. Similar wildfire conditions likely existed in Pleistocene Siberia and were likely to have been an aid to hunters waiting in ambush for game fleeing the fires. This scene also shows a large grave near Khotyk believed to have been excavated by Professor Messerschmidt in 1742 of the Second Academic Expedition to Kamchatka (Lbova and Khamzina 1999). He allegedly found only a few horse bones as the grave had been looted at some earlier time. Left to right: Olga Pavlova, Nicolai Ovodov, and Ludmila Lbova (CGT neg. Khotyk 7-8-03:14).

to one surface or small area. We also did not count most whole pieces that lacked perimortem damage of any sort. These included almost all loose teeth and foot bones. Unlike other sites, especially Ust-Kan, Kamenka contained many teeth and cranial parts. The Kamenka assemblage, like all other sites, was curated in many plastic bags and whatever sorts of cardboard boxes Lbova and her student workers had available. Our working area was limited to one small table. Hence, our study proceeded by opening a bone bag, making and recording observations, doing macrophotography, returning the pieces to the bag, and then moving on to another bag. We were unable to attempt any inter-bag conjoining. Moreover, each bag might contain not only various species, various skeletal elements, but also differing provenience designations, such as I1, K4, I10, K9, no number, and so forth. The basis for bag contents was partly year of excavation, but also unknown reasons.

The faunal collection of Kamenka has been analyzed by M. Germonpré and published (Germonpré and Lbova 1996): mammoth (*Mammonteus primigenius*), one piece; horse (*Equus cf. Caballus*), 191 pieces; Asiatic wild ass, onager (*Equus*

*Hemionus*), one; rhinoceros (*Coelodonta antiquitatis*), ten; camel (*Camelus* sp.), three; bighorn deer (*Megaloceros giganteus*), two; Mongolian gazelle, zeren (*Procapra guturoza*), 229; wild sheep (*Ovis ammon*), seven; bison (*Bison priscus*), 17; cave lion (*Panthera spelaea*), one; horned ungulates (*Bovidae* gen. sp.), seven; indeterminable mammals, 1532.

## Findings

**1 Provenience.** Artifacts from Component A were recovered at a depth of about 9–10 m from the present ground surface. They were found in good stratigraphic and spatial association with the faunal remains. Out of a much larger number of pieces recovered by Lbova, we examined 552 (Table A1.1, site 8). These are about half the total studied by Germonpré and Lbova (1996), whose taphonomic work we will be comparing wherever possible. We could see no reason not to pool all of the Component A pieces, and excluded Components B and C because of apparent stone tool differences and small sample sizes. While we recorded all of the provenience information on each individual piece of bone – such as year, component, square – in some cases year or section were not identified. Informal inspection of the provenience information failed to turn up any obvious spatial or temporal pattern for any of our variables.

**2 Species.** We examined 552 Component A pieces, with gazelle being the most frequently recognized species (57.1%) (Table A1.2, site 8). There were small amounts of horse (4.7%), and even lesser amounts of bison, camel, rhinoceros, and wolf. Less-specific pieces and those that were unidentifiable made up the remainder. Significantly, there was no direct or indirect evidence of hyenas. The range of fauna analyzed by Germonpré and Lbova (1996) is mentioned above.

A statistical comparison of the frequencies of our identified species with the 2009 pieces (all archaeological components) studied by Germonpré and Lbova (1996:40) produced a very significant difference ($\chi^2=70.0$, 12 d.f., $p<0.001$). The difference was caused by: (1) our under-identification of horse, because we did not study most unmodified whole bones; (2) we identified many more examples of gazelle (*Procapra*), attributable to Ovodov's many years of paleontological experience; and (3) we had many fewer unidentified pieces, again probably due mainly to Ovodov's experience plus our minimal size limitation that excluded from study pieces smaller than 2.5 cm in diameter. Given our different study objectives (perimortem taphonomy versus standard taphonomy) and sampling methods, the significant difference is due mainly to our recognition of more gazelle remains.

There are two obvious differences in the species found at Kamenka and our other sites. Most notable is: (1) the absence of gazelle in all other sites except Varvarina Gora, which is only 17 km from Kamenka and was occupied during a similar time period; (2) hyenas were absent at Kamenka.

**3 Skeletal elements.** Our element sample size was 552 pieces (Table A1.3, site 8). Pieces identifiable as long bone but not specific bones were the most numerous (12.7%), followed by pieces of rib (11.4%). Both classes are readily identifiable, but then so are cranial elements, which taken together make up 12.5% of our total. There were no conspicuously missing elements, although given the ease with which vertebrae can be recognized, the large number per individual, and their generally good archaeological preservation, there seems to be a deficiency in their expected occurrence. Fragmented skeletal elements are generally easier to identify than are species.

The skeletal element comparison with Germonpré and Lbova (1996:41) produced a highly significant difference ($\chi^2 = 478.1$, 18 d.f., $p < 0.001$). The main difference is due to Germonpré and Lbova's identification of many more long bones ("unidentified diaphyses") and unidentifiable pieces. The difference between our respective unidentifiable pieces is attributable to Ovodov's years of experience, and our exclusion of pieces <2.5 cm. However, as with our respective species counts, overall the element identifications are reasonably similar.

On the whole, Kamenka displays nothing out of the ordinary with respect to its skeletal element frequencies compared with our other sites. Only one oddity stands out, which was the occurrence of four pieces of penis bone (0.7%). This element occurred elsewhere only in nearby Varvarina Gora, and in the faraway mountainous Altai caves of Ust-Kan and Razboinich'ya Cave. The latter has the highest frequency of penis bones in our entire series – 5.4%. According to Davis (1987:44), male equids, rodents, and carnivores have penis bones.

**4 Age.** Our sample size for age assessment was 552 pieces (Table A1.4, site 8). Adults were represented about five times more often than sub-adults, assuming that the pieces of unknown age were in roughly the same 5:1 ratio. One *Procapra* ulna with osteophytic proliferation around the proximal end and a healed antemortem fracture of the shaft obviously belonged to an adult. Germonpré and Lbova do not report on age.

Comparing our other human-caused bone samples of comparable size with Kamenka, Mal'ta, Varvarina Gora, and Volchiya Griva have similar adult/sub-adult ratios. Most of the others have far fewer sub-adults – for example, Bolshoi Yakor, Denisova, Kaminnaya, Ust-Kan, Ust-Kova, etc. This age ratio difference is not solely attributable to age-dependent bone hardness and calcification, because preservation is good in these cave and open sites, among others. Seasonal hunting techniques, early-in-life deaths due to health and nutrition (Volchiya Griva), and heavier bone processing of younger animals by humans and carnivores alike are potential factors influencing the inter-site age ratios.

**5 Completeness.** We studied 552 pieces for completeness (Table A1.5, site 8). Our whole pieces (8.7%) are mainly those that had some kind of processing stigmata. Those with one remaining anatomical end make up 37.5%. Pieces lacking both anatomical ends were in the majority (53.8%).

By combining our two incomplete categories we could statistically compare "breakage" occurrence with Germonpré and Lbova. The comparison showed no significant difference ($\chi^2 = 0.2$, 1 d.f., $p > 0.5$). Hence our respective whole/breakage ratios are effectively the same, namely 9.0% whole, 91% broken. There is nothing especially unusual about Kamenka compared with our other assemblages. A frequency of roughly 50% lack of both anatomical ends lies near the mean of our other sites.

**6 Maximum size.** The average maximum diameter for the 550 pieces that were either whole or if broken were unaffected by postmortem breakage is 6.5 cm, and the range is 2.9 cm to 20.7 cm (Table A1.6, site 8). While Germonpré and Lbova (1996:42) provide a percentage histogram for fragment length, we are unable to convert it to precise values for statistical comparison. The shape of the histogram is highly skewed, but the median is clearly 3–5 cm, with what we estimate to be about 55% of their unstated total. The histogram shows that their second most frequent class had a class interval of 0–2 cm, and a frequency of about 20%. Hence, their median and our mean values are probably not significantly different, given our different criteria for measurement inclusion.

The Kamenka mean piece size is, as was the case for completeness, similar to the combined mean (9.0 cm) for our other assemblages. Comparing the Kamenka size range values with undamaged long-bone lengths provided by Vera Gromova (1950: table 27), shows that the former upper limit is about the same as the values for Saiga antelope, but smaller than those of sheep (Gromova has no information on gazelles).

**7 Damage shape (Figs. 3.28–3.30).** Damage form was assessed in all 552 Kamenka pieces (Table A1.7, site 8). Most frequent are long bone fragments (36.9%), long bone splinters (11.4%), and long bone flakes (11.2%). Overall, damage-form occurrence correlates, of course, with skeletal element occurrence. Long bone mass (volume or weight) probably exceeds that of other categories such as crania, vertebrae, flat bones, and irregularly shaped bones.

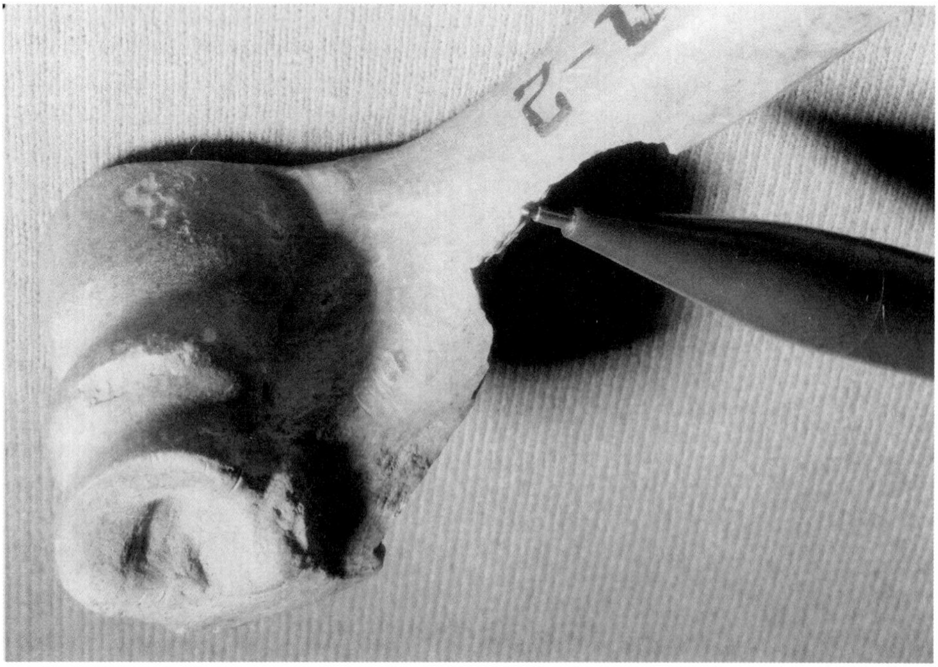

**Fig. 3.28** Kamenka bone damage. A *Procapra* humerus with a big chop notch (at pencil) and cut marks on joint. Fragment length, 7.0 cm. Layer A (CGT neg. MBSC 7-4-03:5).

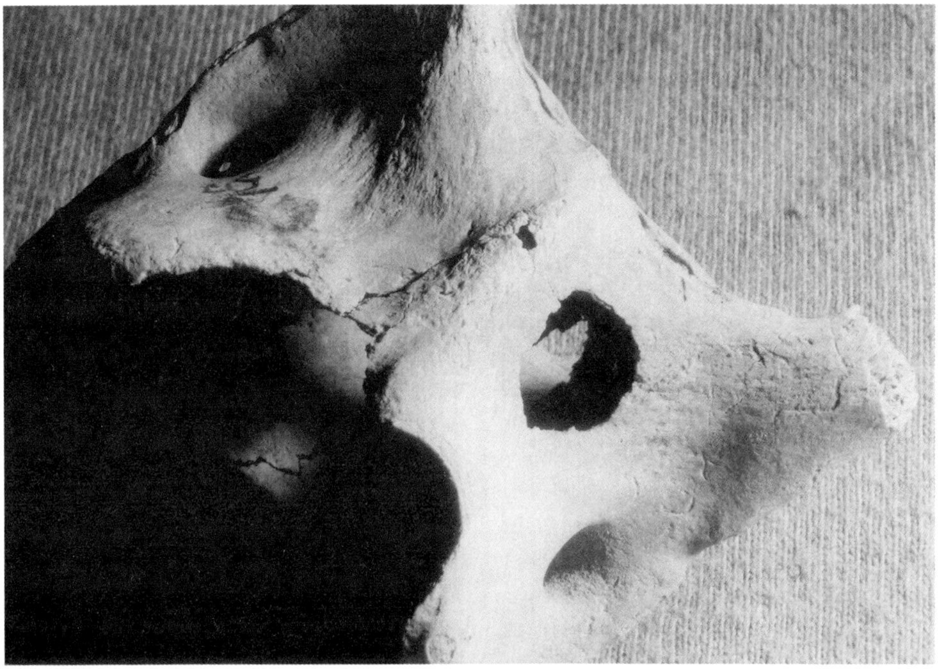

Fig. 3.29 Kamenka bone damage. A punched hole 1.0 × 0.8 cm in a *Procapra* first cervical vertebra, 5.0 cm maximum diameter. Layer A (CGT neg. MBSC 7-4-03:2).

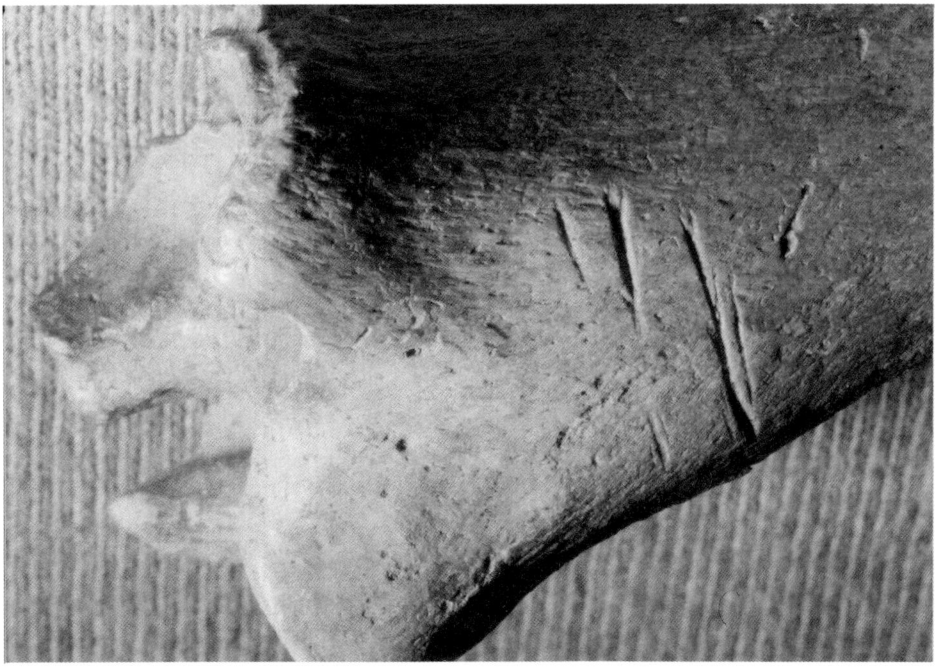

Fig. 3.30 Kamenka bone damage. *Procapra* humerus with cut marks near joint. Side B. Actual width of image is 3.3 cm. Layer A (CGT neg. MBSC 7-4-03:5).

Germonpré and Lbova (1996:42) consider damage form as a variable they call "traces." Traces include gnawing marks, root marks, burning, cut marks, percussion marks, cancellous blocks, and bone splinters. They recognized that Kamenka had a large number of narrow and variably long pieces that look like splinters. They report that 10.5% of their 1928 pieces are splinters (202 splinters). As mentioned, we recognized 11.4% of our total (552) as being splinters. There is no significant difference in our findings ($\chi^2 = 0.35$, 1 d.f., $p > 0.05$).

These authors identified a Kamenka damage form they called "cancellous blocks," which tallied at 3.3% (64 / 1928). We saw several of these but did not count their occurrence. However, cancellous blocks most likely are derived in the main from vertebral bodies, ends of long bones, and the cancellous bone around the pelvic acetabulum ("pelvis center"). Combining our occurrences of vertebral bodies (1.6%), long bone epiphyses (0.4%), and pelvis centers (2.7%) totals 4.7%, a value similar to Germonpré and Lbova's 3.3%. Among rare damage variants, we came across a single case of both horn cores of an adult *Procapra* cranial vault having been pounded off at the horn–vault junction, the purposes of which we speculate had been to get the 9.8 cm wide vault piece into a cooking bag, or to use the horns for some sort of tool, or both.

Another rare damage variant was the occurrence of a seemingly punched 1.0 × 0.8 cm circular hole in the corpus of an adult *Procapra*'s first cervical vertebra.

In comparison with our other assemblages, Kamenka does not stand out as unusual in any of the damage types. While the Kamenka percentage of long bone splinters is among the assemblages with the most splinters, sites such as Bolshoi Yakor (25.1%), Dvuglaska (15.1%), and Ust-Kan (16.9%) have more.

**8 Color.** Color was assessed in all 552 pieces (Table A1.8, site 8). Most of the pieces (98.9%) were scored as ivory in color. Postmortem breakage shows that Kamenka bone coloring is almost entirely surficial. The remaining 1.1% were classified as brown in color. These were pieces that probably had been burned, because the color seems to have been present in the interior of each piece, judging from a few tiny breaks we made on existing fractured edges. We use color as the main identifier of burning. In addition, pieces that we classified as burned had almost no polishing, which we suggest was due to: (1) the pieces having fallen accidentally into a fire before they could acquire polishing; or (2) burned pieces not retaining polishing through time as well as unburned pieces. Hence, a lack of polishing reinforces our view that brown and black colored pieces are the result of burning.

We ignore surficial staining in our color identification, whereas Germonpré and Lbova (1996:41) considered staining, reporting that "colour varies from a dark greyish brown to a lighter brown, but some bones exhibit a light grey hue." One of their "traces" categories is burned fragments, which were rare (1.3%) and small (<5.0 cm long). We agree that most of the burned pieces are smaller than 5.0 cm, although we did identify three probably burned pieces that varied from 7.0 cm to 10.9 cm in length. A comparison of our respective burning frequencies (ours: 1.1%) shows no significant difference ($\chi^2 = 0.23$, 1 d.f., $p > 0.05$).

Compared with our other assemblages, Kamenka is generally without distinction, unless only open archaeological sites are compared. Then, Kamenka does stand out because of its very low amount of deep discoloration. Faunal remains from the open sites at Bolshoi Yakor and Ust-Kova have much lower frequencies of ivory colored pieces. We do not understand the cause of this difference. Lbova (personal communication, July 2003) suggested that rapid burial by shifting sand may have played an important role in these open-site color differences.

**9 Preservation.** All 552 pieces were assessed for quality of preservation (Table A1.9, site 8). Most pieces were ivory hard and excellently preserved (94.4%) – they could not be scratched with a fingernail. The remainder were chalky or intermediate, that is both chalky and ivory in hardness. One rare quality variant, a *Procapra* rib fragment, had a high-quality ivory external surface and a chalky and root-track-riddled lung-facing surface. Presumably the burial position of this piece had the chalky side facing upward, toward the ground surface.

Germonpré and Lbova (1996:41) similarly found that preservation was generally excellent. Using a different method to assess preservation, they found 94.1% had no or only minor weathering. While we are unsure of all the ways bone is degraded to a chalky state – surface weathering being certainly an important factor – we are in complete agreement with their assessment of preservation, including their suggestion that the Kamenka Component A assemblage was on the whole exposed on the ground surface or near the surface for only a short time before deeper burial. We would add that shallow frozen ground would have aided preservation before and after the relatively short summers.

Inspection of bone quality in each of our other assemblages (Table A1.9, site 8) quickly reveals that most cave deposits such as Denisova and Kaminnaya have excellent bone preservation, whereas open sites have assemblages with greater amounts of chalky bone (Kara Bom, Malaya Seeya, Mal'ta, and so forth). But there are exceptions other than Kamenka. These include Krasny Yar and Kurla (while Boisman is an open site with very little chalky bone, it is much younger than all the rest, and it contains large amounts of shell that helps buffer bone-damaging soil acids). It is also evident that open sites containing mainly mammoth remains (Shestakovo and Volchiya Griva) exhibit very poor preservation. Factors involved in this poor preservation probably include bone characteristics such as the relatively thin cortex of mammoth skeletal elements, the longer periods of time these very large bones remain unburied due simply to their size, and the contextual conditions of their locations that might include only modest sources of wind-blown or water-deposited overburden, trampling, and other physical and mechanical hazards to preservation.

**10 Perimortem breakage.** All 552 pieces could be assessed for perimortem breakage (Table A1.10, site 8). Most pieces of the Kamenka assemblage had perimortem breakage (94.7%). When this value is coupled with piece size (Table A1.6, site 8) it is clear that very extensive carcass processing took place in the Kamenka camps. This severe bone reduction and fragmentation has been viewed by a number of workers as indirect evidence of cooking for the purpose of extracting as much nutrient as possible from a carcass, especially fat-rich, high-energy marrow found in the interior of many skeletal elements. As we will mention time and again, we propose that cooking was done in hide or other sorts of containers holding water, fresh bone fragments with adhering flesh, globs of marrow, and heated stones, i.e., bag-boiling. An animal's gutted thorax and abdominal cavity could also serve as a container for the stone-boiling of itself.

A comparison of our perimortem breakage findings with those reported by Germonpré and Lbova (1996:41) shows no significant difference ($\chi^2 = 0.21$, 1 d.f., $p > 0.05$).

The Kamenka perimortem breakage incidence is much like that of our other assemblages, except for the lower occurrence in the open paleontological sites (Krasny Yar, Shestakova, Volchiya Griva). In contrast, the paleontological cave sites such as Razboinich'ya have a great deal of perimortem breakage.

**11 Postmortem breakage.** All 552 Kamenka pieces were examined for postmortem breakage (Table A1.11, site 8). Identifiable postmortem breakage is largely associated with general excavation problems and specifically the difficulty of extracting crumbly mammoth bone from its matrix. Postmortem breakage is rare (1.1%) in the Kamenka assemblage. This is attributable to high-quality excavation, excellent preservation, a fine-grained homogeneous matrix, and the absence of easily fragmented mammoth remains in Component A. However, modified mammoth tooth that had been made into a bracelet or diadem artifact was recovered (Lbova 2000:176), and one non-artifact piece of mammoth was identified by Germonpré and Lbova from Component B. These authors report finding no examples of "recent" bone fractures, which we presume to mean postmortem breakage.

Compared with our other assemblages, Kamenka has a very low frequency of postmortem breakage. Of the fully archaeological sites, only Kurla has a lower value.

**12 End-hollowing.** Only 240 Kamenka pieces could be evaluated for end-hollowing due to the fact that an anatomical end is generally required to assess the occurrence of this variable. Of these pieces, only 0.8% exhibit end-hollowing. These two pieces also have polishing on both ends, and one has tooth dints and small notches in a perimortem fracture. We consider end-hollowing to be a useful but not infallible criterion for carnivore damage.

Germonpré and Lbova report that carnivore gnawing is rare in their study of the Kamenka assemblage ($n = 26$, 1.6%); however, they do not report how they recognized it. Since one of the primary goals of our study is to define the signature of carnivore perimortem damage, we suggest that our respective studies are very similar in this inference, but ours is based on the use of multiple variables, end-hollowing being just one. Hence, we have 31 / 550 pieces (5.6%) that exhibit multiple criteria (excluding perimortem breakage), which we believe represent damage done by carnivores. This multiple-variable issue will be discussed in detail later.

End-hollowing is much more frequent in several of our other assemblages, especially those with direct or strong indirect indications of hyena presence. A return to Table A1.2 will show which sites had hyena bones (direct evidence), while Table A1.21 provides the frequencies for bone fragments with acid erosion (indirect evidence). It is well known that even domesticated carnivores sometimes swallow pieces of bone. If not regurgitated these pieces of bone can show some degree of digestive erosion (Davis 1987:26–27). The occurrence of end-hollowing and a few pieces of acid-eroded bone, plus other criteria of carnivore bone damage, strongly indicates that carnivores were utilizing some of the remains of butchering and cooking left by the human inhabitants of Kamenka. Ovodov's identification of one wolf bone in the Kamenka assemblage points to that species as having been responsible for the end-hollowing. Germonpré and Lbova also identified a bone of a large carnivore as that of a lion. Both species may have occasionally scavenged the refuse of the seasonally departed human hunters and their families, although modern wolves and hyenas are better known for scavenging decayed and dried carcasses than are lions (van Lawick and van Lawick-Goodall 1970).

**13 Notching.** We could assess notching on 549 pieces, of which 9.8% were so modified (Table A1.13, site 8). Notching can be done by hyenas, as testified to by their common occurrence on bones recovered from Razboinich'ya Cave, and by humans, where some abrasions caused by a hammer stone or anvil stone are in direct association with a notch. The majority of the notched Kamenka pieces had only one notch (6.0%). There were no pieces with more than four notches. Of the notched pieces there were 23 (42.6%) that had neither carnivore nor chop marks. There were 17 (31.5%) notched pieces with some carnivore damage (end-hollowing, tooth scratches, tooth dints, heavy polishing, erosion) but no chopping. Lastly, 14 (25.9%) notched pieces had only chop marks. There were no notched pieces with both chop marks and carnivore damage. Hence, notching by itself is not a sure sign of either hyena or human bone processing at Kamenka.

Germonpré and Lbova (1996:43) found a small number of percussion marks ($n=21$; 1.1%) on long bone shafts, most of which were associated with spiral fracturing. They attribute these marks to human bone processing for marrow extraction when found in association with marks attributed to knives, burins, and scrapers. Some of their percussion marks could be due to carnivore notching, just as some of our notching appears to have been done by humans. In our early examinations of the Razboinich'ya assemblage we soon abandoned spiral fracturing as a uniquely human-caused attribute because there were so many pieces that had to have been spirally fractured by hyenas.

The occurrence and multiplicity of notching at Kamenka is not exceptional compared with our other assemblages (Table A1.13). Notching is most frequent in those sites with direct or strong indirect evidence of hyenas or other carnivores having used the caves, including Razboinich'ya, Proskuryakova, Maly Yaloman, Okladnikov, and others. Notching is least frequent in open archaeological sites.

**14 Tooth scratches.** We could assess tooth scratching in 541 pieces (Table A1.14, site 8). Only 2.8% exhibited scratching. There is a wide range in the number of scratches per piece, with eight pieces having more than seven scratches. The largest number of scratches on a single piece is 30. While one or even possibly two scratches might have been caused by some agency other than carnivore chewing – such as trampling – several scratches on a piece are most probably the result of carnivore chewing. This is especially so when their occurrence on a piece is substantially parallel, variable in width and depth, and ends in a kind of squiggle. Looking at co-occurring damage in the eight pieces with more than seven scratches we find that all but one have tooth dints as well. The one piece with 40 scratch marks has 25 tooth dints. At least one of the eight had very fine scratches, suggesting a scavenger of fox-like size.

Comparing only our tooth scratching category with the "gnawing marks" category of Germonpré and Lbova (1996:42), we found no significant difference between our respective frequencies ($\chi^2=0.23$, 1 d.f., $p>0.05$).

Compared with our other assemblages, Kamenka tooth scratching is at the low end of the range (Table A1.14), which generally characterizes open sites. Cave sites have the highest frequencies of tooth scratching.

**15 Tooth dints.** We could assess 544 Kamenka pieces for tooth dints (Table A1.15, site 8). Like tooth scratching, tooth dinting is also uncommon (3.3%). The range of dints per piece is also wide. Both are strongly related, but not so much as to have to look upon them as a single variable. While there are 14 Kamenka pieces with both dinting and scratches, there are four pieces with only dinting, and two pieces with only scratches. As with tooth scratches, Kamenka, being an open site, has a low frequency of pieces with dints, whereas cave sites generally have higher dint frequencies.

**16 Pseudo-cuts.** We could assess 546 pieces for pseudo-cuts (Table A1.16, site 8). Pseudo-cuts were very rare (0.4%). When present there were only one or two pseudo-cuts on each of two pieces. Both pieces also had tooth scratches and dints. One of the pieces also had 20 tooth scratches and five tooth dints, the very fine and delicate size of both classes suggestive of chewing by a fox-sized carnivore. Pseudo-cuts exist as a variable because they were first identified in the faunal remains from Razboinich'ya Cave. The cave had no evidence of having ever been inhabited by prehistoric humans. Hence, what appeared to be cut marks had to have been caused by carnivore bone chewing. While an occasional piece with actual cut marks might have been picked up by hyenas scavenging one or another of the human sites in the valley below, and brought back to Razboinich'ya, we doubt this ever happened because pieces with burning and cut and chop marks at joints were not found in the Razboinich'ya assemblage. Abrasions were, however, present, but they could have been due to trampling or skittering of bone on the gritty limestone cave floor.

The rare pseudo-cut marks at Kamenka are similarly rare in other open sites. They are generally more common in the cave sites.

**17 Abrasions.** We were able to assess 544 pieces for abrasions (Table A1.17, site 8). Only one piece (0.2%) had abrasions. This was a piece of a horse long bone with a set of ten abrasion grooves. The piece also had nine cut and two chop marks.

Abrasions are generally rare in all of our assemblages. It would seem that most bone breakage was done with the aid of non-abrasive hammers and anvils, such as another piece of bone, antler, or wood.

**18 Polishing.** We were able to assess 540 pieces for polishing (Table A1.18, site 8). Of these, 90.0% had polishing on one or both ends (24.8%), in the middle (4.0%), or on the ends and middle (61.3%). We are uncertain how this polishing forms. It is possible it develops by some natural chemical or physical post-depositional process. On the other hand, prehistoric polishing of mainly ends of human bone fragments has been proposed by White (1992), and verified by Turner and Turner (1999), to have likely occurred when freshly smashed pieces of bone were stirred in a water-filled pottery cooking pot with a slightly gritty interior. This treatment helps extract marrow fat as well as loosen and cook adhering pieces of muscle tissue. Perhaps the Kamenka polishing happened when pieces were abraded by heated stones placed in water-filled hide or tightly woven fiber bags for the same marrow-extracting objective.

Compared with our other assemblages, the frequency of Kamenka polishing is at the higher end of the range, but not exceptionally so (Table A1.18). Because polishing is found in our hyena cave assemblages, carnivore bone chewing must be another mechanism producing this condition. This is especially so when polishing of a piece is extensive and severely rounded.

**19 Embedded fragments.** We were able to assess 547 Kamenka pieces for embedded fragments (Table A1.19, site 8). Of these, 1.5% exhibited this condition. In our preliminary work we found embedded fragments largely associated with carnivore damage, and most notably in large tooth dints. In the course of time we found a few instances of embedded fragments associated with what were unquestionably hammer percussion indentations and notches. Kamenka has three pieces with embedding but no chop or cut marks, four pieces with embedding and chop or cut marks, and one piece with embedding and digestive erosion. The latter, while definitely eroded, lacks the greasy or slippery feel of hyena stomach bones, which allows for the possibility for some form of soil-dependent damage.

Compared with our other assemblages the frequency of Kamenka pieces with embedded fragments is relatively low (Table A1.19). The cave sites, as expected, have relatively more embedding, but there are notable open site exceptions such as Kurla and Ust-Kova.

**20 Tooth wear.** We had only 12 tooth-bearing pieces in the Kamenka series, hardly enough for any comments other than very limited suggestions about age. The greater the amount of wear in an assemblage indicates a generally older series of individuals than in an assemblage with less wear. Kamenka seems to have few young individuals at 8.3% (Table A1.20, site 8).

Had the very small sample sizes in Table A1.20 been larger, we would feel much more secure in pointing out that sites such as Boisman II and Kaminnaya, and others, regardless of antiquity differences, apparently have more young individuals than did Kamenka.

**21 Acid erosion.** There were 550 pieces that could be assessed for the occurrence of digestive damage (Table A1.21, site 8). Kamenka has a low frequency of erosion (1.3%), and of these seven pieces, six have only limited erosion and lack the slippery-feeling and scoured surface of stomach bones associated with hyena deposits. It may be that the seven stomach bones represent less-digested pieces swallowed by wolves, other carnivores, or perhaps even the Kamenka humans themselves. The sizes of these pieces were: 2.5, 3.0, 4.3, 4.5, 4.5, 5.5, and 5.7 cm, the diameters of which are all well within the range of stomach bones found in the hyena caves. If these weakly eroded pieces represent regurgitation that limited their time in the digestive track, we would expect the larger ones, not all seven, to have had such short acid exposure. Studies of dried human feces from various archaeological localities have shown considerable ingesting of bone fragments by humans (see, for example, Clary 1984, Heizer 1967).

Kamenka is on the low end of the range for the occurrence of digestive damage (Table A1.21). The highest frequencies are associated with hyenas.

**22–24 Rodent gnawing, insect damage, and human bone.** No examples of these variables were found in the Kamenka assemblage we examined (Tables A1.22, site 8 to A1.24, site 8).

**25 Cut marks.** A total of 546 pieces could be evaluated for cut marks. Of these, 9.5% had one or more cut marks (Table A1.25, site 8). Most frequent were pieces with only one incision (3.8%). Of the pieces with the greatest number of cut marks, one is a long bone fragment 11.4 cm in length with 16 cut marks; a splinter 8.5 cm long has 20 cut marks that may have been produced during excavation; a 7.5 cm long distal portion of a *Procapra* humerus has 24 cut marks; and a 12.8 cm horse long bone with 64 cut marks is of special interest because the cutting tool left multiple parallel lines as if a unifacially chipped flake had been drawn across the surface like a scraper or plane rather than like the slicing action of a knife.

A comparison of our cut mark frequency with that of Germonpré and Lbova shows a very significant difference ($\chi^2 = 111.3$, 1 d.f., $p < 0.001$). We identified about ten times more pieces with cut marks (52 / 546) that did Germonpré and Lbova (19 / 1928). The difference is attributable to the many more years of experience in perimortem taphonomy by Turner (Turner and Turner 1999; and elsewhere) than by Germonpré and Lbova.

Compared with our other assemblages, the occurrence of Kamenka cut marks is near the middle of the range. The Kamenka frequency (9.5%) is much like that of nearby Varvarina Gora (8.8%).

**26 Chop marks.** The Kamenka sample had 548 pieces that could be assessed for chop marks. Of these, 8.6% had chop marks (Table A1.26, site 8). The majority (6.2%) of these had only one chop mark. Compared with our other assemblages Kamenka is near the middle of the range for chop mark frequency.

## Discussion

The Kamenka assemblage has several interesting characteristics. Foremost of the perimortem taphonomic damage is the large amount of intentional bone breakage that is readily inferred as having been intentional human processing. In addition, the absence of direct or indirect evidence of hyenas suggests that these animals were not a problem for the Kamenka people as we believe they were elsewhere, especially in the Altai. Despite the breakage, the Kamenka bone assemblage is remarkably well preserved, especially for an open site. Frozen ground, rapid deep burial, and minimal humic acid are three of the several possible conditions that we suggest favored high-quality preservation. Perimortem bone damage was caused by natural and human agencies, including minor amounts of root etching, surface weathering, and some soil leaching. The amount of chalky or denatured bone is low (see section on site 27, Varvarina Gora, for contrast). Minor amounts of carnivore chewing are evident. As far as we can tell, rodents and insects played no role in the perimortem damage signature.

There is considerable human cutting, chopping, and skeletal element smashing. Because there is so much element breakage, fragmentation, and polishing, whereas there is so little burned bone, it is reasonable to hypothesize that there had been cooking, and that it was done mainly by bag- or wooden bowl-boiling of crushed bone and adhering tissue. The type of evidence needed for hypothesizing roasting of meat is entirely lacking. Evidence for roasting would include burned ends of bones with unburned mid-sections where greater amounts of muscle would have protected the bone from scorching, higher frequencies of burned pieces, especially skulls and other parts with minimal overlying protective soft tissue, and much greater archaeological evidence for roasting, including cooking pits, charred sticks, and other such features known from ethnographic descriptions of roasting (Turner and Turner 1999:24–39).

Having the opportunity to compare our observations with those made by an independent team has been extremely valuable and informative. The work by Germonpré and Lbova (1996) for the most part strongly agrees with our study. Where there are significant differences they are clearly due to differing standards, procedures, or experience, not error.

The Kamenka assemblage broadly resembles our other open site assemblages. Where there are notable differences, the presence of hyenas elsewhere is the major cause. The Razboinich'ya Cave hyena damage signature is absent at Kamenka.

# 9 Kaminnaya Cave

## Background

Kaminnaya Cave was first test excavated during the IAE expedition to the Altai, led by Nicolai Ovodov in 1969 (Fig. 3.31) (Okladnikov and Ovodov 1972, Ovodov 1972, 1973c, 1973d). Its name means "Fire-place." The site is located in a small pine-forested tributary valley to the Annui River in the limestone Altai Mountains at 51°17′ N, 84°28′ E (Vasili'ev *et al.* 2002). The cave faces south onto, and is only 4 m higher than, a swift-flowing, clear, and cold mountain-born artesian stream just 22 m from the entrance. This stream also passes through a small settlement named Karakol, which is within walking distance of Kaminnaya. Its muddy road has wandering pigs, dogs, cows, horses, and chickens, typical of main streets in Siberian villages. The village has an interesting mix of traditional- and Russian-style structures. Some of the villagers operate a maral deer preserve, fenced in the forested mountains that rise immediately behind Kaminnaya Cave. Male deer are rounded up for food and their horns are sold to Chinese traders, who use them for human male sexual enhancement. Like Denisova Cave, Kaminnaya has a spacious entrance: 17 m wide and 5.5 m high (Derevianko *et al.* 1998d:59). However, Kaminnaya differs in having had in the past an active water course that left well-rounded cobbles and boulders mixed in with the non-hydrologic cave deposits (Derevianko and Dergacheva 2002, Derevianko *et al.* 2002d, 2002e). Kaminnaya's floor area is about 120 $m^2$. A distinctive feature of this cave location, as compared with other karst caverns of this region, is its very low position above the nearby stream. The meaning of the word "Kaminnaya" refers to its hearth-like shape. Local people simply refer to it as "Stone Cave." Kaminnaya has a record of *active* history – bioturbation, geological changes, and a history of discontinuous human occupation.

According to archaeological thinking in 1982, when Ovodov made the first test excavation of Kaminnaya Cave, such a low position of the cave would exclude much chance of finding any Paleolithic remains. Ovodov tested about 12 $m^2$ to a depth of 80 cm near the cave entrance. He found only fragments of pottery and ashes, which seemed to confirm the archaeological wisdom of that time. However, shortly after this Yuri V. Grichan dug a little deeper, only some further 15 cm, whereupon he encountered tell-tale loose loamy Pleistocene sediments containing bones of large mammals and Paleolithic stone tools. Beginning in 1983 and continuing to 1990, Grichan (1987, Derevianko and Grichen 1990) conducted intensive excavation of Kaminnaya, mainly

Fig. 3.31 Kaminnaya Cave mouth. Where the students are seated, there had been a talus slope that Grichan excavated into just outside the cave's drip line. Ovodov and Pavlova (left) are standing at about the lowest level of the cave deposits, which are behind them. The original test by Ovodov is the dark vertical hole in the upper center. Ovodov and Pavlova are being photographed for a Russian documentary film (CGT color Kaminnaya 7-10-99:0).

in the fronting talus slope. After a brief hiatus, Sergey V. Markin took over and resumed excavation supervision in 1994, following the completion of his work at Okladnikov Cave. As of today, most of the 15 m long fronting talus slope has been excavated to boulder-covered bedrock. Holocene sediments contained nine cultural layers, within which has been found the burial of a Neolithic female richly decorated with fine jewelry, and a large number of ceramic vessels of different time periods, and other Neolithic and more recent artifacts (Markin 2000). The age of the Holocene cultural layers has been determined by radiocarbon dates (Derevianko *et al.* 1999b). The Pleistocene sediments all over the excavated area have been disturbed by geological processes connected with seismic activity in this locality (Derevianko *et al.* 1998e).

As will be noted in the following section on provenience, our sample may contain a small number of unidentifiable and identifiable bones belonging to Holocene times, as well as pieces of animals that died many millennia before any cave deposit was laid down. The Pleistocene deposits were considerably disturbed by rock falls from the ceiling due to seismic activity in the region. For Layers 11a and 14b there are eight carbon-14 dates ranging from 10 000 BP to 15 000 BP (Derevianko *et al.* 1999b). More recently, Yaroslav Kuzmin and American associates at the University of Arizona, Tucson, have obtained much older carbon-14 dates for nearly the same Kaminnaya Cave levels using accelerated mass spectrometry (AMS). For Level 10b a rhinoceros tooth fragment provided a date of 39 400 BP (AA-38043). For Level 11b an AMS date of 41 000 BP (AA-38044) was obtained. A standard carbon-14 date for Level 11b was 12 150. Kuzmin suggested that the rhinoceros teeth were temporally unrelated, having been in actuality ancient fossil specimens that the more recent Pleistocene folk had carried into Kaminnaya. We can add to this taphonomic assessment by noting that our identification of probable hyena presence might also, or instead, have been the introducers of the rhinoceros specimens. Splitting a single specimen and dating the two halves by AMS and the standard carbon-14 technique produced more concordant dates (Markin *et al.* 2001).

There are some very early and questionable TRL dates. Layer 15 TRL age is 200 000 BP. The layer contains unquestionable stone artifacts (S. Markin, personal communication, July 8, 1999).

The present authors visited Kaminnaya together in 1987, guided by Yaroslav Molodin, then field supervisor for the Denisova Cave excavations, and again together in 1999, while Markin's student field crew was working at Kaminnaya (Figs. 3.32, 3.33). The

**Fig. 3.32** Kaminnaya Cave, screening. A large spring originates far up-slope and its clear cold water flows below the mouth of the cave and past the field camp, through Karakol village, and eventually joins the Annui River. The high mountain snowpack source of this spring might have flowed through Kaminnaya cave off-and-on during the Pleistocene. Whatever the spring's history, it is an excellent source of clean water for the needs of the archaeologists at Kaminnaya, including, as shown here, wet-screening of cave fill. We observed this wet-screening procedure at other sites, and suspect that some of the very minor fine bone scratches occurred during this cleaning process, as well as polishing of stone artifact cutting surfaces, which would cast doubt on wear analysis (CGT color Kaminnaya 7-10-99:37).

**Fig. 3.33** Kaminnaya screened midden. Serghei Markin (far left) explains to Olga Pavlova the wet-screening and subsequent sorting procedures. The mounds of debris on the plastic sheeting are mostly light gravel that did not pass through the wet-screening stage. Each pile after drying will be sorted by a student or professional archaeologist using a small tool to pick out tiny bone and stone tool manufacturing debitage. Most larger artifacts and faunal remains are found during excavation. The senior author estimated that the anthropologically linked recovery level was about 2.0 mm in maximal diameter – far below our minimal level of 2.5 cm (CGT color Kaminnaya 7-8-99:22).

original excavators are shown in Figs. 3.34 and 3.35. As will be discussed later, the Razboinich'ya hyena cave is a steep two-hour and 2 km hike southwestward into the mountains high above Kaminnaya.

## Findings

**1 Provenience.** All of the 771 Kaminnaya pieces we studied came from the deep talus fronting the cave entrance (Fig. 3.31). The stratigraphy in this area includes modern, Neolithic, Paleolithic, and natural geological materials. As much as possible, the Neolithic horizons were excluded from our study; however, ancient stratigraphic disturbances and post-excavation loss of field bag labels and unlabeled bags contributed to our decision to treat all our studied pieces as one temporal and spatial unit. Certainly the vast majority of pieces are from Upper Paleolithic times. Where we were able to make stratigraphic comparisons, we found no meaningful differences in the types of perimortem damage, and in most instances these comparisons lacked species, age, and skeletal

**Fig. 3.34** Odyssey. Chopping a morel deer leg. Nicolai Ovodov is chopping a leg of deer by firelight for the Kaminnaya camp's special Saturday evening meal. Sergei Markin obtained the freshly butchered meat in Karakol village, whose residents manage a vast morel deer farm, fenced by high barbed-wire, in the nearby mountains. The meat was boiled in spring water, potatoes, salt, and onion into a watery stew. Because there are so very few Paleolithic fragments of burned bone that would suggest roasting, we think that boiling like this was also the principle cooking method in ancient times. Notice that Ovodov is using a small log as a chopping block, also possibly done in Paleolithic times. The iron axe head is exactly the same shape and size as one the senior author excavated from the Russian period of an archaeological site in the eastern Aleutians of Alaska. It is illustrated in Turner and Turner (1974) (CGT color Kaminnaya 7-10-99:unnumbered).

**Fig. 3.35** Yuri Grichan. Co-discoverer of the Kaminnaya cultural deposits, Grichan (right) has long been ill and unable to continue the arduous excavation work that he conducted for several years at Kaminnaya Cave. In this photograph with the senior author at the Institute of Archaeology and Ethnography, Grichan had largely recovered from the stroke-related paralysis he suffered for several years. He died in the winter of 2012 (OVP color IAE 6-23-06:x).

element information due to small piece size. Our Kaminnaya assemblage makes up 8.7% of our 9360-piece grand total (Table A1.1, site 9).

**2 Species.** Of the 352 Kaminnaya bones studied by Irina Foronova (Derevianko *et al.* 1998c, 1998d), 17 species were identified: gray wolf (*Canis lupus*), fox (*Vulpes vulpes*), bear (*Ursus* sp.), badger (*Meles meles*), cave hyena (*Crocuta spelaea*), cave lion (*Panthera spelaea*), horse (*Equus cf. caballus*), an Asiatic wild ass (onager) (*Equus hemionus*), rhinoceros (*Coelodonta antiquitatis*), red deer (*Cervus elaphus*), roe deer, roe (buck) (*Capreolus capreolus*), moose, elk (*Alces alces*), bison (*Bison priscus*), wild yak (*Poephagus baicalensis*), saiga antelope (*Saiga tatarica*), goat, ibex, markhor (*Capra sibirica*), and wild sheep (*Ovis ammon*). In Ovodov's (1997b) view the assemblage is quite a typical "soup set" of Paleolithic hunters of the Altai, as well as of cave hyenas who "dine in their underground canteens."

Of the 771 Kaminnaya pieces we studied, the most commonly identified groups are: indeterminable (71.4%), big mammal (8.1%), and horse (3.6%) (Table A1.2, site 9). Compared with the pooled assemblage averages, Kaminnaya has fewer bear, bison, gazelle, goat-sheep, mammoth, and reindeer, and twice the number of indeterminable pieces. Skeletal element breakage is severe in Kaminnaya Cave.

**3 Skeletal elements.** Of 771 Kaminnaya pieces, the common skeletal elements are: long bone (37.0%), unknown (27.5%), toe (6.5%), metapodial (5.8%), mandible (3.9%), and rib (3.9%) (Table A1.3, site 9). Compared with the pooled assemblage averages, Kaminnaya has fewer mandibles, vertebrae, ribs, humeri, and more long bone and unknown elements.

**4 Age.** Only 2.2% of 771 Kaminnaya pieces could be identified as sub-adult (Table A1.4, site 9). Compared with the pooled assemblage, Kaminnaya has fewer sub-adults and twice as many pieces of unknown age.

**5 Completeness.** This variable could be assessed in 714 Kaminnaya pieces. Postmortem breakage is fairly high. There are 6.4% whole bones, 13.6% with one anatomical end, and 80.0% with no anatomical ends (Table A1.5, site 9). The pooled assemblage averages are, in the same order, 10.9%, 30.0%, and 59.1%, all of which shows that the Kaminnaya mammal assemblage is considerably damaged.

**6 Maximum size.** Out of 771 Kaminnaya pieces, the mean is 6.2 cm, and the range is 2.4 cm to 25.0 cm (Table A1.6, site 9). Compared with the pooled assemblage, Kaminnaya's mean and range are somewhat less, indicating greater damage. Looking at the whole long bone lengths for horse provided by Vera Gromova (1950: table 27) shows that the upper range limit for Kaminnaya is generally lower than her lower range limits.

**7 Damage shape.** The most common forms of identifiable Kaminnaya bone damage in 557 pieces are: long bone flakes (38.4%), long bone fragments (25.7%), long bone splinters (9.7%), phalanx butts (5.2%), mostly whole (4.5%), and undamaged bone (3.8%) (Table A1.7, site 9). Compared with the pooled assemblage, Kaminnaya has more long bone flakes and long bone splinters, and fewer rib pieces and undamaged bones.

**8 Color.** All 771 Kaminnaya pieces could be assessed for color, the majority of which are ivory colored (96.6%) (Table A1.8, site 9). Black (burned) pieces are common (3.0%) relative to the pooled assemblage, where they constitute only 0.8% of the total.

**9 Preservation.** Nearly all (99.6%) of the 770 Kaminnaya pieces are ivory hard (Table A1.9, site 9). The remainder (0.4%) are chalky. Compared with the pooled assemblage averages, Kaminnaya has outstanding preservation quality.

**10 Perimortem breakage.** The vast majority (94.3%) of 770 Kaminnaya pieces exhibit perimortem breakage (Table A1.10, site 9). This value is greater than the pooled assemblage average.

**11 Postmortem breakage.** Less than 5% (4.6%) of the 767 Kaminnaya pieces have postmortem breakage (Table A1.11, site 9). This postmortem breakage value is three times less than that of the pooled assemblage.

**12 End-hollowing.** Out of 628 Kaminnaya pieces, 1.9% have end-hollowing (Table A1.12, site 9). Compared with the pooled assemblage, Kaminnaya has less end-hollowing.

**13 Notching.** While the presence of end-hollowing suggests carnivore activity at Kaminnaya, the amount of notching suggests even more. Out of the 768 pieces, 10.8% have one to more than seven notches per notched piece, most of which have only one

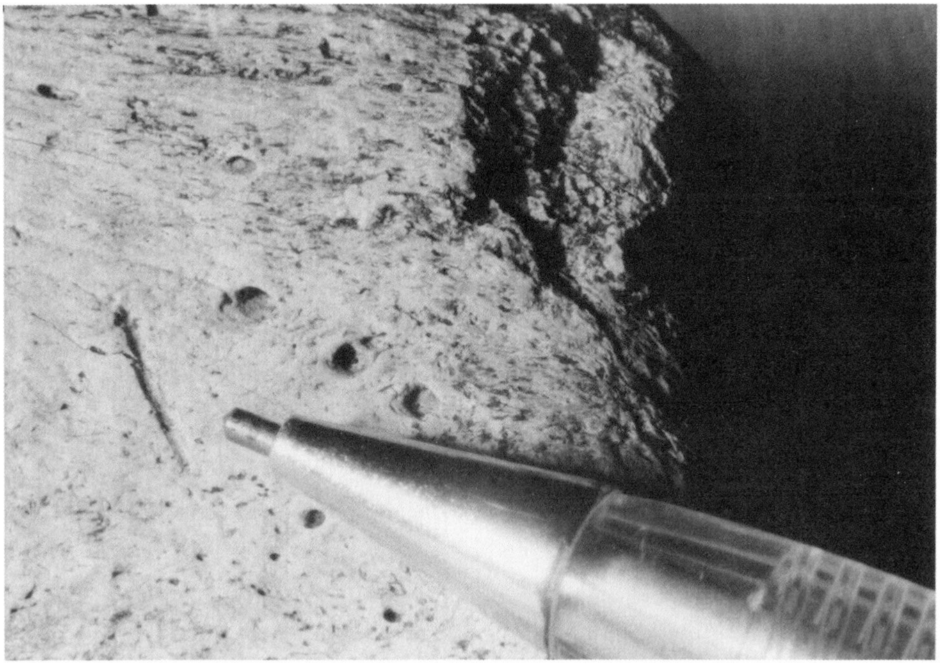

Fig. 3.36 Kaminnaya bone damage. *Equus* calcaneus with 0.6 cm long pseudo-cut (1985, layer 14) (CGT neg. IAE 6-29-00:12).

notch (7.2%) (Table A1.13, site 9). Compared with the pooled assemblage average, Kaminnaya has only slightly less notching.

**14 Tooth scratches.** This unquestionable sign of carnivore perimortem activity is present in 15.6% of 763 Kaminnaya pieces (Table A1.14, site 9). Compared with the pooled assemblage average, Kaminnaya has somewhat fewer pieces with scratching, but not significantly so.

**15 Tooth dints.** Like tooth scratching, tooth dinting is a useful perimortem damage indicator of carnivore presence. Among the 762 Kaminnaya pieces, nearly one-quarter (24.1% ) have from one to more than seven tooth dints per piece (Table A1.15, site 9). One dint per dinted piece is most common (6.2%). Kaminnaya has nearly the same frequency of dinted pieces as the average of the pooled assemblage.

**16 Pseudo-cuts (Fig. 3.36).** Out of 763 Kaminnaya pieces, there are 25 (3.3%) with pseudo-cuts (Table A1.16, site 9). This frequency is slightly less than the pooled assemblage average.

**17 Abrasions.** As in most of our studied sites, abrasions are uncommon. Kaminnaya has in 763 pieces only 12 with abrasions (Table A1.17, site 9). This frequency (1.6%) is almost identical to the average for the pooled assemblage.

**18 Polishing.** A frequency of 78.3% polishing occurs in 770 Kaminnaya pieces (Table A1.18, site 9). By location it is: end (13.4%), middle (1.2%), and end-middle (63.8%). In contrast, the pooled assemblage averages are: total (66.4%), end (19.8%), middle (1.6%), and end-middle (45.1%). These values suggest that carnivore activity at Kaminnaya was substantial.

**19 Embedded fragments.** Out of 770 Kaminnaya pieces, 2.9% have from one to more than seven embedded fragments (Table A1.19, site 9). Compared with the pooled assemblage average, Kaminnaya has insignificantly fewer pieces with embedding.

**20 Tooth wear.** Kaminnaya has 23 teeth, the wear of which suggests that 30.4% were young at the time of death (Table A1.20, site 9). Compared with the pooled assemblage average for young individuals, Kaminnaya's frequency seems reasonable.

**21 Acid erosion.** The frequency of acid erosion in 770 Kaminnaya pieces is 17.4% (Table A1.21, site 9). This is greater than the pooled assemblage average. It is also greater than the Razboinich'ya hyena cave frequency (7.3%), but less than that of Maly Yaloman (48.0%). As previously suggested, Kaminnaya had a substantial carnivore presence. The high frequency of acid-eroded pieces suggests that these carnivores included hyenas.

**22 Rodent gnawing.** Kaminnaya is one of only eight assemblages in this study to have pieces of bone that had been gnawed by rodents. Out of 771 pieces, five (0.6%) had rodent incisor chiseling marks (Table A1.22, site 9). Kaminnaya is nearly identical to the pooled assemblage average for rodent gnawing.

**23 Insect damage.** Even less common than rodent gnawing, insect damage occurs in only three of our assemblages. Kaminnaya has one example of insect damage (0.1%) in 769 pieces (Table A1.23, site 9).

**24 Human bone.** No Pleistocene human bones or teeth have so far been discovered at Kaminnaya, although a spectacular female burial has been recovered in the Neolithic deposits (Markin 2000).

**25 Cut marks.** The frequency of Kaminnaya pieces with cut marks is fairly high (11.4%; 88 / 769 pieces) (Table A1.25, site 9). The number of cut marks per cut piece ranges from one to more than seven, with more than seven being the most common number (3.6%). Compared with the pooled assemblage average, Kaminnaya has a slightly higher frequency of cutting.

**26 Chop marks.** Out of 770 Kaminnaya pieces, 3.4% have one to more than seven chop marks, one per piece being the most common number (Table A1.26, site 9). The occurrence of chopping in Kaminnaya is slightly less than the pooled assemblage average.

## Discussion

As in other Altai cave and open archaeological sites, both human and carnivore activity have been identified in the perimortem taphonomy of Kaminnaya Cave. We infer that much of the carnivore bone damage was done by hyenas. In addition, as a significant contributor to general bioturbation, the use of Kaminnaya Cave by hyenas requires that serious consideration be given to the possibility that its stratigraphy was disturbed to a significant extent. Such disturbance could, of course, blur the distinction between Middle and Upper Paleolithic assemblages. If sufficient mixing had taken place, a model of local evolution (Derevianko 2005:13) could be based on what might seem like clear-cut stratigraphic grounds, when in fact migration into the Altai by anatomically modern Upper Paleolithic humans and their replacement of the Mousterians might have been the actual sequence of events. Basing a local evolutionary hypothesis on stratigraphic trait frequency changes in an environment that seemingly supported an abundant hyena population is a substantial scientific risk.

In addition to the probable bioturbation, Kaminnaya has in lower deposits cobbles and small boulders that are rounded. This means that the cave had at some time been an active underground river. In sum, with such a complex taphonomic history, Kaminnaya would be a poor choice on which to create a chronology and history of humans in the Altai Pleistocene.

Kaminnaya presented some unusual specimens outside our 26 traits. In general, phalanges were not damaged by Paleolithic Siberians, but this occurred at Kaminnaya (Fig. 3.37). Another oddity is a specimen with severe periodontal disease, almost certainly an old and probably weak animal (Fig. 3.38). Fig. 3.39 shows the most extreme plant root damage seen during this study. Fig. 3.40 shows what Grichen believed was a bone tool. If so, it would be one of a very few pieces of bone that might be thought of as

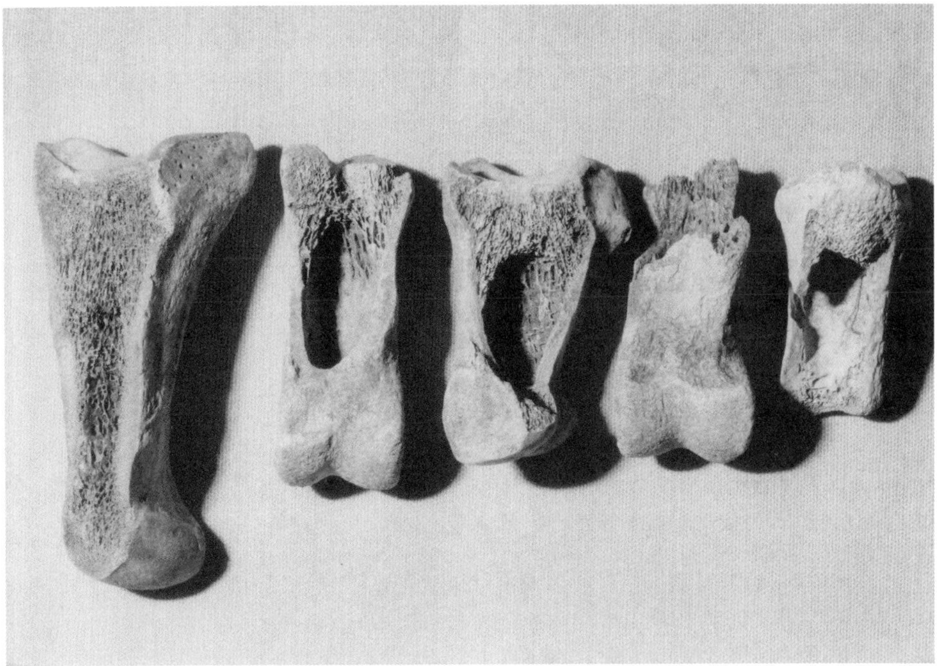

**Fig. 3.37** Kaminnaya bone damage. Cracked open phalanges. Left specimen is 9.5 cm in length (CGT neg. IAE 7-4-00:26).

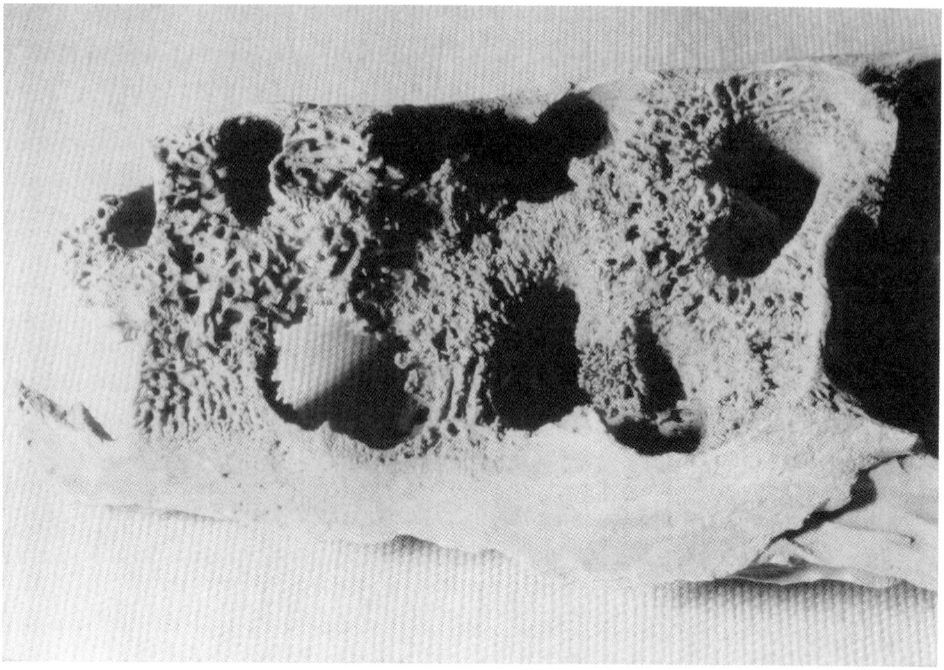

**Fig. 3.38** Kaminnaya periodontal disease. This socket-bearing maxillary fragment shows the pitting and bone erosion characteristic of periodontal disease. This 1985 specimen is one of only a few pieces of bone evidencing disease or trauma. On the basis of what we have seen in these various faunal collections, it would be hard to defend the position that human and non-human hunters preyed mainly on weak and sick game. Young prey may have been the major kill objectives. Actual width of image is 6.0 cm (CGT neg. IAE 7-4-00:27).

**Fig. 3.39** Severe plant root damage. None, or very little, of the original bone surface remains. Specimens like this were eliminated from our study because there is no way to be certain that cut marks or other perimortem damage were once present. Specimen excavated by S. Markin, Kaminnaya Cave, 1994 (CGT neg. IAE 6-29-99:21).

**Fig. 3.40** Kaminnaya artifacts. When these stone flakes and bone fragment were photographed in 1987, Grichan thought that they represented the oldest artifacts from Kaminnaya Cave. Scale is 15 cm (CGT color IAE 6-4-87:36).

**Fig. 3.41** Kaminnaya artifacts. According to Markin, new but very questionable dates for Kaminnaya suggest the site dates to the middle Pleistocene. These artifacts are claimed to be younger than 250 000 BP. In a brief examination the senior author could not distinguish any meaningful difference between these pieces of stone and those of Fig. 3.40. Scale is 10 cm (CGT color Kaminnaya 7-8-99:33).

**Fig. 3.42** Kaminnaya artifacts. Stone flakes and cores from Layer 15 that Markin claims are more than 200 000 years old. Again, there is no readily determined difference between these artifacts and those of the two previous figures. In all three cases, the kinds of cut marks on bone illustrated in the previous Kaminnaya photographs could have been made by any of these flakes. Scale is 10 cm (CGT color Kaminnaya 7-8-99:30).

having been tools. Fig. 3.41 shows stone artifacts of reputed great age. Fig. 3.42 shows more stone artifacts for which great age is proposed. Since bone tools are generally absent in our assemblages, such a tool might suggest a different tribe of people, which among other faunistic considerations has led Ovodov (1973b) to become concerned about the origins of Altai humans.

# 10 Kara-Bom

## Background

Kara-Bom (Black Cliff) is a workshop open site found among a rocky outcrop at the base of a steep foothill (Fig. 3.43) located at an elevation of 1800 m above sea level in the Elovskaya Valley, mountainous central Altai. A year before he died, Okladnikov discovered and tested the site in 1980 (Okladnikov 1983). It was found as a result of exposure due to road construction. Excavations were later directed by A. P. Derevianko and V. T. Petrin from 1987 to 1993 (Derevianko *et al.* 1993, 1994, 1997b, 1997c, 1998f, 1998g, 2000c, 2001c). Site maps, site photos, and various artifacts are newly illustrated by Derevianko and Rybin (2005). Kuzmin and Orlova (1998:6) place Kara-Bom at 50°10′ N, 86°40′ E.

Cultural deposits are overlain by slope diluvium. The thickness of the cultural layers, which included seven Paleolithic living surfaces, was determined by geologist Sergei Nicolaev to be 7.0 m. Derevianko *et al.* (2000c) provide both carbon-14 dates and electron spin resonance (ESR) dates, the latter being much older than the former for stratum M2. Vasili'ev *et al.* (2002:521–522) list two dates for the Mousterian assemblage in stratum M1, which are >42 000 and >44 000, both being Accelerator Mass Spectrometry (AMS) dates. For the Upper Paleolithic assemblages in Layers 3–6, the dates listed range from 30 990 BP to 43 300 BP. All are AMS dates. On the basis of stone tool manufacturing technology, Lbova (2002:131) suggests a date of 43 000 BP for Kara-Bom Layer 6.

Led by Michael Shunkov, two of the authors, Turner and Pavlova, visited Kara-Bom, situated within sight of the Altai village of Yelo, on July 18, 2006, along with John W. Olson, University of Arizona, Tucson, and others. Interestingly, the Russian map (Respooblinka Altai, 1:500 000) the senior author has used to plot the locations of sites visited in the Altai places Kara-Bom at ca. 50°45′ N, 85°35′ E, which is slightly different from the GPS-based location given above. Kara-Bom is protected by a high outcropping of very laminar metamorphosed dark gray, fine-grained limestone, whose use for tool production would inherently produce blade-like artifacts. Shunkov commented that even better stone has been found outcropping atop a nearby hill. Hence, the presence of blade tools found in the cultural levels of this workshop could just as easily be attributed to local stone having natural lamellar characteristics as to technical advancement (Figs. 3.44–3.46). Shunkov also explained that some of the cultural deposits were "mixed," while the remainder retained good stratigraphy. Most of the limited amount of faunal remains came from the mixed area. Shunkov also remarked that the occurrence

Fig. 3.43 Kara-Bom site. Speaking in a light drizzle, Michael Shunkov explains the stratigraphy of this open site, first tested in 1980 by A. P. Okladnikov. Disturbance by hyenas was very apparent. Note the tabular stone outcrop in the upper right. Natural blade tools could be derived from such material without any of the core preparation needed to make Upper Paleolithic blade tools (CGT color Kara-Bom 7-18-06:x).

of blade tools at Kara-Bom proved continuity in the Altai. He noted that Mott and Kozlowsky (2005) proposed blade tool manufacturing began in the Altai before it did in Europe.

Faunal remains were few relative to the large area and volume of excavation. Ovodov and Vasili'iev (unpublished observations) identified goat, ibex, markhor (*Capra sibirica*), horse (*Equus cf. caballus*), bison (*Bison priscus*) or yak, marmot (*Marmota baibacina*), Asiatic wild ass (*Equus cf. hidruntinus*), cave hyena (*Crocuta spelaea*), rhinoceros (*Coelodonta antiquitatis*), mammut (*Mammonteus primigenius*), cave lion (*Panthera spelaea*), and screw-horned antelope (*Spirocerus* sp.). The vast majority of pieces were unidentifiable. Wrinn (2010), following Agadjanyan and Serdyuk (2005), show an absence of hyenas. We found 108 pieces, one (0.9%) of which was hyena.

## Findings

**1 Provenience.** The 108 Kara-Bom pieces make up 1.2% of the 8813-piece grand total of this study (Table A1.1, site 10). Because this is a very old open site, with poor

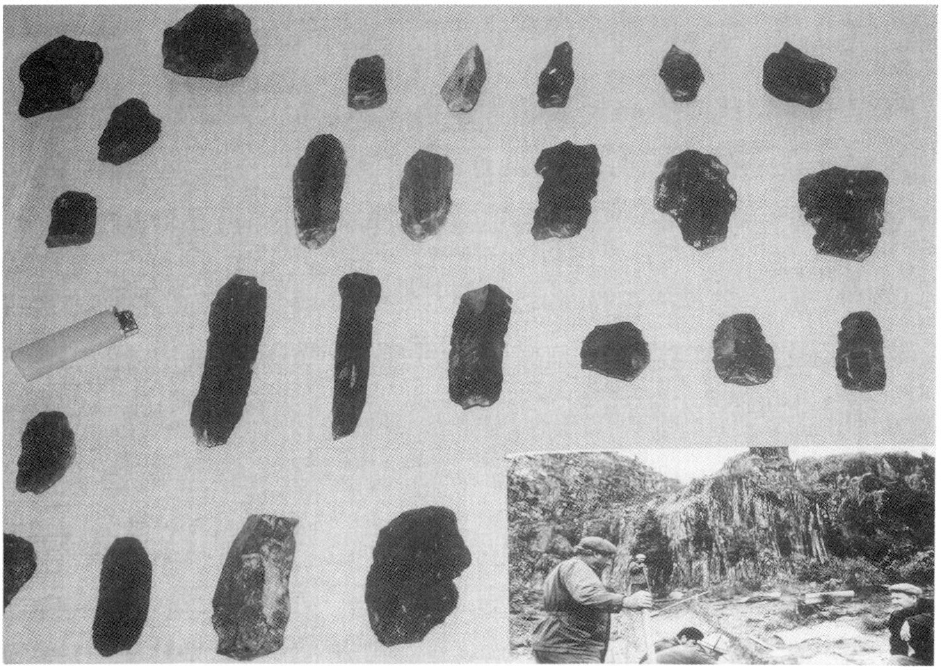

**Fig. 3.44** Kara-Bom artifacts. An exhibit in the former IHPP museum shows some of the stone artifacts found at Kara-Bom. The exhibit photograph shows Okladnikov directing the excavation in 1980 of the open workshop that was believed to be 30 000–40 000 years old. Faunal remains, including those of cave hyenas, came from a largely disturbed part of the site. Butane lighter is 9.0 cm long (CGT color IHPP 2-10-84:1).

**Fig. 3.45** Kara-Bom artifacts. Both flake and blade objects were found at this open workshop site. Scale is 15 cm (CGT neg. IAE 7-30-03:37).

**Fig. 3.46** Kara-Bom blade artifacts. Both objects are knife-like in shape, but only the larger one has any retouching. The larger specimen is 11.6 cm long; shorter one is 8.5 cm (CGT neg. IAE 7-30-03:33).

preservation, we assume that our bone sample dates to the more recent part of the human use of the site.

**2 Species.** The most common groups in the 108 Kara-Bom pieces are: indeterminable (45.4%), big mammal (23.1%), marmot (11.1%), medium mammal (6.5%), and hare (3.7%) (Table A1.2, site 10). Big and very big animals, including bison, horse, hyena, mammoth, and rhinoceros, are also represented, but in very low frequencies. Because their big bones are more likely to be preserved than those of smaller animals, we doubt whether the marmots and hares are actually of Pleistocene age. Compared with the pooled assemblage averages, Kara-Bom has no bear, goat-sheep, reindeer, or roe deer; fewer horse and mammoth; and more big mammal, marmot, medium mammal, and indeterminable.

**3 Skeletal elements.** Out of 108 Kara-Bom pieces, the most common skeletal elements are: unknown (25.9%), long bone (25.9%), rib (11.1%), toe (6.5%), mandible (4.6%), and vertebra (3.7%) (Table A1.3, site 10). Compared with the pooled assemblage averages, Kara-Bom has more ribs, long bones, unknowns, and no scapulae.

**4 Age.** Out of 108 Kara-Bom pieces, there is only a 1.8% sub-adult representation (Table A1.4, site 10). Compared with the pooled assemblage average, Kara-Bom has fewer sub-adults.

**5 Completeness.** Among Kara-Bom's 108-piece assemblage, there are 8.3% whole bones, 25.9% with one anatomical end, and 65.7% without any anatomical end (Table A1.5, site 10). These frequencies are similar to those of the pooled assemblage.

**6 Maximum size.** Postmortem breakage limited the Kara-Bom sample for size measurement to 24 pieces. Of these, the mean is 6.6 cm, the range is 3.0 cm to 24.8 cm (Table A1.6, site 10). Compared with the pooled assemblage values, both the mean and range in Kara-Bom are slightly lower. The limited species information precludes making any meaningful comparisons with Vera Gromova's (1950: table 27) whole long bone length measurements.

**7 Damage shape.** The most common types of damage in the 108 Kara-Bom pieces are: long bone flakes (37.0%), long bone fragments (17.5%), long bone splinters (9.3%), mostly whole (9.2%), undamaged (7.4%), medial ribs (5.5%), and phalanx butts (3.7%) (Table A1.7 site 10). Compared with the pooled assemblage averages, Kara-Bom has more long bone flakes, long bone splinters, and mostly whole; and fewer long fragments.

**8 Color.** Most of the Kara-Bom pieces are ivory colored (102 / 108; 94.4%), and only 5.6% are brown (Table A1.8, site 10). There are no black (burned) pieces. Compared with the pooled assemblage, Kara-Bom has more ivory colored pieces, and fewer of the other colors.

**9 Preservation.** As expected for an open site, Kara-Bom has many chalky pieces – 32.4% of the 108 (Table A1.9, site 10). This frequency is almost twice that of the pooled assemblage average. The most similar of the open sites are Malaya Seeya (15.3% chalky pieces) and Mal'ta (50.0% chalky).

**10 Perimortem breakage.** Kara-Bom has 88.3% perimortem breakage (91 / 103 pieces) (Table A1.10, site 10). This frequency is close to the pooled assemblage average.

**11 Postmortem breakage.** Rather like other open sites, Kara-Bom has a relatively frequent occurrence of postmortem breakage (28 / 106; 26.4%) (Table A1.11, site 10). As expected, this frequency is greater than the pooled assemblage average.

**12 End-hollowing.** Only 42 Kara-Bom pieces could be assessed for end-hollowing, which occurs in 4.8% of the set (Table A1.12, site 10). This frequency is about half that of the pooled assemblage average. Nevertheless, it helps document the presence of carnivores.

**13 Notching.** Almost identical to above, the frequency of Kara-Bom notching is 4.7% in 106 pieces (Table A1.13, site 10). Each has only one notch. Compared with the pooled assemblage average, Kara-Bom has about three times less notching.

**14 Tooth scratches.** This indicator of carnivore activity is present in 10.9% of 92 Kara Bom pieces (Table A1.14, site 10). The number of scratches per piece ranges from one to more than seven. Compared with the pooled assemblage average, Kara-Bom has about half the frequency of tooth scratching.

**15 Tooth dints.** Of 89 pieces that could be assessed for dinting, 7.9% have from one to more than seven dints per piece (Table A1.15, site 10). The pooled assemblage average is about three times greater.

**16 Pseudo-cuts.** No example of pseudo-cutting was identified in 90 Kara-Bom pieces (Table A1.16, site 10).

**17 Abrasions.** Out of 90 Kara-Bom pieces, one (1.1%) has six abrasion grooves (Table A1.17, site 10). The very low frequency is almost identical to the pooled assemblage average.

**18 Polishing.** There is 75.8% polishing in 99 Kara-Bom pieces. These are located on the end (7.1%), middle (2.0%) and end-middle (66.7%) (Table A1.18, site 10). The amount of polishing in the pooled assemblage average is not much different, although there is more end-polishing than in Kara-Bom.

**19 Embedded fragments.** Out of 105 Kara-Bom pieces, 2.9% have one embedded fragment each (Table A1.19, site 10). The frequency of embedding in the pooled assemblage average is very similar, although multiple embedded fragments up to more than seven do occur.

**20 Tooth wear.** Kara-Bom has only five teeth to use for evaluating tooth wear. Of these, two (40%) suggest young individuals (Table A1.20, site 10). This value is much like the pooled assemblage average for young individual occurrence.

**21 Acid erosion.** Out of 105 Kara-Bom pieces, 6.7% have acid erosion (Table A1.21, site 10). This is about the same as the pooled assemblage average (5.9%), and similar to the Razboinich'ya hyena cave acid erosion frequency (7.3%). As we have pointed out previously, the evidence for hyena presence should serve as a warning that stratigraphic disturbances might have taken place as a consequence of the digging and burrowing behavior of these large scavenging and hunting animals.

**22 Rodent gnawing.** One of the 104 Kara-Bom pieces shows rodent gnawing (1.0%) (Table A1.22, site 10). This rare occurrence is nearly the same as the pooled assemblage average.

**23–24 Insect damage and human bone.** There are no examples of these variables in 104 Kara-Bom pieces.

**25 Cut marks.** Out of 89 Kara-Bom pieces, 4.5% have cut marks. These range from one to more than seven cuts per piece (Table A1.25, site 10). The pooled assemblage average is slightly greater. Given the nearby availability of stone that would readily allow blade tool production, we considered whether cut marks at Kara-Bom might differ in some manner compared to those found at sites where flakes were the primary cutting tool. Using a ×20 hand lens, we could see no differences, at least none as apparent as cut marks produced with obsidian blades compared with other less glassy stone types.

**26 Chop marks.** Chop marks in 89 Kara-Bom pieces are slightly more frequent than cut marks, at 6.7%. The number of chop marks also range from one to more than seven per piece (Table A1.26, site 10). The pooled assemblage average is nearly the same.

## Discussion

Kara-Bom is considered to be a major Siberian Paleolithic archaeological site because of its apparently very early occurrence of stone blade production (Derevianko *et al.* 2001c, Okladnikov 1983, Otte and Kozlowski 2001). Assessing the contribution of stone type to blade production has apparently not been done. While the perimortem bone damage includes a number of good examples of human cutting and chopping, substantiating the stone artifact evidence for human occupancy, there is also good osteological evidence for cave hyena and other carnivore presence. This evidence includes identified hyena skeletal remains, as well as seven pieces with acid erosion that we feel is due primarily to the bone collagen extracting power of the hyena gastro-intestinal tract, not damage caused by soil chemistry. This fact, and its corollary of possible stratigraphic disturbance due to hyena digging and burrowing behavior, should be kept in mind when reconstructing the evolution of blade production in northeast Asia. As mentioned elsewhere, burrowing disturbances could mix discontinuous, non-evolutionary, stratigraphic deposits into what would appear to be continuous or local evolution deposits.

It is particularly interesting that the interpretation of the deposits and stratigraphy by Derevianko and Rybin (2005) makes no mention of bioturbation. They say only that the site might have been impacted by hydrological disturbance, erosion from flow of glacial melt water, and gravitational slope erosion (Derevianko and Rybin 2005:236–237).

We can only wonder what sort of relationship existed between the Kara-Bom humans and the cave hyenas. Were the hyenas simply scavenging Kara-Bom refuse when humans were absent? Or might they have driven the few human occupants away to get at fresh meat they carried to the workshop after one or more successful hunts? It is unsettling, if not in fact terrifying, to think about how those people might have reacted to a pack of noisy cave hyenas circling the Kara-Bom camp in the dark following sunset – circling with the intent of rushing in to grab scraps of food or whatever else their powerful jaws might clamp onto. A Kara-Bom human infant comes too readily to mind in this speculative scenario. Speculative or not, as will be discussed, modern hyenas do attack humans and carry off younger individuals.

# 11 Kirkalinskaya Cave

## Background

Located in the Altai Mountains, this cave is named after the nearby Kyrkala River. It is mentioned in a short paper by A. M. Marinina (1966) that dealt with karst geology of the Kaysla and Sarala river basins. In Ovodov's small test trench in the middle of the cave a few large, well-preserved pieces of charcoal were found in the loose sediments at a depth of approximately 50 cm below the cave floor surface. If the charcoal was the product of human activity in the cave, it is the only evidence that Ovodov recovered of such activity, making Kirkalinskaya Cave a strong candidate for being a paleontological site. The species that he could identify, and the number of skeletal elements for each include: hare (1), marmot (1), badger (7), bear (50), cave hyena (11), lynx (3), wolverine (1), horse (3), Asiatic wild ass (1), rhinoceros (12), red deer (1), moose (2), bison (18), wild yak (1), goat-ibex (6), and wild sheep (4).

## Findings

**1 Provenience.** There are only four pieces in the Kirkalinskaya collection, making it 0.04% of our 8813-piece grand total (Table A1.1, site 11). We make no comparisons for this tiny assemblage, which was found on the surface within the cave near its entrance. There obviously had been a larger field collection, as seen in the above listing; however, this additional material could not be located in the IAE zooarchaeological storage area.

**2 Species.** All four pieces are goat-sheep (Table A1.2, site 11).

**3 Skeletal elements.** There is one piece each of the following: antler-horn, humerus, radius, femur (Table A1.3, site 11).

**4 Age.** One of the four pieces is of sub-adult age (Table A1.4, site 11).

**5 Completeness.** Two of the Kirkalinskaya pieces have one anatomical end, two have no anatomical ends (Table A1.5, site 11).

**6 Maximum size.** The four pieces have a mean size of 12.6 cm, and a range of 5.2 cm to 19.6 cm (Table A1.6, site 11).

**7 Damage shape.** One of the four pieces is a long bone flake, one is a phalanx segment, one is a phalanx butt, and one is mostly whole (Table A1.7, site 11).

**8 Color.** Each of the four Kirkalinskaya pieces is ivory colored (Table A1.8, site 11).

**9 Preservation.** All four pieces are ivory in hardness (Table A1.9, site 11).

**10 Perimortem breakage.** All four pieces have perimortem breakage (Table A1.10, site 11).

**11 Postmortem breakage.** One of the four pieces has postmortem breakage (Table A1.11, site 11).

**12 End-hollowing.** Two of the four Kirkalinskaya pieces have end-hollowing (Table A1.12, site 11).

**13 Notching.** One of the four pieces has notching. Six notches are present on the piece (Table A1.13, site 11).

**14 Tooth scratches.** Three of the four pieces have tooth scratches. One piece has one scratch, one has five scratches, and the other has more than seven scratches (Table A1.14, site 11).

**15 Tooth dints.** Only one piece has no dinting. One piece has four dints, one has five, and one has more than seven (Table A1.15, site 11).

**16 Pseudo-cuts.** One of the four Kirkalinskaya pieces has one pseudo-cut (Table A1.16, site 11).

**17 Abrasions.** None of the four pieces has abrasions (Table A1.17, site 11).

**18 Polishing.** Two of the four pieces have end-polishing, and two have end-middle polishing (Table A1.18, site 11).

**19 Embedded fragments.** None of the four pieces has embedded fragments (Table A1.19, site 11).

**20 Tooth wear.** There are no Kirkalinskaya teeth.

**21–23, 25–26 Acid erosion, rodent gnawing, insect damage, cut marks, and chop marks.** None of these variables are present in the four Kirkalinskaya pieces (Tables A1.21, site 11 to A1.26, site 11).

**24 Human bones.** Human remains were recovered from this cave site, five pieces. The cause of damage is uncertain (Fig. 3.47).

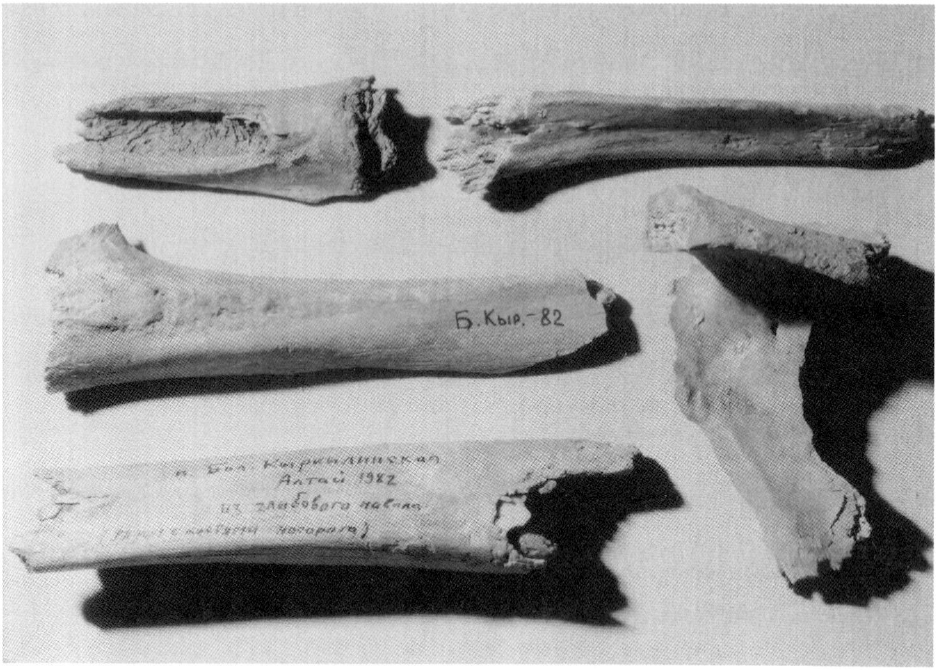


**Fig. 3.47** Kirklanskaya Cave human remains. Although the perimortem damage to these five pieces of human long bone and scapula fragment is attributed to animal-caused activity, the carnivore damage signature is minimal at best, and the human damage signature is absent. We have no explanation for the damage to these bones that likely are all from a single adult individual of indeterminable sex. Length of tibia fragment in lower left is 14.6 cm. Script on the tibia says: "From the boulder's pile with rhinoceros bones, Altai, 1982." The association with rhinoceros bones suggests a Pleistocene age for these remains (CGT neg. IAE 8-11-99:28A).

## Discussion

Despite having only a tiny sample size, the Kirkalinskaya pieces do show good signs of carnivore damage, although we are unable to identify which carnivore was responsible. There is no perimortem evidence for human activity, although a few pieces of human bone were found.

# 12 Krasny Yar

## Background

Krasny Yar (Red or Beautiful Gully) is a deeply buried river bank Pleistocene palentological deposit that accumulated before the overlying near-vertical 30 m bluff – made up of sandy loams, loess-like loams, sand, and silty sedimentary deposits – was laid down (Fig. 3.48). The locality is at 55°02′ N, 82°55′ E on the right bank of the Ob River, 17 km down-river from the main Novosibirsk railroad station. It is the type locality for the geological section of the Ob River Valley in the Novosibirsk region (Saks *et al.* 1978).

Sergei K. Vasiliev has repeatedly collected at the locality every year, starting in 1978 when he was in the sixth grade. He has recovered more than 2800 whole bones and fragments belonging to 20 mammalian species, all of which will be the basis for his Doctor of Science dissertation. Even with his frequent visits and studies at Krasny Yar, where he collects newly exposed bones before they can be washed down-river, it is apparent that the substantial abrasion and polishing of the largely whole bones means that they had washed out from somewhere else up-river, and deposited during the later Pleistocene at the present co-mingled bone bed. Hence, none of the Krasny Yar fossil remains represent *in situ* deaths, and water transport could have been a major contributor to the perimortem and postmortem taphonomy. Not being an archaeological site, there has been no excavation, only shoreside surface collecting (Fig. 3.49).

Vasiliev divides the collection into three groups based mainly on preservation. The best preserved, and presumably the youngest, has been carbon-14 dated between 28 000 and 33 000 BP. The second group is dated to Kazantsev times, as suggested by preservation and the morphological features of some species. The most poorly preserved pieces are considered to be the oldest, with a tentative date of ca. 80 000 BP.

Vasiliev (1995, 2002) has to date identified the following species and specimen numbers in levels designated 6 and 4: hare (*Lepus timidus*) L6 (1), L4 (2); marmot (*Marmota baibacina*) L6 (1), L4 (1); beaver (*Castor fiber*) L6 (0), L4 (11); gray wolf (*Canis lupus*) L6 (1), L4 (9); brown bear (*Ursus arctos*) L6 (1), L4 (8); wolverine (*Gulo gulo*) L6 (0), L4 (1); badger (*Meles meles*) L6 (0), L4 (1); cave hyena (*Crocuta spelaea*) L6 (10), L4 (1); cave lion (*Panthera spelaea*) L6 (6), L4 (10); mammut (*Mammonteus primigenius*) L6 (6), L4 (10); horse (*Equus ex, gr, gallicus*) L6 (401), L4 (443); rhinoceros (*Coelodonta antiquitatis*) L6 (65), L4 (151); big-horn deer (*Megaloceros giganteus*) L6 (0), L4 (160); red deer (*Cervus elaphus*) L6 (24), L4 (113); elk

Fig. 3.48 Krasny Yar locality. Situated near the water's edge of the Ob River, 15 km north (right bank, down-river) of Novosibirsk, this paleontological site is a classic example of taphonomic burial processes that interested I. A. Yefremov (Fig. 1.24). Krasny Yar has been watched over and collected by paleontologist Sergei Konstatinovich Vasilev for several years. He has amassed hundreds of animal bones that had washed into the Pleistocene river shore site. He recently discovered a nearly complete mammoth skull. In this image, fossil mammoth specialist Irina V. Foronova looks for additional Krasny Yar fossils along the narrow beach at the base of the high loess bluffs. While Krasny Yar is obviously a paleontological site, its bone damage was overwhelmingly done by mechanical means – water tumbling, ice crushing, trampling, surface weathering, sand and water polishing mixtures, and other mechanisms prior to deposition. Its perimortem taphonomic damage is markedly different from the paleontological bone accumulations such as Razboinich'ya hyena cave. The ultimate source of the Krasny Yar bones would likely have been natural deaths near to and up-river, deaths like the one illustrated in Figs. 2.5– 2.8 that might have washed into the Ob River from lateral streams and rivers such as the slow-moving muddy right bank Inya River between Academgorodok and Novosibirsk (CGT neg. Krasny Yar 7-7-00:3).

(*Alces* sp.) L6 (13), L4 (111); caribou (*Rangifer tarandus*) L6 (12), L4 (0); bison (*Bison priscus*) L6 (103), L4 (1080); saiga (*Saiga tatarica*) L6 (4), L4 (3). The collection is housed at IAE.

## Findings

**1 Provenience.** Described above. The 203 Krasny Yar pieces represent 2.3% of our 8813-piece grand total (Table A1.1, site 12).

**Fig. 3.49** Participants of the Krasny Yar visit. Clockwise from the top: Nicolai Ovodov, Olga Pavlova (dark hat), Irina Foronova, student, Vlamir Zykin, student, L. Orlova (CGT color Krasny Yar 7-7-00:25).

**2 Species.** The most common species identified in the 203 Krasny Yar pieces are: elk (26.6%), horse (18.2%), small deer (7.9%), lion (7.9%), bear (7.4%), wolf (6.4%), bison (5.9%), beaver (5.4%), saiga antelope (4.4%), big deer (3.9%), reindeer (3.0%), and rhinoceros (3.0%) (Table A1.2, site 12). Compared with the pooled assemblage averages, Krasny Yar has more bear, beaver, bison, big deer, small deer, elk, horse, lion, rhinoceros, and wolf; fewer mammoth and reindeer; no big mammal, *Capra*, gazelle, goat-sheep, hyena, roe deer, and indeterminable. The high degree of specific identification in Krasny Yar has two explanations: First, completeness at Krasny Yar is especially high, even for a paleontological site. Second, Sergei Vasiliev, in consultation with Nicolai Ovodov, has spent several years studying this assemblage for his dissertation, so a lot of time has been spent on identification and classification.

**3 Skeletal elements.** The most common elements in the 203-piece Krasny Yar assemblage are: mandible (30.0%), metapodial (15.3%), humerus (12.8%), tibia (7.9%), radius (5.9%), vault (4.4%), and femur (4.4%) (Table A1.3, site 12). These frequencies do not take into account a large number of foot bones we did not score because they had no identifiable perimortem damage other than polishing. Compared with the pooled assemblage averages, Krasny Yar has more pieces of vault, mandible, humeri, tibia, and metapodial; fewer vertebrae and ribs; and no long bone and unknown elements.

**4 Age.** Out of 203 Krasny Yar pieces, 6.4% belong to sub-adults (Table A1.4, site 12). Compared with the pooled assemblage average, Krasny Yar is very similar; however, it has about 30% more positively identified adults than the pooled assemblage.

**5 Completeness.** Bone preservation in a buried aquatic context is often very good. Krasny Yar is no exception. Out of the 203-piece assemblage, 23.2% are whole, 58.6% have one anatomical end, and only 18.2% have no anatomical ends (Table A1.5, site 12). Compared with the pooled assemblage averages, Krasny Yar is clearly more complete: 10.9% whole, 30.0% one end, 59.1% no ends.

**6 Maximum size.** Piece size correlates with completeness, as the Krasny Yar mean (19.7 cm) and range (3.1 cm to 43.5 cm) readily demonstrate (Table A1.6, site 12). Compared with the pooled assemblage size average (9.0 cm), the Krasny Yar mean is twice as big. Nevertheless, there is considerable piece size reduction. This can be appreciated by comparing the Krasny Yar maximum size values with the ranges for undamaged long bone length provided by Vera Gromova (1950: table 27). In the cases of the most common species found at Krasny Yar (elk, horse, lion, bear, bison), the mean piece size is less than the lower end of the ranges of all five species' major long bones.

**7 Damage shape.** This variable was not scored in the Krasny Yar assemblage due to its entirely different aquatic taphonomy, where most of the bone alteration was likely related to water transport.

**8 Color.** The majority (53.5%) of the 202 Krasny Yar pieces are brown in color, presumably due to water-borne stain. Only 36.1% are ivory colored. The black colored pieces are not due to burning – the cause is a very dark surficial stain of an unknown substance (Table A1.8, site 12). Compared with the pooled assemblage averages, Krasny Yar has half the frequency of ivory colored pieces, and five times more brown colored pieces. Was it a mineral-rich groundwater setting that is the common factor to the assemblages that have a frequent occurrence (>20%) of brown colored bone?

**9 Preservation.** Unlike other open sites in our study, Krasny Yar has a high frequency of ivory hard pieces (194 / 203; 95.6%) (Table A1.9, site 12). Compared with the pooled assemblage average, Krasny Yar is significantly higher in quality.

**10 Perimortem breakage.** Out of 196 Krasny Yar pieces, 60.3% have perimortem breakage (Table A1.10, site 12). Compared with the pooled assemblage average, Krasny Yar has much less perimortem breakage. Presumably this is due to the near absence of identifiable carnivore and human processing.

**11 Postmortem breakage.** Like our other open sites, the 200 Krasny Yar pieces exhibit considerable postmortem breakage (38.5%) (Table A1.11, site 12). Weathering on the ground surface before burial, and subsequently after re-exposure, are the main reasons. Compared with the pooled assemblage average, Krasny Yar has twice as much post-mortem breakage.

**12 End-hollowing.** This variable is surprisingly frequent in the 185 Krasny Yar pieces (30.8%) (Table A1.12, site 12). Compared with the pooled assemblage average, Krasny

Yar has significantly more end-hollowing. At first we thought all this end-hollowing must have been caused by water transport, but our scores for other presumed carnivore indicators are equally frequent. The only conclusion that makes taphonomic sense is that, indeed, carnivores had chewed on some of the carcasses at times and places at Krasny Yar and possibly also up-river when the dead animals still possessed some degree of edible tissue or bone marrow. Given the open bone bed context of Krasny Yar, we should also allow that some of the "carnivore" damage could have been done by bone-chewing, mineral-seeking herbivores.

**13 Notching.** There is considerable notching in the 200 Krasny Yar pieces (18 / 200; 9.0%). The number of notches per notched piece ranges from one to six (Table A1.13, site 12). Compared with the pooled assemblage average (14.9%), Krasny Yar has somewhat less notching.

**14 Tooth scratches.** Tooth scratches are correlated with notching, and the 31.3% scratched pieces in 195 Krasny Yar pieces is no exception. The number of scratches per piece ranges from one to more than seven, with the latter being the largest of the scratched groups (11.8%) (Table A1.14, site 12). Compared with the pooled assemblage average, Krasny Yar has considerably more pieces with tooth scratching.

**15 Tooth dints.** Dints are also correlated with notches as well as tooth scratches. The 24.7% dinting in 194 Krasny Yar pieces meets this expectation. The number of dints ranges from one to more than seven, with the latter being the largest class (8.8%) (Table A1.15, site 12). Compared with the pooled assemblage average (25.6%), Krasny Yar is nearly the same.

**16 Pseudo-cuts.** This condition was judged to have occurred in 5.5% of 200 Krasny Yar pieces. The number of pseudo-cuts per piece ranges from one to four, with one pseudo-cut per piece being the largest class (3.5%) (Table A1.16, site 12). Compared with the pooled assemblage average, Krasny Yar has very nearly the same frequency of pieces with pseudo-cuts.

**17 Abrasions.** Given the riverine depositional setting, we expected to find a high frequency of abrasions, or at least randomly distributed striations. Such was not the case. In the 201 Krasny Yar pieces, abrasions occurred in only 1.5% of the assemblage, all being more than seven in number (Table A1.17, site 12). One of these pieces has more than 100 abrasion striations. Compared with the pooled assemblage average, Krasny Yar is almost exactly the same.

**18 Polishing.** If abrasions did not meet our expectations, then polishing did, because 93.0% of the 200 Krasny Yar pieces are polished. End-polishing occurs in 5.5%, middle in 3.0%, and middle-end in 84.5% of the assemblage (Table A1.18, site 12). Compared with the pooled assemblage for total polishing, Krasny Yar has considerably more. Because there are other sites in this study with large amounts of polishing – for example Varvarina Gora (96.0%) – which we attribute to carnivore and/or human processing, the question that soon arose in our examinations was: What caused the Krasny Yar polishing? Was it mainly water/sand abrasion? Was it mainly carnivore? Was it both? Unfortunately, we never found a way to answer these questions.

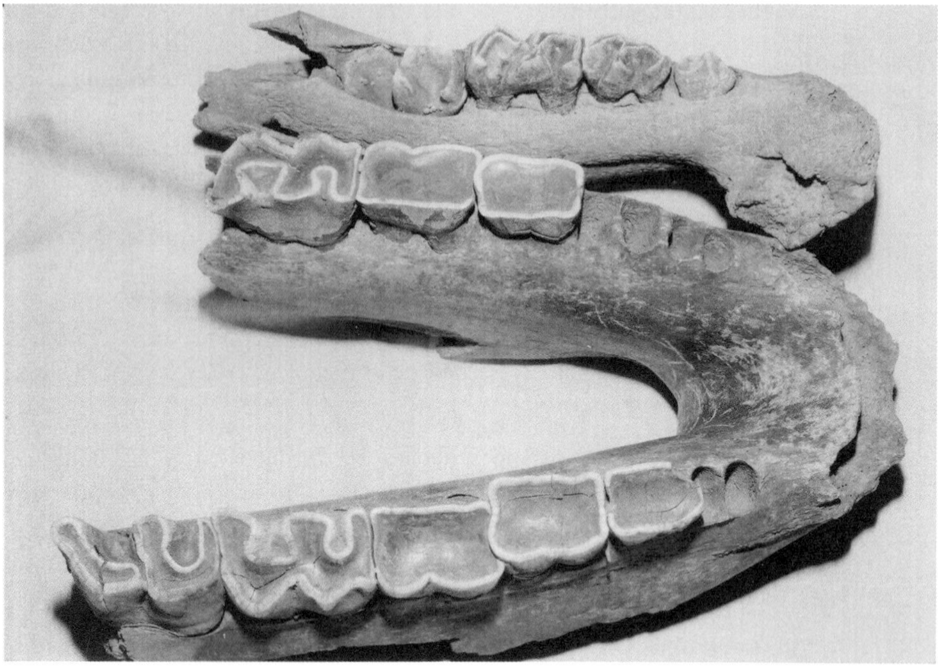

**Fig. 3.50** Krasny Yar animals. Mandibles of an old wooly rhinoceros (Coelodont) and a sub-adult (top). The adult jaw is 27.0 cm in length. Breakage appears to be postmortem (CGT neg. IAE 6-21-00:29).

**19 Embedded fragments.** Out of 203 Krasny Yar pieces, 1.5% have 1–3 embedded fragments (Table A1.19, site 12). Compared with the pooled assemblage average, Krasny Yar has fewer pieces with embedded fragments.

**20 Tooth wear (Figs. 3.50–3.57).** Krasny Yar has the largest tooth wear sample size in our study. Of the 54 specimens, we judged 20.4% to be young individuals at the time of their deaths (Table A1.20, site 12). Compared with the pooled assemblage average for young individuals, Krasny Yar has about half the number. Since we feel that this study has identified so many factors that influence the occurrence or non-occurrence of young individuals in our various contexts, we have no way to tell whether one interpretation of this difference is better than any other. We can be certain that transport and time on the surface (Fig. 3.58) contributed to survivorship as illustrated and discussed in Chapter 2 (small calf vs. large bull).

**21 Acid erosion.** Of the Krasny Yar pieces, 203 were evaluated for acid erosion. Not a single specimen was identified (Table A1.21, site 12). At issue here is differential survival between large and small pieces of bone and teeth in a high-energy riverine context, as well as differential beach collecting without the aid of screens. We feel that both factors have influenced the size survivorship and related taphonomic characteristics of the Krasny Yar assemblage, probably more so than even in other sites where screening had been minimal. In any event, it was our expectation that one or two acid-eroded pieces would show up in the collection because of the carnivore

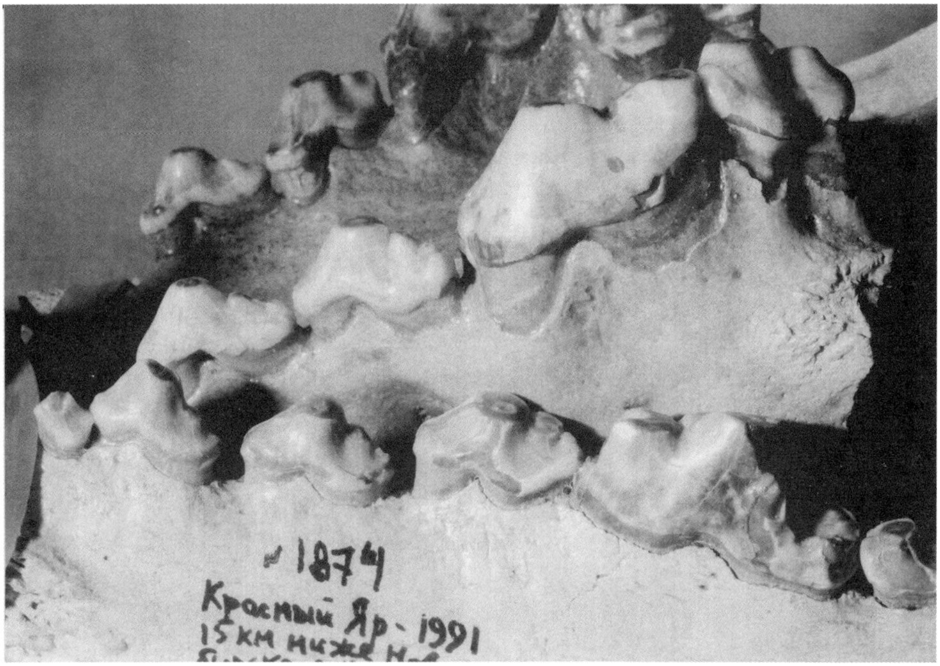


**Fig. 3.51** Krasny Yar animals. Wolf (*Canis lupus*) jaws illustrating cusp wear pattern with tips worn flat. These flat surfaces have sharp chisel-like borders that under some conditions could act like a burin. It is believed that carnivore teeth with this sharp-bordered cusp tip wear were responsible for some of the pseudo-cuts (CGT neg. IAE 7-8-02:7A).

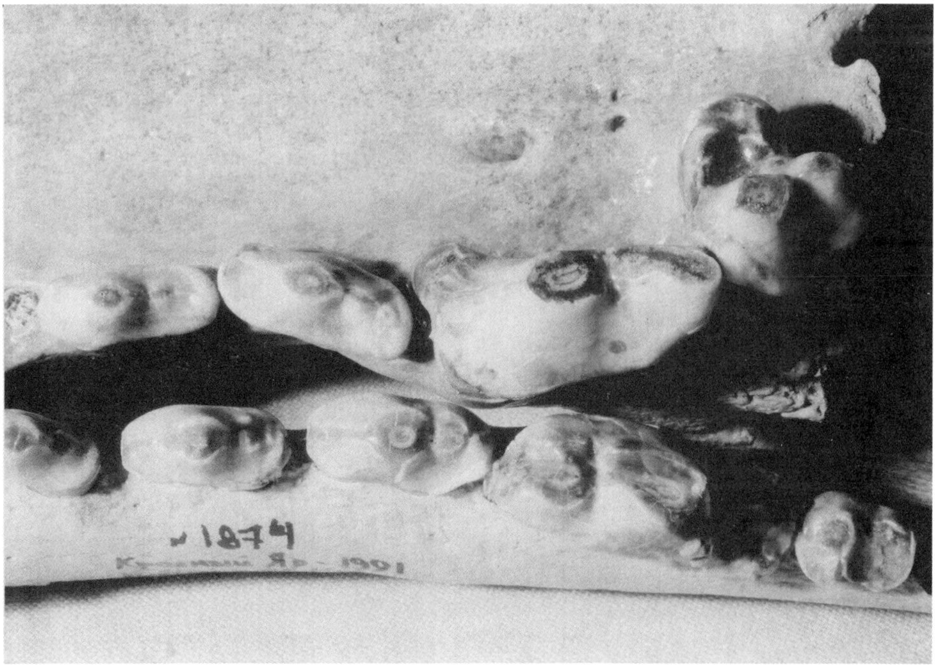


**Fig. 3.52** Krasny Yar animals. Another view of wolf tooth wear to show the flat wear on the cusp tips and the associated sharp borders (CGT neg. IAE 7-8-02:5A).

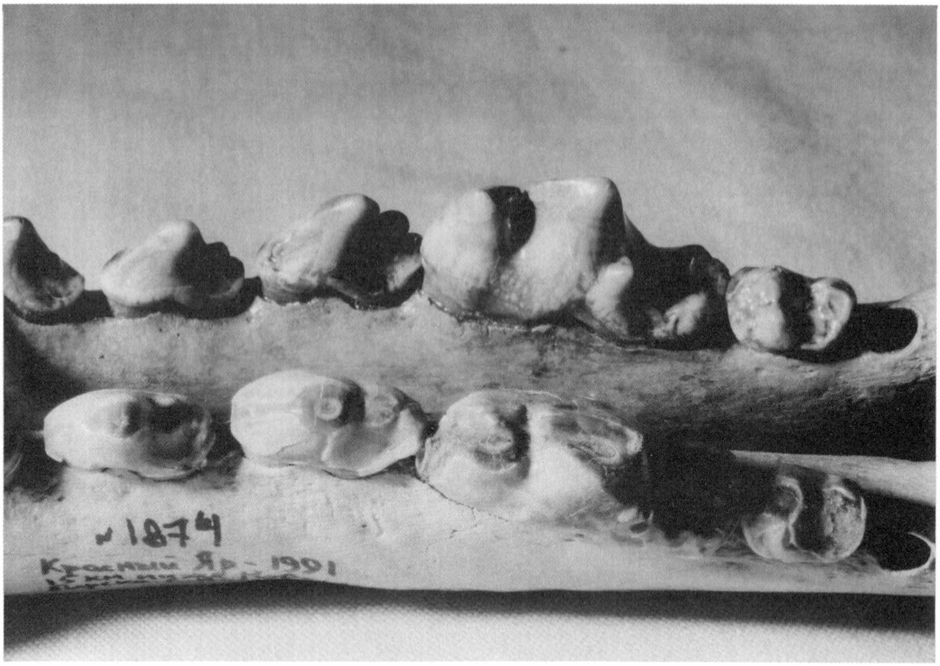


**Fig. 3.53** Krasny Yar animals. A final example of wolf tooth cusp tip wear in an older animal (bottom) in contrast to absence of wear in a sub-adult (top). The unworn teeth do not have the chisel-like cusp tips (CGT neg. IAE 7-8-02L:11A).

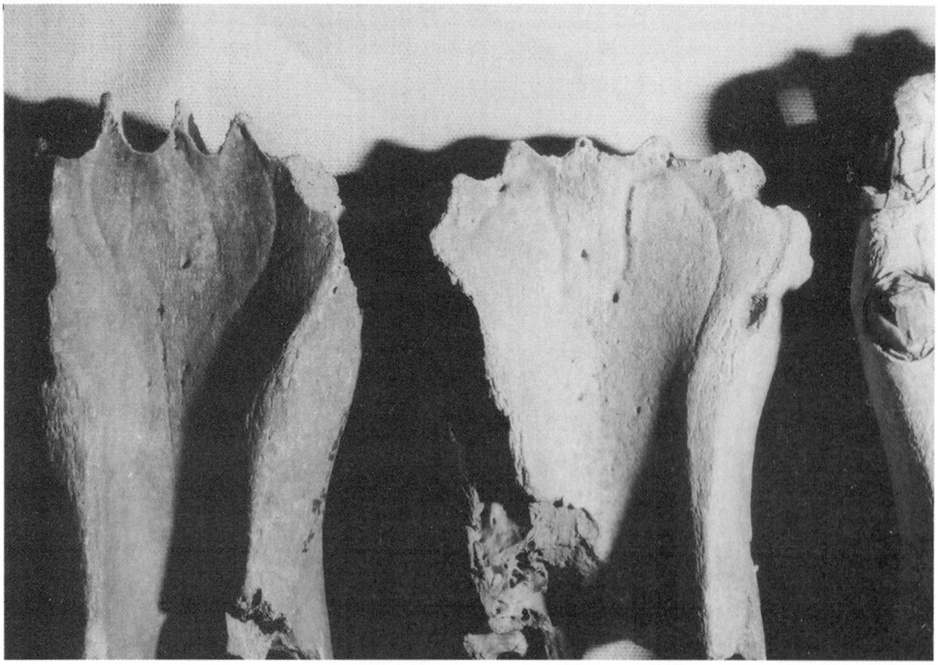

**Fig. 3.54** Krasny Yar animals. Horse mandibles showing variation in the occurrence of a canine tooth. Breakage caused by tumbling, ice, or other mechanical means. There are no signs of carnivore damage. Almost all of the Krasny Yar bone damage was mechanical (CGT neg. IAE 6-20-00:3).

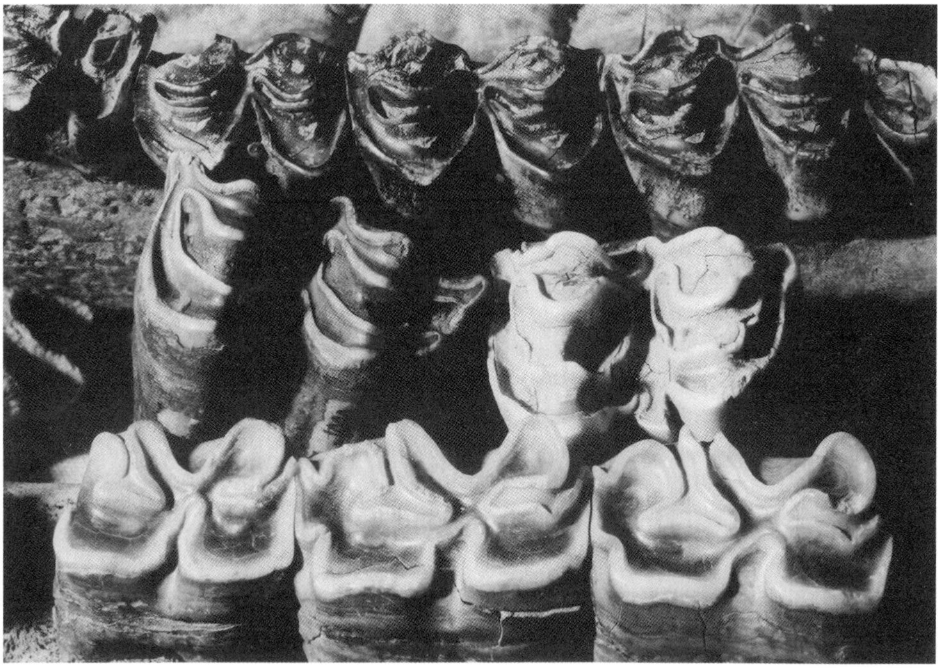

Fig. 3.55 Krasny Yar animals. Teeth of *Equus* (bottom), *Bison* (middle), and *Cervus* (top) are abundant at Krasnoi Yar. Tooth breakage is minimal, suggesting death site and burial site may not have been far apart. Although depositional rates may never be calculated for this site, the abundance of herding animals of large size suggests game was available, if not abundant, for human hunters and carnivores alike. Nevertheless, competition to make a kill, and then to retain the carcass, may have been keen (CGT neg. IAE 6-20-00:2).

damage and the presence of actual hyena in the species inventory. Such was not the case.

**22 Rodent gnawing.** Like Boisman II, rodent gnawing is relatively common in the 204-piece Krasny Yar assemblage, amounting to 4.9%. This is much more frequent than in the pooled assemblage (0.5%) (Table A1.22, site 12). We have no trouble imagining small rodents of the river, lake, or slough shore habitat scampering day or night throughout much of the year across, and occasionally gnawing on, the large bony carcasses partly buried here and there in the muddy banks of their resting places. Among all of our site contexts, that of Krasny Yar is intuitively best for rodent presence.

**23 Insect damage.** Like rodent damage, the contextual conditions we imagine existed at Krasny Yar, or shortly before the assemblage accumulated at Krasny Yar, could have also been beneficial for insect exploitation of the bony remains. This expectation seems realized since a very low but relatively high frequency of insect damage was identified for 203 pieces of the Krasny Yar assemblage (0.5%) (Table A1.23, site 12). The pooled assemblage average is similar.

**24 Human bone.** No human bone has been found at Krasny Yar.

**Fig. 3.56** Krasny Yar animals. *Equus* mandibular teeth showing no perceptible weather-cracking or other postmortem damage. This horse was buried shortly after the animal died. Presumably, all or most of the carcass floated to the Krasnoi Yar locality, where it was soon covered by silt and sand (CGT neg. IAE 6-20-00:9).

**25 Cut marks.** Two pieces (1.0%) out of 202 seem to have cut marks, the identification strengthened by the fact that both have more than seven incisions (Table A1.25, site 12). This frequency of cut pieces is much less than the pooled assemblage average, but given that Krasny Yar is a full-blown paleontological site, the occurrence of cutting is of no small interest relative to the Pleistocene prehistory of this open, flat region of Siberia. Both pieces are well preserved so they perhaps belong to Vasiliev's younger group. One piece is a 6.7 cm long caribou metapodial fragment that has 15 cut marks and 11 abrasion grooves. It also has end-hollowing, four tooth scratches and ten tooth dints, making the cut marks suspect. The other piece is a 12 cm long fragment of a horse mandible with 16 cut marks. It lacks end-hollowing and tooth dints, but has one tooth scratch, making it also suspect but much less so than the caribou metapodial. Fossil ivory is much sought after for carving by Alaskan natives. Perhaps the two Krasny Yar cut pieces represent similar actions. There might be a few more cut pieces if the damage we judged to be pseudo-cuts actually were cut marks (Figs. 3.59–3.61).

**26 Chop marks.** There are no chop marks on any of the 202-piece Krasny Yar assemblage.

Figs. 3.62–3.66 illustrate damage by uncertain causes.

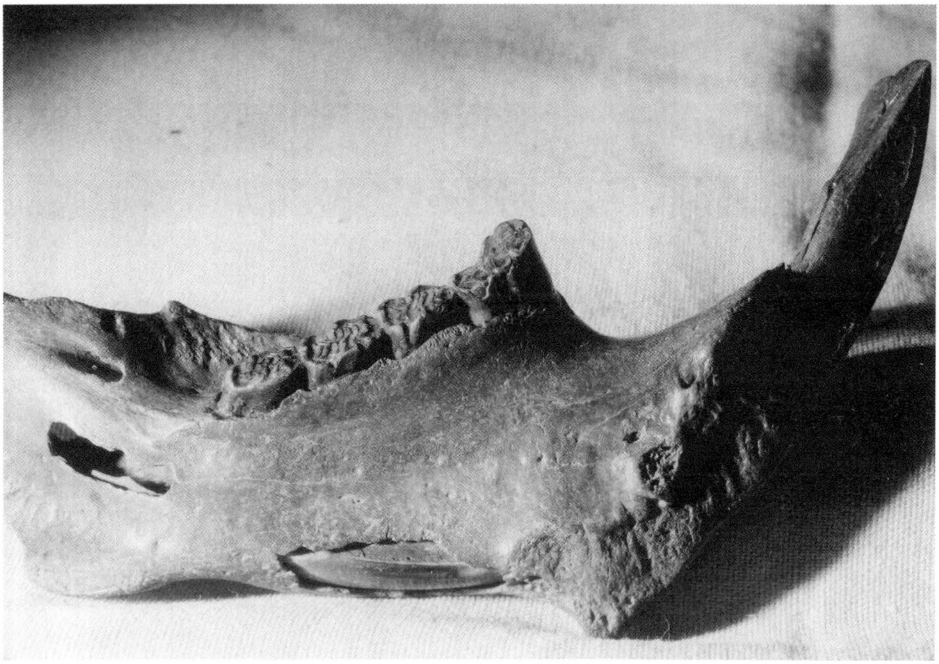

**Fig. 3.57** Krasny Yar animals. A beaver (*Caster fiber*) mandible. A stream with nearby trees, not the mighty Ob River, would be the expected habitat of this creature. It shows very little postmortem damage. What little there is exposed the unerupted portion of the incisor that runs nearly the entire length of the jaw. The jaw is 11.9 cm in length. Perhaps it floated to Krasny Yar from as far south as the Inya River or from one of the many sloughs and streams in between (CGT neg. IAE 6-20-00:18).

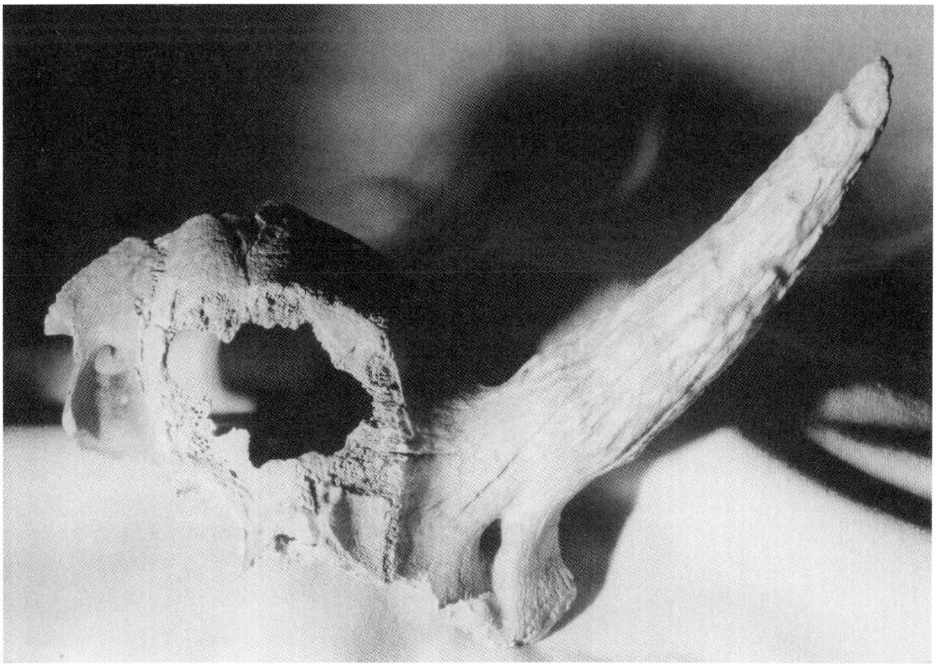

**Fig. 3.58** Krasny Yar animals. The skull base and horn core of a saiga antelope shows considerable breakage and chewing on the horn core. It clearly had been exposed to scavengers before burial at Krasny Yar. Horn is 10.0 cm long (CGT neg. IAE 6-23-00:9).

**Fig. 3.59** Krasny Yar animals. Cave lion (*Panthera spelaea*) ulna with 0.3 cm long pseudo-cut at pencil point (CGT neg. IAE 6-23-00:10).

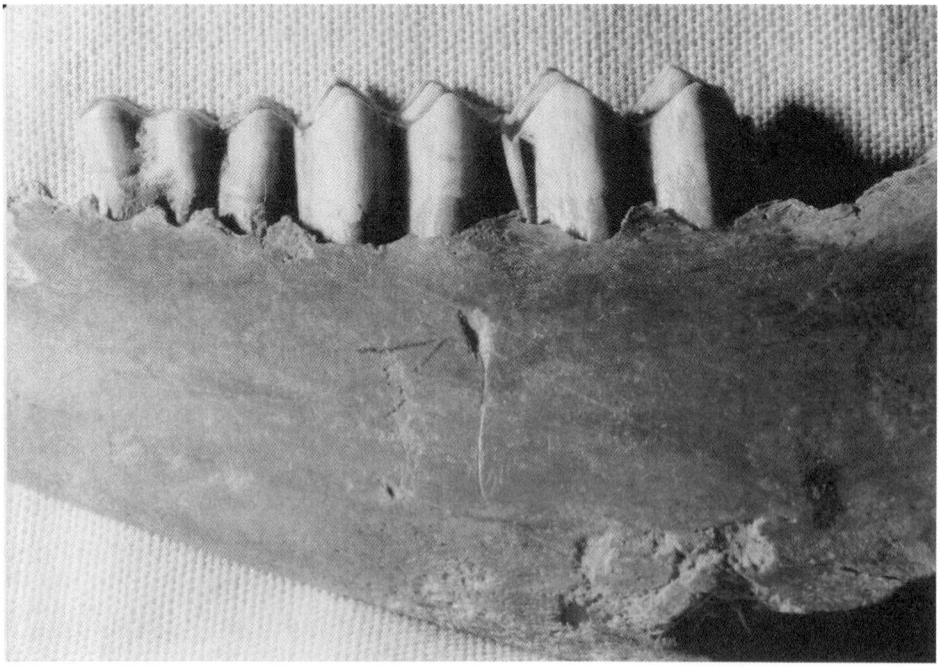

**Fig. 3.60** Krasny Yar animals. Saiga antelope mandible with 0.3 cm long pseudo-cut near the alveolar border and chewing damage to the inferior border of the ramus (CGT neg. IAE 6-23-00:6).

**Fig. 3.61** Krasny Yar animals. Pencil points to a 0.9 cm long pseudo-cut on a bear femur. Nearby tooth scratches are partly obscured by the lettering (CGT neg. IAE 6-21-00:23).

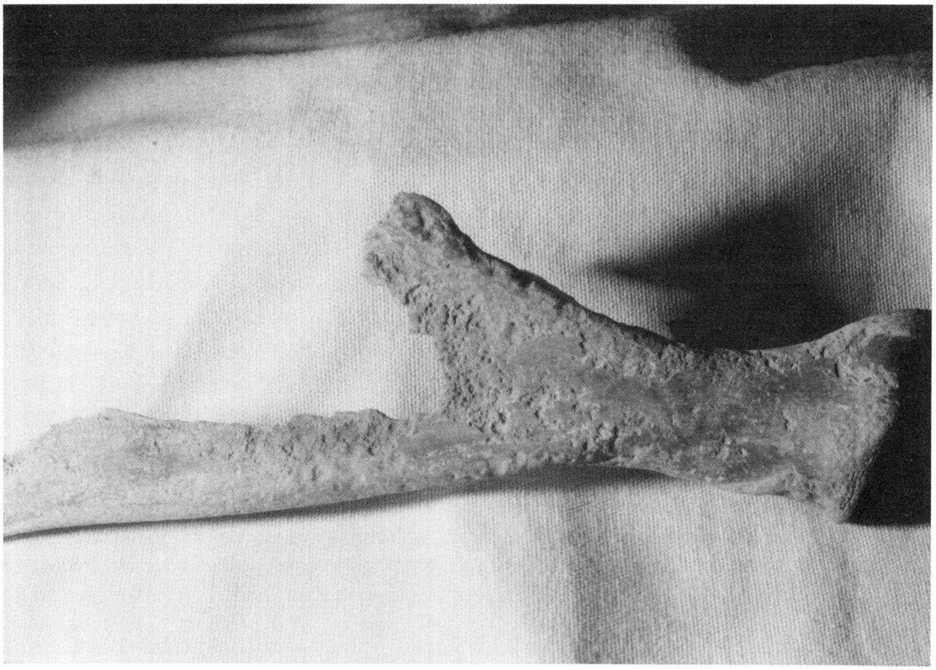

**Fig. 3.62** Krasny Yar animals. A saiga antelope left scapula shows pitting and erosion damage to its internal surface. The cause is unknown. Some root damage is possible. Maximum length is 16.0 cm (CGT neg. IAE 6-23-00:5).

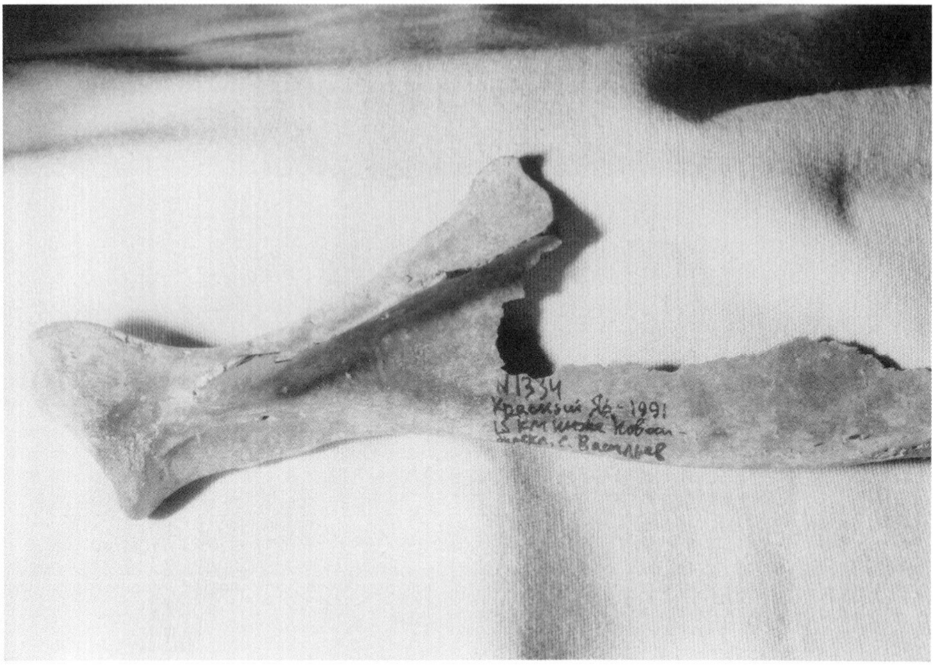


**Fig. 3.63** Krasny Yar animals. The external surface of the scapula shown in Fig. 3.62 shows no pitting or erosion. This peculiar difference was seen in many pieces of bone that had come from open sites (CGT neg. IAE 6-23-00:3).

## Discussion

This river-deposited bone accumulation provides a paleontological comparison for the archaeological sites, just as do the hyena caves. Deposition, post-depositional re-exposure, and collecting procedure have each probably affected the condition of the assemblage. Selection favoring larger-sized pieces over smaller ones must have occurred for each of the three processes, with moving water (river current and waves lapping against the bluff most of the year, and ice scouring during spring break-up) carrying off small bones and pieces of bones. By the time Vasiliev made his many collecting visits to Krasny Yar, many of the smaller items had been removed or reburied by the swiftly flowing Ob River, or simply missed because of their near invisibility in the gray-brown beach silts.

In his discussion of wolf-pack kills on frozen lakes, Haynes (1982:279) suggests that:

> On interior lakes and sloughs that are not disturbed by high energy drainage currents, bones left on the ice by predators tend to float on ice rafts during spring thaw, rather than simply dropping to the bottom where the ice is not thick or the water too shallow to float it, some remains may settle gently into the bottom during the thaw, forming discontinuous beds atop earlier skeletal deposits.

The extensive faunal deposits at Krasny Yar, including large animals like mammoth, may have ice-rafted to the site, as well as having accumulated by other natural processes. Our damage observations do not shed much light on which of several depositional processes

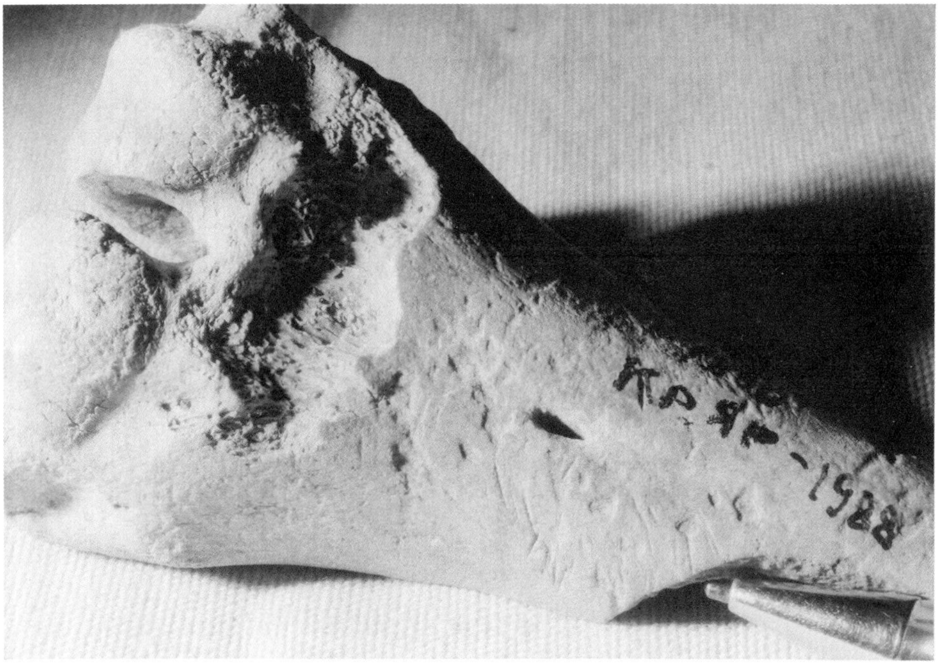


**Fig. 3.64** Krasny Yar animals. The striations at the pencil point on this 7.0 cm long reindeer metapodial were judged to be cut marks, although there is a chance they are pseudo-cuts. Uncertainty in the identification of the few cases such as this arose fewer than one in 100 times (CGT neg. IAE 6-21-00:25).

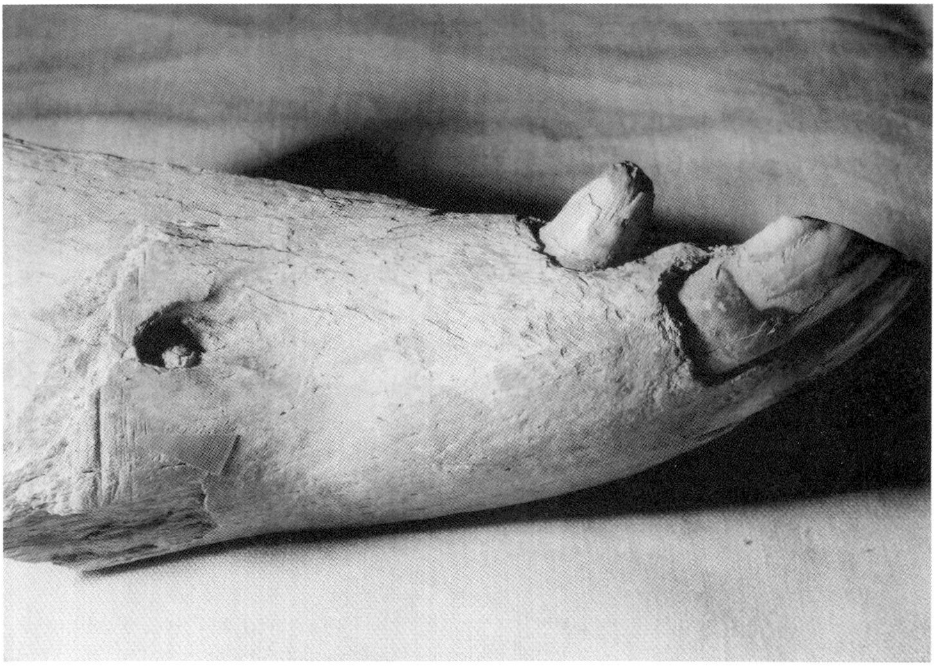

**Fig. 3.65** Krasny Yar animals. There is no doubt about these striations being cut marks. The longest cut on this horse mandible (specimen 3328) is 1.6 cm (CGT neg. IAE 6-20-00:5).

**Fig. 3.66** Krasny Yar animals. A rhinoceros distal humerus with a 2.0 cm long cut mark. Total bone length is 29.5 cm (CGT neg. IAE 7-4-00:33).

were mainly involved in the accumulation, although ice-rafting into a slough or still embayment is an attractive idea.

While the content of Krasny Yar is patently of importance for paleoecological studies, it is the perimortem damage that is our main interest in the assemblage. Of this damage, most unexpected was the substantial amount attributable to carnivores. We cannot identify exactly when in the death history of these animals the carnivores were active; however, two scenarios seem equally possible: First, the damage represents processing at the time of death, that is, initial and early perimortem. In this instance all large carnivores are equally good candidates for the damage – lions, hyenas, wolves, etc. The various carcasses in various degrees of completeness were then transported an unknown distance by water over an unknown period of time to Krasny Yar. Also unknown are the geomorphological conditions that favored the accumulation of the bone bed. At the time bone accumulation occurred at Krasny Yar the site might well have been a slough or a sand bar, like many along the Ob River today. After thousands of years the bones are re-exposed as the Ob River washes away the sediments that had buried them at Krasny Yar.

The second model envisions the carnivore damage having occurred in later perimortem death history. This model offers no explanation for the animal's deaths, but sometime after dying their carcasses are chewed on by scavenging carnivores, which in accord with the damage signature would likely implicate hyenas to some degree. Transport again occurs where bones are deposited at Krasny Yar to be buried and later re-exposed.

We have tried to estimate from the amount of polishing and abrasion how far the assemblage traveled before reaching its burial site at Krasny Yar. Judging from the amount of bone polishing in various of our open sites, where movement must have at best been only a few meters and trampling of little consequence, the Krasny Yar location could have been very near to the death sites of the assemblage, say within a kilometer or so. If this estimate is at all reasonable, then the geomorphological setting of Krasny Yar might have been like that of a lake shore or abandoned river loop rather than the high-energy river bank condition of today. To put it another way, Krasny Yar might have been a freshwater muddy slough counterpart to the tar traps at the La Brea animal pits in southern California. However one interprets the taphonomy of Krasny Yar, it marvelously illustrates the sort of concerns about deposition and association that Yefremov (1940) expressed when he proposed his definition of taphonomy.

# 13 Kurla I

## Background

Kurla I (named after a river that flows into northwest Lake Baikal) is an open site 20 km west of Severobaikalsk village, at approximately 55°39′ N, 109°21′ E. It is one of at least six sites named Kurla discovered in 1975 by Irkutsk geologists. They are located in an area of exposed slope layers that correspond with the 6–8 m terrace of Lake Baikal. Kurla I–III were excavated by Peter E. Shmygun in 1976 and 1977 (Shmygun 1981, Shmygun and Sizikov 1977). Three cultural horizons were identified. The cultural layers produced 54 bone tools, 9770 stone tools, including 45 burins, 40 scrapers, four screblos and screblo-like objects, ten adze-like tools, and stone debitage. The earlier horizon (24 000 BP) had bone foreshafts slotted for side blades; the later horizon (ca. 14 000 BP) had non-toggling harpoon heads. Other details of the Kurla work are in Goryunova *et al.* (1976), Yendrikhinsky *et al.* (1978), Ivaniev *et al.* (1981), Shmygun (1978), Shmygun and Filippov (1982), Shmygun and Yendrikhinsky (1978), and summarized by Derev'anko *et al.* (1998). The collection is housed at the Irkutsk State University Laboratory of Archaeology, Irkutsk.

Vasili'ev *et al.* (2002:527) have no carbon-14 date entry for Kurla I, although they provide dates for Kurla III (24 060 BP to 13 160 BP) and Kurla VI (14 150 BP). L. N. Ivaniev and A. A. Khamzina identified reindeer, red deer, snow sheep, roe deer, hare, sable, and rodents in the faunal remains (Yendrikhinsky *et al.* 1978). N. D. Ovodov (unpublished notes) identified faunal remains from Shmygun's Kurla I second and third cultural horizons, and the second horizon of Kurla II: hare (*Lepus tanaiticus*), 334 bones; red deer (*Cervus elaphus*), one; caribou (*Rangifer tarandus*), 178; bighorn sheep (*Ovis nivicola*), 138; bison/yak, one; gray wolf (*Canis lupus*), one; wolverine (*Gulo gulo*), four; polar fox/fox, two; Baikal seal (*Pusa sibirica*), 16. It is noteworthy that Ovodov made these identifications from pieces of bone and teeth that were largely whole, whereas he was unwilling or unable to make species identifications for the 72 small and fragmentary pieces examined herein.

## Findings

**1 Provenience.** We examined the Kurla I collection on September 28–29, 2000. The 72-piece assemblage makes up 0.2% of our 8813-piece grand total (Table A1.1, site 13). All 72 pieces were excavated from the third cultural horizon in 1976.

**2 Species.** Out of 72 Kurla I pieces, the only species is reindeer (Table A1.2, site 13). This frequency is many times greater than the pooled assemblage average. Our Kurla sample is deficient in all other animal groups.

**3 Skeletal elements.** The most common skeletal elements in the 72 Kurla pieces are: toe (30.5%), mandible (16.7%), scapula (12.5%), metapodial (11.1%), antler-horn (4.2%), and rib (4.2%) (Table A1.3, site 13). Compared with the pooled assemblage, Kurla I has more mandibles, scapulae, metapodials, and toes; and fewer ribs, humeri, long bones, and unknown pieces.

**4 Age.** There are no sub-adults in the 72 Kurla I pieces (Table A1.4, site 13). Compared with the pooled assemblage average for sub-adults, Kurla I has a deficiency that may reflect the time of year when this reindeer assemblage was formed.

**5 Completeness.** The 72 piece Kurla I assemblage is relatively incomplete as it has only 1.4% whole bones, 62.5% with one anatomical end, and 36.1% with no anatomical ends (Table A1.5, site 13). Compared with the pooled assemblage averages, Kurla I is in about the same general condition.

**6 Maximum size.** The above incompleteness is mirrored in the much reduced size of the 72 Kurla I pieces, the mean of which is 4.3 cm, with a range of 2.1 cm to 9.5 cm (Table A1.6, site 13). Compared with the pooled assemblage the individual Kurla I pieces are small. We infer this to mean maximal nutrient extraction. Comparison of the maximum size values of Kurla I with the undamaged long bone lengths provided by Vera Gromova (1950: table 27) shows that not even the upper limit reaches the lower end of the undamaged range.

**7 Damage shape.** The most common forms of damage in the 72-piece Kurla I set are: phalanx butts (29.2%), long bone fragments (25.0%), cracked open phalanges (12.5%), and mandibular condyles (6.9%) (Table A1.7, site 13). Compared with the pooled assemblage, Kurla I has fewer long bone flakes and splinters and undamaged bones; and more cracked-open phalanges, phalanx butts, and condyles. There are no ribs.

**8 Color.** Almost all (98.6%) of the 72 Kurla I pieces are ivory colored. The 1.4% black pieces are burned (Table A1.8, site 13). Compared with the pooled assemblage averages, Kurla I has more ivory colored and black pieces, but no other colors.

**9 Preservation.** As with color, almost all (95.8%) of the 72 Kurla I pieces are ivory hard. Only 4.2% are chalky (Table A1.9, site 13). Compared with the pooled assemblage, Kurla I has more ivory, and fewer chalky pieces.

**10 Perimortem breakage.** Every one of the 72 Kurla I pieces has perimortem breakage (Table A1.10, site 13). This is greater than the average in the pooled assemblage. Intensive carcass processing is evident in the Kurla assemblage, as it was in Bolshoi Yakor I, the other deer kill site in the northern Lake Baikal region.

**11 Postmortem breakage.** None of the 72 Kurla I pieces exhibits postmortem breakage (Table A1.11, site 13). This is much less than the average for the pooled assemblage. Postmortem breakage was very low also at Bolshoi Yakor I (2.7%).

**12 End-hollowing.** There are no examples of end-hollowing in the 72 Kurla I pieces (Table A1.12, site 13). This is less than the frequency in the pooled assemblage, but, as previously pointed out, very much like Bolshoi Yakor I (0.004%).

**13 Notching.** Out of the 72 Kurla I pieces, two (2.8%) have notching. Both have two notches (Table A1.13, site 13). The low frequency of Kurla I notching is five times less than the pooled assemblage average. Bolshoi Yakor I has about twice the number of notched pieces, but all have only one or two notches per piece.

**14 Tooth scratches.** There are only two (2.8%) of 71 Kurla I pieces with tooth scratches. One has one scratch, the other has more than seven (Table A1.14, site 13). This frequency is seven times less than that of the pooled assemblage average (20.6%).

**15 Tooth dints.** Three (4.2%) of the 72 Kurla I pieces have tooth dinting. The three pieces have two, three, and five dints (Table A1.15, site 13). This frequency of dinting is six times less than the pooled assemblage average.

**16 Pseudo-cuts.** Kurla I has one piece with two pseudo-cuts (Table A1.16, site 13). The pooled assemblage average is three times greater. Pseudo-cuts, along with the previous four indicators of carnivore activity, suggest that there was some, but not much, carnivore presence at Kurla I.

**17 Abrasions.** One (1.4%) of the 72 Kurla I pieces has abrasions. The abrasion contains more than seven striations (Table A1.17, site 13). This frequency of abrasions is almost exactly the same as the pooled assemblage average.

**18 Polishing.** Slightly more than half (51.4%) of the 72 Kurla I pieces exhibit polishing: 18.1% end, 1.4% middle, and 31.9% end-middle (Table A1.17, site 13). The average total polishing in the pooled assemblage is not greatly different from Kurla I.

**19 Embedded fragment.** The occurrence of embedded fragments in the 72 Kurla I pieces is unexpectedly high (6.9%). Of these, the number of embedded fragments per piece ranges from two to more than seven (Table A1.19, site 13). The pooled assemblage average is slightly less.

**20 Tooth wear.** Kurla I has no teeth for wear assessment.

**21–24 Acid erosion, rodent gnawing, insect damage, human bone.** None of the 72 Kurla I pieces shows any sign of these considerations.

**25 Cut marks.** As with completeness and maximum size, the Kurla I assemblage shows much processing by its large amount of bone cutting. Out of the 72 pieces, 34.7% have from one to more than seven cut marks per piece. The most frequent number of cuts is more than seven (11.1%) per piece (Table A1.25, site 13). In contrast, the pooled assemblage average for cut marks is more than four times less. In fact, Kurla I has the greatest frequency of cut marks in all of our assemblages.

**26 Chop marks.** As with cut marks, completeness, and maximum size, the number of chopped pieces in the 72-piece Kurla I assemblage is very high (30.6%), suggesting extensive processing, and again, the highest occurrence in our study. The number of chop marks per piece ranges from one to more than seven, with one chop per piece being the most common number (Table A1.26, site 13). There are more than five times the number of chopped Kurla I pieces than in the pooled assemblage average.

## Discussion

The Kurla I perimortem damage provides a remarkably clear taphonomic picture of what happened to the bone assemblage. First, the assemblage exhibits a great deal of human processing by the extensive cutting and chopping, and further nutrient extraction by the small piece size, the amount of perimortem bone breakage, and average piece completeness. The Kurlans' objective seems unmistakable – the maximum extraction of nutrients. This, in turn, along with bone quality, hints at winter supply preparation (there is nothing in the age, tooth wear, or epiphysis variables that helps or detracts from this suggestion). Because of the very good overall bone quality, we suggest that many pieces in the assemblage were trampled into the camp's soft, damp soil by humans before ground freezing set, limiting carnivore scavenging. Second, some carnivore presence is recorded in the small amount of tooth scratching, dinting, and other carnivore indicators.

The amount of carnivore presence is like that of Neolithic Boisman II, which we assumed was caused by the Boisman dogs, whose middle-sized bodies and jaws would approximate those of wolves. We propose that the Kurla I damage was done by wolves, unless future regional excavations show that the Kurlans could have had dogs. For the time being we envision Kurla I as having been a repeatedly used open camp site at or near where a number of reindeer and other animals were killed. Sometime after the Kurla people left their Lake Baikal camp, middle-sized carnivores like wolves or small bears worked over the bone refuse left scattered in the camp. After these two "episodes," very little of the bone refuse remained exposed on the ground surface, judging by the small amount of chalky bone. Embedded in what may well have been permafrost for thousands of years, the Kurla I assemblage remained suspended in as much an inert state as the bones left in the cold, dry limestone caves of the Altai Mountains.

# 14 Malaya Seeya

## Background

Malaya Seeya (Small Seeya, named after nearby Seeya village) is a Paleolithic hillside open site located in the eastern ridges of Kuznetsk Ala-Tau, a mountainous and forested region of the middle Yenisei drainage, Khakasia. It is situated 500 m from the left bank of the Bely Iyus River and 35 m up-slope. Kuzmin and Orlova (1998:8) locate the site at 54° 50′ N, 89°42′ E. Figs. 3.67–3.70 illustrate the environment around Malaya Seeya, including caves in the vicinity that were used by hyenas. Malaya Seeya was first discovered and tested by Nicolai Ovodov in 1974 after he learned from brickmakers in the area that they had found bones and artifacts in the hillside clay deposits that they were mining for brickmaking (Fig. 3.71). About 500 $m^2$ of the site had been destroyed in the mining operation. Ovodov's testing continued in 1975, turning up some 583 Paleolithic stone tools and manufacturing debris, plus a few bone artifacts. It was subsequently more extensively excavated by V. E. Larichev in 1978, whose two seasons of fieldwork turned up more artifacts and 4800 pieces of bone, from which Ovodov identified 13 species. Larichev's bone samples produced carbon-14 dates of 34 500 BP and 34 420 BP (Vasili'ev *et al.* 2002:523). Currently, Malaya Seeya is being further excavated by Yu. P. Kholyushkin and additional carbon-14 dates considerably younger have been derived from bone and charcoal. Vasili'ev *et al.* (2002:523) list dates as ranging from 29 450 BP to 20 300 BP. This large range in age dates and the hillside slope of the cultural level have contributed to the controversy about Malaya Seeya – namely, did it have multiple occupations? A number of Quaternary geologists have visited and studied the site, including V. M. Muratov and S. M. Tseitlin, Moscow; S. L. Troitsky, Novosibirsk; and V. P. Chekha and A. F. Yamskikh, Krasnoyarsk. The authors together visited the testing in progress in July, 2000. We saw an *in situ* artifact that Kholyushkin had exposed at a depth of about 1 m in a 3 m deep test pit dug into light brown, sticky, loamy clay that showed no signs of horizontal stratification. Additional information about Malaya Seeya can be found in Abramova *et al.* (1991), Chekha and Ovodov (1992), Larichev (1978), and Lisitsyn (1997). As suggested above, the site is in need of a comprehensive geological and archaeological analysis.

The 1974–1975 fauna excavated and identified by Ovodov (Chekha and Ovodov 1992) included: marmot (*Marmota baibacina*), two pieces; hare (*Lepus* sp.), 32; brown bear (*Ursus arctos*), five; Asiatic wild ass, onager (*Equus hemionus*), three; horse (*Equus

**Fig. 3.67** The Malaya Seeya landscape. View of the mountainous setting of Malaya Seeya, out of sight to the left. The Seeya River rushes into the distance. View taken from the limestone cliff at the opening to Tokhasky Grotto, a hyena den 83 m vertically above the river. We include this and the three following photographs to make the point that late Pleistocene hyenas were present in the Malaya Seeya area (CGT color Tokhasky 7-22-00:19).

*caballus*), 49; onager aut hors, 39; rhinoceros (*Coelodonta antiquitatis*), eight; mammut (*Mammonteus primgenius*), three; red deer (*Cervus elaphus*), 30; Mongolian gazelle (*Procarpa gutturoza*), one; wild sheep (*Ovis ammon*), three; goat, ibex, markhor (*Capra sibirica*), one; wild goat-sheep, 131; reindeer (caribou) (*Rangifer tarandus*), 290; reindeer, wild sheep/goat, 4054; bison (*Bison priscus*), 113. Clearly, the reindeer are the most commonly represented species. N. M. Ermolova, who studied the faunal remains excavated in 1976 by V. E. Larichev, also found reindeer to occur most frequently. In addition, she identified wolverine and fox.

## Findings

**1 Provenience.** Our 144 pieces make up 1.6% of our 8813-piece grand total. They make up a 2.0% sample of the specimens excavated by Larichev.

**2 Species.** The most common groups in our 144 pieces are: goat-sheep (66.7%), reindeer (25.0%), and horse (8.3%) (Table A1.2, site 14). Compared with the pooled assemblage averages, Malaya Seeya has no big mammal, bison, hyena, mammoth, and indeterminable; and more goat-sheep, horse, and reindeer.

**Fig. 3.68** Tokhasky Grotto. From the mountain summit to the Seeya River is a nearly vertical drop in this massive limestone cliff. Ovodov tested this cave in 1975, finding hyena, horse, bison, and rhinoceros bones (CGT color Tokhasky 7-22-00:20).

**Fig. 3.69** Inside Tokhasky Grotto. About the size of Dvuglaska cave (Fig. 3.16), this cavern differs in being difficult to access and hard to reach water from. These conditions – difficult access and no water – match those of the Razboinich'ya hyena cave. Left to right: Nicolai Ovodov, Elena Popkova, Nicolai Martynovich, and Olga Pavlova to the right of the 1 × 2 m test pit (CGT color Tokhasky 7-22-00:18).

**Fig. 3.70** Proskuryakov Grotto. Another hyena cave high above the Seeya River. Tested also in 1975 by Ovodov, a few stone flakes, hyena, bison, horse, and other Pleistocene animal bones, and a carbon-14 date of 46 000 BP were obtained. Left to right: Olga Pavlova, Nicolai Ovodov, and Nicolai Martynovich (CGT color Proskuryakov 7-21-00:16).

**Fig. 3.71** Malaya Seeya excavation. This open hillside site was originally tested in 1975 by Nicolai Ovodov (left) after learning from brickmakers that they had found bones and stone artifacts while digging clay. During our visit it was being re-examined by Yuri Kholyushkin (right). Sticky loessic clay underlies the darker near-surface Holocene layer. A large stone flake remains *in situ* near the top of the white stick that Ovodov holds in his right hand. Situated high above and back from the Seeya River, the site location is reminiscent of the Dry Creek site in Alaska, which had a good view overlooking the river valley it was in, was in a relatively warm location, and probably had regular hillside winds that would have blown mosquitoes away (CGT color Malaya Seeya 7-21-00:17).

**3 Skeletal elements.** The most common skeletal elements in the 144 Malaya Seeya pieces are: long bone (50.0%), metapodial (17.4%), rib (9.7%), and mandible (4.2%) (Table A1.3, site 14). Compared with the pooled assemblage, Malaya Seeya is low or lacking in humeri, radius, foot, toe, and unknown; and has more long bone and metapodials.

**4 Age.** There are only 0.7% sub-adults represented in the 144 Malaya Seeya pieces (Table A1.4, site 14). This is several times less than the pooled assemblage average.

**5 Completeness.** Malaya Seeya is a very incomplete assemblage by this criterion, there being no whole bones, 9.0% with one anatomical end, and 91.0% with no anatomical ends (Table A1.5, site 14). Compared with the pooled assemblage, Malaya Seeya completeness is much lower.

**6 Maximum size.** The 144 Malaya Seeya pieces have a mean size of 8.3 cm and a range of 3.2 cm to 30.8 cm (Table A1.6, site 14). These values are nearly identical to those of the pooled assemblage. Compared with undamaged long bone lengths provided by Vera

Gromova (1950: table 27) for goat-sheep and reindeer, Malaya Seeya's upper range limit is similar. Undamaged horse long bones are generally larger.

**7 Damage shape.** This variable was not scored for Malaya Seeya because it was being formulated at the time of our examination in 1999.

**8 Color.** For an open site, the 144 Malaya Seeya pieces include somewhat fewer ivory colored specimens (90.3%) than do other open sites. It also has four black pieces (2.8%) and some brown and white (Table A1.8, site 14). Compared with the pooled assemblage, Malaya Seeya has almost the exact same number of ivory, and three times the number of black and brown colored pieces. Because it is an open site subject to staining by soil conditions, we think all the brown and black pieces are stained.

**9 Preservation.** Like bone from other open sites, the 144 Malaya Seeya pieces have a relatively low occurrence of ivory hard specimens (84.7%), and a relatively large occurrence of chalky pieces (15.3%) (Table A1.9, site 14). Compared with the pooled assemblage, Malaya Seeya has slightly more ivory, and slightly fewer chalky pieces.

**10 Perimortem breakage.** Almost all of the 144 Malaya Seeya pieces have perimortem breakage (Figs. 3.72–3.74). Only 0.7% do not (Table A1.10, site 14). Compared with the pooled assemblage, Malaya Seeya has more perimortem breakage.

**11 Postmortem breakage.** The amount of postmortem breakage in the 144 Malaya Seeya pieces is 11.1%, which is somewhat less than the average for the pooled assemblage (Table A1.11, site 14).

**12 End-hollowing.** There is almost no end-hollowing (0.7%) in the 144 Malaya Seeya pieces (Table A1.12, site 14). This is more than ten times less that the average for the pooled assemblage.

**13 Notching.** More than 10% (11.3%) of the 144 Malaya Seeya pieces have 1–3 notches. One notch is most common (Table A1.13, site 14). Compared with the pooled assemblage average, Malaya Seeya has somewhat less notching.

**14 Tooth scratches.** Not quite 5% (4.2%) of the 144 Malaya Seeya pieces have 2–4 tooth scratches (Table A1.14, site 14). Compared with the pooled assemblage average (20.6%), Malaya Seeya has almost five times less scratching.

**15 Tooth dints.** Similar to scratches, Malaya Seeya has 4.9% of its 143 pieces with tooth dints. There are from one to more than seven dints per piece (Table A1.15, site 14). Compared with the pooled assemblage average, Malaya Seeya has five times less dinting.

**16 Pseudo-cuts.** There is only one piece with pseudo-cuts (0.7%) in the 144 Malaya Seeya pieces. It has six cuts (Table A1.16, site 14). Compared with the pooled assemblage average, Malaya Seeya has much less pseudo-cutting. Thus, this variable and the other four that are carnivore indicators show that carnivores had damaged some of the Malaya Seeya bone refuse, but not very intensively. We feel that this modest amount of damage better fits the damage pattern suspected of wolves or small bears rather than hyenas.

Fig. 3.72 Malay Seeya bone refuse. This open hillside site produced a large fauna collection, although preservation was not good, as was the case for a contextually similar site far to the east, Varvarina Gora. These three of several storage boxes contain artiodactyl pieces, collected and identified by Ovodov in 1975. Many, if not most, of the pieces of bone from this and all other collections examined lacked specific labeling. This is because most of the faunal remains recovered were tiny unidentifiable fragments, even to the expert anatomical eye of Nicolai Ovodov. Many of the fragments we were interested in for perimortem damage were smaller than could be identified. Pencil is 11.5 cm in length (CGT color IAE 8-13-99:35).

**17 There are no abrasions in the 144 Malaya Seeya pieces.**

**18 Polishing.** Slightly more than 80% (81.7%) of the 144 Malaya Seeya pieces are polished, which includes 13.4% end, 2.1% middle, and 66.2% middle-end (Table A1.18, site 14). Compared with the pooled assemblage average, Malaya Seeya has about 15% more polished pieces, most being middle-end types.

**19 Embedded fragments.** Of the 144 Malaya Seeya pieces, 4.2% have one or two embedded fragments (Table A1.19, site 14). Compared with the pooled assemblage average, Malaya Seeya is almost identical.

**20–24 Tooth wear, acid erosion, rodent gnawing, insect damage, and human bone.** None of these variables are present in the 144-piece Malaya Seeya assemblage.

**25 Cut marks.** Malaya Seeya has 13.9% of its 144 pieces showing from one to more than seven cut marks (Table A1.25, site 14). Compared with the pooled assemblage average (7.6%), Malaya Seeya has almost twice the number of cut pieces.

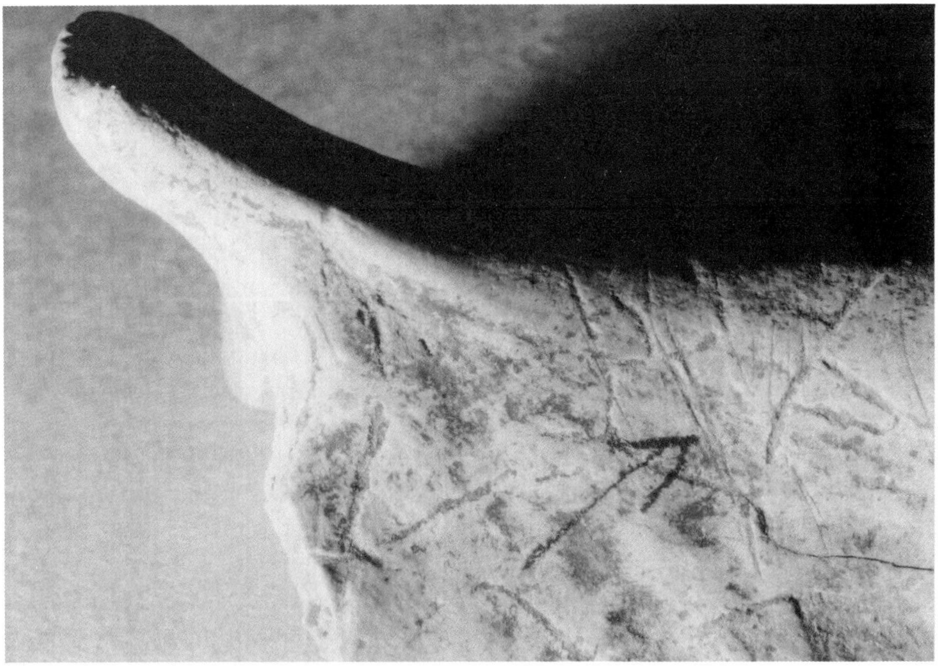

Fig. 3.73 Malaya Seeya bone damage. Butchering cut marks near reindeer mandibular condyle, plant root damage is moderate. Actual width of image is 3.3 cm (CGT neg. IAE 8-11-99:32A).

Fig. 3.74 Malaya Seeya bone damage. Goat-sheep-capra scapula with cut marks at arrow. Actual width of image is 3.3 cm (CGT neg. IAE 8-11-99:29A).

**26 Chop marks.** Only one (0.7%) of the 144 Malaya Seeya pieces had been chopped. This single piece has two chop marks (Table A1.26, site 14). Compared with the pooled assemblage average, Malaya Seeya has less chopping.

## Discussion

The perimortem taphonomy of Malaya Seeya is simple and generally corresponds to what occurs in our other open, human camp sites, most notably in the relatively large amount of chalky pieces. Although we know that hyenas roamed in the vicinity – for example, they used Proskuryakova Cave and Tokhasky (a hyena cave we visited but did not formally work up the few bones of hyena, bison, and rhinceros), both a few kilometers down the Seeya River – the lack of acid-eroded pieces coupled with the moderate amount of carnivore damage is probably best attributed to medium-sized animals such as wolves and possibly small bears. We see nothing in the perimortem taphonomy of Malaya Seeya that helps to explain why the camp was established where it was. Today, despite being in a forested setting, the site seems rather exposed to the elements. Being above or away from the river, along the banks of which likely prowled large predators such as bears and wolves, if not cave hyenas, might be the major reason. Plagues of mosquitoes along the Seeya River might also have favored camping uphill, as well as providing a look-out for game along the river. The site reminds the senior author of the conditions of the hillside site called Dry Creek in Alaska.

# 15 Mal'ta

## Background

Mal'ta is an open site 86 km south of Irkutsk at 52°50′ N, 103°32′ E. It is named after the village beneath which the Upper Paleolithic site was discovered. The typically unprosperous Siberian farming village sits on a flat, rolling plain, that today overlooks the Belaya River (Fig. 3.75). According to Mal'ta expert German Medvedev (personal communication, September 27, 2000) at the time of the late Pleistocene occupation of the Mal'ta site, the river did not exist; instead there were a series of lakes below the site and its companion site called Buret' on the opposite shore of one lake (Lipnina *et al.* 1995). As they do today, villagers in the 1920s, digging in the ground to construct cold vegetable cellars or plant gardens, sometimes encountered bones of large animals. In the winter of 1927–1928 the Mal'ta village teacher-librarian sent word of these discoveries to the Irkutsk Historical Museum, which responded by having a young research fellow named M. M. Gerasimov check out the report. As a youth he developed interests in paleontology, archaeology, osteology, and related fields, so the Mal'ta discoveries fired up his eclectic interests in ancient Siberians. His findings would turn out to be among the most important in Siberian prehistory, as important as his developments in forensic anthropological reconstructions of human faces (Conant 2003, Gerasimov 1955).

The finds were confirmed and Gerasimov began in 1928 several years of archaeological testing (Gerasimov 1931, 1935, 1940, 1958) (Figs. 3.76–3.78). What his excavations eventually discovered included the now world famous Siberian human female bone figurines and bone carvings of water fowl, and other ornaments (Figs. 3.79–3.80), all in association with large amounts of mammoth bone refuse, bones of other animals, stone blade production, a large amount of stone tool manufacturing refuse, and outlines of dwellings and hearths, special analyses of which have been carried out by Gromova (fauna, 1941), Ermolova (fauna, 1978), Kimura (stone tools, 1997, 2003), Turner (human teeth, 1990a), and others. The remains of two young children interred with many ornamental beads topped off the remarkable discoveries made by Gerasimov up to the time he stopped excavating in the late 1950s (Figs. 3.81–3.83). Gerasimov was a man of remarkable talents. In addition to his high-quality excavations, he was also a gifted anatomical artist. He founded the Laboratory of Plastic Reconstruction in Moscow, which still leads the world in historical and forensic facial reconstructions (Conant 2003). Work at the Mal'ta site continues to this day by Gerasimov's student, Doctor of

**Fig. 3.75** Mal'ta–Buret' locality. View from Mal'ta looking across Belaya River toward Buret' village in the distance. Originally, both archaeological sites were camps on opposite shores of a Pleistocene lake. Individuals unknown (CGT color Mal'ta 9-20-00:7).

Historical Sciences, German Medvedev, and his wife Ekaterina Lipnina (Lipnina and Medvedev 1993), also a Doctor of Science. Judging from the Paleolithic remains, the prehistoric Mal'tans were as well off economically if not better than are the modern villagers.

More than 21 000 pieces of bone were recovered in Gerasimov's various excavation seasons. Because of inadequate storage facilities and irresponsible curation, not all of these bones have been preserved. Vera Gromova inventoried the bones collected 1928–1937. Collections made subsequently have been studied by Gromova (1941), Ermolova (1978), and Khenzykhenova and Shushpanova (2001). Fifteen species of larger mammals have been identified by these workers in the Mal'ta faunal assemblage: hare (*Lepus cf. timidus*), gray wolf (*Canis lupus*), fox (*Vulpes vulpes*), Arctic fox (*Alopex lagopus*), brown bear (*Ursus arctos*), wolverine (*Gulo gulo*), cave lion (*Panthera spelaea*), horse (*Equus cf. caballus*), rhinoceros (*Coelodonta antiquitatis*), mammut (*Mammuthus primigenius*), red deer (*Cervus elaphus*), caribou (*Rangifer tarandus*), big-horn sheep (*Ovis nivicola*), wild sheep (*Ovis ammon*), and bison (*Bison proscus*). No cave hyena remains have been identified, although Ovodov knows of one bone of this creature having been found not far away in the Baikal region. In addition, two kinds of pika (steppe and northern), suslik, a few kinds of small voles, even-toed lemmings, and steppe lemmings were found. Patently, this inventory of species does not reflect the complete resource base at Mal'ta, like so many other

**Fig. 3.76** Mal'ta site. Rain-drenched conference-goers inspect the ongoing excavation at Mal'ta. The crowd was participating in the 2001 International Conference: Current Scientific Problems of the Eurasian Paleolithic Dedicated to the 130th Anniversary of the Discovery of the First Paleolithic Site in Russia, "The Military Hospital," in 1871, held in Irkutsk (CGT color Mal'ta 8-4-01:9).

**Fig. 3.77** Mal'ta stratigraphy. German Medvedev, in trench, explains the recent findings to the Irkutsk conference-goers (CGT color Mal'ta 8-4-01:18).

**Fig. 3.78** Mikhail Mikhailovich Gerasimov. Copy of a photograph of the original excavator of the Mal'ta site. He later founded the Laboratory of Plastic Reconstruction in Moscow, where this photocopy was taken (original photographer unknown, CGT color LPR 1-6-81:33).

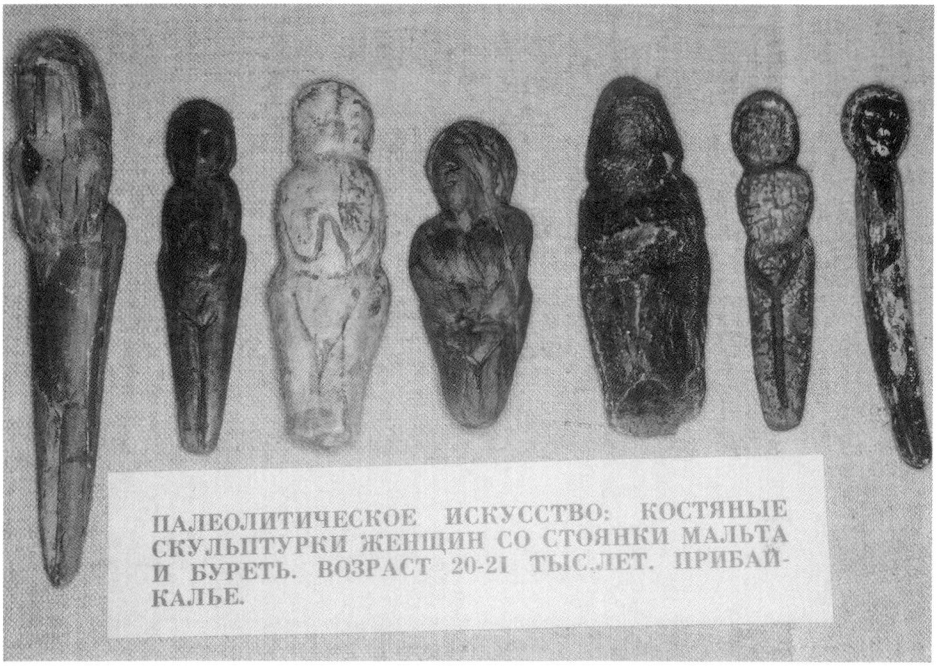


**Fig. 3.79** Mal'ta artifacts. Most renowned of Gerasimov's finds at Mal'ta are carvings of humans and birds. The exhibit caption for these replicas in the former IHPP museum reads: "Paleolithic art: bone figurines of women from the sites of Mal'ta and Buret'. Age: 20–21 thousand years. Cis-Baikal area." Some researchers feel that one or more of the figurines has Mongoloid characteristics. However, the senior author has spent hours studying these objects and has been unable to decide if they as a group are more or less Mongoloid than Europoid. For reasons explained in the text he favors a European affiliation. Artistically they have correspondences to the Upper Paleolithic European "Venus" figurines (CGT color IHPP 6-18-87:34).

Siberian archaeological faunal inventories, because of the lack of information on bird, fish, and other potential dietary items.

Carbon-14 dates suggest a relatively short period of occupation, although there are two old outlier dates based on bone from Stratum 6 (43 100 BP) and "gravel" (41 100 BP). Most of the more than 15 dates tallied by Vasili'ev *et al.* (2002:526–527) cluster around 21 000 BP – for example, 14 750 ± 120 BP (GIN-97); 21 600 ± 170 BP (GIN-8475); 21 700 ± 160 BP (OxA-619). Medvedev (1998) has proposed for various reasons, including the large amount of reindeer bone refuse, that Mal'ta was a repeatedly used, seasonally occupied, reindeer hunting camp. The Mal'ta collection is curated in the Laboratory of Archaeology, Russian Academy of Sciences, Irkutsk State University. There is an extensive literature on Mal'ta, beginning with Gerasimov and continuing to the present day: Tseitlin (1974), Kimura (2003), Medvedev (1998), Medvedev *et al.* (1996, 1998a, 1998b), Turner (1990a), and many others. Derev'anko *et al.* (1998) summarize some of this information prepared by Medvedev. The present authors visited the site together in 2000 and 2001.

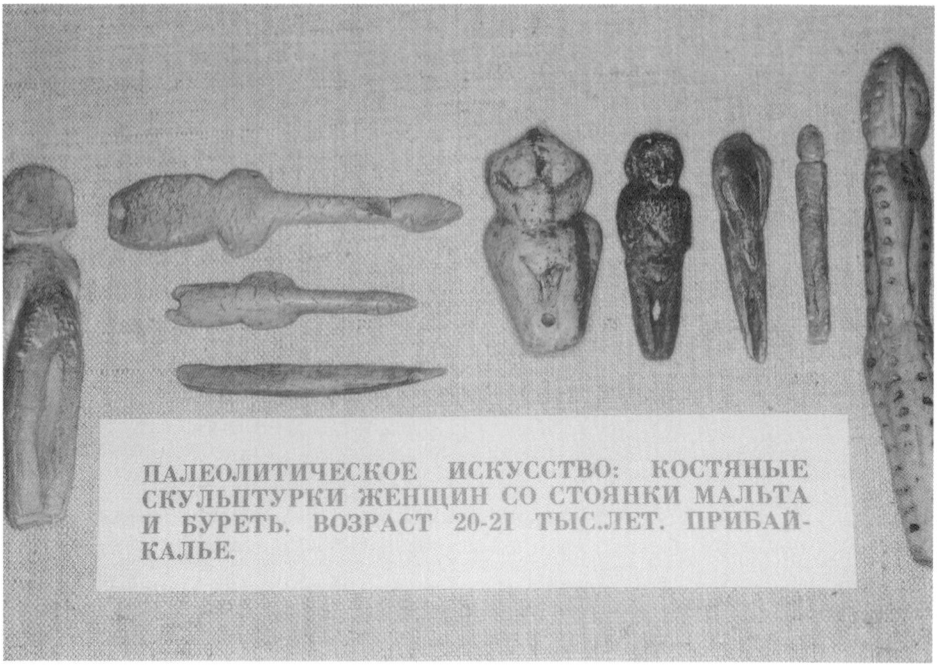


**Fig. 3.80** Mal'ta artifacts. Additional human figurines and birds(?). Exhibit caption reads the same as in Fig. 3.79. One could consider these marvelous objects as examples of perimortem bone damage, but hardly in the sense of accidental damage such as cut marks or tooth scratches (CGT color IHPP 6-18-87:36).

**Fig. 3.81** Location where Gerasimov found the two Mal'ta children. German Medvedev (right) stands on the spot where the two children were buried, 2 m or so below the present ground surface. Korean Paleolithic archaeologist Yung-Jo Lee takes notes (CGT color Mal'ta 8-4-01:29).

**Fig. 3.82** Mal'ta teeth. Unerupted upper permanent teeth of the older Mal'ta child. Unlike the previously illustrated bone figurines, the dental crown morphology has proven to be more useful in racial identification. Modern European teeth are characterized by simplification, that is, weak expression or non-occurrence of several traits. Northeast Asian and all Native American teeth are just the opposite, with strong expression and frequent occurrence. The two teeth in the center of the top row are central incisors. Like European teeth these show weak (left) and no occurrence (right) of shoveling. Asians, on the other hand, have strong and frequent incisor shoveling. The tooth in the upper left is a lateral incisor. It is considerably smaller than the central incisors, a characteristic of European teeth. Asians usually have the central and lateral incisors about the same size. The upper molars in the lower row are in several ways more similar to Europeans than to Asians (CGT color IEL 1-13-81:21).

## Findings

**1 Provenience.** We examined 186 pieces of Mal'ta bone, which is 2.1% of our 8813-piece grand total (Table A1.1, site 15). All of the pieces came from the more recent excavations by Medvedev, Lipnina, and associates in the occupation area(s). None of Gerasimov's material could be located. Two pieces without provenience were excavated in 1991, the other 184 are about equally from precisely defined units excavated in 1992 and 1996–1998.

**2 Species.** The most common groups in the 186 Mal'ta pieces are: mammoth (49.5%), indeterminable (18.3%), big mammal (10.7%), reindeer (9.1%), and bison (4.3%) (Table A1.2, site 15). Compared with the pooled assemblage averages, Mal'ta has no bear, gazelle, goat-sheep, hyena, or roe deer; fewer horse and indeterminable; and more big mammal, mammoth, and rhinoceros.

Fig. 3.83 Unerupted lower permanent teeth of the older Mal'ta child. The two large teeth are first molars. Like those of modern Europeans, both teeth lack cusp 6, the protostylid, and the deflecting wrinkle. These traits are common and pronounced in Northeast Asians and Native Americans. Two inferences can be proposed: (1) the Mal'ta child was more closely related to modern and Upper Paleolithic Europeans than to Mongoloids; (2) the Mal'ta people were not ancestral to Native Americans (CGT color IEL 1-13-81:20).

**3 Skeletal elements.** The most common elements in the 186 Mal'ta pieces are: rib (29.6%), vertebra (22.6%), long bone (9.7%), unknown (6.4%), foot (5.4%), metapodial (4.8%), and toe (4.8%) (Table A1.3, site 15). Compared with the pooled assemblage averages, Mal'ta has fewer pieces of mandible, long bone, humerus, and unknown; and more vertebra and rib.

**4 Age.** Out of the 186 Mal'ta pieces, 15.0% are sub-adult (Table A1.4, site 15). This value is almost twice the frequency of the pooled assemblage average.

**5 Completeness.** The 186-piece Mal'ta assemblage has 17.2% whole bones, 31.7% with one anatomical end, and 51.1% with no anatomical ends (Table A1.5, site 15). These frequencies are similar to those of the pooled assemblage averages.

**6 Maximum size.** The 186-piece Mal'ta assemblage has one of the three largest mean dimensions (16.5 cm) and ranges (3.0 cm to 82.5 cm) of all our studied samples (Table A1.6, site 15). All three contain considerable amounts of mammoth bone. As expected, both the Mal'ta mean and range are greater than the pooled assemblage averages for these statistics. Looking at the lengths of whole long bones provided by Vera Gromova (1950: table 27) for mammoth, reindeer, and bison, the Mal'ta upper range limit is about what she reports for the upper limit of mammoth, but much greater than her bison and reindeer values.

**7 Damage shape.** Out of 185 Mal'ta pieces, the following are the most common types of damage: long bone fragment (17.8%), proximal rib (13.0%), undamaged (11.9%), vertebral spine (11.3%), long bone flake (9.7%), mostly whole (8.1%), and vertebral body (4.3%) (Table A1.7, site 15). Compared with the pooled assemblage averages, Mal'ta has more vertebral bodies and spines, proximal and distal ribs, and mostly whole bones; and fewer long bone flakes, fragments, splinters, and phalanx butts.

**8 Color.** Almost all (99.0%) of the 186 Mal'ta pieces are ivory colored. There is one white and one black (completely burned) piece (Table A1.8, site 15). A fragment of reindeer pelvis (field specimen 118, 1998) is heavily covered all over with manganese "flowers" that also fill root tracks that are common on one surface and rare on the other. Compared with the pooled assemblage averages, Mal'ta has 10% more ivory colored pieces and no brown pieces. White and black are nearly equal in both groups.

**9 Preservation.** Like other open sites, the 186-piece Mal'ta assemblage has a high frequency (50.0%) of chalky pieces (Table A1.9, site 15). This is due in part to the many mammoth pieces that generally do not preserve well because of their relatively thin outer cortex. Compared with the pooled assemblage, Mal'ta has more than twice the frequency of chalky pieces. This high value, along with the amount of mammoth bone, suggests that the site was open to long perids of weathering, or that much of the bone was collected from the ground surface elsewhere and carried to the site.

**10 Perimortem breakage.** Out of 162 Mal'ta pieces, 75.9% have perimortem breakage (Table A1.10, site 15). This value is somewhat less than the pooled assemblage average.

**11 Postmortem breakage.** More than one-quarter (28.5%) of 179 Mal'ta pieces have postmortem breakage (Table A1.11, site 15). Compared with the pooled assemblage average (17.8%), Mal'ta has about 10% more postmortem breakage. As with quality, some of this difference is attributable to the crumbly nature of weathered mammoth bone.

**12 End-hollowing.** The small amount of end-hollowing (4.3%) in 186 Mal'ta pieces (Table A1.12, site 15), as we have suggested previously, probably was caused by middle-sized carnivores such as wolves and possibly small bears. Compared with the pooled assemblage average, Mal'ta has less end-hollowing. Another consideration is that weathered chalky bone is of little interest to carnivores seeking bone grease.

**13 Notching.** As with end-hollowing, notching is also uncommon (3.2%) in the 186 Mal'ta pieces (Table A1.13, site 15). Compared with the pooled assemblage average, Mal'ta has almost five times less notching. We consider this consistent with an inference of small- and middle-sized carnivore presence.

**14 Tooth scratches.** Tooth-scratched pieces are also infrequent (2.8%) in 178 assessable Mal'ta pieces. The number of scratches on the five pieces ranges from one to more than seven (Table A1.14, site 15). Compared with the pooled assemblage average, tooth scratching at Mal'ta is seven times less frequent.

**15 Tooth dints.** There is a low frequency (1.7%) of dinting in 178 Mal'ta pieces. Dints per piece in each of the three pieces ranges from two to more than seven (Table A1.15, site 15). Compared with the pooled assemblage average, Mal'ta dinting is many times less frequent.

**16 Pseudo-cuts.** There are no pseudo-cuts in 178 Mal'ta pieces.

**17 Abrasions.** Almost no (0.6%) abraded specimens occur in the 177 observable Mal'ta pieces. The solitary piece has more than seven abrasion lines (Table A1.17, site 15). Both the pooled assemblage average and Mal'ta are similar for abrasion occurrence.

**18 Polishing.** Out of 181 Mal'ta pieces, 74.6% have some type of polishing. By location, the occurrence of polishing is on the end (34.8%), middle (1.7%), and middle-end (38.1%) (Table A1.18, site 15). Compared with the pooled assemblage averages, polishing is about the same, except for the greater amount of Mal'ta end polishing.

**19 Embedded fragments.** As with most of our samples, pieces with embedded fragments are uncommon (2.7%) in the 186-piece Mal'ta assemblage. The number of embedded fragments in each of the five pieces ranges from one to more than seven (Table A1.19, site 15). Compared with the pooled assemblage average, Mal'ta is similar.

**20 Tooth wear.** The Mal'ta collection had only one tooth usable for tooth wear consideration. It belonged to a sub-adult (Table A1.20, site 15).

**21 Acid erosion.** Mal'ta has three pieces out of 186 that we judge to be acid eroded (1.6%) (Table A1.21, site 15). This frequency is like that of other sites that we consider to have had a medium-sized carnivore presence, and a few acid-eroded pieces – for example, Kamenka and Varvarina Gora. The eroded pieces include a mostly whole, 4.5 cm deer-sized astragulus (field specimen 169, 1998); a whole 3.7 cm reindeer toe (field specimen 287, 1998); and a 7.2 cm fragment of unknown species and unknown element (field specimen 116, 1998). Compared with the pooled assemblage average, Mal'ta has three times less acid erosion.

**22–23 Rodent gnawing and insect damage.** There are no examples of these variables in our Mal'ta collection.

**24 Human bone.** No human bone was found in the Mal'ta faunal assemblage we examined. However, as noted at the beginning of this section, two human children had been recovered by Gerasimov. These have only been partly described (Alexeev 1998, Turner 1990a, Haeussler and Turner 2000 ). They are curated in the Hermitage Museum, St. Petersburg.

**25 Cut marks.** Out of 182 Mal'ta pieces, 11 (6.0%) have cut marks. These range from one to more than seven cuts per piece (Table A1.25, site 15). Compared with the pooled assemblage average, Mal'ta has about the same frequency of cut pieces.

**26 Chop marks.** There are a few more chopped pieces (16; 8.7%) out of 183 Mal'ta pieces than there are cut pieces. Chops per piece range from one to five (Table A1.26, site 15). Compared with the pooled assemblage average, Mal'ta has more chopping, which is attributable to the large amount of very big mammal bones in the assemblage, mammoth in particular.

## Discussion

The Mal'ta perimortem taphonomy picture is straightforward. The assemblage represents the accumulated bone refuse of meals left scattered by humans in the site, and bone collected elsewhere for dwelling construction and fabricational purposes. The large amount of mammoth bone is biased by the great size of these huge animals. We could identify no perimortem processing that would help decide if the mammoths had all been victims of hunting, or if their bones were of previously dead animals simply dragged back to the Mal'ta camp for construction rather than nutritional purposes. In the whole assemblage we failed to identify any signs of roasting. We favor the collecting scenario because of weathering. At some time of the year, medium-sized carnivores were in the Mal'ta camp. We presume these carnivores were wolves and that they were scavenging the site in the absence of its human occupants. This absence may have been seasonal over any number of years, and repeated carnivore visitations are a reasonable hypothesis. Given that cave lion has been identified in the Mal'ta faunal remains, there is a possibility that cave hyenas might also have been present, although our perimortem taphonomy is unsupportive of the possibility. Lastly, given that dog remains have been reportedly found at Afontova Gora at about the same time period, and a dog skull of much greater age has been found at a hyena cave in the Altai Mountains (Razboiich'ya Cave), some of the Mal'ta carnivore damage might have been done by dogs.

# 16 Maly Yaloman Cave

## Background

Maly Yaloman Cave (Little Yaloman, a nearby river) is located in the northern slopes of the Terektinsky Ridge, Ongudai region, Gorny Altai. According to Kuzmin and Orlova (1998:7) it is at 49°80′ N, 86°30′ E. The south-southwest-facing cave is 27 m above the level of the Maly Yaloman River, which runs below the cliff that contains Maly Yaloman Cave. The cave mouth is 5 m wide, and at its highest 2.6 m tall. Its length is about 35 m, trending upward so that the end is 11 m above the floor level at the entrance. There is ca. 60–70 $m^2$ of usable floor area, with the entrance platform having some 15 $m^2$ of living space. It is a cave site that had both human and hyena occupants in late Pleistocene times (Fig. 3.84) (Table A1.1, site 16). Vasili'ev *et al.* (2002:523) list one carbon-14 date of 33 350 BP, derived from charcoal found in Layer 3. There is an additional date of 24 130 (SO AN-2404) based on the bone of a yak or bison.

In 1983, hydrologist A. M. Maloletko (Alexeeva and Maloletko 1984) made the first test excavation, a pit about 1.5 $m^2$ near the entrance. His osteological findings included a human tooth (subsequently lost; personal communication from E. V. Alexeeva to N. D. Ovodov), hare, brown bear, cave hyena, red wolf, ass, small horse, bison, Siberian mountain goat, and sheep (Alexeeva and Maloletko 1984). In 1986, Maloletko, Ovodov, and assistants dug deeper in the original test pit in order to better understand the stratigraphy, get charcoal for dating purposes, and to add to the faunal collection. At that time they fully mapped the cave. identified three hearths of differing ages, and found a few artifacts. From this work they identified 28 species of mammal and 16 species of bird, the latter identified by N. Martynovich. In 1988 Valery T. Petrin, leader of the IAE North Asiatic archaeological expedition, expanded the Maly Yaloman excavation to some 45 $m^2$ (Derevianko and Petrin 1988, Derevianko *et al.* 1998h).

Combining the 1986 and 1988 collections, there are: hare (*Lepus timidus*), five pieces; cape hare (*Lepus capensis*), 38; marmot (*Marmota baibacina*), 21; gray wolf (*Canis lupus*), 59; fox (*Vulpus vulpus*), 138; red wolf (*Cuon alpinus*), three; manul (*Felis manul*), one; lynx (*Felix lynx*), one; snow leopard (ounce) (*Uncia uncia*), eight; cave hyena (*Crocuta spelaea*), 33; horse (*Equus cf. caballus*), 50; rhinoceros (*Coelodonta antiquitatis*), 15; red deer (*Cervus elaphus*), one; roe deer (*Capreolus capreolus*), three; goat, ibex, markhor (*Capra sibirica*), 520; wild sheep (*Ovis ammon*), one; and wild yak (*Poephagus baikalensis*), 46. These species, as well as the small mammals and birds, give Maly Yaloman a pronounced steppe quality (Ovodov *et al.* 2003).

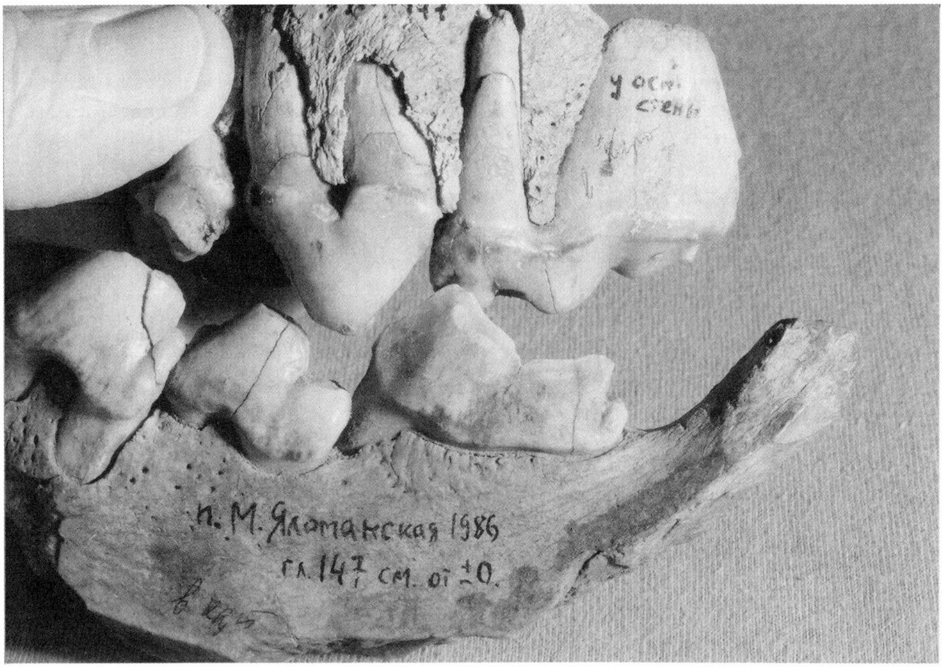


**Fig. 3.84** Maly Yaloman hyena. There is much more evidence of this cave having been used by carnivores than by humans in the Pleistocene. These young adult jaws, found in 1986, Layer 3–5 (147 cm), have chewing damage believed to have been due to cannibalism. Upper jaw is 7.5 cm in length (CGT neg. IAE 8-5-03:34).

## Findings

**1 Provenience.** Cave deposits discussed above.

**2 Species.** Of the 178 pieces we examined, *Capra* (55.1%), hyena (9.5%), goat-sheep (8.4%), fox (7.9%), horse (4.5%), and wolf (3.9%) are the most commonly represented species (Table A1.2, site 16). With the exception of goat-sheep and horse, these species occur more frequently in Maly Yaloman than in the pooled assemblage. Although yak is represented by only 2.2% of the 178 pieces, Maly Yaloman and Kaminnaya (0.3%) are nevertheless the only two of our 30 assemblages where this species has been recognized in this study or previous studies.

**3 Skeletal elements.** Our sample of 178 pieces contained most of the skeletal elements listed in Table A1.3 (site 16). Not represented are the cranial base, hyoid, femur, fibula, epiphysis, and penis bone. Most common are the mandible, teeth, foot, and toe bones. Compared with the averages in the pooled assemblage, Maly Yaloman is unexceptional for the occurrence of most skeletal elements. Only the mandible, teeth, foot, and toe bones occur in greater numbers, and unknown pieces are less common than the averages of the pooled assemblage.

**4 Age.** Adults make up 75.3% of our 178 pieces (Table A1.4, site 16). This value is greater than the 63.0% adults in the pooled assemblage, although sub-adults and unknowns are nearly the same. Although hyena remains were found in Maly Yaloman, the degree of destructiveness for age determination is about half that of Razboinich'ya.

**5 Completeness.** Our 178 pieces are about two-thirds incomplete – that is, one anatomical end (28.1%) and no anatomical ends (39.3%) (Table A1.5, site 16). These values are similar to the averages in the pooled assemblage. Maly Yaloman has more whole bones (32.6%) than does Razboinich'ya (8.6%).

**6 Maximum size.** While the mean maximum size (4.9 cm) of the 177 Maly Yaloman pieces is less than that of the pooled assemblage, the two sets of ranges are nearly the same. There is essentially no difference in size statistics between Maly Yaloman and Razboinich'ya (Table A1.6, site 16). Comparing the whole long bone lengths of sheep-goat, hyena, and fox provided by Vera Gromova (1950: table 27) shows that the Maly Yaloman upper range limit (28.1 cm) is very similar to that of sheep-goat, very slightly larger than that of hyena, and much larger than that of fox.

**7 Damage shape.** The most common form of damage includes long bone fragments (13.5%) and tooth-bearing skull parts (10.7%). Undamaged bones (14.6%) and mostly whole bones (14.6%) were the most common of the damage categories (Table A1.7, site 16). Compared with the pooled assemblage, Maly Yaloman tooth-bearing parts, undamaged bones, and mostly whole bones are more common. Long bone fragments are less common. Compared with Razboinich'ya, Maly Yaloman underwent much less damage.

**8 Color.** Most (96.1%) of the 178 Maly Yaloman pieces are ivory in color, with only 3.9% being brown (Table A1.8, site 16). There are no black pieces. These values are greater for ivory, and less for brown compared with the pooled assemblage. They are similar to the color frequencies of Razboinich'ya.

**9 Preservation.** Quality of preservation is quite high in the 178 pieces of the Maly Yaloman assemblage (Table A1.9, site 16). Ivory (96.6%) pieces far outnumber chalky (1.7%) or intermediate (1.7%) conditions. The pooled assemblage has less ivory and more chalky representation. Razboinich'ya is much like Maly Yaloman.

**10 Perimortem breakage.** Three-quarters (74.7%) of the 178 Maly Yaloman pieces exhibit perimortem breakage (Table A1.10, site 16). This is somewhat less than what turns up in the pooled assemblage (84.4%), even more so compared with Razboinich'ya (92.4%).

**11 Postmortem breakage.** There is very little postmortem breakage in the Maly Yaloman assemblage (178 pieces, 4.5%) (Table A1.11, site 16). This is about the same as the average for the pooled assemblage, and less than what occurs in Razboinich'ya. This hints at the possibility that hyenas introduced less surface-scavenged bone into Maly Yaloman that at Razboinich'ya.

**12 End-hollowing.** Only 116 pieces could be assessed for end-hollowing (Table A1.12, site 16). It was found in 9.5% of the set, the value of which is almost identical with the average for the pooled assemblage and nearly the same as found at Razboinich'ya.

**13 Notching.** All 178 Maly Yaloman pieces could be assessed for notching (Table A1.13, site 16). Nearly one-quarter (23.0%) had notching. Of these, the most frequent number of notches per piece was one (10.1%), three (5.1%), and two (4.5%). There is more notching in the Maly Yaloman assemblage than in the pooled assemblage average, but almost exactly the same amount (23.6%) as found in Razboinich'ya.

**14 Tooth scratches.** Of 175 Maly Yaloman pieces, 28.6% exhibit tooth scratching (Table A1.14, site 16). These occur in roughly the same frequency from one to more than seven scratches per piece. There is a higher frequency of scratches in Maly Yaloman than in the average of the pooled assemblage (20.6%), but less than in Razboinich'ya (37.1%).

**15 Tooth dints.** As above, we have 175 pieces, of which 32.6% have tooth dints (Table A1.15, site 16). This is more than the average for the pooled assemblage. The most frequent number of dints in Maly Yaloman is more than seven (10.2%). Razboinich'ya has more pieces with tooth dints (52.8%).

**16 Pseudo-cuts.** A total of 6.3% of 174 pieces have pseudo-cuts (Table Al.16, site 16). The most frequent number per piece is one (2.3%). The occurrence of Maly Yaloman pseudo-cut pieces is nearly like the average for the pooled assemblage. Razboinich'ya has fewer pieces with pseudo-cuts (3.1%).

**17 Abrasions.** Abrasions occurred in only 1.2% of 173 Maly Yaloman pieces (Table A1.17, site 16). This is almost identical to the pooled assemblage average, and only slightly more than was found in Razboinich'ya (0.5%).

**18 Polishing.** Polishing is very common in Maly Yaloman, there being 89.3% in 178 pieces (Table A1.18, site 16). Polishing occurs on the end (12.9%), middle (1.1%), and both end and middle sections (75.3%). Compared with the pooled assemblage average, Maly Yaloman has about three times more polished pieces. It has more than was found at Razboinich'ya (59.4%).

**19 Embedded fragments.** This feature could be assessed in 177 Maly Yaloman pieces (Table A1.19, site 16). A total of 2.3% had embedded fragments. This is less than the pooled assemblage average. It is about three times less than occurs in Razboinich'ya (8.8%).

**20 Tooth wear.** Tooth wear was scored in 34 Maly Yaloman pieces (Table A1.20, site 16). Of these, 11.8% were judged to be young individuals. This is much less than the pooled assemblage average and Razboinich'ya (36.0%).

**21 Acid erosion (Figs. 3.85–3.86).** Nearly half (48.0%) of 177 Maly Yaloman pieces are acid eroded (Table A1.21, site 16). This is many times more frequent than the pooled assemblage average and the value obtained for Razboinich'ya (7.3%). Even had no hyena skeletal elements been found in Maly Yaloman, the very high frequency of acid erosion (the highest in our entire study) would still signal considerable use of the cave by hyenas.

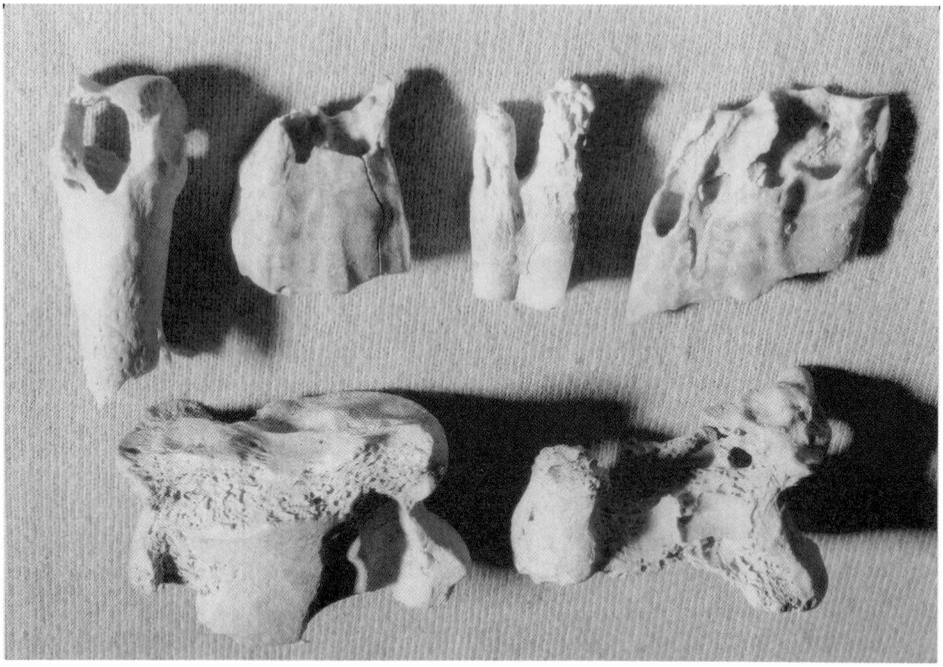

**Fig. 3.85** Maly Yaloman hyena digestive damage. Pieces of *Capra* teeth, toes, astragalus, and uncertain skeletal element that had been partly dissolved by hyena gut secretions (CGT neg. IAE 8-5-03:22).

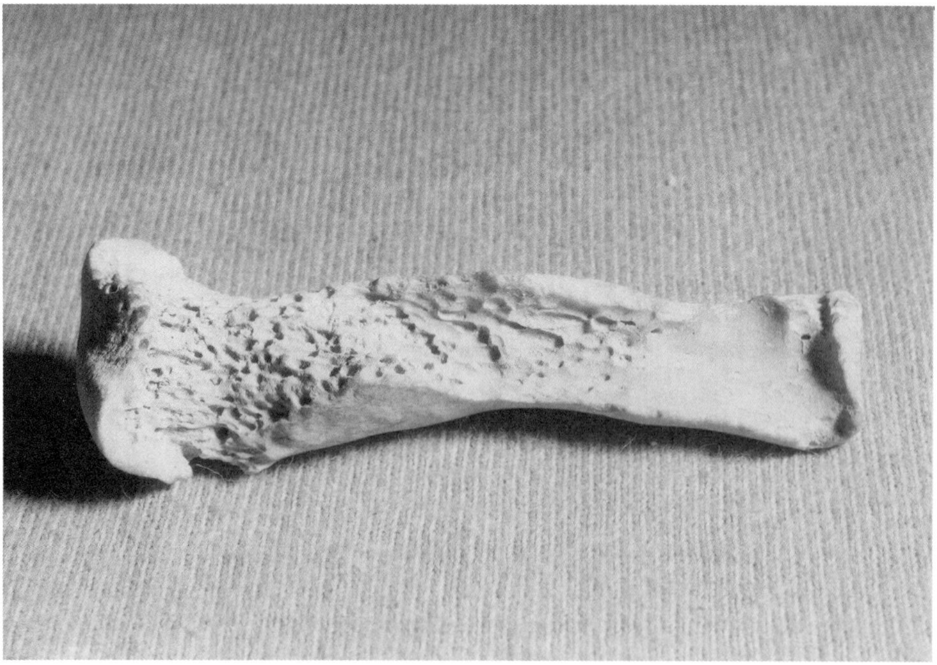

**Fig. 3.86** Maly Yaloman hyena digestive damage. *Capra* phalanx, 5.2 cm in length (CGT neg. IAE 8-5-03:17).

Fig. 3.87 Maly Yaloman bone damage. *Capra* mandible fragment with chewing and polishing, even on roots. Maximum diameter is 4.6 cm (CGT neg. IAE 8-5-03:15).

**22–23 Rodent gnawing and insect damage.** No signs of either form of damage were found.

**24 Human bone.** A human tooth was reportedly found, but lost sometime before our study.

**25 Cut marks.** Of 174 Maly Yaloman pieces, 2.3% have cut marks (Table A1.25, site 16). This is less than the pooled assemblage average. Razboinich'ya has no cut marks. One Maly Yaloman piece has about 200 very fine cut marks. One central portion of a yak pelvis recovered in 1988 that Ovodov had stored in his Krasnoyarsk apartment, which we examined in August 2006 and therefore not included in the sample discussed herein, had heavy chewing damage on the remaining mesial and distal ends and very fine cut marks near the acetabular region. The authors were unable to identify the sequence of human and carnivore damage to this yak pelvis because no superpositioning of cuts and tooth scratches could be recognized.

**26 Chop marks.** Chopping is very rare (0.6%) in our 174 Maly Yaloman pieces (Table A1.26, site 16). This rare occurrence is less than the pooled assemblage average. As with cut marks, Razboinich'ya has no chopped pieces.

## Discussion

While stone artifacts were found at Maly Yaloman, making it by definition an archaeological site, the presence of hyena and wolf skeletal elements documents the use of the cave by these species of carnivores. Moreover, the perimortem taphonomy of the faunal remains shows quite clearly that the cave was not used exclusively by humans. There is sufficient similarity between the carnivore bone damage characteristics of Maly Yaloman and Razboinich'ya to infer a greater overall hyena presence than that of humans (Fig. 3.87). This taphonomic inference is supported by the low amount of cut and chopped pieces of bone, as well as by the absence of black (burned) colored bone. We suspect that most of the faunal remains were carried into the cave by hyenas and other carnivores, not humans, regardless of the multiple hearths that burned at differing times.

# 17 Mokhovo Mine 1

## Background

Mokhovo Mine 1 (meaning of name uncertain, possibly "moss," mokh) is located near the Tom River in the Kemerovo coal-mining region of southern West Siberia. Kuzmin and Orlova (1998:6) place it at 54°40′ N, 86°60′ E. The site contains artifacts and broken bone 38 m deep in Middle Pleistocene deposits that are overlain by younger units, including the Kedrov geological suite (Derevianko *et al.* 1992). It was discovered by geologists S. V. Nikolaev and A. N. Zudin, who have spent many years studying the region's Quaternary deposits (Nikolaev and Markin 1990). Paleozoic rock exposed by the "old Kedrov" river may have attracted Middle Pleistocene people to the locality for its resource of stone for tool production. Additional artifacts were recovered *in situ* by Sergey Markin in 1998 (Derevianko *et al.* 1992, Zudin *et al.* 1983). It had been an open site consisting of mainly bison and human components, which together suggest it had been a kill site, or a location where boggy conditions contributed to the death of various large animals. Some deposits of faunal remains in the Kuznetsk Basin coal-mining district where Mokhovo is located are 1.8 million years old. I. V. Foronova (2000:137, 2001b) refers to Mokhovo as a "premousterian" site that is "probably one of the earliest in Western Siberia." She also reports that damage caused by water transport and tumbling is minimal and carnivore damage is abundant in these deposits (Foronova 2001a). She has identified the trogontherii elephant (*Mammuthus trogontherii*), wolverine (*Gulo cf. dchlosseri*), horse (*Equus mosbachensis*), reindeer (*Rangifer* sp.), and bison (*Bison aff. priscus*). Bone preservation is typical for the Kedrov suite. They are dark brown in color due to the high iron content of the soil, and somewhat mineralized, with some pieces having caliche-like adhesions. Orlova attributes some bone breakage to human butchering and breakage activity. She emphasizes that two fragments of an elephant long bone have obvious signs of processing. Kuzmin and Orlova (1998:6) list Mokhovo Mine 1 as having a carbon-14 date of 30 330 ± 445 BP. There is some confusion as to how the carbon-14 date fits into the middle Pleistocene context (Derevianko *et al.* 1990).

## Findings

**1 Provenience.** Only 12 pieces of bone were studied, making this one of our smallest assemblages. It is only 0.1% of the 8813-piece grand total (Table A1.1, site 17).

Foronova (2000:137) reported that 19 pieces were found, and on the basis of the morphology of the mammoth tooth parts she suggested a date of Mindel-Riss time, that is 350 000+ BP.

**2 Species.** Half of the 12 pieces are bison, and half are indeterminable (Table A1.2, site 17). Foronova (2000:137) reported the same ratio, but with the addition of fragments of mammoth tooth. Foronova classified the bison as *Bison priscus*.

**3 Skeletal elements.** A bison tooth, two ribs, a long bone, two metapodials, humerus, femur, and four unknowns make up the 12-piece Mokhovo Mine 1 assemblage (Table A1.3, site 17).

**4 Age.** There are no sub-adults represented in the 12 pieces.

**5 Completeness.** There are no whole bones; five pieces have one end; and seven have no anatomical ends (Table A1.5, site 17).

**6 Maximum size.** The mean is 10.2 cm, and the range is 6.2 cm to 21.0 cm for the 12 Mokhovo Mine 1 pieces (Table A1.6, site 17).

**7 Damage shape.** There is one tooth fragment, one long bone flake, nine long bone fragments, and one piece of a proximal rib (Table A1.7, site 17).

**8 Color.** All 12 of the Mokhovo Mine 1 pieces are ivory colored (Table A1.8, site 17).

**9 Preservation.** All 12 of the Mokhovo Mine 1 pieces are ivory hard (Table A1.9, site 17).

**10 Perimortem breakage.** Seven of the eight Mokhovo Mine 1 pieces that could be scored have perimortem breakage (Table A1.10, site 17).

**11 Postmortem breakage.** Seven of the 11 Mokhovo Mine 1 pieces that could be scored have postmortem breakage (Table A1.11, site 17).

**12–13 End-hollowing and notching.** There are no examples of these two variables in the 12 Mokhovo Mine 1 pieces.

**14 Tooth scratches.** There is one Mokhovo Mine 1 piece (the bison tooth) that is scratched. This piece has three tooth scratches (Table A1.14, site 17).

**15 Tooth dints.** The same bison tooth showing scratches also has ten tooth dints (Table A1.15, site 17). Foronova (2000:138) reports that seven pieces have "dints," but she does not define what she means by the term.

**16–17 Pseudo-cuts and abrasions.** Neither of these considerations appears in the 12 Mokhovo Mine 1 pieces (Table A1.16, site 17–A1.17, site 17).

**18 Polishing.** Four Mokhovo Mine 1 pieces have end-polishing, and six have end-middle (Table A1.18, site 17).

**19–24 Embedded fragments, tooth wear, acid erosion, rodent gnawing, insect damage, and human bone.** There are no examples of these variables in the 12 Mokhovo Mine 1 pieces.

**25 Cut marks.** There is one out of the 12 Mokhovo Mine 1 pieces that has apparent cutting in addition to perimortem breakage. This 12.1 cm long bison femur fragment has five cut marks on the distal articular surface. Each cut in the nested set is less than 5 mm long and each is very faint (Table A1.25, site 17).

**26 Chop marks.** There are no chop marks on any of the 12 Mokhovo Mine 1 pieces.

## Discussion

This extremely small bone sample, of what might represent a bison kill, surprisingly provided useful evidence – for both carnivore and human perimortem processing. The carnivore damage is slight, in line with our thinking about medium-sized animals such as wolves. The human processing (cutting) frequency of 8.3% is almost the same as that of the pooled assemblage average (7.6%). Such a nice fit could easily make one become a believer in the statistical reliability of small samples. The one bison femur fragment with cut marks and perimortem breakage helps support the presumed association of faunal remains and stone artifacts – a remarkable association in light of Foronova's 350 000 BP age estimate based on the evolution of proboscidean tooth morphology.

# 18 Nizhneudinskaya Cave

## Background

History does not record why or how the scholar Erik Laxman was drawn to the very isolated Nizhneudinskaya Cave in 1788, nor why he prepared a map of it. Located in the rugged foothills of the eastern Sayany Mountains, 264 m above the steep right bank of the Uda River, the cave is at 54°28′30″ N, 98°58′48″ E, approximately 50 km south of the town of Nizhneudinsk (the source of the cave's name). The town is located on the old east–west road that first crossed all Siberia, its paved modern counterpart, and the Trans-Siberian railway about half way between Krasnoyarsk and Irkutsk, the distance between which is 1050 km (630 miles). We mention these details to emphasize how very remote the cave was in 1788. In 1875 the paleontologist I. D. Chersky (1876) made a small excavation in the cave and recovered several bones of Pleistocene mammals. A distinctive feature of his assemblage was the retention of soft tissue on some bones – and even pieces of mummified rhinoceros skin – due to the dryness and low temperature of the cave. Unfortunately, his collection was destroyed by fire before it could be described when the wooden building that housed the Irkutsk Museum burned down in 1879. Nizhneudinskaya Cave was revisited in 1930 and 1980 by Moscow paleontologists M. G. Prokhorov, E. A. Belyaeva, and V. S. Slodkevich (1936), and the Irkutsk geologist, V. M. Filippov. The cave has two main areas (Malaya Nizhneudinskaya and a branch called Devil's Hole that is a natural trap) and two major strata: a lower one that consists of dense layered clay sediments devoid of organic inclusions; and an upper one that was rich in Pleistocene mammalian remains. Nizhneudinskaya is strictly a paleontological site, as no evidence of human use has been reported by any of its investigators (Melkheyev 1965).

The largest number of bones collected in the cave belonged to brown bears. Chersky calculated that he had bones of no fewer than 50 individual bears. In 1930 at least 28 more individuals were added to the list. Other species that have been identified include: hare (*Lepus timidus*), gray wolf (*Canis lupus*), fox (*Vulpes vulpes*), red wolf (*Cuon alpinus*), wolverine (*Gulo gulo*), sable (*Martis zibellina*), roe deer (*Capreolus capreolus*), saiga (*Saiga tatarica*), goat, ibex, and markhor (*Capra siberica*). Mummification followed the death of a wolverine and a sable. There are, in addition, remains of northern pika, shrew-mouse, voles, and bats – all of which, like the two mummies, probably died of natural deaths in the cave.

In the central part of the Uda River basin there seems to have been in late Pleistocene times a large bear population, or so it would seem from the number found in

Nizhneudinskaya Cave, whose deaths may have resulted from competition for hibernation sites and other causes (Ovodov 1970, 1977a). Conflict or cannibalism is reflected in carnivore damage to several of the disarticulated bear limb bones and lower jaws. It is puzzling as to why so many bears were attracted to the cave, since its temperature is always below freezing, while other caves in the area explored by the above scientists have above-freezing temperatures but lack bear remains.

## Findings

**1 Provenience.** Ovodov collected the animal remains from the floor levels of this near-vertical trap. He recovered 36 pieces, which make up only 0.4% of our 8813-piece grand total (Table A1.1, site 18).

**2 Species.** Of the 36 Nizhneudinskaya pieces, 94.4% are bear, and 5.6% are indeterminable. Presumably, the latter are also bear (Table A1.2, site 18). Compared with the pooled assemblage, the Nizhneudinskaya collection lacks representatives of all other species.

**3 Skeletal elements.** The most common elements in the 36 pieces are: mandible (13.9%), rib (19.4%), fibula (5.6%), toe (11.1%), scapula (5.6%), and humerus (19.4%) (Table A1.3, site 18).

**4 Age.** More than one-third (36.1%) of the 36 Nizhneudinskaya pieces are of sub-adults. This is the highest frequency of sub-adults in our study and one that is very difficult to explain (Table A1.4, site 18). Seemingly, bear cubs must have crawled into the animal trap opening, slipped into the lower chamber, and were unable to crawl back out, remaining there until their death by starvation. Compared with the pooled assemblage average, Nizhneudinskaya has almost five times greater sub-adult representation. Clearly, this is a unique Siberian paleontological site.

**5 Completeness.** Given the just-mentioned death scenario, it is somewhat surprising to find that the completeness profile is unexceptional. Of the 36 pieces, there are 8.3% whole, 30.6% with one anatomical end, and 61.1% with no anatomical ends (Table A1.5, site 18). This is very similar to the pooled assemblage averages. Since there is no evidence of human presence in Nizhneudinskaya, its ratio of 1:3:6 may well be the "natural" ratio of completeness.

**6 Maximum size.** The 36 pieces have a mean of 10.9 cm and a range of 4.7 cm to 28.5 cm (Table A1.6, site 18). Compared with the pooled assemblage mean and range, Nizhneudinskaya is again unexceptional. Looking at whole long bone lengths for bear provided by Vera Gromova (1950: table 27) shows that the upper range limit for Nizhneudinskaya fits well within her ranges for bear.

**7 Damage shape.** The most common damage forms in the 36 Nizhneudinskaya pieces are: long bone fragment (33.4%), medial rib (22.2%), tooth-bearing (13.9%), long bone flake (8.3%), undamaged (8.3%), and mostly whole (5.6%) (Table A1.7, site 18). Compared with the pooled assemblage averages, Nizhneudinskaya has more tooth-bearing pieces, long bone fragments, and medial ribs; and fewer long bone flakes and splinters and phalanx butts.

**8 Color.** Being a cave assemblage, it is not surprising that all 36 of the Nizhneudinskaya pieces are ivory colored (Table A1.8, site 18). Compared with the pooled assemblage, Nizhneudinskaya has more ivory colored pieces.

**9 Preservation.** Again, as a cave assemblage, all of the 36 Nizhneudinskaya pieces are ivory hard (Table A1.9, site 18). This is more than the pooled assemblage average.

**10 Perimortem breakage.** Almost all (94.4%) of the 36 Nizhneudinskaya pieces have perimortem breakage (Table A1.10, site 18). This is more than the pooled assemblage average.

**11 Postmortem breakage.** None of the 36 Nizhneudinskaya pieces has postmortem breakage (Table A1.11, site 18). This is considerably less than the pooled assemblage.

**12 End-hollowing.** There is a large amount of end-hollowing in the Nizhneudinskaya assemblage (44.4%; 16 / 36) (Table A1.12, site 18). Compared with the pooled assemblage, Nizhneudinskaya has almost five times more end-hollowing. The trapped and starving young bears must have been cannibalizing the dead or dying of their own kind.

**13 Notching.** The 36 Nizhneudinskaya pieces have a great deal of notching (41.7%), as we would predict for a carnivore presence. The number of notches ranges from one to five, with one being the most frequent number per piece (22.2%) (Table A1.13, site 18). Compared with the pooled assemblage, Nizhneudinskaya has more than twice the number of notched pieces.

**14 Tooth scratches.** Almost half (48.6%) of 35 Nizhneudinskaya pieces have tooth scratches. They range in number from one to more than seven scratches per piece, one being the most common number (11.4%) (Table A1.14, site 18). Compared with the pooled assemblage, Nizhneudinskaya has more than two times the number of scratched pieces.

**15 Tooth dints (Fig. 3.88).** More than half (69.4%) of the 36 Nizhneudinskaya pieces have tooth dints. The number of dints per piece ranges from one to more than seven, with the latter being the most common number (22.4%) (Table A1.15, site 18). Compared with the pooled assemblage, Nizhneudinskaya has more than twice the number of dinted pieces.

**16 Pseudo-cuts.** As we predict for a carnivore context, Nizhneudinskaya exhibits some pseudo-cuts (4 / 36; 11.1%). The number per piece is one (8.3%) or two (2.8%) (Table A1.16, site 18). Compared with the pooled assemblage, Nizhneudinskaya has more than twice the number of pieces with pseudo-cuts.

**17 Abrasions.** One of the highest frequencies (5.6%) of abrasions in this study occurs in the 36-piece Nizhneudinskaya collection. The two pieces each have more than seven abrasion striations (Table A1.17, site 18). They are more like tiny "mat burns" or a "skinned knee" than the slippage associated with deeper hammer stone and anvil abrasions. The two

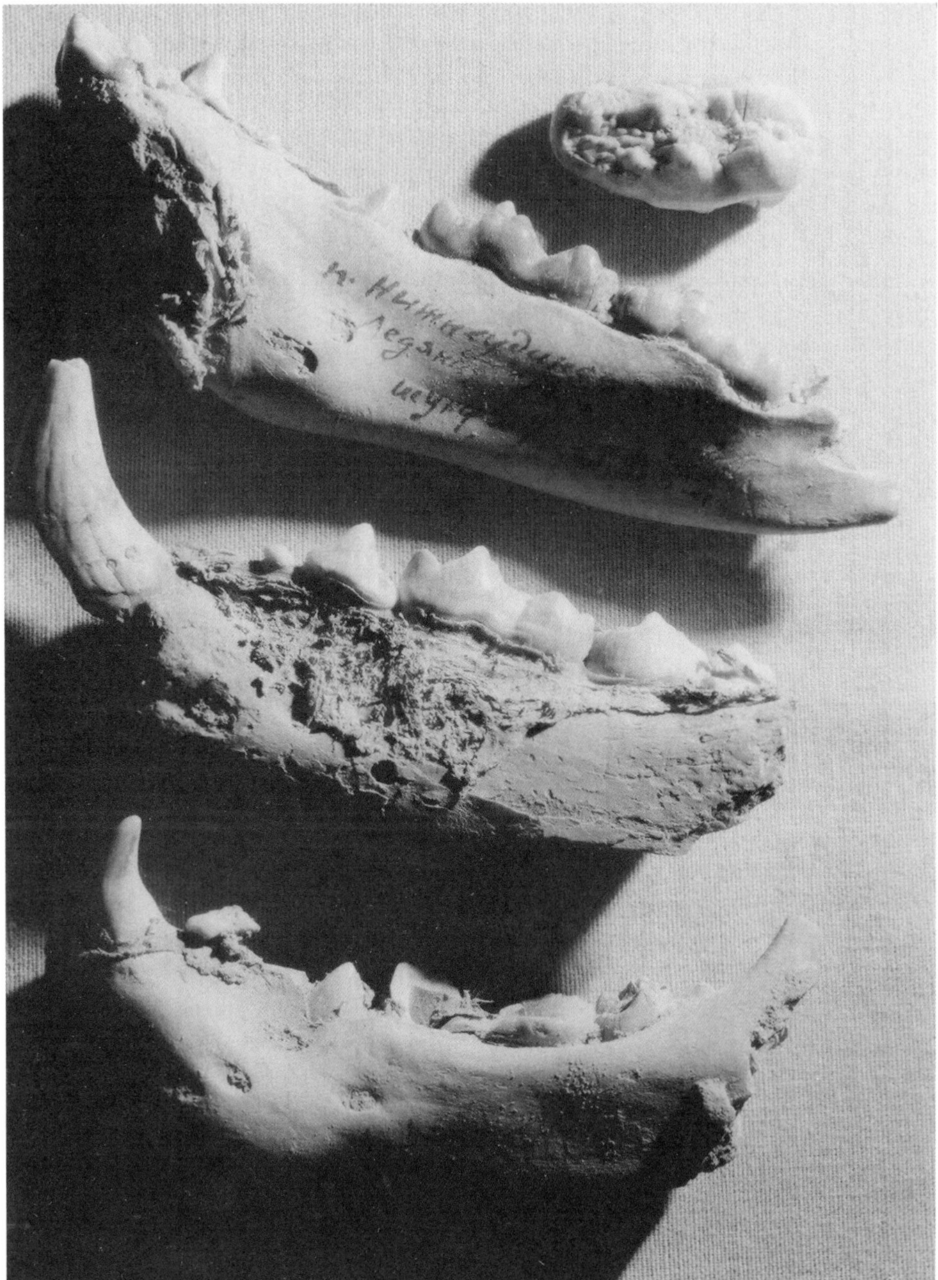

**Fig. 3.88** Nizhneudinskaya animal trap. Mandibles of sub-adult bears showing cannibalistic damage (CGT neg. IAE 7-5-01:25).

pieces must have grated upon a granular stone surface when accidentally stepped upon. Compared with the pooled assemblage average, Nizhneudinskaya has more pieces with abrasions.

**18 Polishing.** There is 72.2% polishing in the 36 Nizhneudinskaya pieces. By location the amount is: end (61.1%), middle (0.0%), and end-middle (11.1%) (Table A1.18, site 18). Compared with the pooled assemblage, Nizhneudinskaya has about the same amount of polishing.

**19 Embedded fragments.** The 36 Nizhneudinskaya pieces have the highest frequency (33.3%) of embedding in this study, a fact supportive of our view that embedding is strongly correlated with carnivore bone processing. The number of embedded fragments per piece ranges from one to more than seven, with one and three both having the highest frequency (11.1%) (Table A1.19, site 18). Compared with the pooled assemblage, Nizhneudinskaya has many more embedded pieces.

**20 Tooth wear.** All six of the Nizhneudinskaya tooth-bearing pieces have very little crown wear, which we believe indicates young animals (Table A1.20, site 18).

**21–26 Acid erosion, rodent gnawing, insect damage, human bone, cut marks, and chop marks.** None of these variables are present in the 36-piece Nizhneudinskaya collection.

## Discussion

Nizhneudinskaya Cave contained a pure paleontological bone assemblage that resulted from young bears slipping into this animal trap and not being able to get out. Their remains show perimortem damage caused by the animals trying to stay alive by cannibalizing the remains of other young bears that fell in at the same or some earlier time. Hence, like Razboinich'ya, Nizhneudinskaya provides a pristine perimortem signature of carnivore damage, which in this case was due to the feeding efforts of young bears rather than hyenas. The amount and types of processing damage are similar in the two assemblages with two exceptions. First, the Nizhneudinskaya damage was qualitatively less severe even if quantitatively similar. We can only illustrate the difference in photographs since we were never able to develop practical scales for destructiveness within a variable other than number (e.g., number of dints per piece). Second, Nizhneudinskaya has no acid-eroded pieces. While this absence may have been somehow related to the captive state of the young bears, we prefer to look upon acid erosion as caused primarily by hyenas. Thus, Nizhneudinskaya, despite its small sample size, is a powerful contributor to identifying the damage signature of Pleistocene and Holocene Siberian carnivores.

# 19 Okladnikov Cave

## Background

Okladnikov Cave is located in the northern foothills of the Altai Mountains at approximately 650 m above sea level. The south-facing limestone cave is about 14 m above a side valley meadow floor of the Sibiryachikha River (Fig. 3.89), a tributary of the Anui River. The cave is about 1 km from a small village named Sibiryachika (Fig. 3.90). Kuzmin and Orlova (1998:6) place the cave at 51°40′ N, 84°20′ E. The Russia–China border is relatively close, only some 60 km due south. Originally named after the just-mentioned nearby village, the site was renamed by Academician Anatoly P. Derevianko in honor of his predecessor, the first director of the now-named Institute of Archaeology and Ethnography in Academgorodok, who also possessed the title Academician. That some structure or archaeological monument should be named after Okladnikov goes without saying, given his distinguished career. This career has been masterfully described by A. K. Konopatsky (2001, 2009), and in less detail but equal respect by R. S. Vasilievsky (1999), although as will be seen in the following text, perhaps this particular cave is not the most appropriate one to bear his name.

The cave was first examined in 1984 by Derevianko and Yaroslav I. Molodin. Test excavation began the same year and continued until 1987, first under the field supervision of V. T. Petrin, who worked mainly in the entrance area, followed by Serghei V. Markin, who carried out the interior investigation. The two-branched, tube-shaped cave has a cap-like overhang entrance ledge that protects the nearly level entrance stone bench that is 8 m wide, 2 m high, and 4.2 m deep. The narrow, low, pitch-black tunnels extend into the hill for about 35 m. Compared with the grotto entry, the rest of the cave looks like a cold, wet burrow. Additional information is readily available in English (Derev'anko *et al.* 1998). On the basis of our personal experience in the cave, we regard these tunnels as uninhabitable, and whatever cultural remains found in them resulted from animal bioturbation of the outside living area.

One hundred meters away and slightly higher on the hill is Sibiryachikha VI, an animal cave in which a child's humerus was found by Ovodov in 1985. He found it with Pleistocene faunal remains during his test excavation of this small cave. In Ovodov's view this cave would have served better for human use than Okladnikov Cave, but he encountered no evidence of human occupation. In August 2006 we examined this sub-adult humerus shaft, which shows evidence of small carnivore or vole chewing. A small

Fig. 3.89 Okladnikov Cave. Archaeological field camp. The south-facing cave entrance is visible in the distant limestone outcrop in the upper right of the photograph. Left to right: Jacqueline Turner, Olga Pavlova, crew members, and Sergei Markin at far right. The road from Denisova camp to Okladnikov camp was very muddy, as the Soviet-era jeep shows (CGT color Okladnikov 6-13-87:32).

portion of the distal end remained unchewed. It is kept temporarily in Ovodov's Krasnoyarsk apartment and home office, a not uncommon practice since Russian scholars do much of their work and writing at home. Ovodov had originally identified in 1985 the child's upper arm bone among the animal remains he found in Sibiryachikha VI. The present authors, having no experience with morphology or age-related robusticity in fossil human osteological remains, are unable to say whether it represents an anatomically modern human or a more archaic variety, nor can we identify precise age or sex, although we can say that there was no obvious pathology such as osteomyolitis or fracture. As for finding human remains in Okladnikov Cave, Ovodov related in August 2006 a curious story about human bone fragments and teeth from this site that have created so much controversy that we review the discussion at the end of this section.

One summer during the period that Petrin was field supervisor for the Okladnikov Cave excavations, ten large bags, 25 kg each or more, of washed bone were trucked back to IAE for identification. Ovodov had time to examine the contents of only one bag that winter, and in the faunal content he identified ten pieces of human bone and teeth. These obviously had not been recognized *in situ* during excavation, which suggests that there is a strong possibility that the other nine large bags of bone contained additional human material. The really curious part of the story is that a decision was made to take the bags of bone back

**Fig. 3.90** Okladnikov Cave entrance. Jacqueline Turner (left) and Olga Pavlova stand on the 40 m long "cap" area that makes up the cave's ledge entrance. Sibiryachikha village is in the distance at the upper right (CGT color Okladnikov 6-13-87:14).

to the Altai field camp and study them there the following summer. At some point in time after they arrived at the institute, all the large bags of bone were misplaced, forgotten, or lost. The amount of missing bone must have numbered in the thousands of pieces, as it is common for a small 20 × 30 cm plastic bag to contain several hundred small pieces. This loss is, of course, significant to late Pleistocene Siberian paleontology and physical anthropology, especially in light of the controversy stirred up by the two Okladnikov teeth that were saved by Ovodov's careful examination. We emphasize *careful* because in the million or so pieces that we have examined during this project, we never found a piece of human bone or tooth fragment that had been missed by Ovodov.

As elsewhere in the Altai, thin forests of conifer and birch today blanket the upper hillside slopes near Okladnikov Cave. The foothills and valley bottoms are carpeted with grasses and numerous species of annual and perennial flowering plants, neatly trimmed by the grazing of the gentle village dairy cattle. Logging for village house and barn construction has increased the area of meadow plant growth. With help from Molodin, the senior author, along with his late wife, Jacqueline, and Olga Pavlova, made a one-day visit to Okladnikov Cave on June 13, 1987. Markin was the field supervisor, and he led the senior author into the low tube-like, narrow depths of the ring gallery to where a final small shallow excavation was being completed. The very narrow, low ceiling, barrow-like damp and pitch-black condition of this part of the cave seemed a most unlikely place for human occupation. The small amount of archaeological stone and bone debris then being recovered seemed at the time to have probably been shifted by animal activity from its original place of deposit nearer the entrance. There were abundant carnivore tooth marks on the bone fragments. This bioturbation by cave-using animals would be seen repeatedly in other Siberian cave sites. Refuse concentration indicates that most of the human use of the cave was immediately outside, on the ledge (called the "cap") of the protective rock shelter from which one enters the cave proper. Nearly 4000 stone items (tools and debitage) have been recovered from this ledge area and the talus slope immediately below. Various analyses of the Okladnikov Cave finds have been done by Ovodov (1987b), Derevianko and Markin (1992), Derevianko et al. (1998a), Ivleva (1990), Kulik and Markin (2003), Martynovich (1998), and others. No differences in tool types could be recognized in the various Pleistocene stratigraphic levels. One item of concern not mentioned in these reports is the fact that the large (25 kg or more) bags of faunal remains mentioned above were never studied. The few human bones and teeth identified by Ovodov were transferred to T. Chiksheva, who in turn loaned them to Academician V. P. Alexeev, Moscow. By a clandestine arrangement, the senior author, accompanied in 1987 by A. K. Konopatski and Jacqueline Turner, briefly examined in Alexeev's Moscow apartment the teeth from this collection, leaving them in the possession of Alexeev's wife, Tatyana Alexeeva. Following the death of Alexeev, the Okladnikov Cave human remains have been transferred to a German laboratory for DNA analysis, and at some point in time the teeth were examined by E. Sphakova. The paper trail for the transfer and storage of these few bones and teeth would not be acceptable in modern forensic practice. We have not been able to locate any photographs or catalog numbers of these Okladnikov Cave bones and teeth. As of this writing, their location is unknown.

Vasil'ev *et al.* (2002:521) inventoried the published carbon-14 dates for the Mousterian assemblage at Okladnikov Cave (Layers 1–3). The range is from >16 210 BP to 43 300 BP. These extreme values are based on bone specimens from Layer 3. There is a Layer 1 date of 33 500 BP. The senior author carried back to the United States a midshaft segment of a bovid leg bone excavated by Markin from near the bottom of the Okladnikov Cave midden. On his July 1987 return to the United States, Turner requested a uranium series date of the bone by the US Geological Survey, Menlo Park, California. This was generously provided by J. L. Bischoff. The date Bischoff obtained was ca. 33 500 BP, well in line with the carbon-14 dates from Okladnikov Cave, published 11 years later by Derevianko *et al.* (1998i), and with other radiocarbon dates for the late Pleistocene occupation of the Altai. Vasili'ev *et al.* identify no Upper Paleolithic dates for Okladnikov Cave. Temporally and typologically (dental morphology) the humans who used the cave were Mousterian culture-bearing Neanderthal-like people.

## Findings

**1 Provenience.** Table A1.1 (site 19) shows that we examined a total of 168 pieces, making the Okladnikov assemblage 1.9% of our total study of 8813 pieces. Our sampling of the Okladnikov faunal collection was carried out in 1999, 2000, and 2003 at the Institute of Archaeology and Ethnography, Novosibirsk. A large portion of our sample has no identifiable-level provenience because of the way it was curated after excavation.

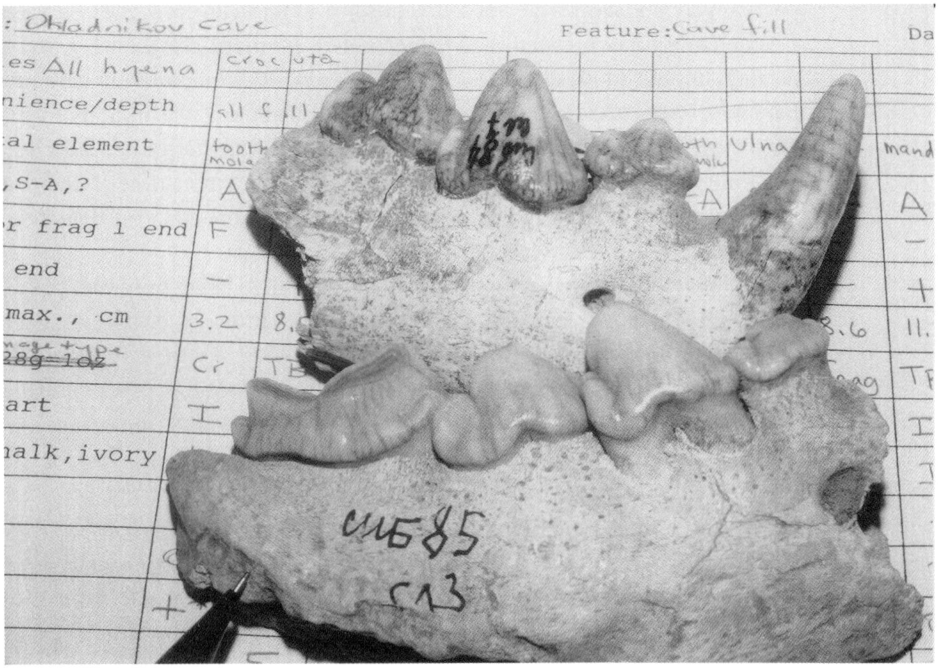


Fig. 3.91 Okladnikov Cave hyenas. Chewed young adult hyena mandible fragments from Layer 3 (1985). The pencil points to a tooth notch and polishing. The lower piece is 10.5 cm in length. The site was originally named Sibiryachikha after the nearby village. Okladnikov never had anything to do with this site (CGT color Krasnoyarsk 8-3-00:3).

As will be seen, everything we had read in the many archaeological publications dealing with Okladnikov Cave failed to prepare us for the abundant evidence of bone damage by large carnivores, even though the senior author had personally seen carnivore-damaged bone *in situ* in 1987. Hyenas seemingly used the cave as much if not more so than did the late Pleistocene Mousterian people.

**2 Species.** The identified bones belonged to animals widely spread in Late Pleistocene timesare: marmot, gray wolf, fox, bear, cave hyena (Fig. 3.91), horse, wooly rhinoceros, reindeer, wild sheep, and bison. There are a few bones that belonged to beaver, red wolf, and cave lion.

Table A1.2 (site 19) shows the near-dozen species that Ovodov could identify in 168 pieces studied in this project. It should be noted that the bone fragments used in this study are generally smaller and much less complete that those used for paleontological species identification. Goat-sheep (16.7%), cave hyena (14.3%), and big mammal (10.1%) are the most frequent groups identified. Indeterminate was our largest class (36.3%). Each of these is more common compared with our pooled assemblage averages. Okladnikov Cave has the highest frequency of rhinoceros of all of our assemblages, and its frequency of hyena pieces slightly exceeds even that found at Razboinich'ya (11.7%). Derevianko and Markin (1992:83) report that hyena remains were recovered from Levels 1–6, whereas bear and wolf were not found in some of these levels. Level 7 had only horse and rhinoceros bone (we found also *Ovis*/*Capra*). None of our Level 7 had cut marks or burning, but had tooth scratches, dints, and other signs of carnivore damage. We suggest that Level 7 also had hyena presence. Hence, Okladnikov was used by hyenas for most if not all of its stratigraphic history. Although the Mousterian stone artifacts found at Okladnikov Cave (Derevianko and Markin 1992) demonstrate without question that people used the shelter, the continuous presence of hyenas also demonstrates without question that the human occupancy was discontinuous.

The zooarchaeological remains are numerous, there being minimally 6773 elements of large- and medium-sized mammals representing minimally 20 species, each of which is widely distributed in the late Pleistocene of the Altai region. Ovodov and Martynovich (2004) have identified hare (*Lepus timidus*), cape hare (*Lepus capensis*), marmot (*Marmota baibacina*), beaver (*Castor fiber*), gray wolf (*Canis lupus*), fox (*Vulpes vulpes*), red wolf (*Cuon alpinus*), brown bear (*Ursus arctos*), badger (*Meles meles*), cave lion (*Panthera spelaea*), cave hyena (*Crocuta spelaea*), horse (*Equus cf. caballus*), rhinoceros (*Coelodonta antiquitatis*), mammut (*Mammonteus primigenius*), red deer (*Cervus laphus*), roe deer (*Capreolus capreolus*), reindeer (*Rangifer tarandus*), wild sheep (*Ovis ammon*), goat, ibex, markhor (*Capra sibirica*), and bison (*Bison proscus*). One unexpected faunal identification for Okladnikov Cave was a giant Pleistocene frog (Gutieva and Chkhikvadze 1990).

**3 Skeletal elements.** Of 168 pieces, the most common Okladnikov skeletal elements are long bone (32.1%) and mandible (8.9%) (Table A1.3, site 19). Unknown elements make up 16.1%. The near-absence of teeth in Table A1.3 is due to our not studying loose teeth unless they evidenced perimortem damage. However, we are unable to explain the near-absence of skull elements as is the case with several of our assemblages. Compared with our pooled assemblage averages, most elements occur in similar frequencies. Although Okladnikov has fewer ribs and more long bone pieces, Okladnikov and Razboinich'ya are remarkably similar in their frequencies of nearly all skeletal elements, so much so that we propose that Okladnikov Cave was occupied as frequently by hyenas as by humans.

**4 Age.** Of 168 Okladnikov pieces, 8.3% are sub-adult, and 42.9% are adult (Table A1.4, site 19). The remainder are of questionable or unknown age. The Okladnikov sub-adult frequency is nearly identical with our pooled assemblage average, but the Okladnikov adult frequency is less. Age representation at Okladnikov and Razboinich'ya are quite similar.

**5 Completeness.** Of 168 Okladnikov pieces, 70.8% have no anatomical ends, 23.8% possess one end, and 5.4% are more-or-less complete (Table A1.5, site 19). These values differ somewhat from the averages of our pooled assemblage, indicating greater perimortem damage in the Okladnikov sample. However, the Okladnikov frequencies are rather like those of Razboinich'ya.

**6 Maximum length.** We measured 180 Okladnikov pieces, obtaining a mean of 7.9 cm and a range of 2.1 cm to 26.9 cm (Table A1.6, site 19). These and the other statistical values are quite similar to those of our pooled assemblage. Compared with Razboinich'ya, the Okladnikov statistics indicate slightly more size variability, even though the means are very similar. Looking at the Okladnikov upper range limit for goat-sheep and hyena and the undamaged bone lengths for these species provided by Vera Gromova (1950: table 27) shows considerable similarity, although the lower limits are very different.

**7 Damage shape.** We classified damage form in 655 pieces from Okladnikov Cave (Table A1.7, site 19). The sample for this variable is not comparable because it had been pulled out of the larger sample for identifying species.

**8 Color.** Most of the 168 Okladnikov pieces are ivory colored (90.5%; Table A1.8, site 19). The 6.5% pieces of light brown color are most likely soil-stained. They came from various layers, both shallow and deep, so color here does not correspond well with time. The 2.4% black colored pieces unquestionably had been burned, as probably was a single black-brown specimen. All of the black or black-brown pieces have perimortem breakage but no sign of carnivore damage. The black specimens include: (1) a 2.9 cm long adult? sheep-goat phalanx from Layer 3; (2) a 5.2 cm long adult?, species?, long bone fragment from Layer 2; (3) a 2.1 cm long adult?, species?, vertebral fragment from Layer 2; (4) a 2.3 cm long adult, species?, long bone fragment from Layer 2. The probably burned brown-black piece is an 8.2 cm long adult?, "big" species, long bone fragment from Layer 3. These five pieces are useful evidence for fire, but not cooking. Because of their small size, we consider them to have burned accidentally at different times by either falling or being discarded in a fire hearth, or having a fire started over them while they were surface trash.

The color ratios are similar to our pooled assemblage values, with Okladnikov having almost exactly the same frequency of ivory colored pieces. Okladnikov differs from Razboinich'ya only in its having black (burned) pieces, which are absent in the latter site.

**9 Preservation.** Ivory-like hardness characterizes the preservation of the 168 Okladnikov Cave pieces (73.2%; Table A1.9, site 19). The 3:1 ratio of ivory to chalky conditions seems reasonable for a site whose excavation occurred within and outside the protective limestone entrance overhang and interior cave walls. Fully open sites such as Kara-Bom and Ust-Kova have more chalky pieces, while cave interiors such as Ust-Kan have fewer. Compared with our pooled assemblage averages, Okladnikov has a slightly higher percentage of chalky pieces. Compared with Razboinich'ya, Okladnikov has more chalky pieces. Presumably this means that Okladnikov had a weather-exposed bone-littered fronting talus slope.

**10 Perimortem breakage.** Almost 95% of the 168 pieces have perimortem breakage (Table A1.10, site 19). Compared with the Shestakovo and Volchiya Griva open mammoth sites with their low 18.7% and 8.0% perimortem breakage, respectively, a lot more human and carnivore processing transpired at Okladnikov Cave. If we assume that perimortem breakage was done by carnivores when the combination of tooth dints, scratches, end-hollowing, or pseudo-cuts also occurs on a piece, then the amount of carnivore bone processing at Okladnikov Cave is almost twice the amount

(64.6%; 102 / 158 pieces) attributable to humans (35.4%; 56 / 158). This is especially so if we assume that pieces with perimortem breakage, but lacking carnivore damage, resulted from human processing. What's more, there is carnivore tooth damage on most of the pieces lacking perimortem breakage. Compared with our pooled assemblage average, Okladnikov has more perimortem breakage, but nearly the same as Razboinich'ya.

The taphonomic inference from the Okladnikov Cave perimortem damage is that the site had an extensive late Pleistocene carnivore presence. Since hyenas are the only carnivores directly identified in our Okladnikov sample, we assume that they did much of the perimortem bone damage. A further implication is, of course, that hyenas were also responsible for the introduction of some portion of the animal remains found in the site. We feel certain that the Okladnikov Cave faunal remains do not represent only the results of human hunting activity. In fact, we suspect that possibly the majority of the animals represented in our sample were brought to the cave by hyenas and probably by other carnivores not identified in our bone and tooth sample, such as cave lions and wolves. This raises a number of questions about site occupancy, including seasonal use by humans and hyenas, periods of human absence from the locality or region, and possible competition between humans and hyenas for use of the cave site and the small valley that it overlooks.

**11 Postmortem breakage.** Of 168 Okladnikov pieces, 22% have postmortem breakage (Table A1.11, site 19.) This is not much more than our pooled assemblage average (17.8%), which we see as indicating nothing out of the ordinary for the amount of Okladnikov postmortem breakage. Okladnikov has about twice the amount of postmortem breakage compared with Razboinich'ya (10.9%), again not exceptional in our view, given the talus slope context and the rather unmindful method used to transport faunal remains from this site.

**12 End-hollowing.** Of 166 Okladnikov pieces, 17.5% possess end-hollowing (Table A1.12, site 19). This value is almost twice that of our pooled assemblage average, and more than twice that seen in Razboinich'ya (8.1%). We consider end-hollowing to be a robust indicator of carnivore presence based on its extensive documentation in the taphonomic, forensic, and bioarchaeological literature (Lyman 1994), and on our own casual observations of domestic ranch dog and coyote bone chewing. The Okladnikov Cave sample has 29 examples of end-hollowing. This number includes three adult hyena molar and premolar teeth whose roots have been chewed off – a condition we include with end-hollowing because of its damage-type similarity. The frequency of end-hollowing would most likely be greater if more ends of long bones had remained. Most pieces of bone with end-hollowing also have tooth dints, tooth scratches, or pseudo-cuts, or all three conditions (96.6%; 28 / 29 pieces). Only one (3.4%) end-hollowed piece has no dints, scratches, or pseudo-cuts.

**13 Notching.** Fracture plane notching occurs in 39.2% of our 168 pieces (Table A1.13, site 19). The number of notches ranges from one (13.2%) to as many as 15 per piece (0.6%). One to two notches is most common. Compared with our pooled assemblage average, Okladnikov has more than two times as many notched pieces. It even has more than Razboinich'ya (23.6%).

In our pilot study of the Razboinich'ya hyena cave assemblage, we were quickly struck by the frequent occurrence of notching. These numerous pieces suggested that notching was an important characteristic of the powerful bone-cracking capability of these animals. Relative to the very small amount of carnivore damage that can be identified in our two paleontological mammoth sites (Shestakovo and Volchiya Griva), both of which have almost no notching, we still maintain this view. For Okladnikov Cave, the relationship between notching and undoubted tooth damage (dints and scratches) is strong. Of pieces with notching, 96.6% (56 / 58) also have tooth dints, scratches, pseudo-cuts, and/or end-hollowing. Only 3.4% (2 / 58) of the notched pieces lack dints, etc. Like Razboinich'ya, Okladnikov Cave has so much notching at the edges of fracture planes that we cannot help but infer an extensive hyena presence. We believe that when a bone assemblage has high frequencies of tooth dints, tooth scratches, end-hollowing, and fracture plane notching, that a very strong case exists for carnivore perimortem processing. This is what we conclude occurred at Okladnikov Cave. While numerous stone tools and two human teeth prove without any doubt that humans used the cave in late Pleistocene times, perimortem bone taphonomy shows that carnivores used the cave as well, and possibly even more so than the humans.

**14 Tooth scratches.** More than half of our 168 Okladnikov pieces have tooth scratches (57.0%; Table A1.14, site 19). The number of scratches per piece ranges from 1 to 105, with higher numbers being 25, 35, 44, and 55. The most frequent number of scratches per piece is more than seven (23.0%), followed by four (9.1%), two (5.5%) and one (5.5%). As might be expected, more scratches are on larger than on smaller pieces. The specimen with 105 scratches was found in Layer 6. It is a 21.5 cm long wedge-shaped flake of an adult? mammoth leg bone. It also has perimortem breakage, 30 tooth dints, two pseudo-cuts, and polishing all over the piece. It may have a chop mark. We noted at the time of study that it "could have been struck from the long bone shaft by humans, and later chewed on by hyenas." We have since come across flakes of similar size and form that have none of the more diagnostic signs of human processing (cutting, abrasions, etc.). Therefore, this large flake's entire suite of perimortem damage could have been produced solely by a hyena. Compared with our pooled assemblage average, Okladnikov has more than two times the number of pieces with scratches. It also exceeds Razboinich'ya (37.1%) in number of scratches.

**15 Tooth dints.** As with tooth scratches, the Okladnikov Cave sample has more than half of its 167 scorable pieces exhibiting tooth dints (55.7%; Table A1.15, site 19). The number of dinted pieces ranges from those with one (7.8%) to a piece with 70 (0.6%). The most frequent number of dints per piece is more than seven (20.4%), followed by two (9.6%), and one (7.8%). The fragment with 70 dints is a 6.5 cm long goat-sheep metapodial with one anatomical end. It also has end-hollowing, four notches, and six tooth scratches, but, unexpectedly, no definable polishing. Because it was found in Layer 1, the carnivore involved might have been a dog or wolf instead of the ever-suspected Pleistocene hyenas.

Compared with our pooled assemblage average, Okladnikov has more than twice the number of pieces with one or more dints. The Okladnikov dinted piece number is close to that of Razboinich'ya (52.8%).

**16 Pseudo-cuts.** Unlike other sites in this study lacking pseudo-cuts and lacking osteological or dental evidence of hyenas, the Okladnikov Cave assemblage has 12 examples in the 167 pieces (7.2%; Table A1.16, site 19). The number of pseudo-cuts per piece ranges from one (3.0%) to six (0.6%). There are six pieces with either one or two pseudo-cuts. Without exception they all occur on pieces that also have tooth dints, scratches, end-hollowing, or notches. One questionable pseudo-cut also occurs on a piece with tooth dints, notching, and end-hollowing.

Pseudo-cuts are so named because they look very similar to stone tool cut marks. However, at Okladnikov Cave there can be little doubt that pseudo-cuts were produced by carnivore teeth, probably teeth with small newly chipped, burinated cusp tips. When chewing on a bone, the carnivore's chipped tooth may accidentally produce a groove V-shaped in cross-section, not unlike that left by a sharp stone flake or burin.

Compared with our pooled assemblage average, Okladnikov has a slightly greater occurrence of pseudo-cut pieces. It has almost four times the number found at Razboinich'ya (3.1%). Two possibilities come to mind: (1) We have misidentified pieces with true cut marks as having pseudo-cut marks. (2) There was much more hyena and other carnivore activity at Okladnikov Cave than is reported in the literature of this well-studied site.

**17 Abrasions.** Abrasions are believed to be caused by the slippage of a bone being broken open on an abrasive anvil, or by a glancing blow by an abrasive hammer stone employed for the same purpose. There are only two Okladnikov pieces out of 167 that have abrasions (1.2%; Table A1.17, site 19). One has only one minute striation, the other has a set of more than 25 minute parallel striations. Each of these pieces also has tooth dints, and none has cut or chop marks that would favor inferring that these abrasions were produced by human bone processing. At Okladnikov Cave, the association between abrasions and carnivores is stronger than that between abrasions and human bone processing. Abrasions alone cannot be used to argue for their human origin at this site.

Compared with our pooled assemblage average, Okladnikov is nearly identical in its frequency of pieces with abrasions. Razboinich'ya, as expected, has fewer examples of abrasions (0.5%).

**18 Polishing.** Polishing is common in the Okladnikov Cave sample of 167 pieces (73.7%; Table A1.18, site 19). Because it is so strongly associated with other types of bone damage that were undoubtedly caused by carnivore chewing, polishing is considered to be another indication of carnivore presence. This is especially so when polishing occurs on perimortem fractures that are on the end(s) and middle of a bone fragment. There are only four pieces with middle- or middle-and-end polishing that lack carnivore damage. End-polishing alone may also be due to carnivore activity; however, there are more examples that lack such an association (32.3%; 10/31) than is the case for middle- and middle-and-end polishing. Thus, end-polishing alone is not as diagnostic of carnivore damage as are the other forms discussed.

Compared with our pooled assemblage average, polishing of end, middle, and both locations is slightly more common in the Okladnikov assemblage. It is considerably more (73.7%) in Okladnikov than in Razboinich'ya (59.4%).

**19 Embedded fragments.** There are only six instances of fragment embedding in 168 Okladnikov pieces (3.6%; Table A1.19, site 19). The number of embedded fragments per piece ranges from one to three. Each example is associated with some form or another of presumed carnivore damage. Only one piece with embedded fragments is associated with cut marks, but this piece also has tooth dints, scratches, and notching. Either we misidentified the cut marks, or it is a piece that was processed by both a human and a carnivore agent.

Our pooled assemblage average for notched pieces is nearly the same as Okladnikov. Notched pieces occur twice as often (8.8%) in Razboinich'ya.

**20 Tooth wear.** There are 18 Okladnikov teeth or tooth-bearing mandible fragments (Table A1.20, site 19). Three have no tooth wear (16.7%), nine have slight cusp wear but no dentine exposure (50.0%), and six cases have dentine exposure (33.3%). All but one of these teeth belonged to hyenas, including all three cases without wear. Adults (47.1%) and sub-adults (52.9%) are about equally divided among the hyena specimens, all of which have perimortem chewing damage done by presumably some other hyena. As we define age by tooth wear, two-thirds were young individuals.

Our pooled assemblage average has fewer young individuals than Okladnikov, a value nearly the same as in Razboinich'ya (36.0%). There is no severe tooth wear (grades 3 and 4) in Okladnikov or Razboinich'ya, and far fewer in our pooled assemblage.

**21 Acid erosion (Figs. 3.92–3.93).** We consider this form of perimortem damage to be strongly diagnostic of carnivore damage, especially by hyenas when the acid-eroded pieces are large and highly eroded, and when hyena remains are part of the faunal assemblage. Eight Okladnikov pieces out of 168 (4.8%; Table A1.21, site 19) are acid-eroded. Their maximum diameter ranges from 2.8 cm to 6.8 cm. One eroded piece is a tooth root of an adult animal of unknown species. Our sample contains eroded pieces from Layers 2, 3, and 6 – that is, throughout most of the late Pleistocene occupation of Okladnikov Cave.

While Okladnikov has somewhat less acid erosion than does our pooled assemblage or Razboinich'ya (7.3%), we nevertheless have no doubt about the cave having been used often, if not mostly, by hyenas.

**22 Rodent gnawing.** Okladnikov is one of our few sites with rodent gnawing. Two examples were identified in 1.2% of 168 pieces (Table A1.22, site 19). Both specimens came from Layer 3, and both were sheep-goat bones – a sub-adult cervical vertebra with no carnivore or human damage, and an adult scapula with extensive carnivore damage.

Compared with our pooled assemblage average, Okladnikov rodent damage is nearly the same, as is Razboinich'ya (0.8%). There are only eight sites with rodent gnawing, half of which are open sites, and half are cave sites.

**Fig. 3.92** Okladnikov Cave hyena stomach bones. A sample of partly digested bones and teeth found in Layer 6 (1984). Scale is 15 cm (CGT neg. IAE 7-22-99:2).

**Fig. 3.93** Okladnikov Cave stomach bone with cuts. This specimen from Layer 6 (1984) is of special interest. There is a small field of seven cut marks on this 3.1 cm long unidentifiable bone fragment that was also eroded and polished. The inference is not only that a hyena had scavenged refuse left in Okladnikov Cave by the Mousterian inhabitants, but that it and they were in temporal and spatial proximity with one another (CGT neg. IAE 7-22-99:9).

**23 Insect damage.** Like nearly all of our sites, the Okladnikov sample of 168 pieces has no damage that could be attributed to insects.

**24 Human remains.** No pieces of human teeth or bones were present in our study collection. However, as previously discussed, a few sub-adult bones and five teeth had been found before 1987 and reported on elsewhere (Alexeev 1998, Shpakova and Derevianko 2000, Turner 1990a, 1990b, 1990c, see discussion at end of this section).

Five teeth of Paleolithic people from Layers 2, 3, and 7 (Derevianko *et al.* 1998a; Turner 1990a, 1990b, 1990c) were found amid numerous remains of vertebrate animals in the loose cave sediments. Later, one tooth was lost (Shpakova 2001). It should be noted that factually there were essentially more of the human remains than the mentioned ones. In the catalog made by Ovodov after 1987 the following parts of the human skeleton were listed: (1) a fragment of humerus in Layer 1 under the cap; (2) two fragments of the cranial vault and a fragment of metapodia in the same layer but in the grotto; (3) a fragment of humerus bone, calcaneus bone, and two kneepans in Layer 2 under the cap; (4) a part of femur bone in Layer 3 under the cap; and (5) one tooth root in Layer 7 under the vault of room 1 (Ovodov and Martynovich n.d.).

In the cave uphill from Okladnikov cave, Ovodov made a small test excavation and among the faunal remains found a 15 cm long young human humerus of Pleistocene age without its distal epiphysis. Shallow dints left by teeth of a carnivore or perhaps voles are observable on the well-preserved surface.

**25 Cut marks.** Ten Okladnikov pieces out of 168 (6.0%) have cut marks (Table A1.25, site 19). The range is from two to more than 100 cut marks per piece. Most occur on bone fragments without carnivore damage, but four have both cutting and carnivore damage. The cut pieces in our sample originated from Layers 2 and 3.

Compared with our pooled assemblage average of, Okladnikov has a slightly lower frequency of pieces with cut marks. Razboinich'ya has no cutting.

**26 Chop marks.** Six Okladnikov pieces out of 167 (3.6%) have chop marks (Table A1.26, site 19). The range is 1–11 marks per piece. All of the pieces with chop marks also have carnivore damage. They were recovered from Layers 2 and 3.

Compared with our pooled assemblage average, Okladnikov has a slightly lower frequency of pieces with chop marks. Razboinich'ya has no chopping.

## Discussion

The perimortem damage identified at Okladnikov Cave has far more carnivore processing (Figs. 3.94–3.95) than anything we had been led to expect from published archaeological accounts of the site. Clearly, the cave had been used repeatedly for thousands of years by both humans (Fig. 3.96–3.100) and hyenas. The evidence for both comes from their individually distinctive refuse types, signature perimortem damage, and from the skeletal and dental remains of both species. However, the perimortem taphonomy leads us to believe that hyena use of Okladnikov Cave far exceeded that of humans. As for the

**Fig. 3.94** Okladnikov Cave chewed bone. A sample of carnivore-chewed bone fragments from Layer 6. The left piece is 3.2 cm long (CGT neg. IAE 7-22-99:15).

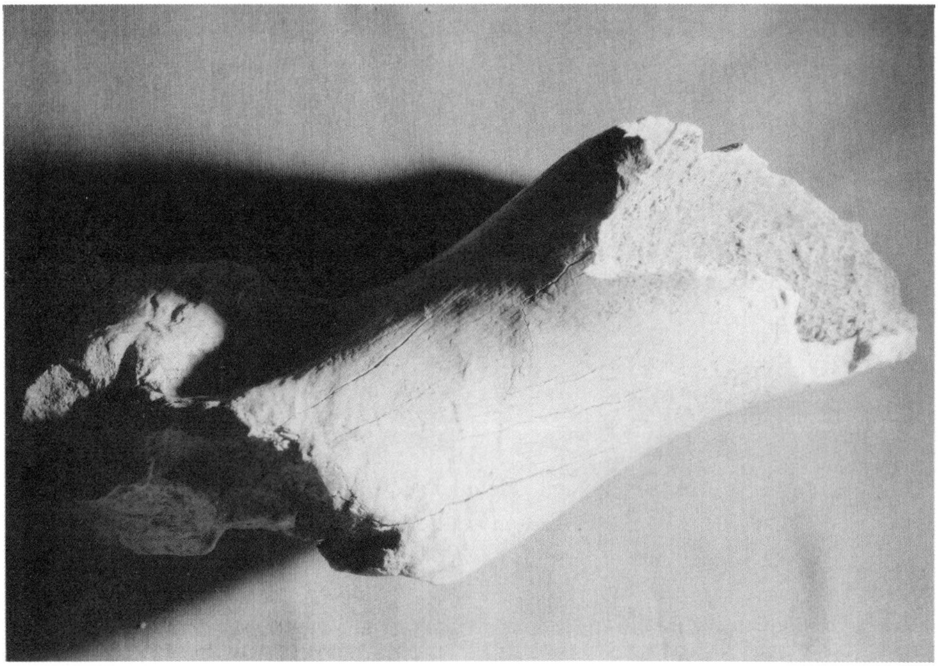

**Fig. 3.95** Okladnikov Cave chewed bone. A carnivore-chewed rhinoceros humerus with end-hollowing and tooth scratches from Layer 7. Specimen is 26.9 cm in length (CGT neg. IAE 7-20-99:36).

**Fig. 3.96** Okladnikov Cave stone artifacts. Mousterian scrapers. These and following artifact photographs illustrate the variability in cutting edges, with no identifiable temporal significance. Scale is 3.0 cm (CGT color IHPP 6-2-87:20).

**Fig. 3.97** Okladnikov Cave stone artifacts. Unifacial pointed objects. Scale is 3.0 cm (CGT color IHPP 6-2-87:23).

**Fig. 3.98** Okladnikov Cave screblos. These knife-like unifacial tools would have left highly variable cut marks, depending on how they were used. Scale is 3.0 cm (CGT color IHPP 6-2-87:22).

**Fig. 3.99** Okladnikov Cave scrapers. There is also a small bifacial tool in the upper left corner. Scale is 3.0 cm (CGT neg. IHPP 6-2-87:25).

**Fig. 3.100** Okladnikov Cave scrapers and points. These Mousterian tools are among the older specimens from the cave, coming from Layers 4–6. Scale is 3.0 cm (CGT color IHPP 6-2-87:18).

archaeological claim that Okladnikov Cave "was used as a hunters' site for long-term occupation" (Derevianko and Markin 1992:207), our perimortem analysis is unsupportive. One or two weeks a year, or perhaps each generation, seems much more likely if we assume that one or two bones per carcass will receive cut marks. If we assume more than two bones were accidentally cut, then the amount of time spent butchering out an animal might mean that the cave was used only a few visitations per century. On different grounds, Ovodov and Martynovich (2004:180) also found the previous quote to be unrealistic. They write:

> [we made] some statistical calculations, based on the time of accumulation of Pleistocene sediments in the cave and the number of tools and [stone] waste, we have come to a paradoxical conclusion: one flake was being produced each 6 years. As P. Volkov (the only specialist in Siberia in wear analysis) thinks, 5 minutes is enough [time] for making one screblo. . . . About 100 flakes would be produced at that. During three to four days of experiments in his laboratory P. Volkov has produced about 10 000 flakes. Most probably, Okadnikov Cave was seldom used by humans and for a short time only.

Ovodov and Martynovich made their calculation based on the recovered 3911 stone artifacts, of which 78% were flakes, as reported by Derevianko and Markin (1992). Boaz *et al.* (2000) reached a similar inference in their favoring of carnivores as the primary destruction agency for bones in the Lower Cave of Zhoukoudian, rather than the result of extensive hominid processing. Regarding a Middle Paleolithic German

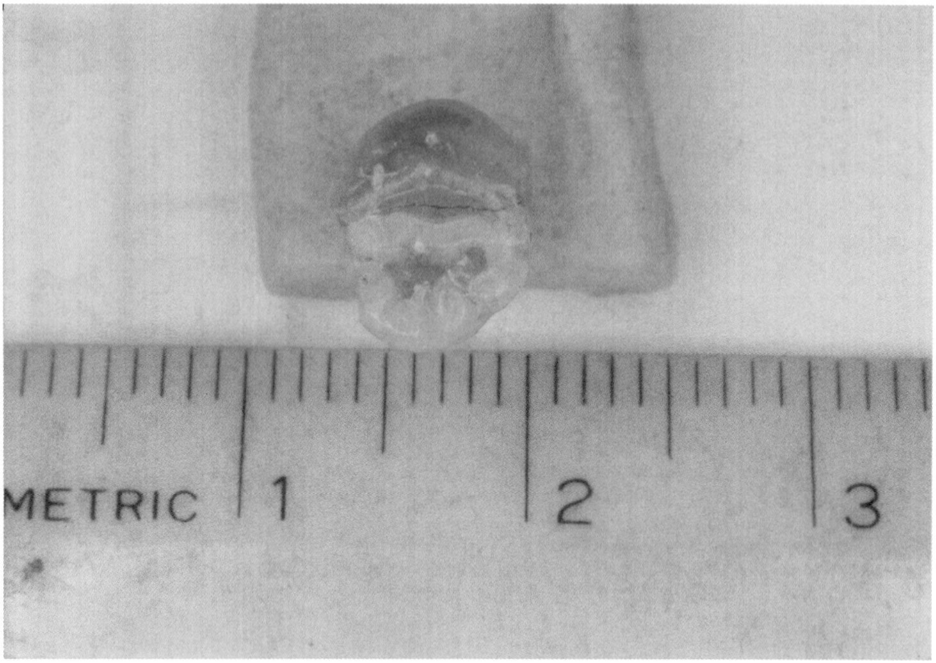


**Fig. 3.101** Okladnikov Cave human tooth. This adult lower left first premolar was found in Layer 3. It is associated with stone tools typologically identified as Mousterian (CGT color Moscow 6-25-87:17).

cave, Diedrich (2011) reviewed its bone accumulation characteristics resulting from hyenas and hominid occupation, an issue we have encountered with the accumulations in Okladnikov Cave.

At an international summer conference held in 2000, at N. I. Drozdov's Kurtak field camp near the Yenisei reservoir, Ovodov gave an unscheduled report on July 29 about some of his paleontological findings with emphasis on hyenas. Relative to our concern about hyena use of Okladnikov Cave, he determined that ca. 16% (944 / >6000) of the pieces of bone were hyena. Compared to a site he called "Hyena Den," where he found ca. 12% (95 / 800) hyena pieces, it would appear that hyenas were common occupants of Okladnikov Cave.

One of the mysteries of the late Pleistocene Siberian archaeological record, at least up to now, has been the nearly complete absence of human skeletal remains, especially remains of whole or nearly whole individuals (Fig. 3.101–3.104). The unique exception is the largely complete older child found by M. M. Gerasimov at Mal'ta. Despite many years of excavation, as well as the excavation of numerous late Pleistocene sites, Siberia has a strikingly sparse record of human skeletal recovery, particularly when compared with anywhere else in the Old World, including Australia and even bone-destructive tropical environments such as Southeast Asia. This discovery of a strong dual presence of humans and hyenas at Okladnidov Cave provides a basis for hypothesizing on empirical grounds as to why there is a near absence of Siberian Pleistocene human remains. If

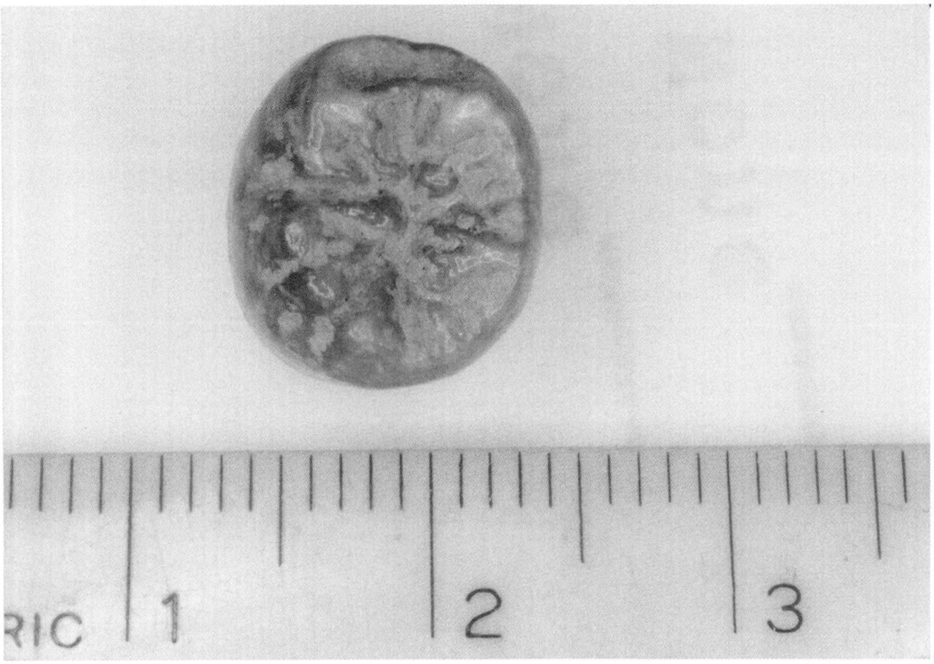


**Fig. 3.102** Okladnikov Cave human tooth. This tooth, found in Layer 2, is considered by the senior author to be a lower left third molar of an individual who was about 12 years old at the time of death. It has no wear. The question that arises from this unerupted tooth is: What happened to the rest of the child's body? Presumably the child died in the cave (CGT color Moscow 6-25-87:16).

**Fig. 3.103** Okladnikov Cave human tooth. This tooth is considered to also be a lower left third molar. It was found in Layer 3. It also has no wear (CGT color Moscow 6-25-87:12).

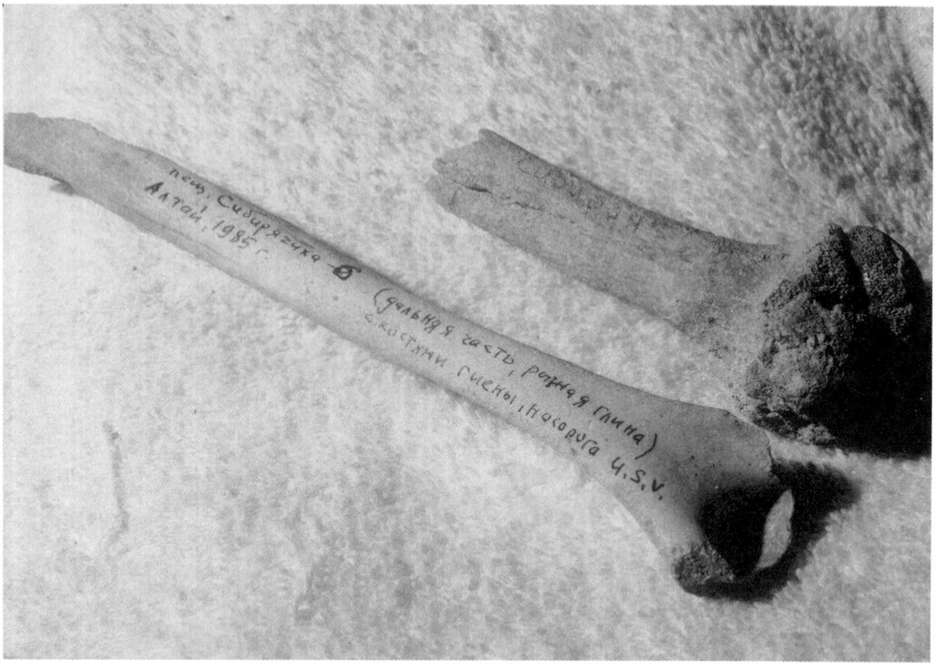

**Fig. 3.104** Okladnikov Cave human arm bones. One of the two humeri in this photograph came from Okladnikov Cave Layer 3 (the shorter piece). The other was found by Ovodov in an animal cave he tested near Okladnikov Cave. The provenience indicated on the animal cave bone reads: "Cave, Sibiryachikha-6, Altai, 1985, remote part, red clay, with bones of hyena, rhinoceros, usv." The humerus from the animal cave appears to be that of a child, around five years of age at the time of death. Ovodov suspects that both arm bones date around 30 000 BP (CGT color IAE 7-27-90:23).

humans had buried their dead in the caves or in the open, we propose it possible, if not likely, that neighborhood hyenas – with their strong scavenging instincts – dug up the buried human corpses, especially those individuals who died in the winter time and would therefore have received only a shallow burial or no burial at all due to frozen ground. Hyenas today dig up relatively shallow Bedouin burials in Israel and carry away body parts to their isolated den sites (Horwitz and Smith 1988). After the end of the Pleistocene, discoveries of Siberian human skeletons are not common, but their recovery is much more frequent than that of the late Pleistocene. Obviously, time and preservation contribute to the amount of human skeletal recovery, but possibly so also does the extinction of the Siberian cave hyena population.

Another thought triggered by the analysis of the Okladnikov assemblage has to do with seasonal or discontinuous use of the cave, since we seriously doubt that the evidence for the presence of both humans and hyenas means they lived together in Okladnikov Cave. Two lines of evidence seem most relevant. First, the small amount of burned bone argues against roasting of game and against a lot of accidental burning from fires kept going for human warmth all winter long. Second, the non-human teeth that were recovered with no wear or very little wear probably represent spring to fall births and deaths. Both humans and hyenas could have been the hunters or the scavengers. In either case, these game animals indicate one or the other species was responsible. The small amount of burned bone suggests humans were not using Okladnikov Cave in the winter, when camp fires would have been needed for warmth. We propose that the spring–fall kills occurred when humans would have been most likely to use the cave. Humans likely abandoned Okladnikov Cave during the winter in favor of forest camps, where fuel supplies for warmth would have been much easier to obtain. On the other hand, hyenas must have over-wintered in locations where they could obtain shelter from deep cold and lengthy winter storms. Our perimortem taphonomy suggests that Okladnikov Cave and other Siberian caves aided hyenas and other carnivores more than humans in the winter.

Finally, a few words on the controversy generated by the Okladnikov teeth, and those from Denisova Cave. As mentioned, the senior author had a very brief opportunity to examine these Altai teeth in Moscow in 1987. There was no comparative material available, neither modern or archaic humans. The few observations of crown non-metric traits that were possible followed the Arizona State University Dental Anthropology System (ASU-DAS) (Turner *et al.* 1991, Scott and Turner 1997). This system is concerned only with the permanent dentition of anatomically modern human populations. It does not consider deciduous teeth for various reasons (for discussion, see Scott and Turner 1997), and the standards were developed using dental casts and actual teeth obtained from around the world. In other words, ASU-DAS rigorously defines anatomically modern human dental variation. In Turner's examination of the Altai Pleistocene teeth, it was patently obvious that these teeth had morphology far outside the range of variation of modern human teeth. Moreover, Turner felt that the morphology was more like that seen in archaic humans of Europe than in the few later Pleistocene archaic teeth of East Asia. Hence, he concluded that the Altai teeth were more like those of Neanderthals than those of modern humans of both Europe and Asia (Turner 1990c:242). In 1998 a review paper by Alexeev was published wherein he concluded that the Altai teeth were archaic but race could not be identified. Then, Derevianko and Sphakova published a reaction to the Turner paper, claiming that his study was wrong – he misidentified a deciduous molar, and was incomplete because he did not include the traits used in the Zoubov dental system. It should be noted that the Zoubov system is used nowhere outside of Russia, a criticism inconsistent with world dental anthropology, whereas ASU-DAS is used widely around the world. As for the alleged misidentification of the deciduous teeth, even if true, it is irrelevant since ASU-DAS does not deal with deciduous teeth. This critical point eluded Derevianko (2005) in his page-long critique of Turner's observations of the Altai teeth (as much space as he devoted to criticizing Yu. Mochanov).

Why the criticism of Turner and the favoring of Sphakova's view that the Altai teeth were modern? Sphakova's experience in dental anthropology was not much more than that of a first-year graduate student. Derevianko (2005) provides the answer both directly and indirectly. Directly, he states explicitly that he believes in multi-regional evolution of modern humans. He rejects the out-of-Africa hypothesis that has become the major world view of modern human origins. He believes that Neanderthals were the ancestors of modern Europeans and Siberians. Why? We reject the possibility of racial prejudice in having a relatively recent African ancestor despite the strong anti-Black sentiment throughout Russia. We suspect that his acceptance of the multi-regional view of modern human origins is colored by the work of he and his associates in the Altai caves, where they have concluded that there is stratigraphic evidence for cultural continuity from the Middle Paleolithic to the Upper Paleolithic, i.e., from Mousterians to anatomically modern humans. They see no cultural break that might indicate incoming modern people to the Altai, bringing with them the hallmarks of Cro-Magnon culture – movable art, symbolic representations, bone tools, considerable formal burials, jewelry, and other cultural characteristics of modern humans. In our view, cultural continuity is like a null hypothesis – it cannot be demonstrated, only, as in geology, can discontinuity be demonstrated. Okladnikov Cave stratigraphy, while appearing as continuous, was probably blurred and homogenized by the bioturbation of hyenas, marmots, wolves, hares, and other burrowing and digging animals that used the cave and fronting talus slope. As will be discussed in Chapter 4, all of the Altai caves that Derevianko and associates – ignoring the open sites that hyenas would have quickly disturbed – use to argue for cultural continuity have a significant hyena/carnivore presence, meaning that bioturbation ruined the stratigraphic virginity. Had there been a major human biological and cultural intrusion into the Altai, or elsewhere in Siberia, it would never be detected given the overwhelming presence of cave hyenas.

So, who was right in the question of dental identification of late Pleistocene humans in the Altai? First, fewer than a dozen teeth is hardly what one needs to settle this question. But, putting sampling concerns aside, according to T. Chicksheva (personal communication, August 23, 2006), preliminary DNA analysis in Germany of the Altai material suggests a "mixed" combination of Mousterian and *Homo sapiens*. What this means is impossible to say since the terms were not defined. But, regardless of terminology, it certainly favors Turner's view that the Altai teeth were not those of anatomically modern humans. How this information will eventually relate to the issue of cultural continuity remains to be seen – hopefully before this book is finished.

In sum, we regard Okladnikov Cave as more of a paleontological site than an archaeological one. Indeed, humans did use this rock shelter on occasion, of that there is no question. These people possessed a Mousterian tool assemblage, and in the senior author's view, based on dental morphology, were more likely to have been closer to European Neanderthals than to anatomically modern humans. Because of the substantial use of this cave by hyenas and their likely bioturbation of stratigraphy, the cave and its cultural and physical anthropological contents is not a good resource for the theoretical discussion of cultural and biological continuity, or multi-regional evolution in Siberia. This is particularly so since it seems to lack an Upper Paleolithic component and dates. It is Ovodov's view, which we share, that the density of the Mousterian population in the Altai was less than that of cave hyenas and much less than that of a number of herbivores.

# 20 Proskuryakova Cave

## Background

Proskuryakova Cave (named by Ovodov after Pavel Stepanovich Proskuryakov, the first director of the Krasnoyarsk Historical Musuem and explorer of Khakasia caves) is located in the middle Yenisei River basin at 54°27′ N, 89°28′ E, in the foothills of the eastern slopes of Kuznek Ala-tau, above the Bely Iyus River. It is not far from the Malaya Seeya site, and 1 km upstream of Yefremkino, the administrative headquarters for the Shirinsky district of Khakasia. The triangular-shaped entry to the south-facing limestone cave is a nearly vertical 70 m above the Bely Iyus River. At the mouth, the cave is 3.5 m high and 4.2 m wide, and its back wall is about 14 m deep. Archival records show that cave sites in the Bely and Cherny Iyus River valleys were being noted as early as 1799. In 1888, these caves were investigated by P. S. Proskuryakov, an archaeologist from St. Petersburg University. He located 18 caves in the valley of the middle course of the Iyus River, and made test excavations in eight (Proskuryakov 1889, 1890), although it cannot be determined from his articles if he tested the cave that Ovodov named in Proskuryakov's honor.

Bones of extinct animals and stone blades on the floor of Proskuryakova Cave were first noticed by Novokuznetsk speleologists S. A. Rybakov and R. A. Tsetel. Learning of their discovery, Ovodov began excavations in 1975 with a team of workers for two field seasons, establishing that the five stone blades were stratigraphically younger than the time when the bone accumulations occurred (Okladnikov *et al.* 1975, Ovodov 1975). The 40 cm thick fragmented bone mass occurred at the border zone between wet and dry sections in the remote back part of the cave. We examined the collection in 1999 and visited the site on July 21, 2000.

Vasili'ev *et al.* (2002:521) list four carbon-14 dates based on bone that range from >40 000 BP to 40 770 BP. Lab reports given to Ovodov have dates of 46 000 BP based on rhinoceros bone (SO AN-848), and three dates based on bison bone of 40 690 ± 1150 (SO AN-1571), 40 595 ± 875 (SO AN-1518), and 40 770 ± 1075 (SO AN-1519).

Ovodov's team turned up remains of at least 47 species, of which there were 25 large- and medium-sized mammals based on 530 pieces. Some joint-sharing pieces were deposited together, indicating articulation was retained after discarding. Hence, the cave taphocenosis was primary, not secondary as at the Krasny Yar river deposits. The taphonomic agency for accumulation was cave hyena and other carnivore activity, as evidenced by 666 of 819 pieces

(81.3%) having some manner of carnivore damage (Ovodov *et al.* 1992a). Ovodov's identifications include: hare (*Lepus timidus*), marmot (*Marmota baibacina*), beaver (*Castor fiber*), gray wolf (*Canis lupus*), fox (*Vulpes vulpes*), brown bear (*Ursus arctos*), badger (*Meles meles*), wolverine (*Gulo gulo*), lynx (*Felis lynx*), ounce, snow leopard (*Uncia uncia*), cave lion (*Panthera spelaea*), cave hyena (*Crocuta spelaea*), Asiatic wild ass, onager (*Equus hemionus*), horse (*Equus cf. caballus*), rhinoceros (*Coelodonta antiquitatis*), mammut (*Mammonteus primigenius*), red deer (*Cervus elaphus*), moose, elk (*Alces alces*), saiga (*Saiga tatarica*), wild sheep (*Ovis ammon*), goat, ibex, markhor (*Capra sibirca*), bison (*Bison priscus*), and wild yak (*Poephagus baicalensis*).

## Findings

**1 Provenience.** Noted above. The 47 Proskuryakova pieces make up 0.5% of our 8813-piece grand total (Table A1.1, site 20).

**2 Species.** The only two groups in the Proskuryakova collection are hyena (51.1%) and saiga antelope (48.9%) (Table A1.2, site 20). Compared with the pooled assemblage averages, Proskuryakova exceeds in these two groups, and is missing all other groups.

**3 Skeletal elements.** The most common elements in the 47 Proskuryakova pieces are: vault (29.8%), antler-horn (25.5%) (Figs. 3.105–3.106), long bone (21.3%), humerus (8.5%), and radius (8.5%) (Table A1.3, site 20). Compared with the pooled assemblage averages, Proskuryakova has more pieces of vault, antler-horn, humerus, and radius; and fewer pieces of vertebra, rib, foot, metapodial, toe, scapula, femur, and unknown.

**4 Age.** Only 4.3% of the 47 Proskuryakova pieces are sub-adult (Table A1.4, site 20). Compared with the pooled assemblage average, Proskuryakova has a smaller representation of sub-adults.

**5 Completeness.** There are no whole pieces in the Proskuryakova assemblage, 55.3% have only one anatomical end, and 44.7% have no anatomical ends (Table A1.5, site 20). Compared with the pooled assemblage, Proskuryakova has a slightly higher frequency of possible ends (27.7%; 26 ends out of 94 possible) than does the pooled assemblage (20.4%; 3231 ends out of 15 798 possible), indicating slightly more completeness as assessed by this variable of "endedness."

**6 Maximum size.** The Proskuryakova mean is 11.9 cm, and the range is 7.1 cm to 18.8 cm (Table A1.6, site 20). The mean is slightly larger than that of the pooled assemblage, but the range is less. Looking at the whole long bone lengths of hyena and saiga provided by Vera Gromova (1950: table 27) shows that the Proskuryakova upper range limit is generally less than the lower range limit for hyena, but fits well with her values for saiga antelope.

**7 Damage shape.** This variable was not scored in 1999 for Proskuryakova.

**8 Color.** All 47 Proskuryakova pieces are ivory colored (Table A1.8, site 20). The pooled assemblage average is less (89.2%) because there are other colors present in the pooled group.

Fig. 3.105 Proskuryakova Cave, saiga antelope. Patterned carnivore damage to crania recovered in 1974. It was these specimens that when first seen in 1984 triggered the senior author's desire to learn more about how carnivores damaged bone. Maximum diameter of specimen in lower right is 15.0 cm (CGT neg. IAE 8-10-99:15).

**Fig. 3.106** Proskuryakova Cave, saiga antelope. Horn core stumps. The cranial vaults and most of horn cores have been chewed off. Whole horn core in the upper left is 15.0 cm in length (CGT neg. IAE 8-10-99:12).

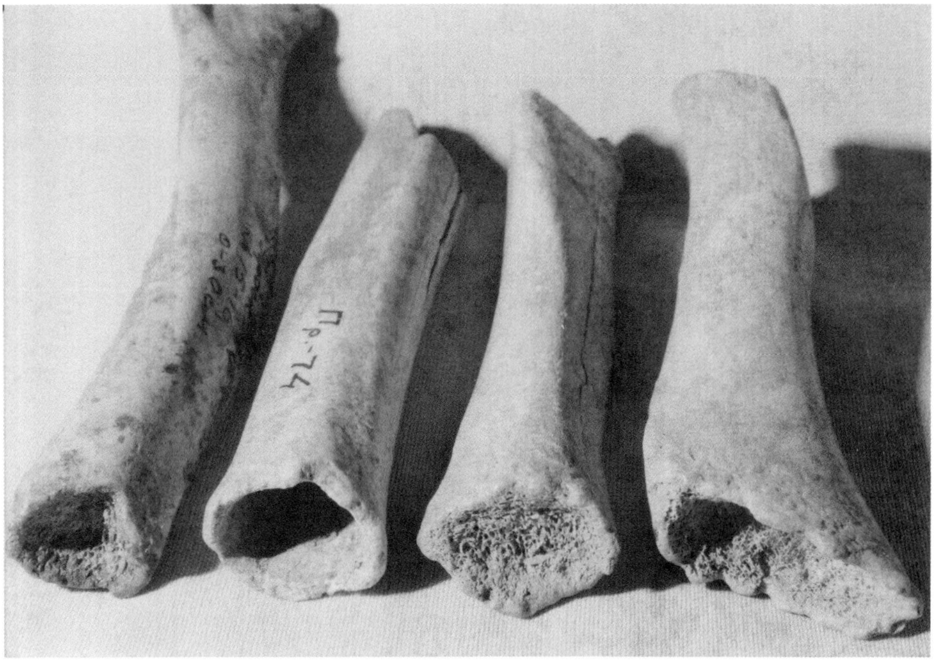

Fig. 3.107 Proskuryakova Cave, end-hollowing. These and most other bone refuse in this cave have one or several elements of the carnivore signature. Cross-sectional diameter of left specimen is 2.3 cm (CGT neg. IAE 8-10-99:20).

**9 Preservation.** There are four (8.5%) chalky specimens among the 47 Proskuryakova pieces (Table A1.9, site 20). This frequency is about half that of the pooled assemblage average.

**10 Perimortem breakage.** All of the 47 Proskuryakova pieces have perimortem breakage (Table A1.10, site 20). This frequency is greater than the pooled assemblage average.

**11 Postmortem breakage.** Six (12.8%) of the 47 Proskuryakova pieces have postmortem breakage (Table A1.11, site 20). This value is somewhat less that the pooled assemblage average.

**12 End-hollowing (Fig. 3.107).** This indicator of carnivore activity is common (29.8%) in the 47 Proskuryakova pieces (Table A1.12, site 20). End-hollowing is three times more frequent in Proskuryakova than in the pooled assemblage.

**13 Notching (Fig. 3.108).** Another carnivore indicator, notching is very frequent in Proskuryakova (40 / 47 pieces; 85.1%) (Table A1.13, site 20). The number of notches ranges from one to more than seven, with three and four notches tied for the most frequent per piece (19.1% each). Compared with the pooled assemblage average, Proskuryakova has five times more notching.

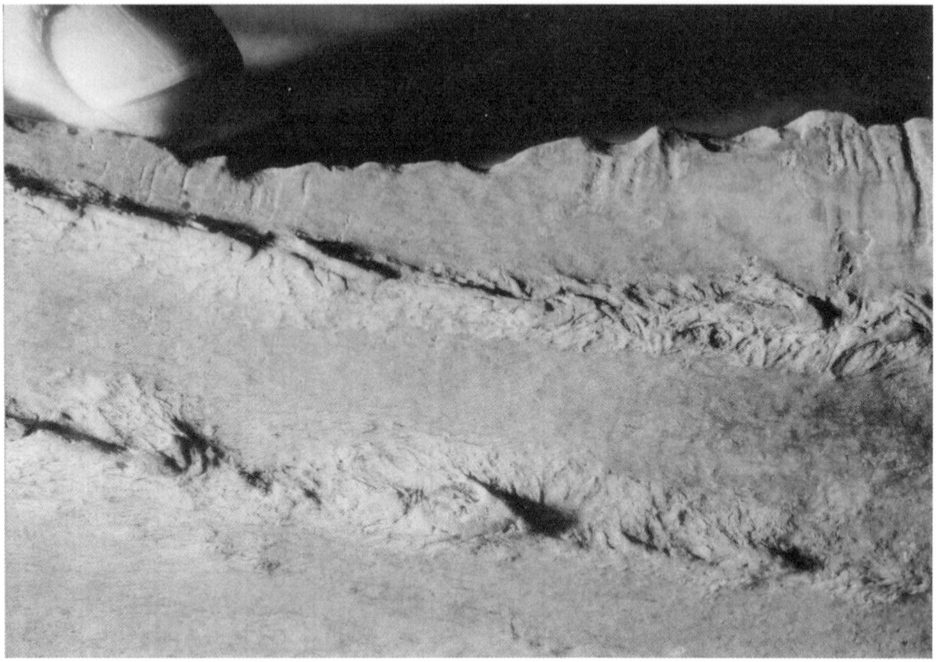

**Fig. 3.108** Proskuryakova Cave, notching. Rhinoceros tibia with human-like notching along edge of fracture. However, closely associated tooth scratches indicate that the notching was most likely caused by carnivore chewing (CGT neg. IAE 8-11-99:16).

**14 Tooth scratches.** Tooth scratches are nearly universal (97.9%) in the 47-piece Proskuryakova collection (Table A1.14, site 20). The number of scratches per piece ranges from one to more than seven. The vast majority (80.8%) has more than seven scratches per piece. Scratches are almost five times less common for the pooled assemblage average.

**15 Tooth dints.** Almost as frequent as scratching, tooth dints are very common (93.6%) in the 47 Proskuryakova pieces (Table A1.15, site 20). The number of dints per piece ranges from one to more than seven, with the latter being the most frequent (55.3%), as was the case with scratches. Dinted pieces are three times less common for the pooled assemblage average.

**16 Pseudo-cuts.** Four pseudo-cuts (8.5%) occur in the Proskuryakova collection.The number of pseudo-cuts per piece ranges from one to four (Table A1.16, site 20). This frequency is almost two times greater than the pooled assemblage average (4.8%). Thus, all of the carnivore indicators up to now occur in relatively or absolutely high frequencies, as would be predicted on the basis of the Razboinich'ya hyena cave baseline collection.

**17 Abrasions.** Three pieces with abrasions (6.4%) are present in the 47-piece Proskuryakova assemblage. The number of abrasion striations per piece ranges from one to three, much lower than is characteristic of abrasion sets made by hammer stone and anvil slippage (Table A1.17, site 20). While abraded pieces are uncommon in the Proskuryakova assemblage, they are even less common in the pooled assemblage.

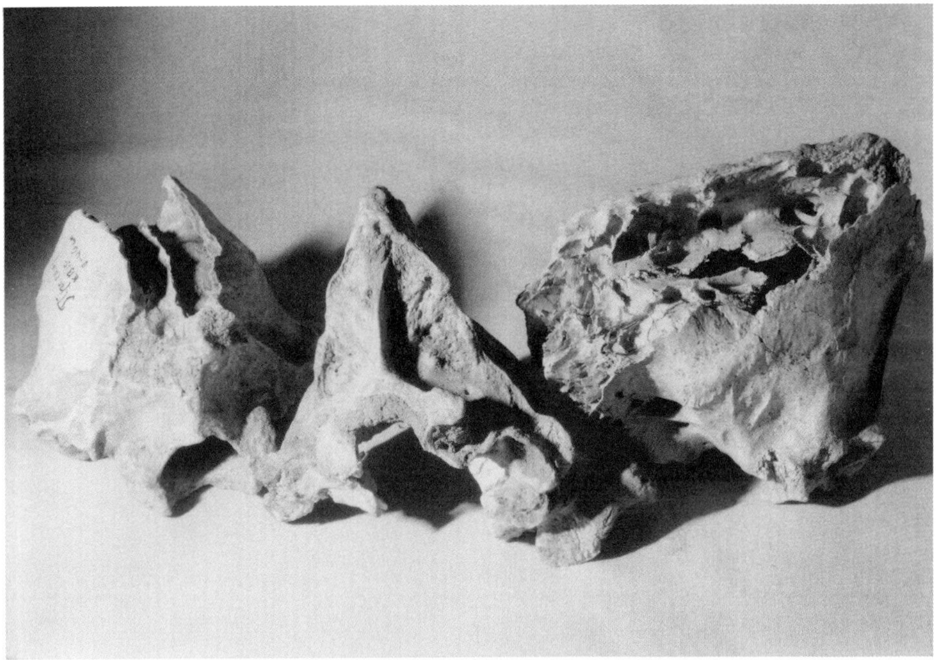

Fig. 3.109 Proskuryakova Cave hyenas. The actual presence of hyena bones as shown with these three cranial vault pieces suggests that most of the carnivore damage was done by these creatures. The left and middle specimens show the foramen magnum, the right specimen shows a crestless vault from the left. Middle specimen is 13.4 cm maximum width (CGT neg. IAE 8-10-99:8).

**18 Polishing.** Polishing is almost universal in the 47 Proskuryakova pieces (97.9%). It occurs on the end (6.4%) and end-middle (91.4%) (Table A1.18, site 20). Polishing is considerably more frequent in Proskuryakova than in the pooled assemblage.

**19 Embedded fragments.** A total of 6.4% of the 47 Proskuryakova pieces have 1–3 embedded fragments (Table A1.19, site 20). This is a little more than the pooled assemblage average, and a little less than the frequency in Razboinich'ya (8.8%).

**20 Tooth wear.** Proskuryakova has only one tooth to use for wear assessment. It belonged to a young animal (Table A1.20, site 20).

**21 Acid erosion.** It is not completely surprising that there are no examples of acid erosion in the Proskuryakova assemblage, despite the presence of hyena bones and the severe perimortem bone damage caused by these carnivores (Table A1.21, site 20). This is because acid-eroded pieces are generally smaller than the smallest pieces in this assemblage, whose incomplete content represents selection and curation primarily for species identification.

**22–26 Rodent gnawing, insect damage, human bone, cut marks, and chop marks.** There are no examples of these variables in the 47-piece Proskuryakova assemblage.

## Discussion

There are numerous statistical similarities between Proskuryakova, Razboinich'ya, and our other carnivore caves. The principle difference between Proskuryakova and the caves occupied mainly by hyenas (Dvuglaska, Maly Yaloman, Razboinich'ya, and Straschnaya) is the absence of acid-eroded pieces in Proskuryakova. We believe this absence was more likely caused by selective collection and curation than by any known taphonomic process, because of the direct evidence of hyena presence (Fig. 3.109).

# 21 Razboinich'ya Cave

## Background

Razboinich'ya Cave (Deserter's Hideaway) is a palentological site located in the southern Siberian Altai Mountains of northwestern Gorny Altai at 51°20′ N, 84°25′ E. The cave's name comes from the fact that three young men resisted being drafted into the Soviet army, running away and hiding in the cave entrance for a long time. All three were eventually found and taken back to Karakol, the village where they originated. One was killed while captive. The other two served in the army during World War II and eventually returned to Karakol (Galina Urusova, Karakol Village school teacher, July 10, 1999, personal communication). Razboinich'ya is relatively high for southern Siberia, about 1350 m above sea level, and situated on the top of a limestone ridge. We attempted to reach the cave in 1987 but were unsuccessful due to bad weather (Fig. 3.110). We were successful in 1999 (Figs. 3.111–3.114). The cave temperature stays near freezing year round. Razboinich'ya is higher and east of the regional administrative town named Ust-Kan. It is 30 km northwest of a large but poor farming village named Chorny Annui, and about 3 km west of, and at a higher elevation than, Kaminnaya Cave, an archaeological cave site situated near the valley floor of the Annui River (Ovodov 1991, 1997a, Ovodov *et al.* 1999). Both caves, Razboinich'ya and Kaminnaya, like many others in these pine-covered rugged limestone mountains, were formed long ago as underground water channels that subsequently dried up, formed an entrance, some time following episodes of mountain-building, and became habitable in the Pleistocene. The west-facing entrance to Razboinich'ya Cave is 5 m high, 2 m wide at the base, and rather hidden by a rocky cliff near the summit of the small Bashckelak Ridge, which is covered with mixed evergreen forest. The cave location is poorly situated for human use – not near water, difficult to reach, dark, cold, and without a warming southern exposure. The cave has a slightly sinuous descending profile whose main course extends about 90 m. There is one near-vertical 3 m drop about half way between the entrance and the cave's end near an enlarged 3 × 10 m chamber. The end is 19 m below the surface at the entrance. Razboinich'ya was filled with 2.5–3.0 m of uncompacted loamy sediments containing limestone fragments, stalactites, and many bones and teeth of mammals and some birds. From May to October the temperature inside the cave ranges from −1°C to 0°C. In addition

**Fig. 3.110** Odyssey. Razboinich'ya Cave: the failed 1987 visit. Heavy rain and deep, muddy tracks convinced our jeep driver that we six (seven counting Slava Molodin's cocker spaniel) should not delay returning to Karakol Village after we reached and lunched at Ovodov's log cabin field station on the mountainside between Kaminnaya and Razboinich'ya caves. In this image we had stopped to inspect a large Bronze Age burial mound called Peschyorkin Log, excavated in 1986 by Molodin and Japanese archaeologists. Afterward we scraped lots of mud off our shoes and boots. Note that the mixed pea and oat crop, winter feed for the local dairy herds, has only recently begun to sprout when this photograph was taken on June 12. Razboinich'ya Cave is in the distant mountains. Left to right: Olga Pavlova, Jacqueline Turner, Nicolai Ovodov, Slava Molodin (CGT color Karakol 6-12-87:3).

to the decomposing rock debris and faunal remains, large numbers of coprolites, particularly of hyenas and voles, were found by Ovodov and associates. A second layer, 20–40 cm beneath the pre-excavation cave floor, was carbon-14 dated at 14 500 BP. A fragment of a porcupine skull was directly AMS dated at 27 600 BP (Ovodov *et al.* 2010).

The excellent organic preservation of all the faunal remains is due to constant low temperature, dryness, total darkness, minimal disturbance, and the alkalinity of the cave, suggesting that the deposits probably began to accumulate around 40 000 to 50 000 years ago. Although the remains of other carnivores were found in the cave, the cave hyena was the most numerous (50%, MNI of 75 as of 1996 – see MNI below); fox was 20% and wolf, 12% (Ovodov *et al.* 1999:66).

There is no evidence of Pleistocene humans having ever *lived* in Razboinich'ya Cave (i.e., no hearths, stone or bone artifacts, or tool manufacturing debris, etc.). However, the

**Fig. 3.111** Razboinich'ya Cave. In 1999, hiking from Kaminnaya camp, we followed a steep trail through tick-infested brush for more than an hour to reach our destination. We rested and ate a lunch snack before entering the cave. The north-northwest-facing entrance can be seen in the center background. Left to right: Nicolai Ovodov, Olga Pavlova, student, and Nicolai Martynovich (CGT color Razboinich'ya 7-9-99:3).

cave may have served more than once as a sacred place or shrine for the Kaminnaya hunters living lower down the mountain, because Ovodov identified 41 very small patches of burning and a few burned bone fragments suggesting ritual ceremony. Oddly, this remarkable discovery of a possible Pleistocene Altai shrine that includes the recovery of a skull of an old dog has received very little professional or public attention until recently. The dog skull was found in association with the skulls of five wolves and four brown bears. Excluding this possible limited ceremonial use of the cave, its main inhabitants were various species of carnivores (hyena, brown bear, wolf, fox), various rodents, and bats. The combined identification efforts of Ovodov, N. V. Martynovich, L. L. Galkina, N. G. Ivleva, and A. Nadakhovsky turned up 50 species of mammals and 34 species of birds. Based on MNI, the Pleistocene animals that seemed to have used the cave the most were mice, moles, voles, pika, and other small mammals. The larger carnivores included the gray wolf (*Canis lupus*, 12%), fox (*Vulpes vulpes*, 20%), corsac fox (*Vulpes corsak*), red wolf (*Cuon alpinus*), brown bear (*Ursus arctos*, rare), cave lion (*Panthera spelaea*, rare); by bone count the most frequent was the cave hyena (*Crocuta spelaea*, 50%). In a more recent paper, Ovodov (1997a) added the Balkalian yak to this list. The excavations of Ovodov and Martynovich (2005) have

**Fig. 3.112** Razboinich'ya Cave. We packed in warm clothing, candles, flashlights, and rope for our inspection of the cave. Inside, the cave is cold, dry, dark, and perceptibly limey and musty. However, fresh air moved through the cave as demonstrated by the small flickering candle to the lower left of Olga Pavlova, who is standing about where the Pleistocene bone deposits began. The candle was placed to light our way through this section of the descending cave. Nicolai Ovodov examines the stalactites that reveal the cave was once wet as well as having been an ancient underground water channel (CGT color Razboinich'ya 7-9-99:10).

produced a very large number of hyenas. On the basis of the occurrence of the distinctive lower molar (right side) they identified an MNI of 137 individual hyenas ranging in age from sub-adults to very old individuals. They estimate that double or triple this number would be the total MNI if the cave had been completely excavated. At least one over-wintering bear seems likely to have died in the cave. Judging from the types and amounts of faunal remains, and the numerous distinctive round, white coprolites containing numerous bone chips found throughout the maximal 3.5 m depth of loamy pale brown Pleistocene deposits, cave hyenas were the major large inhabitants. Their use of the cave ended with their natural extinction, generally estimated to be around 13 000–14 000 BP, or just prior to the end of the Pleistocene. Nearly all the hyena remains abundantly and clearly show indications of cannibalism.

Archaeological exploration of Razboinich'ya began in 1975 and was continued from 1977 to 1991 by Nicolai Ovodov and associates Nicolai Martynovich, Yuri Grichan, Sergei Glinsky, Andrei Sagalaev, and Alexander Smirnov following the discovery of the cave in 1962 by geologists Nicolai K. Moroz, Lev Sandakhchiev, and Vladimir Grober.

**Fig. 3.113** Razboinich'ya Cave. Several meters further into the cave from the previous scene, the floor drops vertically. It is at this point that the deposits of bones, desiccated animal parts, coprolites, and decomposed coprolite soil were their deepest. The deposits are remarkable because every bone represents an animal that walked, crawled, flew, or was carried into the cave. Ovodov and Pavlova (CGT color Razboinich'ya 7-9-99:20).

Other archaeologists who have visited the cave include A. P. Derevianko and V. I. Molodin. Ovodov was the first scientist to dig a test pit in the cave, in 1975, followed by another exploratory test pit that year in Denisova Cave. He dug the 1975 test pit in the grotto-like mid-portion of Razboinich'ya that is 19 m lower than the elevation at the cave mouth. The 2 × 2 m pit was dug through cave fill until bedrock was reached at a depth of 3.5 m. His Razboinich'ya testing quickly recovered a hyena jaw and remains of two cave lions. He was with geologist A. M. Panichev, who initially waited outside; however, upon seeing the remarkable specimens Ovodov brought out, Panichev joined in the test excavation. When Ovodov first entered the cave he found the floor covered with a 10–15 cm deep soft loam that was some 3 m deep at the back of the cave. Hyena coprolites were present throughout the loamy deposit. The only material culture remains in Razboinich'ya are small amounts of food container refuse and rags left on the surface near a fire hearth just inside the entrance by three Russian army draft-evaders during World War II. The abundant presence of Pleistocene hyena remains and the near total absence of all prehistoric human refuse suggest that the cave locality was considered dangerous and largely avoided by humans, as well as having been inconveniently located for human occupation. Danger, fear, and darkness, however, may have played some part

**Fig. 3.114** Razboinich'ya Cave. Back outside, Ovodov is examining the 2 L sample of deposit he collected at the far end of the cave during our visit inside. Various studies have shown that when the late Pleistocene hyenas were the cave's main occupants, the landscape was largely alpine tundra at this elevation, although lower down the mountain, Kaminnaya and other cave site localities were vegetated about the same as today and as in this scene. The tent behind Ovodov is where he and his associates camped when rain or snow made it difficult to hike back to his log cabin field station (CGT neg. Razboinich'ya 7-9-99:14).

in whatever ceremony ancient hunters might have performed >30 000 years ago when a large dog skull was placed upright deep in the cave in the same layer with the skulls of five wolves and four brown bears (Ovodov *et al.* 1999:66). The dog was elderly and seemed to have died a natural death, although it could have been sacrificed by strangulation or blood-letting. We found no cut marks on the base of the skull or elsewhere. It had not been damaged in any discernible way by hyenas or humans. Small bits of charcoal and burned bone were near and beneath the apparent shrine, but not near the dog skull. Dogs may have helped the Kaminnaya Cave dwellers in their hunting of mammoths (Ovodov 1999; Ovodov and Kuzmin 2006).

In all, about 15 $m^2$ of cave floor were excavated to bedrock. Double screening was done, first with a $2 \times 2$ cm mesh, with the filtrate finally washed through netting of a $5.0 \times 1.5$ mm mesh.

A year after the end of our data-collecting phase of study, Ovodov recalled another paleontological cave he had tested (Baryshnikov and Maloletko 1998). He felt that the main large occupants had been wolves. Called Eckina Cave, it is likely Holocene in age (no extinct species, only small animals such as marmot and small deer). We briefly examined the small 50-piece assemblage but did no formal scoring for our perimortem damage variables. The bones had very little damage. There was no acid erosion, no deep dints, very little polishing. The carnivore damage was trivial compared to that found in Razboinich'ya Cave, which leads us to believe that most of the perimortem bone damage present in Razboinich'ya was caused by hyenas, not wolves. This inference is supported by the much greater presence of hyena skeletal remains and coprolites than all other carnivores combined.

The abundance of carnivore bones, not only small but also large, is striking: fox, gray wolf, brown bear, cave hyena (cave lion was also found, but not in the Razboinich'ya sample studied herein). A complete tally of the vertebrate remains recovered during a few years of excavation is more than 73 500 pieces. Of special interest is the large number of bones and teeth of cave hyenas that belonged to 137 individuals at a minimum (Ovodov and Martynovich 2005). According to the faunal classification by G. Baryshnikov (2003), this cave was used by Pleistocene *Crocuta* as a permanent den in which they manifested their cannibalistic habits from time to time.

## The dog skull

Until recently only a few examples of a wolf-like dog have been found in a Paleolithic context of Siberia. The best-known example was excavated from Afontova Gora in the late 1800s by I. T. Savenkov. As we have previously said, Afontova Gora is a multilayered, multi-locality archaeological site(s), with some five cultural horizons embedded in various locations in the riverside deposits of the Yenisei that flows through the city of Krasnoyarsk. The famous Russian natural historian and mammalian osteologist I. D. Chersky thought well of Savenkov's find when he traveled through Krasnoyarsk in July 1885, although he himself did not study the dog remains (Ovodov and Kuzmin 2006). Afontova Gora has produced many stone and bone artifacts, faunal and human skeletal remains, and other information about the Upper Paleolithic inhabitants of the Yenisei basin. In addition, there are at least 11 carbon-14 dates for this site, ranging from 11 330 BP to 20 900 BP. Unfortunately, it is unknown exactly where the Afontova Gora dog remains are today or precisely where they were originally found, or if they were actually deposited during this range of late Pleistocene time. Hence, their age is uncertain, whereas the Razboinich'ya dog has much better contextual control, which places it at more than 30 000 BP. In European Russia, the dog does not appear in the archaeological record until late Magdelanian times, some 18 000 years ago (Golomshtok 1938:221). Outside of Europe claims for an early dog have been made for a find in Israel (Davis and Valla 1978), and the dingo of Australia must have been derived from a wolf population in India.

Ovodov and Kuzmin (2006; Ovodov *et al.* 2011) provide comparative measurements for the Razboinich'ya dog skull and those of some modern wolves. The dog is smaller in most respects.

Dogs are increasingly receiving attention concerning the time and place of their domestication, their aid in human hunting in the late Pleistocene, and genetics (Koop *et al.* 2000,

Morey 2010, Ovodov 1999, Ovodov and Kuzmin 2006, Savolainen *et al.* 2002) following the pioneering dog origin studies of Stanley J. Olson (Olson and Olson 1977, Olson 1985) In addition, we consider the importance of dogs for their probable aid in human entry into the Eurasian high Arctic and the eventual crossing of Beringia (Turner 2002).

One especially critical discussion of early Eurasian dogs is by Susan J. Crockford and Yaroslav V. Kuzmin (2012). Their main point is concerned with distinguishing wolves from dogs. Their concern deals with taxonomy and domestication, biological questions for the most part. They are quite uncomfortable with the term *dog* being applied to late Pleistocene canids on the basis of tooth and skull morphology and dimensions. However, their concern is not ours because of the context wherein the Razboinich'ya dog was found. Regardless of whether it was a domesticated wolf to some degree or a full but runty wolf, it was important to the people who left its skull in the cave. Its relatively small size and old age hints that it was likely cared for for several years by the people who enshrined it in the cave.

There is a wide range of carbon-14 dates for the upper layer of the Razboinich'ya Cave fill. This clearly demonstrates the stratigraphic disturbance associated with carnivores in these southern Siberian caves and open sites – a theme we have repeatedly referred to in this book. The dates range from 670 ± 60 years (SOAN-2012); 7370 ± 40 years (SOAN-2015); 9760 ± 70 (SOAN-2013); to 22 400 ± 320 years (SOAN-2014).

The underlying second layer of red-brown loam, frozen in some places (mummified soft tissue is sometimes preserved), includes remains of cave hyena, brown bear, wolf, and various other large and small mammals. It is in this layer that the complete skull of an adult dog was discovered (Ovodov 1999). The preservation of the dog's skull does not differ from that of other skulls found in the same layer, suggesting that it was not introduced at a later time in the Holocene when dogs are archaeologically well known. Wolf-like but smaller, its presence is more likely linked to ancient cult or shrine activity during the time this layer was forming. Such activity is additionally suggested by the presence of some burned bone and small fragments of charcoal found in this layer deep in the pitch-black cave.

In addition, a brown bear skull and wolf skulls were also found (see Varvarina Gora for possible ritual wolf burial). The nearest Paleolithic site, Kaminnaya Cave, is located 2 km from Razboinich'ya. The lack of perimortem carnivore damage to the dog and bear skulls allows us to infer that it was people, perhaps individuals from the Kaminnaya band, who placed the dog and bear skulls in Razboinich'ya Cave. Such placement of offerings in dark and dangerous places is not without strong analogy in the ethnographic present.

## Findings

**1 Provenience.** For this book we examined and re-examined the Razboinich'ya faunal collection during 1998, 1999, 2000, and 2003 (Table A1.1, site 21). It is our preeminent baseline assemblage for identifying the bone damage signature of late Pleistocene Siberian carnivores. Excluding the modern Russian surface trash found near the cave entrance, we found few differences in the assemblage regardless of vertical stratigraphy or horizontal location; hence, all the faunal remains have been pooled. The deposits span a period of perhaps 40 000 radiocarbon years. Some of the lack of stratigraphic difference is presumably due to the digging behavior common to hyenas of today (see Chapter 4). Altogether we scored 739 pieces out of thousands that were either too small or were undamaged whole bones of complete small animals that probably died of natural causes within the cave. This total is 7.9% of the grand total for our multiple sites.

**2 Species.** Of 638 pieces larger than 2.5 cm assessed for species, those that could not be identified even in a general way were the most frequent (45.1%), followed by fox (31.0%), hyena (11.7%) (Figs. 3.115–3.116), and wolf (4.7%) (Table A1.2, site 21). The high frequency of fox probably represents natural deaths within the cave, because most of these bones and teeth have very little perimortem damage. Had these foxes been carried into the cave as hyena kills, we feel that their perimortem condition would have exhibited much more damage, if not in fact having been almost totally destroyed due to the light and delicate nature of the fox skeleton. Species represented at about 1.0% include bear, big mammal, bird, and goat-sheep. Compared to the unweighted averages for all our assemblages, Razboinich'ya seems meaningfully below average for big mammal, bison, goat-sheep, horse, mammoth, and above average for fox, wolf, and indeterminable. Wolves may have also denned in the cave when vacated by hyenas. Indeterminacy of species undoubtedly reflects both small piece size and the powerful destructiveness of hyena jaws and teeth.

**3 Skeletal elements.** There were 638 pieces studied for skeletal element identification (Table A1.3, site 21). Of those represented by about 5.0% or more, the semi-specific category of long bone was most frequent (26.0%). This was followed by rib (10.2%), vertebra (8.3%), indeterminable (5.6%), penis (5.4%), mandible (4.7%), and humerus (4.7%). With one exception, none of these values, nor the others of lower percentage, are noteworthy when compared to the total in our other site assemblages. The exception is the penis bone. This skeletal element is absent in all but three of our other assemblages, and in each of these the frequency is less than 1.0%. We can think of no explanation for the relatively high Razboinich'ya penis bone occurrence.

**4 Age.** Of the 638 Razboinich'ya pieces scored for age, adults were more frequently identified (52.5%) than sub-adults (7.4%) (Table A1.4, site 21). Compared with our other assemblages, there is nothing out of the ordinary in these values. Pieces whose age could not be determined (40.1%) were similarly unexceptional, although twice as common as the mean for all sites combined.

**5 Completeness (Fig. 3.117).** Of the 638 Razboinich'ya pieces scored for completeness (Table A1.5, site 21), those that lacked both anatomical ends were the most common (62.8%). Those with one anatomical end were next most frequent (28.5%), and whole

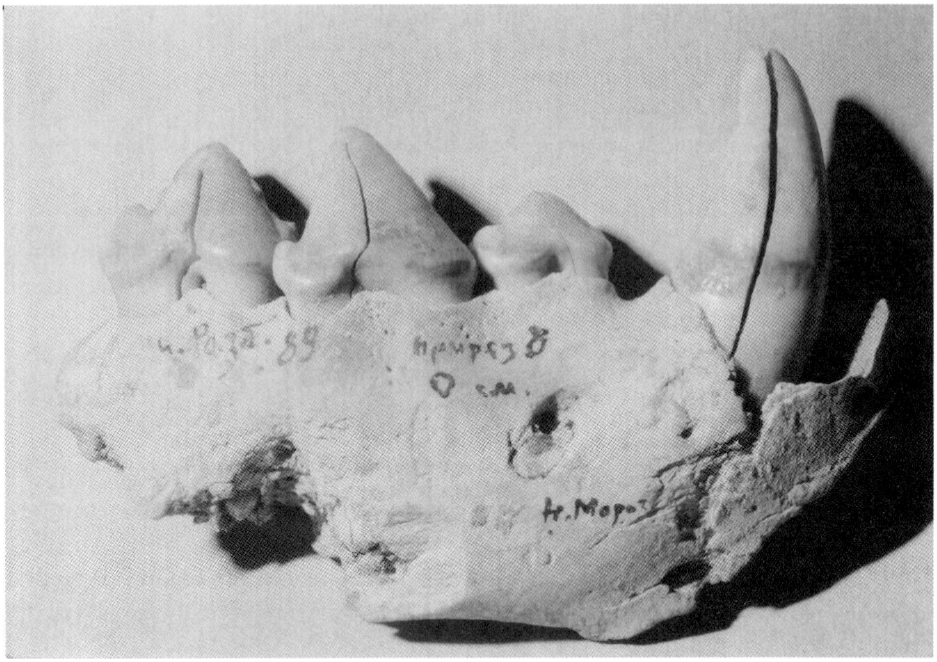

**Fig. 3.115** Razboinich'ya Cave, hyena (1989). A 10 cm long fragment of the right side of a lower jaw that belonged to a young occupant shows shearing and puncturing damage, which signals cannibalism. The tooth dint beneath the first post-canine tooth has small embedded bone fragments. (CGT neg. IAE 7-28-99:32).

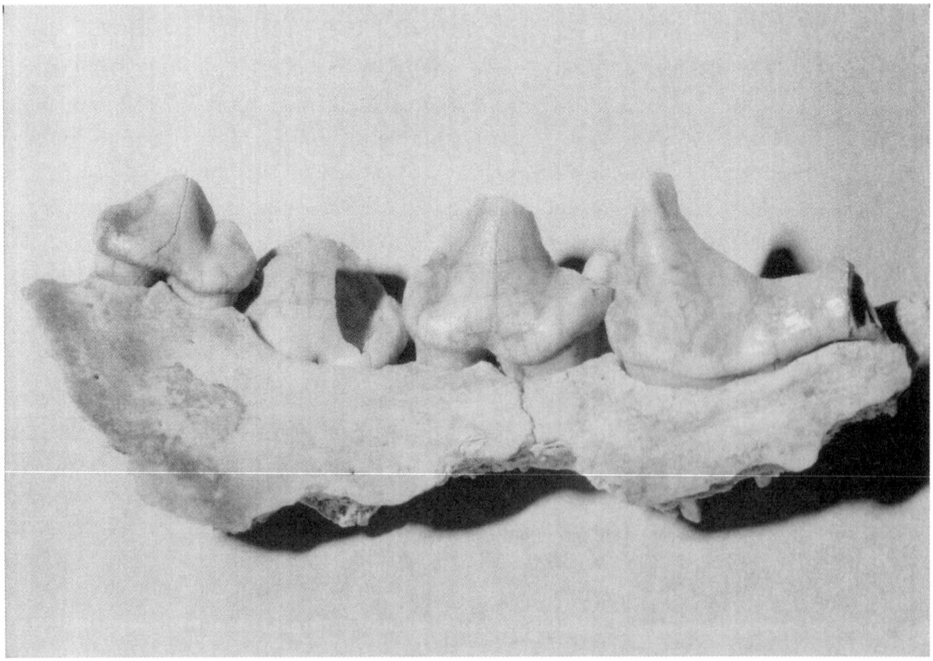

**Fig. 3.116** Razboinich'ya Cave, hyena (1988). Another young cannibalized cave occupant. This right side mandibular fragment is 10.1 cm in length (CGT neg. IAE 7-28-99:36).

**Fig. 3.117** Razboinich'ya Cave, unsorted deposit. A sense of the cave's contents can be gotten from this view of yet-to-be-studied bones, teeth, and hyena coprolites recovered by Ovodov in 1989. Two round, white coprolites are in the lower left corner. The match boxes are 5.0 cm in length and contain tiny teeth and bones. The pencil points to a hyena canine with tooth scratches on its root (CGT neg. IAE 7-6-00:1).

bones were least frequent (8.6%). None of these values are exceptional compared with the other 28 assemblages, being as they are fairly close to the means for each condition in the pooled 30-site assemblage.

**6 Maximum size.** The mean and related statistical values were calculated on the basis of 599 measured pieces that were for the most part greater than 2.5 cm in length of maximal dimension (Table A1.6, site 21). Although there are some statistically significant differences between the ordinal values of the Razboinich'ya pieces when compared with other assemblages, such as Mal'ta and Shestakovo, these significant differences are due solely to the latter two sites having some very large pieces of mammoth bone that produce much larger means and related statistics. On the whole, the Razboinich'ya mean (6.5 cm) and range (2.4 cm to 24.4 cm) has no exceptional size differences with most of the other faunal assemblages. Nor does it differ in any meaningful way from the pooled assemblage averages. Compared with the lengths of undamaged long bones of fox, hyena, and wolf provided by Vera Gromova (1950: table 27), the Razboinich'ya upper range limit (24.4 cm) is greater than all five fox long bones, and about the same as those of the hyena and wolf. The lower range limit of Razboinich'ya is, of course, much smaller than those of these species.

**7 Damage shape.** We assessed the type of damage in 739 pieces. The forms most common are fragments (31.1%), medial rib pieces (15.4%), and flakes (13.4%). Because of the many fragments and flakes in the Razboinich'ya assemblage, the majority of the other damage types, such as "mostly whole," have lower frequencies than the pooled averages. Compared with the pooled averages, the Razboinich'ya assemblage shows more overall damage and processing.

**8 Color.** We could evaluate color in 637 Razboinich'ya pieces (Table A1.8, site 21). Most pieces are ivory colored (92.9%), with some brown (6.4%) and white (0.6%) examples. These values are not much different from the pooled averages, but differ greatly from the open sites like Ust-Kova. Black mineral "flowers" (see Fig. 3.170) occurred on 35 ivory colored Razboinich'ya pieces. We have not found any explanation for how these deposits are formed.

**9 Preservation.** In total, 625 pieces were assessed for quality of preservation (Table A1.9, site 21). Most Razboinich'ya pieces are ivory hard (93.8%). A much smaller number (6.2%) are chalky. These values indicate significantly better preservation at Razboinich'ya than in the pooled assemblage, although there are individual cave sites such as Kaminnaya with even better preservation. Because of the cave's excellent preservation conditions, we presume the chalky pieces represent old surface-weathered bones that hyenas carried into the cave.

**10 Perimortem breakage.** Of 607 pieces, Razboinich'ya has identifiable perimortem breakage in 92.4% (Table A1.10, site 21). This value is more than that of the pooled assemblage. While representing a great deal of bone breakage by carnivores, there are a few archaeological sites with even higher values of perimortem breakage, including Bolshoi Yakor and Varvarina Gora.

**11 Postmortem breakage.** Postmortem breakage was found in 10.9% of 607 Razboinich'ya pieces (Table A1.11, site 21). Much of the damage was caused by the difficulty of archaeological recovery in the pitch-black cave, and in subsequent handling and transport down the mountain and later in the laboratory. Although one-tenth of all pieces show some postmortem damage, the amount is less than the average for the pooled assemblages, and is therefore unexceptional.

**12 End-hollowing.** This damage form is distinctive of carnivores, especially when there are co-occurring tooth dints and tooth scratches. End-hollowing was found in 8.1% of 543 Razboinich'ya pieces (Table A1.12, site 21). The Razboinich'ya frequency is nearly the same as that of the pooled assemblage. Although we expected to find end-hollowing in sites with carnivore remains, we did not know how much to expect. Nor did we expect to see as much end-hollowing in archaeological sites as actually occurred in a few sites, such as Okladinov and Ust-Kan Caves. These two caves, as well as others, must have had a significant carnivore presence from time to time. Nevertheless, most of the archaeological sites have much less end-hollowing than does Razboinich'ya. It seems reasonable to propose that late Pleistocene sites in Siberia exhibiting 5–10% or more end-hollowing should be considered as having had a significant carnivore presence. We will discuss this proposition in more detail in Chapter 4.

**13 Notching.** Of 618 pieces, Razboinich'ya has 23.6% with notching (Table A1.13, site 21). While this value is greater than the average for the pooled assemblage, it is generally like the values found in other paleontological sites. It is greater than occurs in the archaeological sites, with the exception of Okladnikov Cave. Most of the Razboinich'ya pieces with notching have only one notch (76.4%). While there are no Razboinich'ya pieces with more than six notches, notching nevertheless seems to be a characteristic of sites with definite carnivore presence, especially that of hyenas.

**14 Tooth scratches (Fig. 3.118).** Razboinich'ya has a high frequency of pieces with tooth scratches. In 661 pieces, 37.1% have one or more scratches. This value is almost twice that of the pooled assemblage, and much like that of our other paleontological sites (Table A1.14, site 21). There are several Razboinich'ya pieces with multiple scratches; one piece had 50 scratches. If we assume that 30–40% tooth scratching is evidence of a significant carnivore contribution to the perimortem taphonomy of a site, then, once again, Okladnikov Cave would appear to have been used by carnivores as much as by humans.

**15 Tooth dints (Fig. 3.119).** Like the three previous variables, Razboinich'ya shows that tooth dints, as well as scratches, are also a useful indicator of carnivore presence. More than half (52.8%) of 629 Razboinich'ya pieces have one or more tooth dints (Table A1.15, site 21). This value is similar to that of our other paleontological sites. It is greater than the average for the pooled assemblage. There are several Razboinich'ya pieces with multiple dints; 14.0% have more than seven dints. As noted previously, pieces showing tooth dinting are very common at Okladnikov Cave.

**16 Pseudo-cuts.** We examined 638 Razboinich'ya pieces for pseudo-cuts, finding them expressed in 3.1% of pieces (Table A1.16, site 21). This value is slightly less than the average for the pooled assemblage (4.8%). Most of the Razboinich'ya pieces with pseudo-cuts have only one, but a very small fraction (0.2%) have more than seven. Several of our sites have pseudo-cuts. Paleontological sites have the most, but the Denisova archaeological site has the greatest amount (28.6%). We suspect a sampling error for Denisova, although observation error is possible because we studied the Denisova assemblage in the field and not under more ideal laboratory conditions.

**17 Abrasions.** Abrasions are rare in all of our assemblages, although they are slightly more common in the archaeological than in the paleontological sites. Of 623 Razboinich'ya pieces, we identified only 0.5% with abrasions (Table A1.17, site 21). This is less than the average for the pooled assemblage, as might be expected since abrasions are believed to have been caused mainly by bone slippage on a gritty stone surface such as that of a hammer or anvil stone. A grainy limestone cave floor could also have caused trampling abrasions, although much more finely and with fewer lines than those resulting from the glancing impact of stone tools or slippage on a stone anvil.

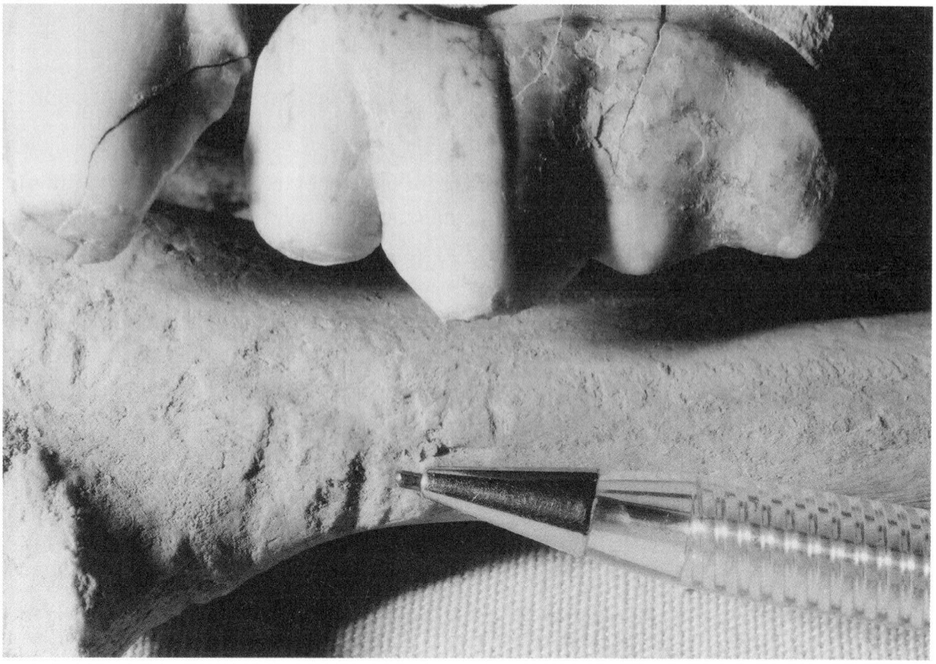

**Fig. 3.118** Razboinich'ya Cave, bone damage. The pencil points to tooth scratches on a hyena ulna. A hyena jaw has been placed at the top to show the burin-like form of the chipped cusp tip above the pencil. It is these chisel-like cusp tips that we believe caused the pseudo-cutting (CGT neg. IAE 7-6-00:4).

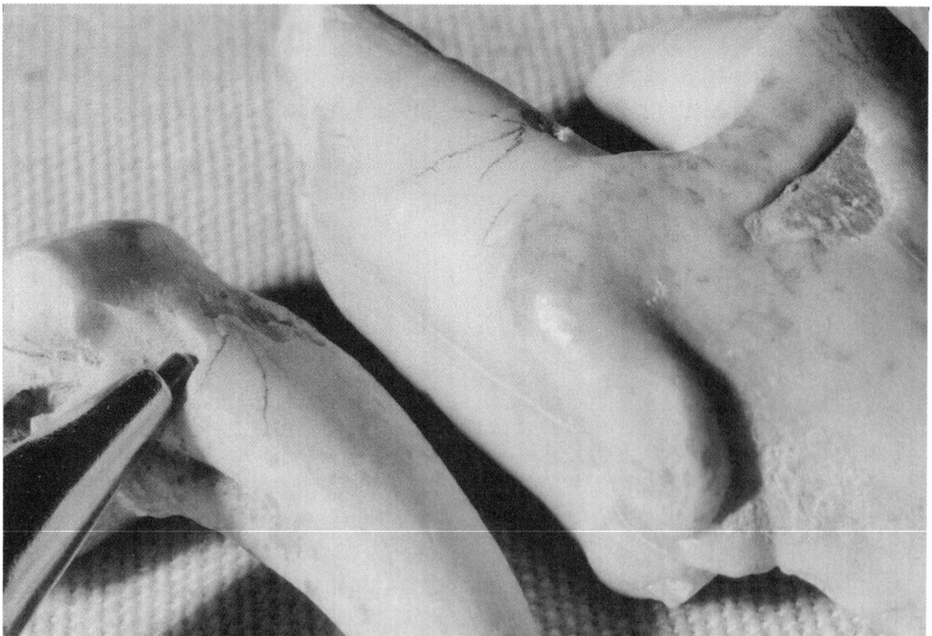

**Fig. 3.119** Razboinich'ya Cave, tooth damage. Arrows point to tooth dints on the roots of two hyena teeth, both dints being near the crown–root border. Cannibalism is suggested. Note also the area of developmentally defective enamel on the upper tooth in the upper right corner of the photograph (CGT neg. IAE 7-6-00:5).

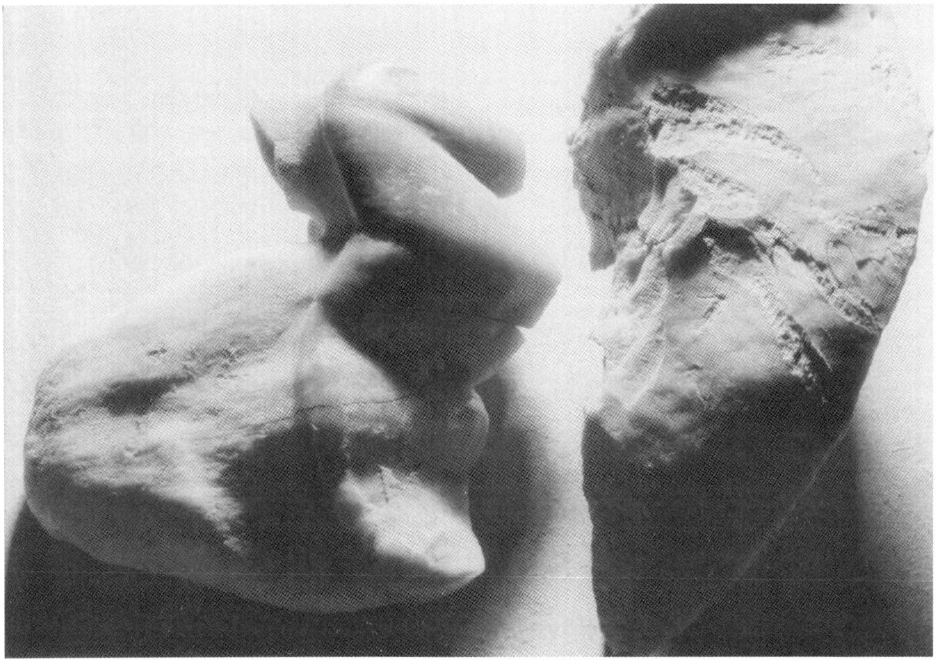

**Fig. 3.120** Razboinich'ya Cave, bone damage. On the right is a hyena cranial fragment with wide tooth scratches. A hyena molar from the same excavation unit is placed near to illustrate how the worn cusps could have scratched the bone in a chisel-like fashion (CGT neg. IAE 8-3-99:18).

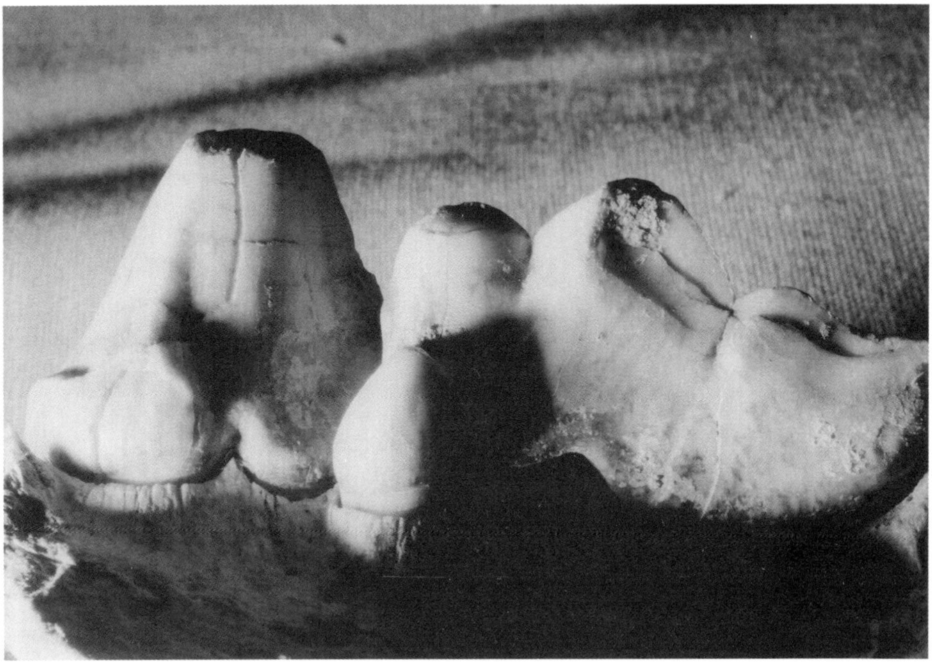

**Fig. 3.121** Razboinich'ya Cave, hyena teeth. Worn and burinized tips of maxillary teeth (CGT neg. IAE 7-6-00:14).

Fig. 3.122 Razboinich'ya Cave, stomach bones. These specimens from the 1977 excavation season were found 15 m from the entrance, along with other cave midden. The largest piece is 4.8 cm in length (CGT neg. IAE 7-28-99:28).

The number of abrasion grooves or lines ranges from one to more than seven in the Razboinich'ya assemblage.

**18 Polishing.** We originally assumed that polishing would be more common in sites with carnivore presence than in archaeological contexts. Razboinich'ya and several other assemblages showed this assumption to be incorrect. Out of 598 Razboinich'ya pieces, fully 40.6% had no identifiable polishing (Table A1.18, site 21). We further assumed that polishing due to carnivore chewing would occur more frequently on both the end and the middle of a piece than would be the case in an archaeological assemblage. Razboinich'ya has only 27.1% of its pieces with polishing on the middle and one or both ends, compared with the 45.1% average for the pooled assemblage or archaeological sites such as Kamenka (61.3%) or Varvarina Gora (71.2%). However, very heavily polished and highly rounded pieces are impressionistically far more common in paleontological than in archaeological sites. Unfortunately, we failed to make systematic observations on very heavy polishing, so we cannot provide any quantitative information on this perimortem taphonomic condition.

**19 Embedded fragments.** We examined 611 Razboinich'ya pieces for embedded fragments, finding 8.8% (Table A1.19, site 21). This is about twice as many as in the pooled assemblage (4.1%). The Razboinich'ya frequency is generally greater than we found for

**Fig. 3.123** Razboinich'ya Cave, stomach bone. This specimen is 6.1 cm long and 1.8 cm wide, which is larger than the diameter of most of the hyena coprolites from this cave (CGT neg. IAE 7-28-99:8).

**Fig. 3.124** Razboinich'ya Cave, bone damage. An unstudied 1986 collection of bones, teeth, and coprolites that illustrate the size and variety in the cave hyena-age deposits. The earliest deposits in the cave are sterile. The oldest animal remains Ovodov found were those of foxes. Sometime later, hyenas started to use the cave, and did so until near the end of the Pleistocene (CGT neg. IAE 7-29-98:34A).

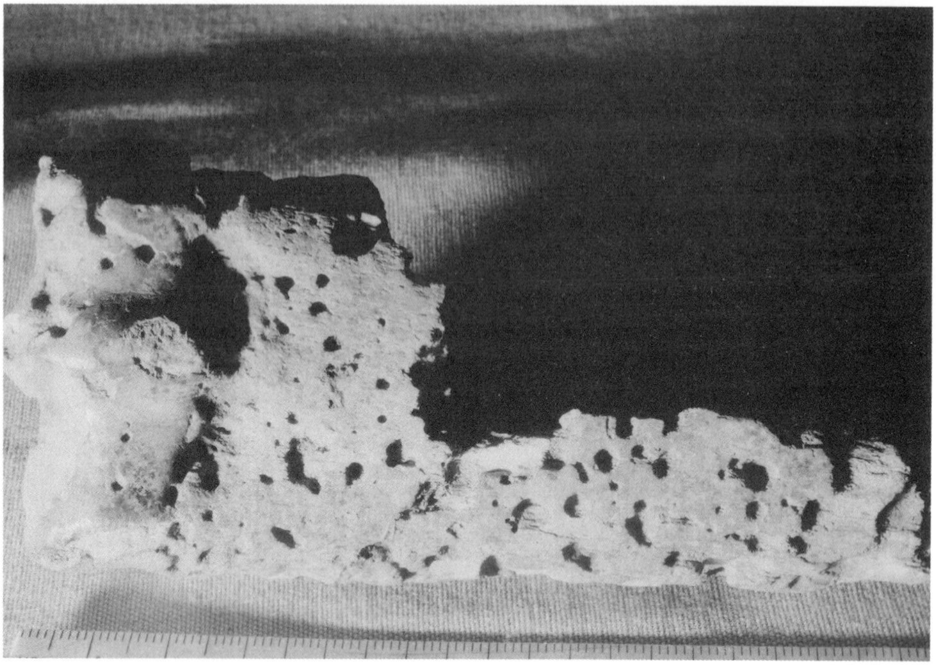

**Fig. 3.125** Razboinich'ya Cave, bone damage. The cause of this unusual pitting damage is unknown. Possibilities that come to mind include insect burrowing (like wood-boring bark beetles), digestive erosion (but the surface is hardly eroded and bone size is too large), and bone cancer. Pitting has ragged borders. Found at a depth of 60 cm. Scale is in centimeters (CGT neg. IAE 7-6-00:19).

the archaeological sites such as Denisova (1.2%), Kaminnaya (2.9%), Ust-Kan (1.1%), but not Ust-Kova (9.5%). Embedded fragments occur slightly more often in most of the paleontological sites. One embedded fragment per piece (5.7%) is the most common condition in the Razboinich'ya assemblage. Five fragments per piece is the largest number occurring in the Razboinich'ya series. These were found in only 0.6% of pieces with embedded fragments.

**20 Tooth wear (Figs. 3.120–3.121).** Only 25 Razboinich'ya pieces were scored for tooth wear, and generally fewer were scored for almost all of the other sites, making this variable quantitatively weak. Nevertheless, about one-third of the Razboinich'ya teeth indicated young individuals, which is close to the average for the pooled assemblage.

**21 Acid erosion (Figs. 3.122–3.123).** This perimortem taphonomic condition may well be the hallmark of carnivore bone processing, especially that done by hyenas. Our examination for acid erosion in 370 Razboinich'ya pieces turned up 7.3% (Table A1.21, site 21), a value close to the average for the pooled assemblage. On the basis of Razboinich'ya Cave, we propose that a frequency of 5–10% acid-eroded pieces indicates the presence of carnivores, especially cave hyenas. Hence, archaeological

**Fig. 3.126** Razboinich'ya Cave, bone damage. Magnification of Fig. 3.125. Here, it can be seen that some of the pits have a basin or container shape. The senior author with others studied one prehistoric Alaskan complete human skeleton with similar but less pitting, concluding that the individual had suffered from multiple myeloma, a form of cancer. Width of image is 3.3 cm (CGT neg. IAE 7-6-00:23).

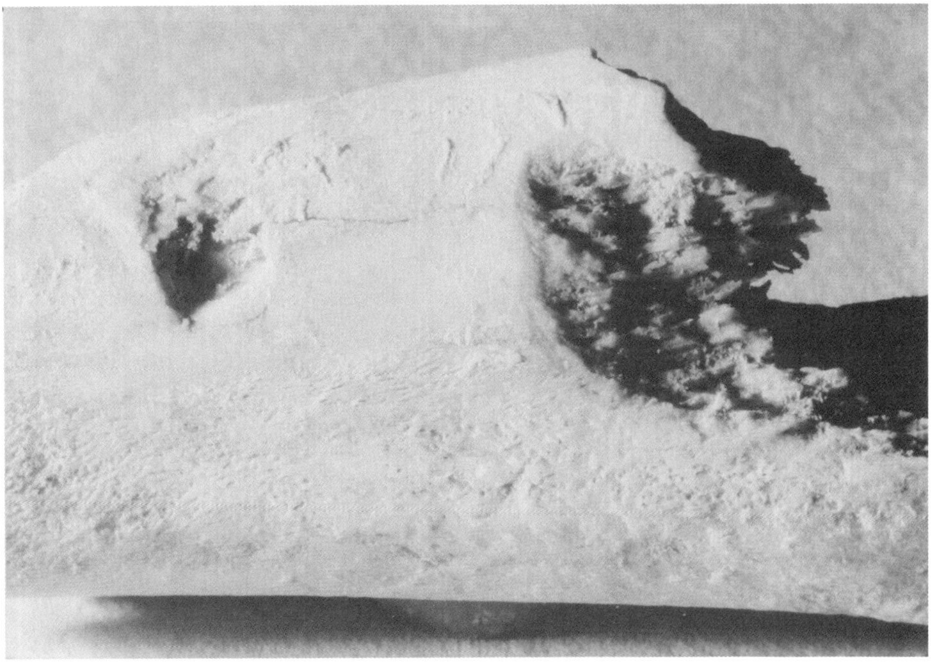

**Fig. 3.127** Razboinich'ya Cave, bone damage. Rib fragment (1980, 2 m deep) with two large tooth dints, each 4 × 5 mm in diameter. Tiny bone fragments are embedded in the left dint (CGT neg. IAE 7-28-99:16).

**Fig. 3.128** Razboinich'ya Cave, bone damage. Tooth notches on fracture lines. Actual width of image is 3.3 cm. This figure illustrates that notches are not always attributable to human bone smashing (CGT neg. IAE 7-28-99:2).

**Fig. 3.129** Razboinich'ya Cave, bone damage. Tooth notch on fracture line. Notch is 1.0 cm wide. Tooth dints are also present, near bottom of photograph (CGT neg. IAE 7-28-99:19).

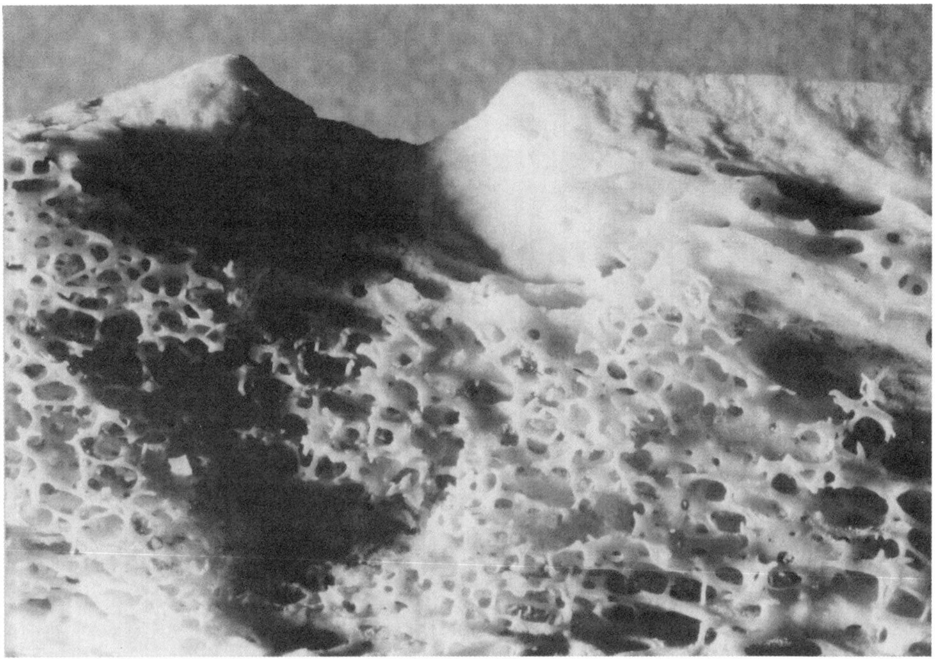

**Fig. 3.130** Razboinich'ya Cave, bone damage. Tooth notch on opposite surface of Fig. 3.129 (CGT neg. IAE 7-28-99:18).

**Fig. 3.131** Razboinich'ya Cave, bone damage. Pseudo-cut (at arrow). Actual width of image is 3.3 cm (CGT neg. IAE 7-6-00:7).

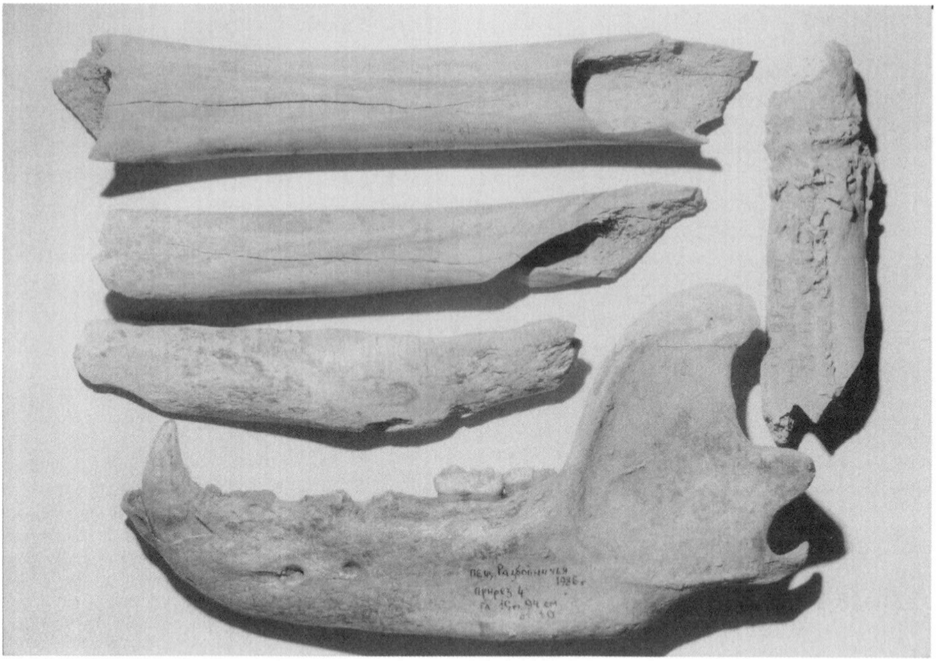

**Fig. 3.132** Razboinich'ya Cave, bears. Ovodov and associates found bones and teeth of brown bears, which had been damaged by hyenas, randomly distributed in the cave deposits. The pieces shown here belonged to more than one animal. Ovodov suggests that bears wintered in the cave. Once they were asleep, the hyena occupants killed and ate them. Note spiral fracturing on right end of pieces second from the top. Jaw is 26.0 cm in length (CGT neg. IAE 7-23-98:30).

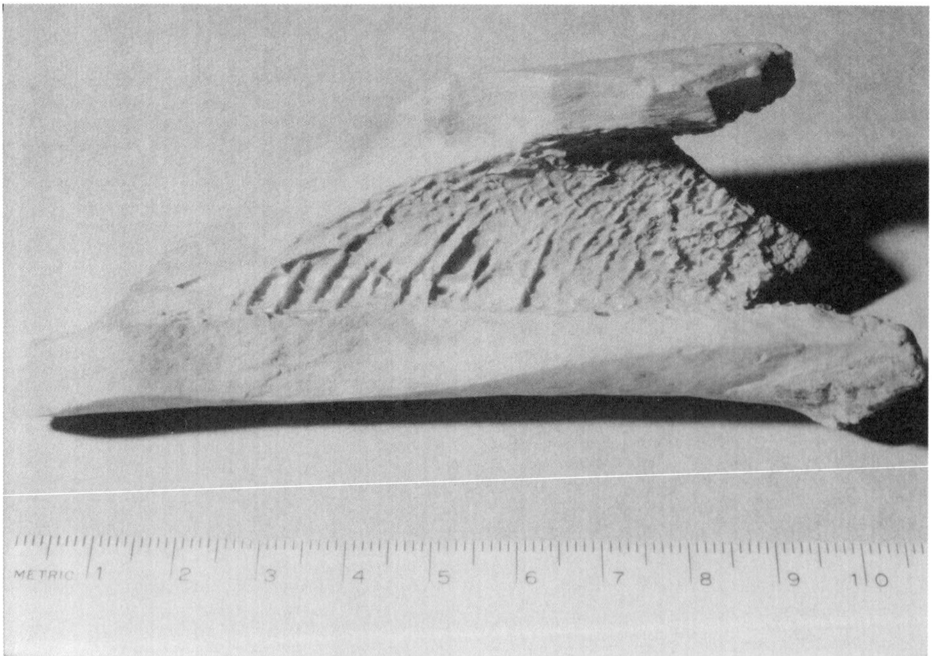


**Fig. 3.133** Razboinich'ya Cave, bear. Spiral fracturing of this bear humerus was done by a hyena. The piece is 10.8 cm in length (CGT neg. IAE 7-28-99:25).

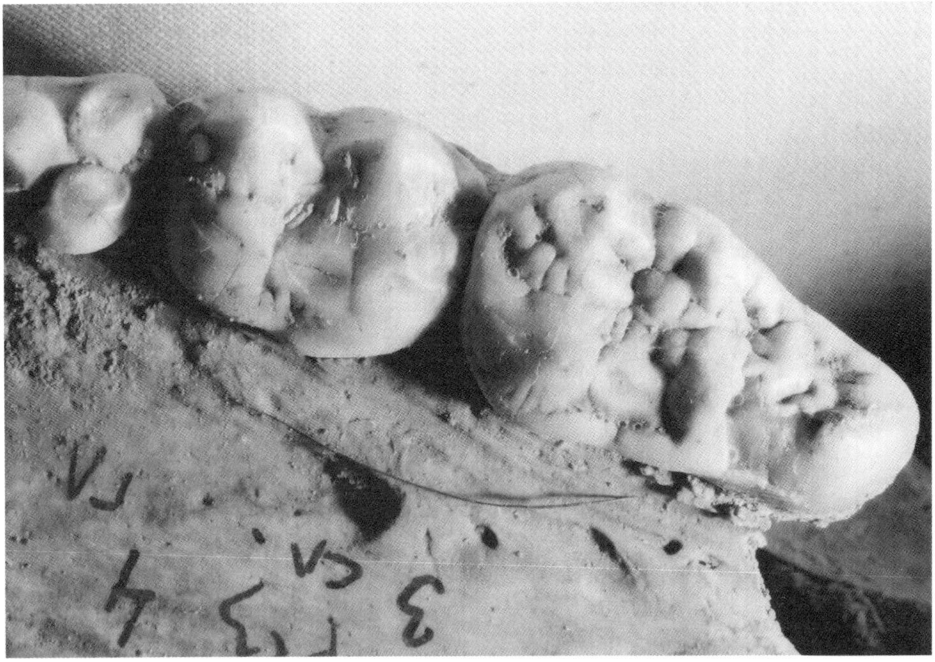

Fig. 3.134 Razboinich'ya Cave, bear. The maxillary teeth show relatively little wear, although the large molar has enamel breakage near its distal end (far right). That tooth is 3.7 cm in length. In contrast to the shearing teeth of hyenas, the bear dentition is better suited for milling, one of several reasons we feel that much of the bone breakage in this study was done by hyenas (CGT neg. IAE 7-8-02:22A).

sites such as Kaminnaya, Okladnikov, and Ust-Kan should be viewed as having also been paleontological sites. As such, stratigraphic disturbances may have been severe (Fig. 3.124).

**22 Rodent gnawing.** Only 0.8% of 611 Razboinich'ya pieces had gnawing incisions. This is nearly the same as the average for the pooled assemblage (0.5%). Rodents were not a meaningful source of perimortem damage in nearly all of our assemblages.

**23 Insect damage.** There was no identifiable damage done by insects in 611 Razboinich'ya pieces, which corresponds to the total or near-total absence in all of our assemblages.

**24 Human bone.** No human bones or teeth were found in Razboinich'ya Cave.

**25 Cut marks.** None of 611 Razboinich'ya pieces had cut marks. This is strong supporting evidence for our inference that Paleolithic humans rarely used the cave in Pleistocene times. As mentioned previously, only on the modern cave floor near the entrance are there any indications of relatively long-term human use of the cave. This evidence includes a fire hearth, cooking refuse, and three pieces of bone with cut marks, one of which was also burned.

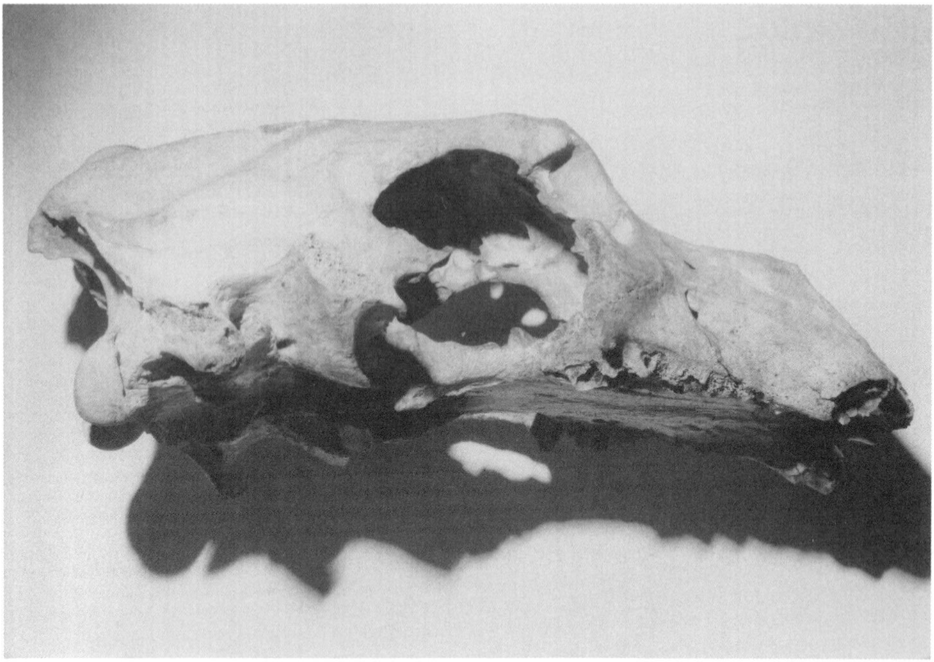

**Fig. 3.135** Razboinich'ya Cave, bear skull, lateral view. Ovodov and associates made a remarkable discovery in 1990. Deep in the cave they found the possible burial of a brown bear skull in Layer 2. This skull of an old edentulous bear with partly healed antemortem snout wounds also had massive perimortem breakage to the right side of its head, behind the right eye and zygomatic process. The damage appears traumatic, like the impact from a blunt weapon. There is no sign of healing or infection on any of the perimortem fractures. Most surprisingly, there is no identifiable carnivore damage. Ovodov proposes that the skull was placed in the cave by humans, possibly the Mousterian inhabitants of Kaminnaya Cave, down below the mountain and an hour away. Skull is 40 cm in length (CGT neg. IAE 7-23-98:25).

**26 Chop marks.** As with cut marks, the only Razboinich'ya piece out of 611 is a mid-twentieth-century bone with a single chop mark. There are a few damaged pieces that do not exactly fit our trait definitions (Figs. 3.125–3.131).

## Discussion

The Razboinich'ya Cave assemblage provides a unique resource for identifying and defining the bone damage signature of late Pleistocene carnivores, especially the cave hyena. The characteristics of this signature include: (1) such severe destruction that identifying species, skeletal elements, and age is impossible for at least 25% of an assemblage; (2) more than half of an assemblage is made up of pieces without any anatomical end; (3) around 50% of a bone assemblage consists of flakes, fragments, and splinters, there being very few whole bones or bones only slightly damaged; (4) perimortem breakage is extensive, and present in at least 90% of an assemblage;

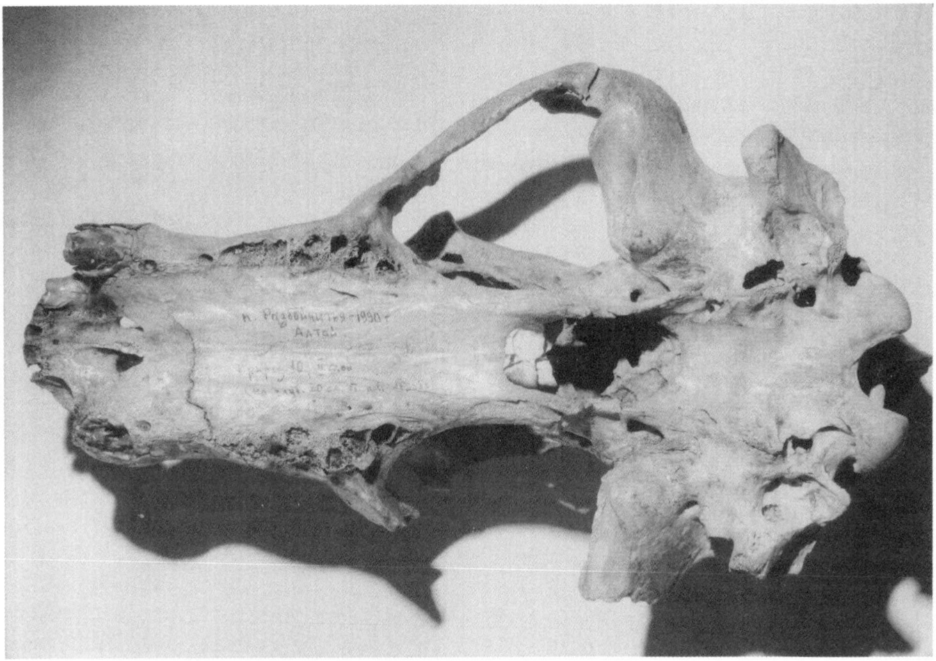

**Fig. 3.136** Razboinich'ya Cave, bear skull, basilar view. This old bear was effectively toothless when it died. The canines had been broken in life (they show wear on the remnant roots). There was inflammation in and around the sockets of the post-canine teeth, and severe fracture of the snout shows clearly in this and the previous figure. Such a break likely occurred in combat or by a kick from a large hoofed herbivore. Careful examination shows not a single tooth scratch, dint, or damage to the fragile bones so easily broken by powerful hyena jaws. The case for humans having placed the skull in the cave is forensically satisfactory (CGT IAE neg. 7-23-98:22).

(5) end-hollowing occurs in at least 5% of an assemblage; (6) notching is frequent, being present on at least 25% of all pieces; (7) tooth scratches occur on at least one-third of an assemblage, and tooth dints are present on 50%; (8) pseudo-cuts can occur; (9) embedded fragments occur around 5–10% of the time; (10) pieces with acid erosion are present at least 5% of the time; (11) a carnivore assemblage has no pieces with cutting, chopping, or burning. As is discussed in our various other sites viewed as archaeological sites, this carnivore signature can easily be recognized when it is present, meaning that some of the faunal remains found in an archaeological site may have been introduced by hunting or scavenging carnivores, not by humans. When the carnivore signature occurs in an archaeological site, then economic reconstruction must consider that some of the game animals may not have been the prey of the human occupants. Conversely, humans may have carried meat-bearing bones of large animals into archaeological caves such as Kamminaya. These bones may have been broken up later by scavenging hyenas following short- or long-term departure by the human inhabitants, of whom Paleolithic hyenas may have been moderately fearful.

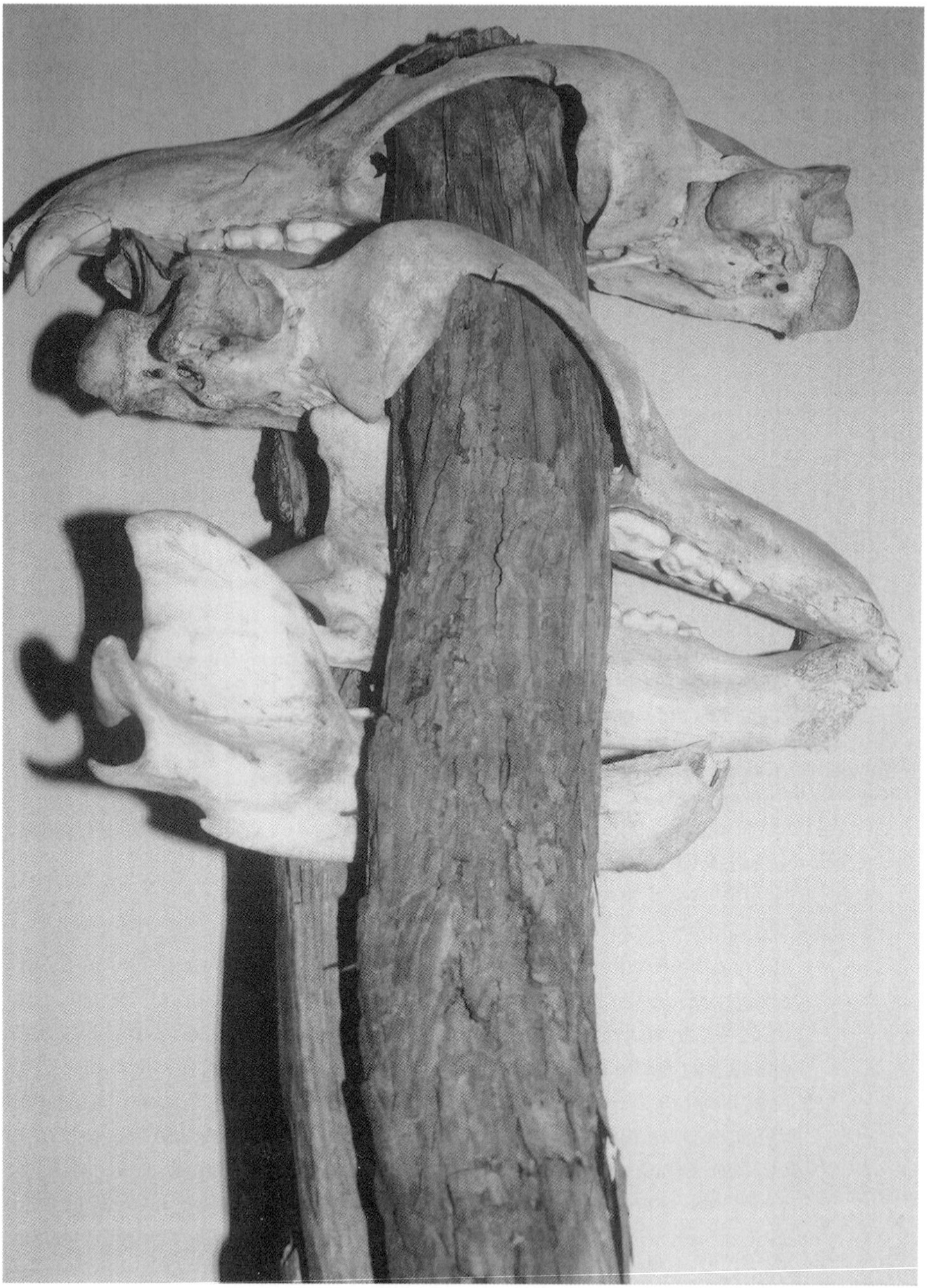

**Fig. 3.137** Bear sacrifice. An exhibit in the former IHPP museum of a Siberian Tungus bear sacrifice and monument pole. In addition to Tungus animism, animal sacrifice is well known ethnographically in far eastern Siberia and adjacent Hokkaido (Jomon-Ainu bear cult, Kamchatkan Koryak dog sacrifice). The Razboinich'ya find suggests that the practice of animal worship and sacrifice might extend back into Mousterian times. The Razboinich'ya bear skull may be the oldest evidence of "animism" in all of Siberia. If so, might there have been shamanism also? (CGT color IHPP 6-18-87:26).

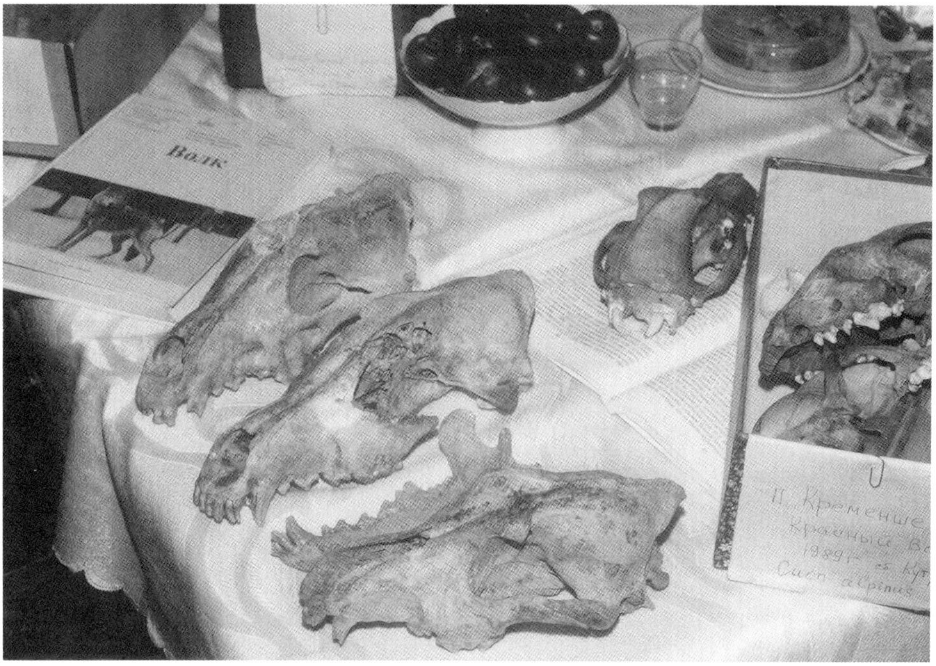


**Fig. 3.138** Razboinich'ya Cave, dog. As if the Ovodov team's find of the ritual bear skull was not enough, they also discovered the skull of what turns out to be one of the most ancient dogs in northern Eurasia (34 000 years old). Ovodov was preparing another scientific paper on the dog when this photograph was taken in the small Krasnoyarsk apartment where he lives with Lena Popkova. Dinner was being served when the senior author hastened to snap a few photographs of the dog, comparative carnivore skulls, books on wolves, a box of hyena skulls, and a small part of a delightful Russian dinner that, like most, terminated scientific study and measurements. Traditionally, Russian scholars conduct much of their work at home, which explains why the specimens were in Ovodov's apartment (CGT color Krasnoyarsk 8-2-00:6).

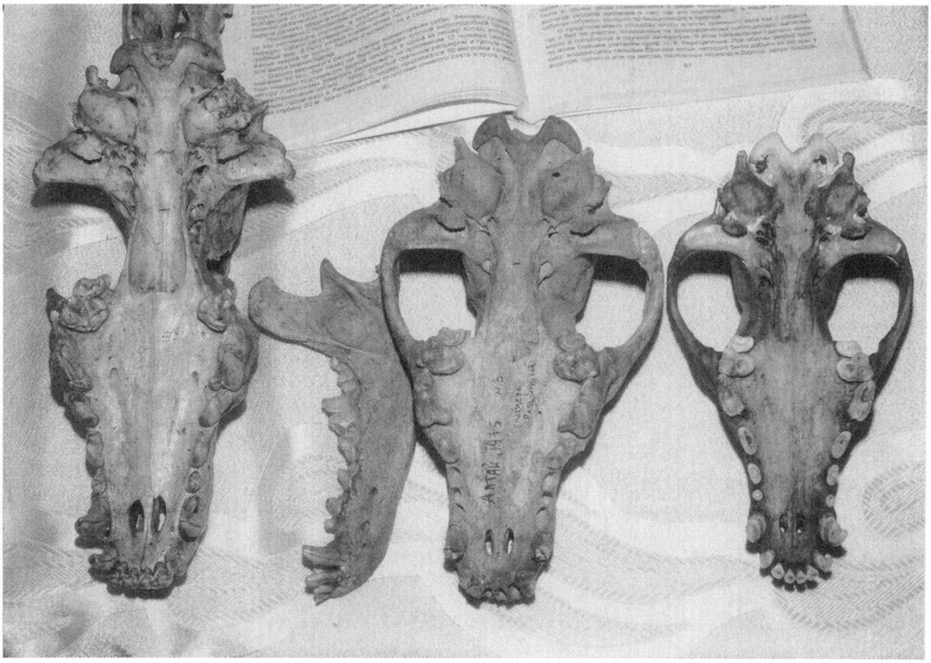


**Fig. 3.139** Razboinich'ya Cave, dog. Basilar view, same positions as previous figure (CGT color Krasnoyarsk 8-2-00:10).

Seasonal occupants of Razboinich'ya Cave included brown bears that Ovodov proposes were likely killed and eaten by hyenas while hibernating. One bear skull seems to have been left in the cave by humans (Figs. 3.132–3.136). A sacrificial bear cult was widespread in historic Siberia (Fig. 3.137), as was dog sacrifice in parts of far northeast Siberia (Figs. 3.138–3.139). Thus, Razboinich'ya Cave provides direct evidence of the carnivore bone damage signature, and indirect evidence of very ancient animism-shamanism.

# 22 Sarala Cave

## Background

Sarala Cave (named after the river Sarala) is located in forest-steppe habitat of the middle Yenisei basin, Khakasia. It is a small limestone cave near the site of Malaya Seeya at ca. 54°50′ N, 89°42′ E. It was tested in 1970 by geologists V. M. Maratov (Moscow), V. A. Panychev (Novosibirsk), vertebrate paleontologist N. D. Ovodov (Novosibirsk), and archaeologist Yu. V. Grichan (Novosibirsk). The remains of horse, mountain goat, and bison were recovered in the 45 $m^2$ area excavated to a depth of 60 cm. Despite the Pleistocene species, no Paleolithic artifacts were found, only some of Bronze Age to the nineteenth-century refuse.

## Findings

**1 Provenience.** The site seems to have been used mainly by animals (Table A1.1, site 22). Our cave fill assemblage consists of only 44 pieces, which make up 0.5% of our grand total of 8813 pieces.

**2 Species.** Table A1.2 (site 22) shows that the most prevalent classes for the 44-piece Sarala assemblage are indeterminable (63.6%), goat-sheep (18.2%), big mammal (9.1%), and horse (4.5%). The pooled assemblage averages show that Sarala is lacking bear, bison, hyena, mammoth, reindeer, roe deer, and is higher for goat-sheep and indeterminable. The small sample size makes these comparisons statistically weak, but not far out of line for other sites in the area or similar habitat.

**3 Skeletal elements.** Sarala's 44 pieces are most frequently represented by unknown (65.9%), long bone (40.9%), rib (9.1%), humerus (4.5%), ulna (4.5%), and metapodial (4.5%) (Table A1.3, site 22). Compared with the pooled assemblage, Sarala has a lower occurrence of the mandible, vertebra, femur, foot, and toe, and more frequent occurrence of long bone and unknown classes.

**4 Age.** There are no identifiable sub-adults in the 44 Sarala pieces (Table A1.4, site 22). This is less than the average for the pooled assemblage. We suspect that this deficiency is due to sample size, not some aspect of seasonal birth time.

**5 Completeness.** The 44-piece Sarala sample has only 4.5% whole pieces, 9.1% with one anatomical end, and a significantly large number, 86.4%, with no anatomical ends (Table A1.5, site 22). The averages for the pooled assemblage have considerably more completeness.

**6 Maximum size.** Despite the above substantial incompleteness, the 44 Sarala pieces have a mean maximum size of 8.8 cm, which is nearly identical to the pooled assemblage mean (Table A1.6, site 22). Sarala's range is about 10 cm less than that of the pooled assemblage. Comparing the upper range limit (21.2 cm) with whole long bones of goat-sheep and horse provided by Vera Gromova (1950: table 27) shows that the Sarala upper limit is less than the lower limit of both groups in Gromova's series.

**7 Damage shape.** The most common of the damage shape classes in the 44-piece Sarala assemblage are long bone flakes (36.4%), long bone splinters (27.3%), long bone fragments (11.4%), phalanx segments (9.1%), phalanx butts (4.5%), tooth-bearing pieces (4.5%), and undamaged pieces (4.5%) (Table A1.7, site 22). Compared with the pooled assemblage, Sarala has more long bone flakes and splinters, phalanx segments, and fewer long bone fragments, medial ribs, undamaged pieces, and mostly whole bones.

**8 Color.** All of the 44 Sarala pieces are ivory colored (Table A1.8, site 22). This frequency is more than the occurrence in the pooled assemblage. There is no indication of Sarala bone burning.

**9 Preservation.** As above, all of the 44 Sarala pieces are ivory hard (Table A1.9, site 22). Also as above, this frequency is much more than the occurrence in the pooled assemblage.

**10 Perimortem breakage.** Nearly all of the 44 Sarala pieces have perimortem breakage (97.7%; 43 / 44) (Table A1.10, site 22). This is considerably more perimortem breakage than the average of the pooled assemblage.

**11 Postmortem breakage.** A relatively low 6.8% of the 44 Sarala pieces exhibit postmortem breakage (Table A1.11, site 22). This is about half of the average for the pooled assemblage.

**12 End-hollowing.** End-hollowing is relatively common in the 44 Sarala pieces (Table A1.12, site 22). Compared with the pooled assemblage average, Sarala's 15.9% is greater by about 5% – not a significant difference.

**13 Notching.** Exactly one-quarter of the 44 Sarala pieces has 1–4 notches, most of which possess only one notch (Table A1.13, site 22). Nevertheless, this frequency is greater than the pooled assemblage average, and is part of the perimortem taphonomic signature for inferring a substantial carnivore presence at Sarala. In the absence of hyena and wolf skeletal and dental remains, the single lion skeletal element recovered from the Sarala excavation suggests that this species may have caused the notching. Compared with the pooled assemblage average, Sarala has nearly a 10% greater frequency of notching, and even slightly more examples of multiple notching.

**14 Tooth scratches.** Almost half (47.7%) of Sarala's 44 pieces have from one to more than seven tooth scratches (Table A1.14, site 22). This frequency is more than twice that of the pooled assemblage average. In fact, Sarala is in the upper one-fifth of sites with the greatest amount of tooth scratching.

**15 Tooth dints.** Dinting could be assessed in 43 Sarala pieces (Table A1.15, site 22). A remarkable 88.4% have from one to more than seven dints. Only one other studied assemblage, Proskuryakova, has a greater frequency (93.6%; 44 / 47 pieces). More than seven dints per piece (23.3%) is the most common Sarala class. The pooled assemblage average is less than Sarala has.

**16 Pseudo-cuts.** Sarala has one of the higher frequencies of pseudo-cuts in our study (9.8%; 4 / 41 pieces) (Table A1.16, site 22). This frequency is almost exactly twice the average for the pooled assemblage, and three times greater than that found in Razboinich'ya, where pseudo-cutting was first identified. While the difference between Sarala and Razboinich'ya may signal a species difference (cave lion vs. hyena) in pseudo-cutting, we are doubtful of whether such is the case. In fact, despite the absence of hyena remains in Sarala, we are inclined to view the site's major carnivore occupants as having been hyenas.

**17 Abrasions.** There were no cases of abrasions in the 44 Sarala pieces (Table A1.17, site 22).

**18 Polishing.** Approximately 80% of the 44 Sarala pieces exhibit polishing (Table A1.18, site 22). Most of this polishing occurs on both the middle and one or both ends of a piece (61.4%), a condition compatible with a carnivore inference. The pooled assemblage polishing average is somewhat less than Sarala, but not significantly so.

**19 Embedded fragments.** Like most other assemblages in this study, Sarala has a low frequency of embedded fragments (4.5%; 2 / 44 pieces) (Table A1.19, site 22). There is one piece with two embedded fragments, and one with four. The pooled assemblage average is nearly identical to that of Sarala.

**20 Tooth wear.** There are only two teeth in the Sarala assemblage that could be assessed for wear, and both were grade 1 (dentine exposed). This degree of wear we consider to be associated with most adults. Small sample size prohibits any meaningful comparisons.

**21 Acid erosion.** None of the 44 Sarala pieces has any sign of acid erosion (Table A1.21, site 22). While this is exceptional in light of the previous indications of carnivore processing, we propose three possible explanations. First, the absence of acid-eroded pieces that we normally find associated with hyena presence may actually indicate that Sarala had not been used by hyenas. Perhaps there was some environmental factor at Sarala that was unfavorable for hyenas. Second, the sample size may be below some sort of "visibility threshold" beyond which acid-eroded pieces show up. In our view, the most likely explanation was excavator error, since the site was excavated in 1970 and the workers were largely unfamiliar with the need for total recovery. Bones that they could not identify had no reason at that time to be saved. Using the pooled assemblage average for expectation of finding acid-eroded

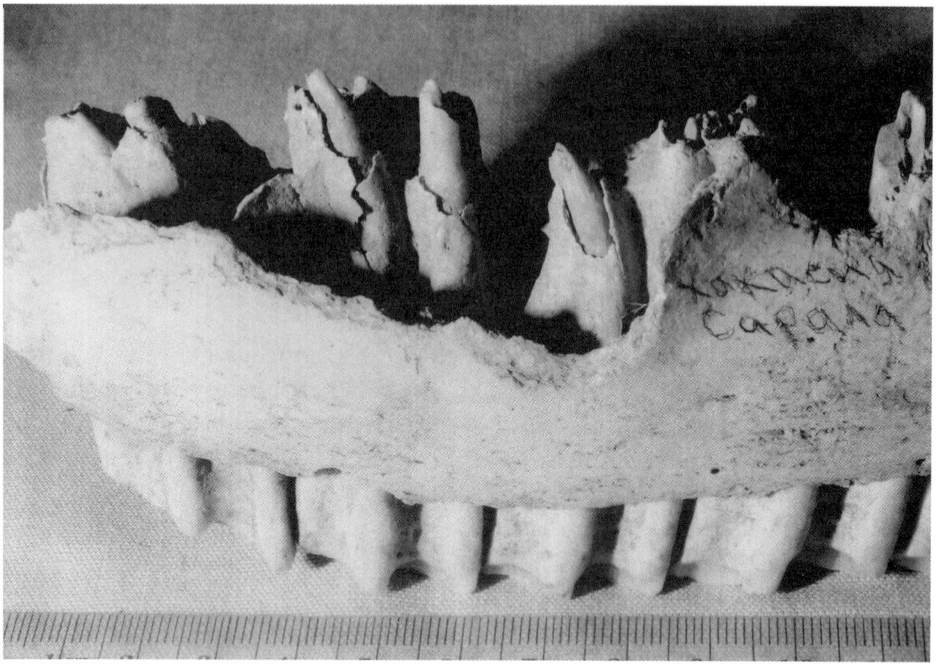

Fig. 3.140 Sarala Cave, bone damage. Horse maxilla (1975). It is difficult to envision how this damage occurred if the carnivore was not a heavy, bone-crushing hyena. There is no suggestion that the damage was done by humans. A mystery piece (CGT neg. IAE 7-16-01:31).

pieces, then on the basis of chance alone we could expect there to have been 2.5 pieces. Not finding any is not out of line with expectation.

**22–25 Rodent gnawing, insect damage, human bone, and cut marks.** There are no examples of these variables in the 44 Sarala pieces.

**26 Chop marks.** There are more than seven chop marks on one of the 44 Sarala pieces (2.3%) (Table A1.26, site 22). This frequency is about half the average for the pooled assemblage. Despite the absence of Paleolithic artifacts, the chopped piece suggests humans briefly took shelter in Sarala Cave.

## Discussion

The perimortem bone damage at Sarala reveals a far greater presence of carnivores that humans. With the exception of one piece with chop marks, there is no other distinctive perimortem taphonomic evidence for humans ever having used the cave. Since there are no hyena bones or teeth, nor any pieces with acid erosion, we are inclined to accept that the carnivore damage was done by a cave lion(s), one skeletal element of which was identified in the faunal remains that were recovered. However, because of the small sample size we are reluctant to exclude entirely the possibility that some of the damage might also have been done by hyenas (Fig. 3.140).

# 23 Shestakovo

## Background

Shestakovo is an open site of 125 000 m$^2$ in area east of Novosibirsk. It is located on a high, steep exposure of the Kita River bank, 500 m down-stream from Shestakovo village (literally, pole village) at 55°54′ N, 87°57′ E. Sergei V. Leshchinskiy (1998, 2000, 2001a, 2001b) and Vasily N. Zenin (Zenin *et al.* 2000b) directed the more recent excavations that over time have produced more than 3000 pieces of bone and teeth, 90% of which are mammoth (*Mammuthus primigenius*). Of an estimated 18 MNI, almost half (44%) are sub-adult, whose good preservation despite bone and tooth immaturity is attributed to their having been buried in a protective muddy setting. These deaths do not represent an epidemic die-off or mass killing by human hunters. Instead, the skeletons are thought by Leshchinskiy (2001a:295) to represent isolated deaths that occurred over many centuries, perhaps often of sickly animals. The concentration of skeletons is attributed to a higher density of living animals in one location than expected on a random basis because of the attractive mineral nutrients at Shestakovo.

Excavations are continuing at Shestakovo (Derevianko *et al.* 1997e, 1999e, 1999e, 2003c, Zenin and Maschenko 2000). During the 12 field seasons of excavation, from 1975 to 1978 and 1992 to 1999, 11 species of large mammals were recovered. They represent natural deaths at what Leshchinskiy calls a "beast solonetz." The term refers to a zoogeological locality where animals came to obtain various minerals and salts containing essential metabolic elements such as calcium, magnesium, and sodium that were concentrated in wet hollows close to mineral-rich bedrock. Gary Haynes (2002b:204) emphasizes the attraction for mammoths at "mineral licks" as being the need for iodine. Leshchinskiy hypothesizes that large herbivores frequented the Shestakovo beast solonetz or "mineral oasis" to obtain the nutrients that were missing in their diets due to the unfavorable mineral-poor acidic soil condition of the late Pleistocene steppe-tundra. In addition to the strong association between the soil chemistry of Shestakovo and the extensive mammoth remains, Leshchinskiy cites Panichev (1990) for examples of earth-eating (lithophagy) by large herbivores. Panichev inferred lithophagy on the basis of finding rock and mineral fragments in large herbivore digestive tracts and coprolites, including the lower gut of the Kirgilyakh mammoth calf. Despite the excellent associations for the mineral oasis–multiple mammoth hypothesis, this portion of the West Siberian Plain has a limited number of fossil localities with one or more mammoths. One is on the Tom River within the city of Tomsk, mentioned in Chapter 2. Two others are on the Ob River. Both are called Krasny Yar. One is in the region of Tomsk. The other is located to the south, a short distance down-river from Novosibirsk and described previously in this chapter (site 12). Another consideration that comes to mind is that Shestakovo might have been an "elephant cemetery." The key to assessing this model would be the perimortem taphonomy expected when live animals trampled the bodies of the dead (Haynes 1988). We would expect perimortem and postmortem bone breakage, gritty sand and rock abrasions, much disarticulation, and random scatter (see especially Haynes 1991, 1988). A final consideration would be microbial poisoning by cyanobacteria as proposed by Thekla Pfeiffer and Andreas Braun for many dead articulated Pleistocene animals found in an ancient lake in eastern Germany (Perkins 2001).

Twenty-one carbon-14 dates for Shestakovo range from approximately 25 660 BP (Layer 24) to 18 040 BP (Layer 17) (Vasili'ev *et al.* 2002:523). This late Pleistocene time range embraces the end of the Kargin warming to the middle of the Sartan cooling period (Leshchinskiy 2001a:295, Zenin *et al.* 2000b).

Late Pleistocene hunters and their families also visited Shestakovo, as evidenced by six blade- and burin-containing cultural layers above the bone-bearing unit. Over the six layers some 1500 artifacts were recovered, many made of bone and some stylistically like those recovered at the Upper Paleolithic sites of Mal'ta and Buret (Derev'anko *et al.* 1998). Zenin (personal communication, July 4, 2001) has found no cut marks on the mammoth bone, but believes that there may be some on the reindeer bone. Leshchinskiy proposes that mammoth ivory was a sought-after fabrication material inasmuch as some pieces were found with notches and splitting, and some ivory pieces had been fashioned into formal tools. He also suggests that humans also consumed the salty and mineral-rich mud and water, although there is no evidence for this suggestion. As Shestakovo is located near the southern boundary of the late Pleistocene West Siberian Plain, Leshchinskiy (2001a:293) proposes that a close relationship existed between the mineral oases, the migration paths of mammoths, and the hunting and travel routes of Upper Paleolithic human groups. In this regard, Zenin and Konovalenko (2001) reported that the nearest possible stone source for a cache of stone artifacts was 220 km away. Because humans could not survive by eating only plant products of this region, they were probably as dependent on animals for food as were Arctic Eskimos at historic contact. Almost 95% of the Eskimo diet was obtained from animal sources. Most of the remaining 5% consisted of watery berries and the sour-tasting lichen "salad" obtained from the stomachs of slaughtered reindeer. For the Paleolithic people of the West Siberian Plain the primary land animals would be those adapted to the tundra-steppe habitat – mammoth, wooly rhinoceros, bison, horse, saiga antelope, polar fox, lemmings, and other species. Leshchinskiy (2001a:298) further suggests that the changes in the late Pleistocene geochemical landscape of northern Eurasia could have played a "crucial role in the extinction of megafauna." The paleontological remains that we studied were housed in two locations. First, there was a large collection in the Institute of Archaeology and Ethnology, Academgorodok, Novosibirsk. This collection was made available for study by V. N. Zenin in July 2001. The second collection was housed in the Department of Palaeontology and Historical Geology, Tomsk State University, Tomsk. This smaller collection was made available by S. V. Leshchinskiy in June 2002.

In sum, Shestakovo is a natural, non-anthropogenic, large-mammal-bone site whose contents accumulated over many centuries of time. According to Leshchinskiy the accumulation was due to the nutrient attraction of the mineral-rich locality. As such, we should expect the perimortem bone damage signature to be very different from that found in sites formed by human occupation. If, as Leshchinskiy suggests, the human utilization of Shestakovo is largely later than the major mammoth-bearing unit, then the frequency of human-produced damage – such as cut and chop marks, burning, etc. – should be extremely low.

## Findings

**1 Provenience.** Since Leshchinskiy has himself written about Shestakovo as a single long-term unit accumulation, we do the same. Table A1.1 (site 23) shows that the 159 Shestakovo pieces make up 1.8% of our 8813-piece grand total.

**2 Species.** Our sampling of the Shestakovo assemblage resulted in identifying five species. The vast majority of pieces are mammoth (87.4%). There were a few pieces of bison, reindeer, wolf, horse, and unknown forms (Table A1.2, site 23). This is less than half of the 11 reportedly identified, because we ignored some whole or nearly whole bones that had no recognizable perimortem damage.

**3 Skeletal elements.** The elements recovered that we studied seem to reflect their anatomical number and hardness (Table A1.3, site 23). Hard, dense foot bones are most common (23.9%), followed by ribs (20.1%), vertebral bodies (15.1%), phalanges (8.2%), and small numbers of other elements. We see nothing in our sample of elements that suggests any special taphonomic events. Missing large heavy bones represent excavation, salvage, transportation, and storage decisions.

**4 Age.** The high frequency of sub-adults mentioned in the beginning is well reflected in the age as based on individual pieces (Table A1.4, site 23). Of the 159 pieces, 30.2% are unquestionably sub-adult. In fact, a mandible of a mammoth fetus has been found at this site (Maschenko and Leshchinskiy 2002). The remainder are adult (plus one unknown). The near absence of age indeterminate pieces is related to the mammoth's large size, their osteological and dental preservability, and the minimal human or carnivore exploitation.

**5 Completeness.** Given that Shestakovo is mainly a mammoth paleontological site without much identifiable human or carnivore involvement in the bone damage, the degree of completeness probably reflects the inherent preservability of mammoth bone. Our impression is that with the exception of the feet, toes, and vertebrae, the other bones possess relatively thin amounts of external compact bone, and very large amounts of spongy interior bone. In other words, mammoth bones are very strong relative to weight, but lack the hard, thick, and strong external cortex structure of smaller mammals such as horse, bison, and rhinoceros. Hence, mammoth bones can be expected to be less complete due to their biomechanical construction. Table A1.5 (site 23) shows that whole bones make up slightly more than half of the 158 pieces (55.3%), whereas pieces with only one anatomical end are 27.2% of this sample. Pieces with no anatomical end make up 16.5%. In other words, about half of the incompleteness can be attributed to the limited preservability of mammoth bone. This "weakness" is compensated by the huge size of these animals, so even if a bone is incomplete, it is still readily identifiable, although its surface may be so eroded that human and carnivore processing may not be preserved or identifiable. The large size of a mammoth would make rapid burial less likely than a smaller animal would experience. As such, mammoth bones would be prone to more weathering and associated breakdown.

**6 Maximum size.** The mean for 155 Shestakovo pieces is 16.1 cm, and the range is 2.4 cm to 68.0 cm (Table A1.6, site 23). As expected for a mammoth paleontological site, these values are significantly different from those of the pooled assemblage, respectively, 9.0 cm, and 3.2 cm to 31.0 cm. Compared with the long bone lengths of mammoth, bison, reindeer, wolf, and horse provided by Vera Gromova (1950: table 27), the Shestakovo upper range limit is greater than the upper limits for all but mammoth.

**7 Damage shape.** The classes of damage form listed in Table A1.7 (site 23) can be attributed mainly to natural breakage, perhaps largely by cryoturbation, wet and dry cycling with concomitant bone-splitting caused by salt or ice crystal growth, and trampling. Thin soil atop bedrock is also involved because protective burial for huge animals would not have been quickly possible. In addition, Leshchinskiy (2012) suggests that breakage and preservation should be expected to occur more in mammoths that had some type of bone disease than in animals that were healthy at the time of their deaths.

More than half (57.7%) of our sample has no identifiable perimortem damage. Heading the list of damage forms are segments (8.8%), shafts (6.3%), vertebral spines (5.7%), long bone flakes (5.0%), and vertebral bodies (5.0%). Types of damage that would be expected from human and carnivore activity are almost non-existent.

Of the damage type that might best be viewed as age-dependent – the separation of the vertebral spines and bodies – this expectation does not hold up because the condition is about equal in sub-adults and adults. On the whole, the Shestakovo damage types are probably best viewed as resulting from the natural causes suggested above. This damage pattern largely excludes carnivorous and anthropogenic agencies.

**8 Color.** Of 159 pieces studied, most are ivory colored (97.5%). Four (2.5%) are brown (Table A1.8, site 23). Unlike cave sites such as Razboinich'ya, where the ivory color is due to the natural color of unstained bone, or the brown-stained bone of open sites such as Bolshoi Yakor and Krasny Yar, the ivory color of Shestakovo bone has two possible explanations. First, it probably reflects the negligible content of organic and mineral-staining materials in the embedding sediments. Second, the external exfoliating surface of most mammoth bones found in the Shestakovo context may not be able to retain soil stain as well as do their teeth. As expected on archaeological grounds, there are no black or white pieces that would indicate, respectively, charring or calcining. Compared with the pooled assemblage, Shestakovo has more ivory colored pieces, and fewer with the other four colors.

**9 Preservation.** Of the two qualitative states recognized for this study, most of the 159 pieces are chalky (88.0%). The remainder are ivory-like (Table A1.9, site 23). This ratio is due not so much to the corrosiveness of the Shestakovo embedding sediments as it is to

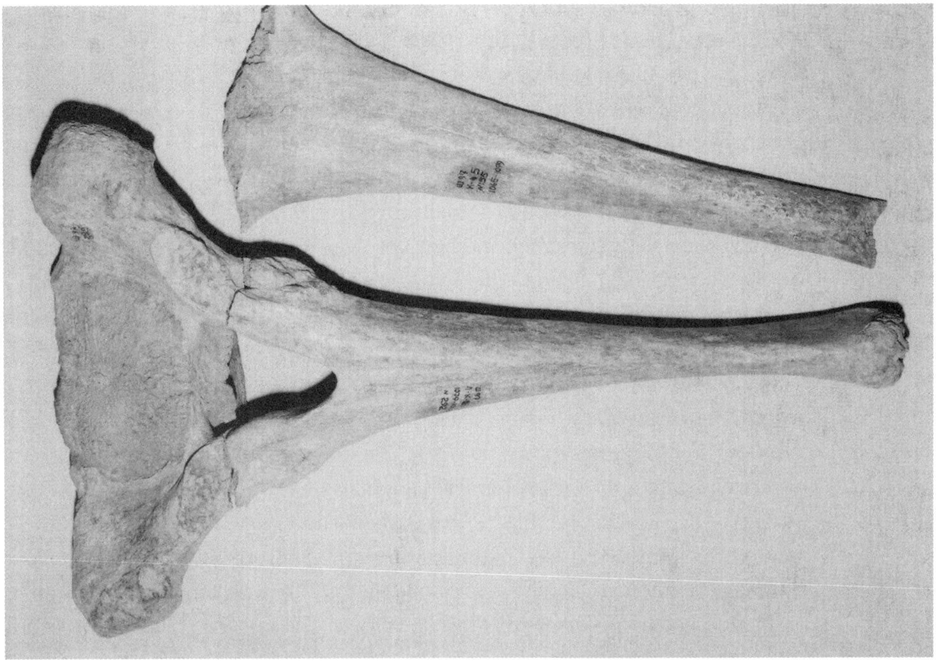

Fig. 3.141 Shestakovo mammoth vertebrae. Spinal process without corpus (upper) has perimortem breakage. There is no identifiable carnivore damage. Upper spine is 34.0 cm in length (CGT neg. IAE 7-5-01:21).

the easily damaged thin cortex of many bones of the mammoth skeleton. Those that have a thicker cortex make up 73.7% of the ivory-like pieces. These are foot bones in general (36.8%; 7 / 19), vertebrae (15.8%; 3 / 19), skull and jaw (15.8%; 3 / 19), and the one astragalus (5.3%) in our sample.

**10 Perimortem breakage.** There is relatively little perimortem breakage identifiable in the Shestakovo assemblage, only 18.7% (Table A1.10, site 23). As discussed above, the archaeological interpretation of Shestakovo as having a minimal carnivore and human presence is supported by several of our perimortem damage features, including perimortem breakage (Fig. 3.141). We suspect that the amount of perimortem breakage was actually higher, but with time sharp fracture edges have eroded and rounded to such an extent that they look like postmortem breaks.

**11 Postmortem breakage.** Postmortem breakage (Table A1.11, site 23) is about two times more frequent (32.5%) than perimortem breakage (18.7%). How much of the postmortem breakage should be attributed to postmortem trampling (if any) in contrast to the easily damaged surface of mammoth bone cannot be determined. However, in comparison with the open paleontological site of Krasny Yar, whose postmortem breakage frequency is 38.5%, the difference is not significant.

**12 End-hollowing.** This feature is commonly seen in contexts where carnivores must have been the principle cause of perimortem damage. We are aware from the senior author's archaeological experience in the Aleutian Islands that wave action on abrasive beaches can produce end-hollowing in long bones of sea mammals. But since all our Siberian sites except Krasny Yar and Boisman II lack taphonomic involvement with high-energy rivers or ocean shores, end-hollowing here is viewed as an exclusive result of carnivore chewing, especially when associated with tooth dints and scratches.

The three examples of end-hollowing at Shestakovo (1.9%; Table A1.12, site 23) are best attributed to carnivore processing. If, as Leshchinskiy suggests, the ground around Shestakovo was wet, muddy, and perhaps even churned and hoof-cratered like the mucky edges of ponds used by domestic livestock, then it is not too difficult to imagine that some carnivores might have preferred to look for something easier to reach, and something more rewarding that a rotting mound of mammoth carcass.

The first example of end-hollowing occurs on a nearly whole vertebra of an unclassifiable animal, but presumably a mammoth. This 9.2 cm wide adult specimen (W375) also has 24 tooth dints. The second example occurs on an unnumbered 15.7 cm long adult mammoth metapodial with only one anatomical end. It also has polishing on the missing end and in the middle of the shaft. The third example is an unnumbered whole 40.8 cm long sub-adult mammoth humerus. Being classified as whole means that the end-hollowing had not completely removed the chewed-upon anatomical end. Alternatively, trampling might have been the cause of the damage to this specimen. The first two pieces with end-hollowing are sure evidence of carnivore presence.

**13 Notching.** As with end-hollowing, notching is a common feature in our carnivore and archaeological assemblages, discussed previously. It is practically non-existent (1.3%; Table A1.13, site 23) in the Shestakovo sample. The two notching examples include specimen 353, a 4.3 cm long adult reindeer metapodial with butt-shaped perimortem breakage. The second specimen, number 358, is also from an adult reindeer. It is a 4.5 cm long astragulus with one notch between the two anatomical ends. Both bones probably belonged to the same individual. The damage to both was most likely caused by humans, since neither has any tooth dints or scratches.

**14–15 Tooth scratches, dints.** We could not find a single tooth scratch or dint in 159 pieces. Multiple tooth scratches and dints are good indicators of carnivore activity. However, given that the surfaces of mammoth bone breaks down relatively fast, except for the nearly solid harder-surfaced mammoth foot bones that have almost no nutrient value for carnivores or humans, we allow that a few tooth scratches may once have been present, but are now eroded away (Fig. 3.142). The same holds for sand and grit scratches that might have been produced with trampling. We found no random surface scratching that would indicate trampling.

**16 Pseudo-cuts.** There is one example of pseudo-cutting (0.6%; Table A1.16, site 23). This is specimen 868, a 25.1 cm long chalky mammoth rib fragment with perimortem and postmortem breakage, and end-polishing. The animal's age at death is indeterminable. There are nine pseudo-cuts present. We consider this piece to be a valid case of carnivore perimortem damage.

**17 Abrasions.** There are no examples of abrasion.

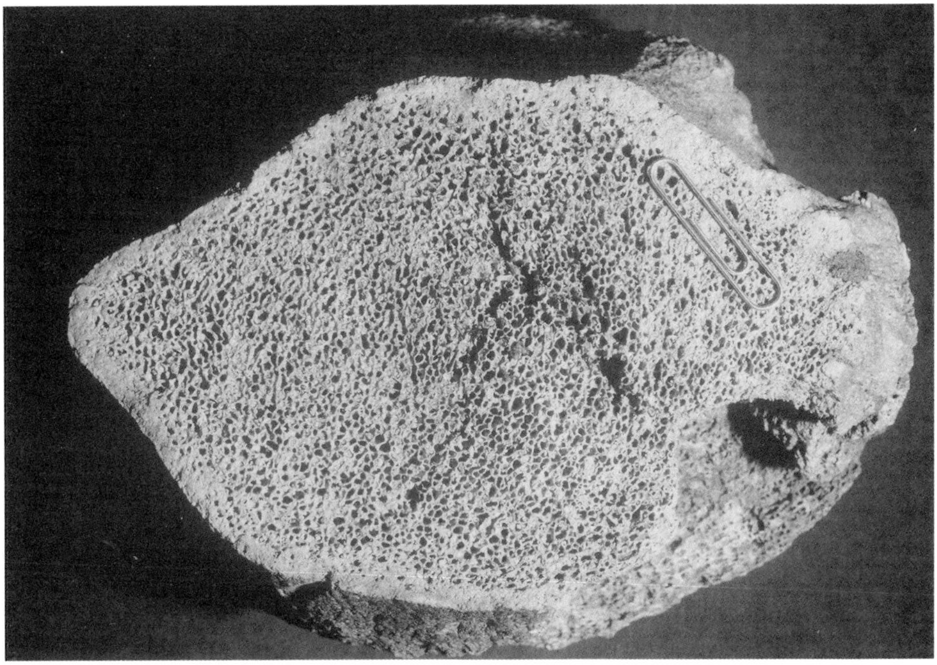

**Fig. 3.142** Shestakovo mammoth long bone cross-section. This long bone fragment with postmortem breakage illustrates how very little cortex there is in a mammoth long bone. It is largely for this reason that bone surface weathering is often pronounced in mammoth remains. Paper clip is 4.8 cm long (CGT neg. TU 6-27-02:11).

**18 Polishing.** Polishing is rare but sufficiently represented in our view to provide some slight evidence of carnivore perimortem damage when judged in conjunction with the above-mentioned end-hollowing and pseudo-cuts. Of 162 pieces, seven have polishing (4.2%; Table A1.18, site 23). (1) Specimen 393 is a dog- or wolf-sized chalky scapula flake, 5.0 cm long with postmortem and perimortem breakage and end-polishing. (2) Specimen 334 is a 6.2 cm long flake of an unidentifiable bone of ivory quality with perimortem breakage and middle-area polishing. (3) An unnumbered piece is the 15.7 cm long mammoth metapodial mentioned under end-hollowing. This specimen has end and middle-area polishing. (4) Specimen 868 is the 25.1 cm long rib fragment discussed under pseudo-cutting. (5) Specimen 155 is a 34.2 cm long chalky mammoth vertebral spine with perimortem breakage and end-polishing. (6) Specimen 306 is a 30.0 cm long chalky mammoth rib fragment with middle-area polishing. (7) Specimen 326 is an 11.4 cm long chalky mammoth rib fragment with middle-area polishing.

The rarity of polishing in this assemblage may also be due to the nature of the burial matrix. Judging from the very fine gray clay adhering to incompletely cleaned fragments, it is unlikely that the faunal remains moved around much. On the other hand, cave deposits probably were more often disturbed – for example, by hyena digging or by hoofed animals using the cave for shelter. Bone movement in a gritty limestone soil probably produces a significant amount of the end-polishing we see in our samples from cave sites.

**19 Embedded fragments.** The only piece (specimen 82) with embedded fragments (0.06%; Table A1.19, site 23) is a 6.2 cm wide chalky sub-adult mammoth vertebral body with perimortem breakage. Inasmuch as there are few other indications of processing by humans or carnivores, perhaps the two embedded fragments resulted from trampling damage. Still, carnivore presence is absolutely certain because of the wolf mandible mentioned next.

**20 Tooth wear.** There are only two tooth-bearing pieces (Table A1.20, site 23). An unnumbered 4.6 cm long adult wolf mandible fragment has grade 1 wear (dentine exposed). Also unnumbered is an 11.4 cm long reindeer mandible fragment with grade 1 wear. Both are young animals.

**21 Acid erosion.** There are no examples of bone fragments having been swallowed and exposed to stomach acid. This is supportive evidence for viewing carnivore presence at Shestakovo as minimal. It also suggests that hyenas, which are highly correlated with bone acid damage, were rare or non-existent in the Shestakovo district. Limited den sites, severe cold, and competition from wolves may have been responsible for their apparent absence.

**22–24 Rodent gnawing, insect damage, and human bone.** There are no examples of these variables in the Shestakovo assemblage.

**25 Cut marks.** One bone flake has two cut marks (Table A1.25, site 23). These occur on specimen 334, a 6.2 cm long flake of ivory color and quality, but of unknown species, age, or skeletal element. The flake had perimortem breakage and middle-area polishing. This small solitary flake is the major non-artifactual evidence for human presence at Shestakovo. It speaks almost with no voice for human activity at Shestakovo, despite the number of artifacts mentioned in the introduction.

**26 Chop marks.** No examples of chopping were identified in the Shestakovo sample of 159 pieces.

## Discussion

Shestakovo is a paleontological open site where mammoths and other animals most likely died of natural causes. There are stone tools present. Because of the general absence of perimortem butchering marks and bone breakage, humans probably utilized the mammoths for their bone and ivory fabricational needs rather than for their dietary needs of meat, fat, and other edible tissue. We note, however, that the outer cortex of mammoth bone is thin and does not preserve as well as the harder and thicker cortex bone of smaller mammals such as the reindeer found here. This biological quality of mammoth bone could have contributed to the loss of surfaces that might once have had cut marks, thus falsifying our inference of non-nutritional carcass utilization.

# 24 Straschnaya Cave

## Background

The word "Straschnaya" in Russian means "terrible." It is the name used by local residents for a cave in the Altai Mountains. When the cave was being excavated, none of the workers thought to ask the local people *why* the cave was so named. Hence, we have no explanation (Okladnikov *et al.* 1973). According to Kuzmin and Orlova (1998:6), Straschnaya Cave is located at 51°75′ N, 83°84′ E. It is carbon-14 dated at >25 000 BP (Vasili'ev *et al.* 2002:521). A. P. Okladnikov and Nicolai Ovodov did the first excavation in Straschnaya Cave in 1969 (Okladnikov and Ovodov 1972, Okladnikov *et al.* 1973, Ovodov 1973a, 1977c), followed by additional stratigraphic analyses by Derevianko and Zenin (1992) and Derevianko *et al.* (1998i). Faunal studies have been carried out by Ovodov (1973a, 1977c), Galkina and Ovodov (1975), and Ovodov and Martynovich (2004). The cave had been used by both hyenas and humans.

Straschnaya is a horizontal cave, 23 m in length. The cave is located on the right bank of the Tigirek River directly before it merges with the Inya River (the basin of the upper Charysh). In ancient times it was repeatedly used by Paleolithic hunters. The main excavation of Straschnaya was made by Ovodov near the entrance. The test pit went through 9.6 m of loose sediments. The second test pit was made in the back of the cave, 18 m from the entrance. It was dug to a depth of 1.9 m. The first test pit had six levels of loose sediments. The second pit had three horizons. Bones of large and small mammals were found all over the cross-section in varying numbers.

Galkina and Ovodov (1975) reported that Layers 1 and 2, and the uppermost zone of Layer 3 (down to a depth of 160–180 cm from the ground surface in the remote part of the cave), were considered to be of Holocene age because of the limited degree of bone fossilization, by the absence of large mammals that went extinct at the end of the Pleistocene, and because of the presence of species characterized by evolutionarily progressive dental qualities. Starting from the middle part of Layer 3, including Layer 4, the loose sediments of Straschnaya are considered to be late Pleistocene because of the presence of *Citellus* sp., *Marmota* sp. with relatively low tooth crowns and primitive molar 3 structure, and the presence of gray voles with features transitional between *Pitymys* and *Microtus* and *Clethrionomys* sp. Their dating inferences were in good agreement with archaeological data (Okladnikov *et al.* 1973), and confirmed by radiocarbon dating based on bones of large mammals. Upper Layer 3 is 25 000 BP (SO-AN SSSR, 785); lower Layer 3 and 4 (depth 4–6 m) more than 40 000–45 000 BP (SO-AN SSSR 786–787).

A comparative study of modern mammals in the local area and those found in Straschnaya Cave, along with pollen analysis made by Novo-Kuznetsk paleontologist L. N. Yefimova, enabled Galkina and Ovodov (1975) to reconstruct the environmental setting of northwestern Altai in the terminal Pleistocene. Straschnaya, being located in the foothills of the Altai, is set in the forested slopes which share a border here with the vast West Siberian Plain and the low hills of eastern Kazakhstan, blanketed with steppe and forest-steppe vegetation. In the late Pleistocene this area was probably covered with forest-steppe where mixed birch, spruce, and pine forests grew in multi-herbal steppes. This inference was based partly on the occurrence of much grass (53–60%) and woody plant (35–42%) pollen in the cave midden, as well as by the existence at that time of such animals as Allactaginae, *Ellobius talpinus*, and *Equus hemionus*, each of which inhabited this area in direct proximity with forest species: *Ursus arctos*, *Alces alces*, *Cervus elaphus*, *Capreolis capreolis*, *Clethrionomys glareolus*, *Clethrionomys rutilus*, and *Clethrionomys rufocanus*. In the final late Pleistocene a drying up of the northwestern Altai foothills seems to have taken place, promoting the development of steppes with cereal/orach/multigrass complexes that are evidenced by pollen spectra with the dominance of grasses (87–96%), and woody plants reduced to 0.4–5.0%. As for mammals at this time, the steppe and meadow (alpine?) species dominate (*Lagurus lagurus*, *Myospalax myospalax*, *Allactaga* sp., Cricetinae, *Equus caballus*, *Bison priscus*), as well as mountainous (*Alticola* sp., *Ovis ammon*) and semi-desert (*Lagurus luteus*, *Equus hemionus*) species.

Forest fauna in Straschnaya's final Pleistocene days are represented by only a small number of species: *Clethrionomys rutilus*, *Ursus arctos*, *Mustela nivalis*, *Mustela erminea*, *Capreolus capreolus* and *Cervus elaphus*. In the beginning of the Holocene some mammals, members of a general Late Pleistocene terracomplex, went extinct: *Crocuta spelaea*, *Mammuthus primigenious*, *Coelodonta antiquitatis*, and *Bison priscus*. Galkina and Ovodov (1975) concluded that there were no continuous tracts of periglacial tundra-steppe in the late Pleistocene of the western Altai. Instead, there was a diversity of biotopes that made possible the co-existence of species such as *Mammuthus primigenius*, *Allactaga* sp., and *Alces alces*. For this reason there was no drastic replacement of the open-type vegetation by a forested condition, which did occur in many other regions of Siberia.

## Findings

**1 Provenience.** The 527-piece Straschnaya assemblage makes up 6.0% of our 8813-piece grand total. Most are probably from the Pleistocene horizons, although we are not absolutely certain because of inadequate specimen labeling.

**2 Species.** The two most common groups in the 527 pieces are: indeterminable (91.1%) and big mammal (6.4%) (Table A1.2, site 24). There is a lot of marmot bone in Straschnaya, but we recorded only the pieces that seemed that they might have been carried in, i.e., those having tiny, fox-sized tooth dints. Ovodov feels that marmots burrowed in the cave, and some died naturally in their burrows, leaving clusters of undamaged bones in largely articulated conditions. The more common groups in the pooled assemblage averages that Straschnaya is lacking or has very low frequencies of include: bear, bison, gazelle, goat-sheep, horse, hyena, mammoth, reindeer, and roe deer.

**3 Skeletal elements.** The most common groups represented in the 527 pieces include: unknown (49.7%), long bone (31.7%), rib (5.3%), and metapodial (4.2%) (Table A1.3, site 24). Compared with the pooled assemblage, Straschnaya has no or fewer pieces of mandible, vertebra, scapula, humerus, radius, femur, foot, and toe; and more long bone and unknown pieces.

**4 Age.** There is almost no (0.6%) sub-adult representation in the 527 Straschnaya pieces (Table A1.4, site 24). This is much less than the pooled assemblage average.

**5 Completeness.** Straschnaya is a very incomplete assemblage as assessed by this "missing end" variable. Of the 527 pieces there is only 0.9% whole, 8.0% one anatomical end, and 91.1% with no anatomical ends (Table A1.5, site 24). Compared with the pooled assemblage, Straschnaya is much less complete. Presumably this reflects its having had hyena and/or other carnivore occupants.

**6 Maximum size.** This variable correlates with completeness. The mean for the 527 Straschnaya pieces is 5.4 cm, the range is 2.5 cm to 16.2 cm (Table A1.6, site 24). These values are about half those of the pooled assemblage. Regardless of what species are represented in the Straschnaya assemblage, its upper range limit is smaller than most of the long bone lengths provided by Vera Gromova (1950: table 27), with the exception of fox.

**7 Damage shape.** The most common forms of damage to the 537 Straschnaya pieces are: long bone flake (71.8%), long bone splinter (17.6%), and irregular (4.2%) (Table A1.7, site 24). Compared with the pooled assemblage averages, Straschnaya has more long bone flakes, splinters, and irregular bones; and fewer long bone fragments, phalanx butts, medial ribs, undamaged, and mostly whole bones.

**8 Color.** All 527 Straschnaya pieces are ivory colored (Table A1.8, site 24). This frequency is greater than the pooled assemblage average.

**9 Preservation.** Only 3.2% of the 527 Straschnaya pieces are chalky. The rest are ivory hard (Table A1.9, site 24). Compared with the pooled assemblage, Straschnaya has almost six times fewer chalky pieces. Information on the amount of time sunlight fell directly into the cave would be useful here, but the low frequency of chalky bone suggests that old ground surface carcasses were not often carried into the cave by scavengers.

**10 Perimortem breakage.** Almost all (98.1%) of the 526 Straschnaya pieces have perimortem breakage (Table A1.10, site 24). This frequency is more than the pooled assemblage average.

**11 Postmortem breakage.** The amount of postmortem breakage in the 527 Straschnaya pieces is 11.4% (Table A1.11, site 24). This amount is somewhat less that the pooled assemblage average. Given the nature of our perimortem damage "universe," we do not see that it indicates any meaningful condition.

**12 End-hollowing.** There is almost no (0.9%) end-hollowing in the 527 pieces (Table A1.12, site 24). Much more end-hollowing occurs in the pooled assemblage average. Here, there is good reason to be curious about the difference, and it is surely due to the low frequency of "whole" bones. The lesson to be learned here is that a multi-variable study may have more highly correlated variables than expected.

**13 Notching.** About 10% (11.4%) of the 527 Straschnaya pieces exhibits 1–7 notches. One notch is the most frequent number per piece (7.8%) (Table A1.13, site 24). Compared with the pooled assemblage average, Straschnaya has slightly fewer notched pieces.

**14 Tooth scratches.** Tooth scratches are fairly frequent (22.9%) in the 526 Straschnaya pieces. The number of scratches per piece ranges from one to more than seven (Table A1.14, site 24). Compared with the pooled assemblage average, Straschnaya has about the same frequency of tooth scratching.

**15 Tooth dints.** The frequency of dinted pieces in the 526 Straschnaya pieces is 31.9%. The number of dints per piece ranges from one to more than seven. One dint is the most common number (9.1%) of dints per piece (Table A1.15, site 24). Compared with the pooled assemblage average, Straschnaya has slightly more dinted pieces. We feel these values are supportive of a significant carnivore presence, and whatever bioturbation it implies relative to human use of this cave.

**16 Pseudo-cuts.** There are six pieces with pseudo-cuts in the 526-piece Straschnaya assemblage. In this 1.1% occurrence, the number of pseudo-cuts per piece ranges from one to a piece that has 22 pseudo-cuts (Table A1.16, site 24). Compared with the pooled assemblage average, Straschnaya has less pseudo-cutting. Nevertheless, Straschnaya has all of the perimortem features we associate with carnivores, and as will be seen, there is also a considerable number of pieces with acid erosion. Taken altogether, the perimortem taphonomy suggests that the cave was more likely a hyena den than a continually used human shelter.

**17 Abrasions.** Out of 526 Straschnaya pieces, there is only one (0.2%) with abrasions. This single piece has four striations (Table A1.17, site 24). Compared with the pooled assemblage average, abrasions are also rare in the Straschnaya assemblage.

**Fig. 3.143** Straschnaya Cave, stomach bones. A 1988 sample of this type of bone damage. Scale is in centimeters (CGT neg. IAE 7-11-01:2).

**18 Polishing.** Of the Straschnaya pieces, 511 could be scored for polishing, which is 69.7%. By location, polishing occurs at: one end 5.7%, middle 1.0%, and end-middle 63.0% (Table A1.18, site 24). The total amount of polishing in the pooled assemblage is similar to the Straschnaya condition, although the locational amounts are different. Without question, carnivores used this cave.

**19 Embedded fragments.** Embedding is rare (0.4%; 2 / 527) in the Straschnaya assemblage. One piece has one embedded fragment, the other has four (Table A1.19, site 24). Compared with the pooled assemblage average, Straschnaya has much less.

**20 Tooth wear.** Only one tooth was available for scoring wear, but was not because it had been swallowed and had acid erosion. It was classified as a stomach bone.

**21 Acid erosion (Figs. 3.143–3.145).** Out of 528 Straschnaya pieces, more than one-quarter (28.3%) have acid erosion (Table A1.21, site 24). This frequency is more than the Razboinich'ya (7.3%) and Dvuglaska (22.1%) hyena caves, but less than Maly Yaloman (48.0%). Hyenas were most likely the principal occupants of Straschnaya Cave.

**22–24 Rodent gnawing, insect damage, and human bone.** There are no examples of these variables.

**Fig. 3.144** Straschnaya Cave, stomach bone. A very large eroded toe bone 7.2 cm in length (CGT neg. IAE 7-5-02:6).

**Fig. 3.145** Straschnaya Cave, tooth. A mature hyena incisor with tooth scratches and polished root tip. Cannibalism. Actual width of image is 3.3 cm (CGT neg. IAE 7-5-02:17).

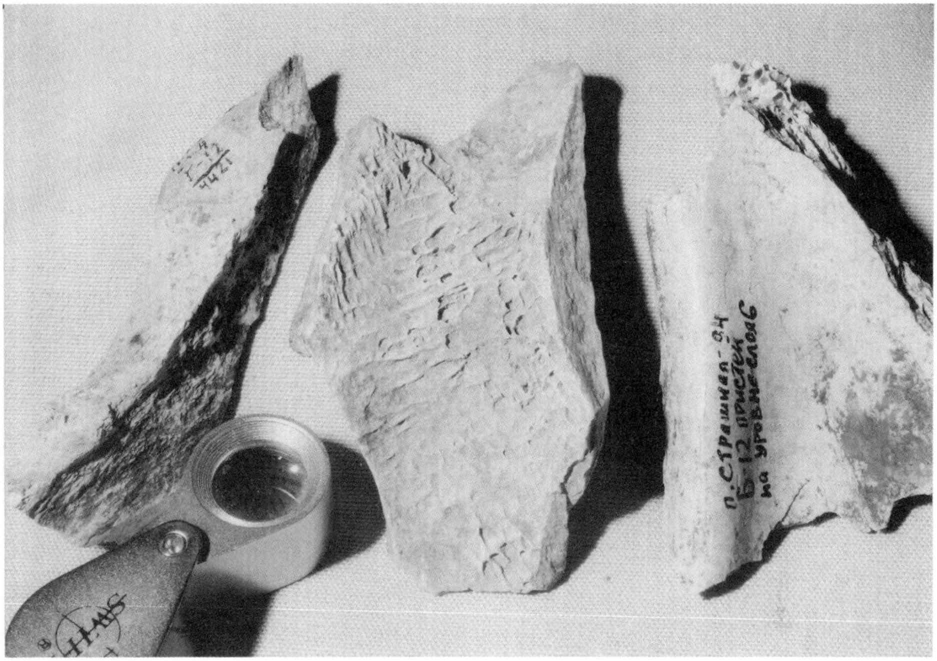

**Fig. 3.146** Straschnaya Cave bone damage. Large bone flakes produced by carnivores, presumably hyenas. Center long bone fragment is 9.0 cm in length (CGT neg. IAE 7-5-02:20).

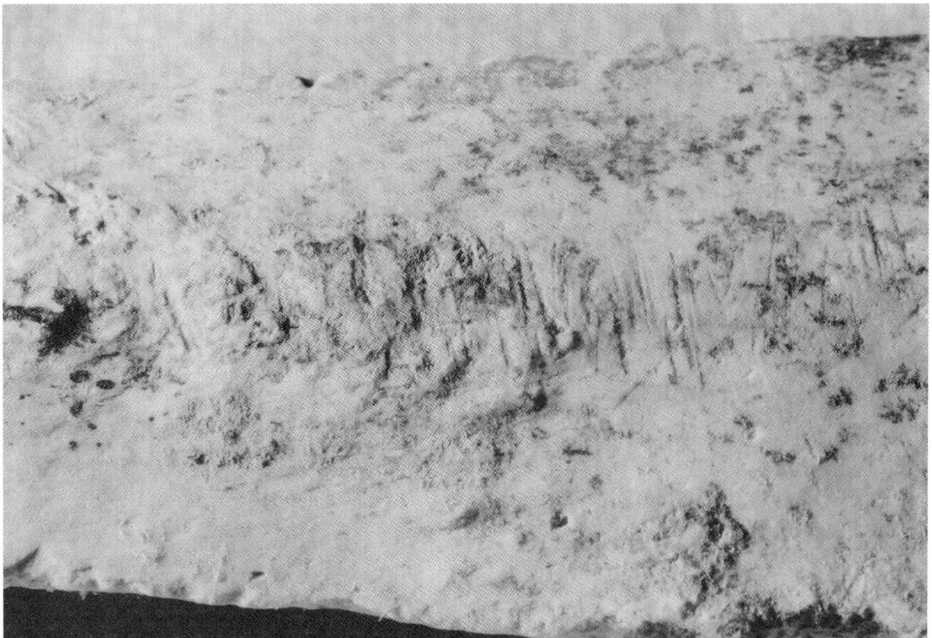

**Fig. 3.147** Straschnaya Cave (1988) pseudo-cuts. Because of the rounded edges of the "cuts," plus other unquestionable carnivore damage to this specimen, the numerous fine parallel striations were judged to be pseudo-cuts. This specimen is an excellent example of the need of a class of damage that allows for uncertainty of identification. Actual width of image is 3.3 cm (CGT neg. IAE 7-11-01:7).

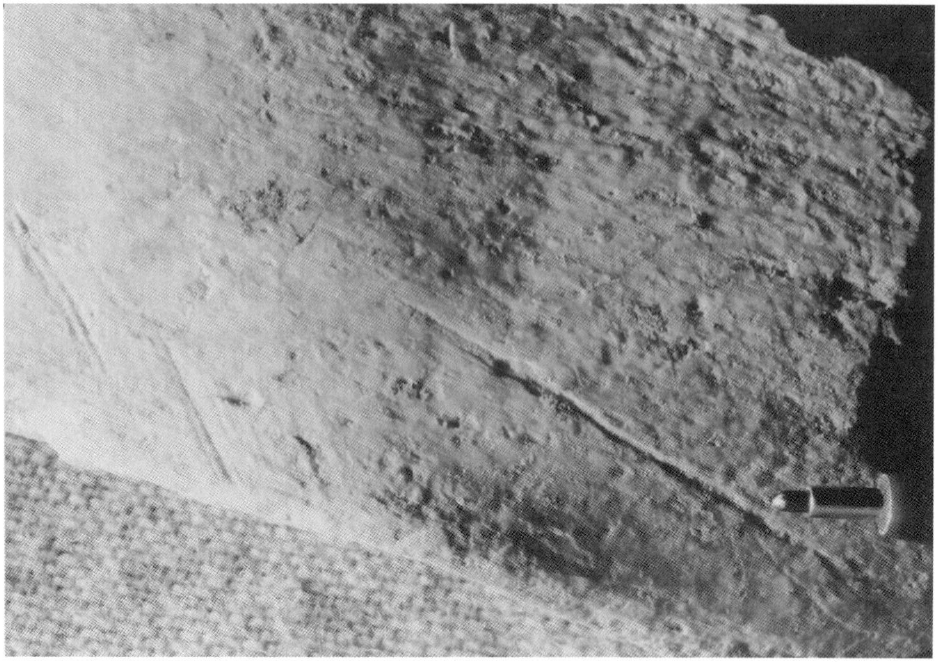

**Fig. 3.148** Straschnaya Cave, cuts and pseudo-cuts. This odd piece has both stone tool cut marks (lower left) and a pseudo-cut mark (lower right at pencil point). There also appears to be some surface erosion, suggesting that the piece had been swallowed but regurgitated some time afterward. Actual width of image is 3.3 cm (CGT neg. IAE 7-16-01:17).

**Fig. 3.149** Straschnaya Cave, stone artifacts. Exhibit of stone objects found in the cave. The text indicates that there is a flake, a pointed screblo, a fragment of a Levallois point, and unidentified specimens. The variability in quality of these artifacts may explain some of the difficulty in identifying some cut marks. Exhibit photograph shows entrance to cave (CGT color IAE 8-5-02:31).

**Fig. 3.150** Straschnaya Cave, "symbolism." This bone fragment, numbered 369, was found in the paleontological collection, seemingly missed in the field or lab sorting of cultural and non-cultural remains. The equally spaced cut marks were probably made as a decorative activity, although a calendrical purpose might alternatively be inferred. The bone fragment is 7.2 cm in length. There are >25 parallel cut marks (CGT neg. IAE 7-5-02:1A).

**25 Cut marks.** Only four Straschnaya pieces out of 527 have cut marks, a frequency of 0.8%. One, three, six, and more than seven are the number of cuts per piece (Table A1.25, site 24). Compared with the pooled assemblage average, cutting is very rare in the Straschnaya assemblage, approaching the total absence of the Razboinich'ya and Dvuglaska hyena caves.

**26 Chop marks.** Out of 527 Straschnaya pieces, only one has been chopped (0.2%). This piece bears only one chop mark (Table A1.26, site 24). Compared with the pooled assemblage average, chopping is extremely rare in the Straschnaya assemblage. Such rarity opens up the possibility of our misidentification.

## Discussion

While Straschnaya is treated by Derevianko *et al.* (1998i) as an archaeological site, the perimortem taphonomy shows that the bone refuse found in the cave has far more damage attributable to carnivores (Figs. 3.146–3.148) than to humans

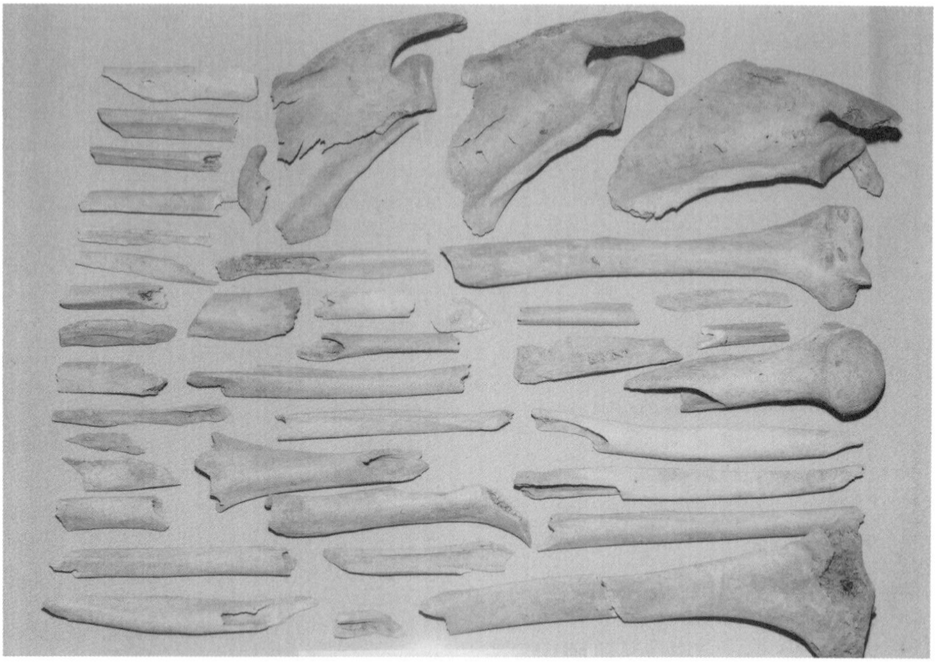

**Fig. 3.151** Straschnaya Cave, human remains. Found by Anatoly Zenin, the damage pattern to these human bones is not a great deal different from other assemblages elsewhere in the world whose damage pattern strongly suggests cannibalism. These remains are thought to be of early Holocene or late Pleistocene age. Whatever their exact age, they post-date or are near to the time of the extinction of hyenas, there being no clear-cut carnivore damage. At least four adults are represented. In addition to the perimortem breakage, there are also some cut marks. Scale is 15 cm (CGT neg. IAE 7-5-01:3).

(Figs. 3.149–3.151). In addition, the large number of acid-eroded pieces suggests that hyenas were the primary large carnivore occupants. If so, then our cautionary warning about hyena disturbances of stratigraphy applies here as well as in the other sites discussed. On taphonomic grounds, Straschnaya is more a paleontological than an archaeological site.

# 25 Ust-Kan Cave

## Background

Ust-Kan is a cave site 3.5 km east of the Altai Republic mountain town and local administrative center of the same name. The cave is located at 50°54′40″ N, 84°48′50″ E (Derevianko *et al.* 2001b:130). The cave opening is about half way up (52 m) a steep, rounded limestone butte set apart in the surrounding steppe plain from nearby forested hills and more distant mountains (Figs. 3.152 and 3.153). Not far from the base of this butte a clear mountain river rushes northward in the Charysh River valley and through the town. Its water eventually reaches the Arctic Ocean via the Ob River.

The 8.3 m wide, 16.7 m deep, and 10 m high cave (Derevianko and Markin 1998:85) was first tested in 1954 by archaeologist Sergei I. Rudenko (1960, 1961). Derevianko and Markin (1998) lay out a brief history and summary of the stratigraphy, recovered materials, cultural affinity, and reconstructed environment based on Rudenko's excavation (Fig. 3.154). They provide no carbon-14 dates for Ust-Kan, nor does the recent inventory and interpretation of Siberian Paleolithic dates compiled by Vasili'ev *et al.* (2002). On the basis of 520 artifacts, Derevianko and Markin conclude that the cave dwellers had a late Mousterian tool kit. There is no physical evidence of the dwellers themselves. Of 1749 pieces of late Pleistocene faunal remains recovered by Rudenko, one-third (532) could be identified as to species – 17 of which were mammalian and 12 avian. Elsewhere, L. I. Galkina and N. D. Ovodov (1975) report that only 12 species of mammals had been identified by Vereschagin. The inventory suggests a combination of species that lived in open spaces and forest, as well as being moderate-temperature and cold-loving, which altogether implied an arid, treeless Central Asian plateau landscape surrounded by mountain forests. This would match to a large degree the botanical landscape today. Large carnivores included cave bear, wolf, and cave hyena. More recently, cave lion has been added to this list as a result of Patrick Wrinn's finding two paw fragments (Postnov and Karavaeva 2005) in the Ust-Kan osteological assemblage of "layer 5, cave 2." Ust-Kan stratigraphy is far from simple (Agadjanyan *et al.* 2002). This condition we propose involves much carnivore bioturbation.

An undisturbed hearth was discovered. It was given special study, including the surrounding rock structure (Postnov and Kulik 2003). It serves as a reminder to us that bioturbation missed some fraction of the midden deposit, but the amount has not yet been estimated.

**Fig. 3.152** Ust-Kan site. The cave is in the upper right corner of this view, 52 m above the valley floor on which the people are standing. An aerial cable and bucket system carries excavated midden down to the edge of the Charysh River, where Nicolai Ovodov (right) explains to student workers his paleontological identifications from previous summer excavations supervised by IAE archaeologist Alexander Postnov. The aerial cable system is a continuous loop of heavy single-strand wire that is propelled by the weight of a filled bucket as it descends from the cave to the river below, where its contents are water-screened. An empty bucket is propelled up to the cave on the returning loop of cable (CGT color Ust-Kan 7-12-99:26).

There is an extensive Russian literature on Ust-Kan, in addition to those cited above. Most are based on the materials recovered by field supervisor Alexander V. Postnov: Derevianko *et al.* (1998j, 1999c, 1999d, 2000a, 2001a, 2001b, 2002a, 2003d, 2004); Postnov and Kulik (1998); Postnov and Karavaeva (2005); among others.

More recent fieldwork led by A. P. Derevianko and supervised by Alexander V. Postnov took place in the late 1990s, continuing into the first years of this century. Our inspection of Ust-Kan was on July 12, 1999 (Figs. 3.155–3.156). On that date, Postnov explained to us his understanding of the archaeology, and of how Ust-Kan was formed – by faulting and earthquake activity, unlike the karst solution caverns elsewhere in the Altai. Postnov could readily demonstrate that some of the troubling past interpretative problems are due to Rudenko's excavation methods. However, as will be shown, these stratigraphic problems very likely began long before Rudenko's excavation. The possibility of Ust-Kan bioturbation is not mentioned by Derevianko *et al.* (2001b). We did not examine the Rudenko faunal finds. Our sample came entirely from the IAE excavations of Postnov and associates.

Fig. 3.153 Ust-Kan site. Of all the Altai cave sites excavated by IAE archaeologists, Ust-Kan is the most difficult to reach, making it both highly defendable and a trap. The difficulty arises from the steepness (notice that the slope at the upper left and bottom right form a right-angled triangle), and the loose, broken rock covering the slope. The climb would have made winter use of Ust-Kan undesirable because of the effort it would have taken to haul firewood up to the cave. Catching their breath are, left to right: Nicolai Martynovich, Olga Pavlova, Alexander Postnov, and a female student (CGT color Ust-Kan 7-12-99:35).

## Findings

**1 Provenience.** The Ust-Kan faunal assemblage is the largest of our fully scored assemblages (Table A1.1, site 25). It was possible to generate two large nearly equal subsets by pooling the pre-2002 pieces ($n = 780$) and comparing their maximum lengths with those Postnov and associates recovered in 2002 ($n = 876$). Most of the latter were stratigraphically mixed due to domestic sheep and goats using the cave for shelter in the winter of 2001–2002, wherein they wrecked the excavated face that had been left exposed for continued work in summer 2002. Most of the disturbed midden came from Levels 3 and 5 (Postnov, personal communication, June 11, 2003). A $t$-test shows only a small statistically significant difference ($t = 2.263$; $0.05 > p > 0.02$) despite the means and standard deviations being very similar (means = 4.9 cm and 4.7 cm; SDs = 1.61 and 1.60). Given the provenience conditions, we do not consider the barely significant difference to be meaningful, and have pooled the Pleistocene faunal remains recovered from all of the IAE excavation years. We do so in light of the relatively shallow cave deposits (generally less than 1.0 m), the sheep disturbance, and the abundant evidence of

**Fig. 3.154** Ust-Kan site. The cave has a sloping floor, so all deposit is subjected to some degree of downward creep. Ust-Kan was first tested in 1954 by S. I. Rudenko, at the place where the student is standing near the distant cave wall. Since then, Anatoly Derevianko and Alexander Postnov have extended the excavation deep into the cave, the back wall of which is out of sight on the right in this figure. Unlike Razboinich'ya Cave, Ust-Kan is situated in highly fractured limestone. Standing near the bucket cable's pulley mechanism, from left to right, are: Ovodov, Martynovich, Postnov, and Pavlova (CGT color Ust-Kan 7-12-99:33).

late Pleistocene hyena and wolf use of the cave that most likely also caused substantial stratigraphic disturbance. The strongly sloping cave floor would not have helped with stratigraphic equilibrium.

**2 Species (Fig. 3.157).** We examined 1654 Ust-Kan pieces. Wolf is the most frequently recognized species (5.5%) (Table A1.2, site 25). In contrast with the 17 (or 12) mammalian taxa reported by Derevianko and Markin (1998:85), the present study identifies only eight. However, the difference is not statistically significant ($\chi^2 = 3.0$; $p > 0.05$). The difference is due mainly to the fact that Ovodov was not able to assist Turner and Pavlova during the period of study of the Ust-Kan assemblage. Turner has long held to an observation policy of scoring a condition as "missing data" rather than guessing.

**3 Skeletal elements.** Of the 1654 Ust-Kan pieces, the most common are: long bone (42.2%) and rib (4.7%) (Table A1.3, site 25). We queried Postnov about the small number of cranial pieces, and he assured us that we had all that were in the faunal sample. The cranial elements had not been separated out for special study, and given the many very small pieces of bone, there can be no question about the carefulness of the field crew in collecting everything possible. Thus, it seems that most game animal heads had not been

**Fig. 3.155** Ust-Kan. View northward (down-river) toward the town of Ust-Kan. There, in a crowded general store, we purchased cheap soup spoons and souvenir scarves for ourselves, and several bottles of vodka to serve as gifts for the adult members of the 20-person summer field crew (CGT color Ust-Kan 7-12-99:34).

**Fig. 3.156** Ust-Kan. View southward (up-river). Four vegetation types can be seen in this view: drought-resistant small steep slope species; steppe; riparian; and mountain forest. On the basis of Ovodov's earlier Ust-Kan faunal identifications, he believes that the landscape shown here would have been similar in the late Pleistocene, although probably somewhat drier and colder. Ust-Kan Cave would have served as a good lookout for Pleistocene human hunters and hyenas alike, but not at the same time (CGT color Ust-Kan 7-12-99:36).

**Fig. 3.157** Ust-Kan, bone sample. Shown is a representative collection of bone refuse from a typical excavation unit (3VG 99 8/31) about 1 $m^2$ and 10–15 cm deep. As can be seen, fragmentation is severe. Whole bones or even largely complete bones are rare in Altai cave sites, making species, age, and MNI identifications quite difficult. Only one piece in this lot could be identified as to genus (a tooth fragment). Most of this damage represents carnivore activity, making hyenas a major factor in Ust-Kan taphonomic reconstruction. Hand lens is 6.3 cm long (CGT neg. IAE 9-20-00:3).

carried up the steep slope to the Ust-Kan cave, or if they had, then perhaps they later somehow slid out and rolled down to the base of the butte. Nowhere else in our study have cranial elements been so conspicuously missing.

**4 Age.** Of the 1654 Ust-Kan pieces, only 1.2% are sub-adult (Table A1.4, site 25). Compared with the pooled assemblage sub-adult average (7.6%), Ust-Kan has many fewer sub-adult pieces. If this difference has any ecological or taphonomic significance, given the cave's complex ancient use and modern exploration, we are unable to say.

**5 Completeness.** Only 2.5% of the 1647 Ust-Kan pieces are whole. Pieces with one anatomical end make up 6.1% of the sample, and those with no anatomical ends form the majority (91.4%) (Table A1.5, site 25). Compared with the pooled assemblage averages, Ust-Kan is much less complete.

**6 Maximum size.** The mean maximum diameter for 1656 pieces without postmortem breakage is 4.8 cm. The range is 2.2 cm to 19.9 cm (Table A1.6, site 25). The Ust-Kan mean is the second lowest is our study. Kurla I has the smallest mean (4.3 cm), but generally speaking all archaeological assemblages were heavily processed. The best single interpretation is that all possible carcass nutrients were being sought. Compared with the pooled assemblage values, Ust-Kan pieces are generally smaller. The limited species identification precludes comparison with Vera Gromova's (1950: table 27) whole long bone length information.

**7 Damage shape.** The most common forms of damage in 1558 Ust-Kan pieces are: long bone flakes (53.5%), long bone fragments (18.6%), and long bone splinters (15.3%) (Table A1.7, site 25). Compared with the pooled assemblage averages, Ust-Kan has more long bone flakes and long bone splinters; and fewer long bone fragments, phalanx butts, medial ribs, undamaged bones, and mostly whole bones.

**8 Color.** Out of 1654 Ust-Kan pieces, ivory is the most common color (94.6%). Nearly all of the brown (2.7%) and black pieces (2.6%) had been unquestionably burned or scorched (Table A1.8, site 25). The sizes of the brown and black pieces are mostly small. Brown pieces average 3.8 cm in diameter; black, 3.0 cm. Their respective ranges are 2.4 cm to 8.5 cm, and 2.3 cm to 8.2 cm. Flakes are the most common damage type in both brown (40.0%) and black (44.4%) pieces. Other than being fragments of long bones, none of the flakes, splinters, or fragments could be identified. Three fragments of small mammals of hare or fox size were burned brown. We do not understand what caused the brown and black color differences, as both are burned thoroughly and the damage forms and sizes are much the same. Given the likelihood that the Ust-Kan butte lacked woody plant cover, fuel would have been hauled up to the cave, as well as game, as documented by the 76 burned pieces of bone, all of which must have been burned accidentally by falling into a fire hearth, or burned as a result of a fire having been built upon cave floor bone refuse. Ust-Kan color is unexceptional compared to most of the other assemblages. There is no evidence of roasting, that is, there are no elements of any size or completeness that are both burned and unburned. Wrinn (2010) reports also identifying burned pieces of bone.

**9 Preservation.** Like other cave sites, ivory hardness is very common, being almost universal (99.4%) in the 1654 Ust-Kan pieces (Table A1.9, site 25). Compared with the pooled assemblage average, the Ust-Kan bone is much better preserved.

**10 Perimortem breakage.** Most of the 1654 Ust-Kan pieces have perimortem breakage (97.6%) (Table A1.10, site 25). As in several of our other assemblages, the few pieces without perimortem breakage are largely foot bones (64.3%), followed by teeth (17.9%), and a few other elements. Compared with the pooled assemblage average, Ust-Kan has 10% more perimortem breakage.

**11 Postmortem breakage.** This damage type is rare in the 1654-piece Ust-Kan assemblage, occurring only in 1.9% of all pieces (Table A1.11, site 25). Compared with the pooled assemblage average, Ust-Kan has less postmortem breakage.

**12 End-hollowing.** One of our carnivore indicators, end-hollowing, occurs in 9.5% of the 1654 Ust-Kan pieces (Table A1.12, site 25). The pooled assemblage average is almost exactly the same as that of Ust-Kan.

**13 Notching.** Another carnivore indicator, notching on a perimortem fracture plain occurs in 9.2% of 1653 Ust-Kan pieces. The number of notches per piece ranges from one to more than seven, with one notch per piece being the most common occurrence (6.8%) (Table A1.13, site 25). Compared with the pooled assemblage average, Ust-Kan has a slightly lower frequency of notching.

**14 Tooth scratches.** Of 1654 Ust-Kan pieces, 9% have tooth scratches. The number of scratches per piece ranges from one to more than seven, with one notch per piece being the most common number (Table A1.14, site 25). Compared with the pooled assemblage average, the occurrence of scratches at Ust-Kan is at the lower end of the intermediate to upper range, well below Razboinich'ya (37.1%), but greater than the open sites. Tooth scratches are strongly associated with bones found in caves.

**15 Tooth dints.** Out of 1653 Ust-Kan pieces, dinting is present in 13.7%. The number of dints per piece ranges from one to more than seven. We found 36 pieces to have more than seven dints (Table A1.15, site 25). Compared with the pooled assemblage average, Ust-Kan has about half the amount of dinting. Dinting often occurs together with scratches, but not so frequently as to have to view both as a single variable. There is one long bone flake with dinting that was probably caused by a hammer stone or anvil stone instead of by carnivore teeth. Another unusual Ust-Kan specimen is a 6.9 × 2.0 cm fragment with four of the largest dints we have seen. They were 6.6, 5.4, 6.3, and 2.2 mm in diameter. The piece also had acid erosion.

**16 Pseudo-cuts.** In our judgment, 65 (3.9%) of the 1654 Ust-Kan pieces have one or more examples of pseudo-cuts (Table A1.16, site 25). The number of pseudo-cuts per piece ranges from one to more than seven, with one per piece being the most common number (1.9%). Compared with the pooled assemblage average, Ust-Kan has a similar incidence of pseudo-cuts. Not all of the Ust-Kan examples need necessarily to have been caused by carnivores. A few may have actually been left by unusual stone tool gouging or chiseling, or left by a bone having been scraped or trampled upon a sharp stone surface. Said another way, pseudo-cuts do occur in sites where hyena and human presence has been identified, *and* can be found in assemblages where the presence of hyenas or other carnivores is mostly absent. In the case of Ust-Kan, hyenas are abundantly evident along with the refuse of human occupation.

**17 Abrasions.** Ust-Kan is similar to all our assemblages in having a very low frequency of abraded pieces – only 2.7% of the 1653 pieces. The number of abrasions per piece ranges from one to more than seven, with the latter occurring most often (1.7%) (Table A1.17, site 25). Of these, there are some interesting conditions: 40 striations on a 4.6 cm long bone flake; 11 striations near a notch, suggesting a hammer stone blow; 28 striations on a 8.8 cm long bone flake with 80 cuts; 27 striations with one cut on a 5.1 cm long metapodial splinter; 55 striations on a 4.4 cm fragment with 35 cuts; 55 striations on a 6.7 cm long bone flake with 19 cuts and one chop mark; and one 7.4 cm long piece of leg bone with scores of near-microscopic striations randomly oriented that are probably the result of sanding with a fine-grained stone. Depending on how the term artifact is defined, this latter piece might or might not be considered an artifact.

**18 Polishing.** The amount of Ust-Kan polishing is more than in several of our other assemblages – 83.2% of 1651 pieces have end, middle, and end-middle polishing (Table A1.18, site 25). The difference does not suggest a simple cave or animal relationship. Instead, we propose that carnivore chewing is involved along with soil effects and possibly bag-boiling. Cooking of some form like bag-boiling with heated stones could have produced polishing due to bone fragments with adhering meat and fat having been stirred against abrasive heated stone surfaces. Ust-Kan, like most of our other sites, has very little charcoal or ash, and thermally altered stones are non-existent. If cooking was done with stone boiling it must have been done on the plain at the base of the steep butte. Compared with the pooled assemblage average, Ust-Kan has about 20% more polishing.

**19 Embedded fragments.** We checked 1652 Ust-Kan pieces for embedded fragments, finding only 1.1%. The number of pieces with embedded fragments ranges from one to more than seven, but they are rare in all cases (Table A1.19, site 25). Compared with the pooled assemblage average, Ust-Kan has almost four times less embedding.

**20 Tooth wear.** There are 13 tooth-bearing pieces in the Ust-Kan assemblage. Of these, 23.1% are considered to be young because of their lack, or limited amount, of crown wear (Table A1.20, site 25). Compared with the pooled assemblage average, Ust-Kan has a lower representation of young individuals.

We came across a complete adult hyena canine, 4.7 cm in length, with the crown split in half from side to side (lingual–buccal). On the root near the crown–root junction, the crack has a small amount of use-wear polishing. This wear indicates that the fracture occurred sometime before the animal died, and that the tooth had been in use between the time of breakage and the hyena's death. We mention this curiosity because a hyena canine is a massive tooth whose shape reminds one of a large-caliber pistol bullet. The force needed to crack this tooth must have been enormous. The break serves as yet another reminder of what hyena jaws are capable of doing.

**21 Acid erosion (Figs. 3.158–3.165).** This variable is strongly associated with carnivores, as has been suggested previously. The high frequency (22.9%) of acid erosion in the 1654-piece Ust-Kan assemblage indicates to us a substantial hyena presence (Table A1.21, site 25). Compared with the pooled assemblage average, Ust-Kan has almost four times the frequency of acid-eroded pieces. With respect to other cave sites known or with a strong likelihood of having had a hyena presence, including Dvuglaska (22.1% acid erosion), Maly Yaloman (48.0%), Razboinich'ya (7.3%), and Straschnaya (28.3%), Ust-Kan could be considered as much a paleontological site as an archaeological site. Stomach bones were very common in Ust-Kan excavation square 7/31, Level 2, suggesting a "toilet area" that had been selected by hyenas for defecation or vomiting.

There are five stomach bones that had also been burned. These are: (1) a 3.3 cm long fragment; (2) a 4.5 cm flake; (3) a 2.8 cm flake, the damage sequence of which seems to have been acid erosion first, followed by the burning; (4) a 2.7 cm long proximal end of an adult hare-sized ulna; and (5) a 2.7 cm one-ended butt of an adult goat-sized toe. In addition, there are stomach bones with cut marks (see cut marks, below).

**Fig. 3.158** Ust-Kan, stomach bones. Sorted from the sample shown in the previous figure, this image shows that there is strong indirect evidence of relatively pronounced hyena use of Ust-Kan Cave. This and several following illustrations of Ust-Kan stomach bones are included to make a visual statement about the marked hyena presence in Ust-Kan, where the impact of examining tabular data (Table A1.28, for example), might not be as much recognized and appreciated. Scale is 15 cm (CGT neg. IAE 9-20-00:4).

**22 Rodent gnawing.** Ust-Kan is one of eight assemblages in this study with rodent-gnawed pieces of bone (Table A1.22, site 25). Of these, Ust-Kan has the lowest frequency (0.5%; 8 / 1654 pieces). Ust-Kan and the pooled assemblage average have the same frequency of gnawing. In all cases, the gnawers were mice or squirrel-sized animals, because the chiseled grooves are smaller in mesio-distal diameter than the incisors of marmots.

**23 Insect damage.** Ust-Kan and two other sites (Kaminnaya and Krasny Yar) have the only assemblages where we attributed damage to insects, all three with less than 1.0% (Table A1.23, site 25). The lone Ust-Kan piece exhibits about 50 tiny holes in a fragment 6.5 cm long. This fragment also has five cut marks and two chop marks. While most of these holes appear to be burrowing holes, we cannot exclude the possibility that some other agency might have been responsible.

**24 Human bone.** No specimens of human bone or teeth have been found at Ust-Kan.

**25 Cut marks.** Out of 1653 Ust-Kan pieces, 18.0% have cut marks ranging from one to more than seven per piece. The most common number is one per piece (3.8%), followed closely by more than seven (3.7%) (Table A1.25, site 25). Compared with the pooled assemblage average, Ust-Kan has twice the frequency of cut pieces.

**Fig. 3.159** Ust-Kan, stomach bones. Another sample, this one from what field supervisor Postnov calls the "disturbed" area where Rudenko excavated. Note the three eroded tooth fragments in the upper left top row. As in the previous figure, note also the relative size uniformity of these pieces. Perforated stomach bones as in the lower left might have been worn as necklace amulets. Scale is in centimeters (CGT neg. IAE 7-30-99:1).

**Fig. 3.160** Ust-Kan stomach bones. Another sample (Layer 2 7–31) (CGT neg. IAE 10-14-00:10).

**Fig. 3.161** Ust-Kan, stomach bone detail. Lower jaw fragment, 5.0 cm maximum diameter (CGT neg. IAE 10-14-00:12).

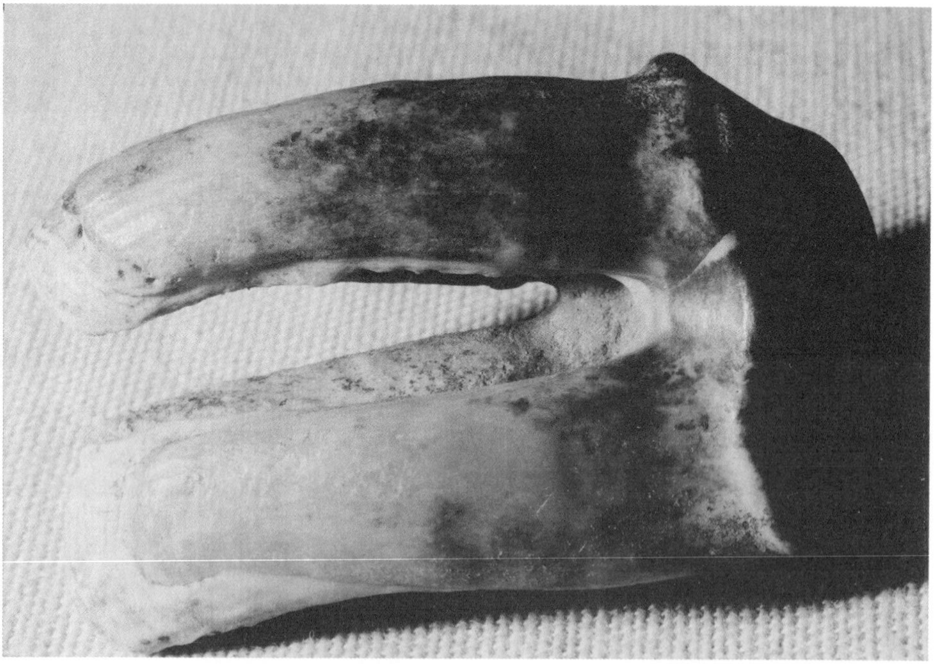

**Fig. 3.162** Ust-Kan, stomach bone detail. Mature hyena tooth that had been swallowed and exposed to digestive erosion by another hyena. The dentine roots and enamel crown are only slightly eroded, but the bone of the adhering socket is largely destroyed. Tooth is 4.0 cm in length (CGT neg. IAE 9-20-00:8).

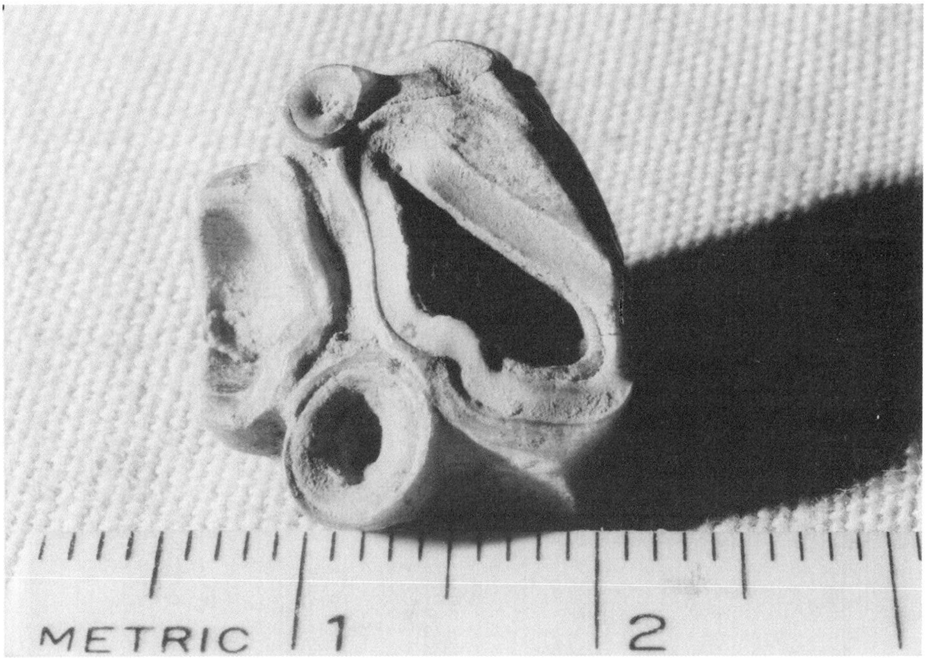


**Fig. 3.163** Ust-Kan, stomach bone detail. Worn hyena tooth crown. View is occlusal, looking into the four cusp pulp cavities. Actual width of image is 3.3 cm (CGT neg. IAE 9-20-00:12).

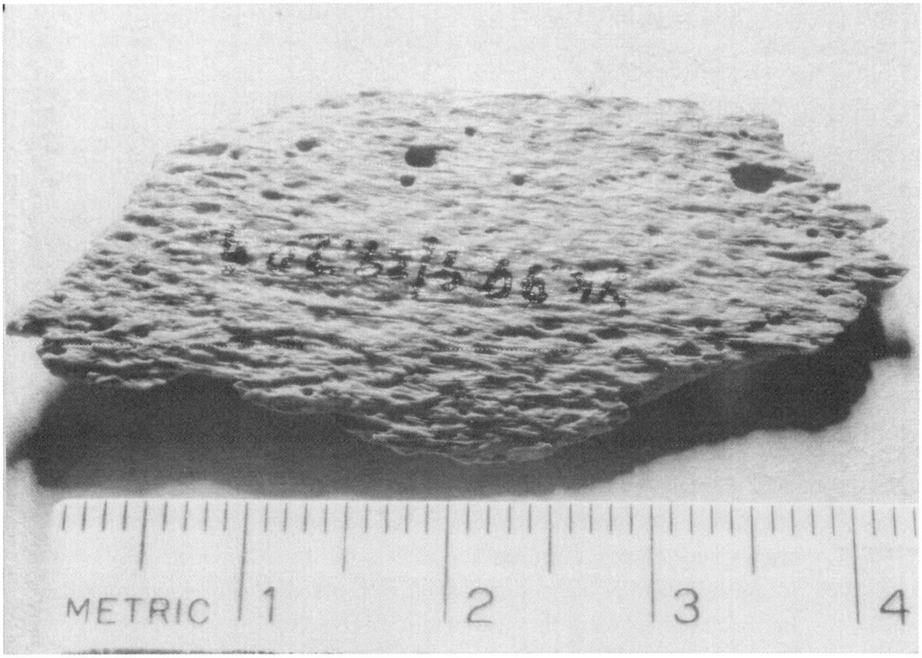


**Fig. 3.164** Ust-Kan stomach bone detail. A piece from deep Layer 8. The entire cortex has been dissolved. These stomach bones are found in the earliest to the most recent Pleistocene deposits of the cave. Length is 4.3 cm (CGT neg. IAE 8-2-99:4).

**Fig. 3.165** Ust-Kan, stomach bone detail. Burned specimen. These burned stomach bones could have been accidently burned by a campfire or were larger burned bones subsequently scavenged by hyenas when humans were absent. Actual width of image is 3.3 cm (CGT neg. IAE 6-18-03:22).

Of pieces with many cut marks, the following is a sample of their size and nature: 15 cut marks prior to breakage, adult long bone splinter, 8.1 cm; 20 cuts, flake, 4.3 cm; 21 cuts, adult long bone flake, 4.1 cm; 28 cuts, adult long bone fragment, 6.3 cm; 29 cuts, adult long bone flake, 4.5 cm; 35 cuts, adult (element?) fragment, 4.4 cm; 40 cuts, adult long bone flake, 5.9 cm; 42 cuts, metapodial flake, 5.5 cm; 80 cuts, adult long bone flake, 8.8 cm.

There are four pieces of special interest. These demonstrate that large carnivores – presumably hyenas – and humans had both processed some of the same pieces: (1) A horse-sized distal humerus fragment 8.9 cm long, polished on its ends and middle, has four cut marks, five tooth scratches, and 12 tooth dints made by a large carnivore. The cut marks are 4, 5, 9.0, and 10.0 mm long. We are unable to determine if the cutting or the chewing occurred first. (2) A 3.3 cm fragment, polished all over, has two cut marks of 2.2 and 2.0 cm long. This piece was subsequently swallowed and partly digested by a hyena. (3) Like the previous piece, a 4.7 cm adult long bone flake, polished all over, with two cut marks, was subsequently swallowed and partly digested by a hyena. This specimen shows very clearly the digestive effects on the cut marks, because their borders are rounded instead of being sharp. (4) A well-polished adult long bone flake, 6.4 × 3.7 cm, has one faint cut mark that just barely survived the digestive erosion of a hyena.

**26 Chop marks.** All 1654 Ust-Kan pieces were evaluated for chop marks. We found that 9.5% had one or more chop marks. The most common number per piece was one

Fig. 3.166 Ust-Kan, bone damage. This specimen was classified as having carnivore tooth dints, chewing, and pseudo-cuts. In light of the burned specimens in the previous figure, the pseudo-cuts might better have been identified as stone tool cut marks. Actual width of image is 3.3 cm (CGT neg. IAE 9-20-00:24).

(5.8%) (Table A1.26, site 25). Compared with the pooled assemblage average, Ust-Kan has almost twice the frequency of chop marks.

Figs. 3.166–3.179 are a gallery of damaged pieces of special interest. This array is meant to illustrate the range and variation of perimortem bone damage. The figure captions explain what we think to be the significance of each piece.

## Discussion

Ust-Kan is our largest assemblage, containing 17.7% of the 9360-piece grand total. As such, it is for statistical sample size considerations the best of our 29 assemblages. For example, the decidedly low frequency of Ust-Kan cranial pieces cannot rationally be attributed to sampling error. Some heads may have been cut off before the remaining carcasses were hauled up the steep climb to the cave entrance, where further butchering and bone breakage can be shown to have occurred within the cave. On the other hand, head parts left in the cave may have been scavenged by hyenas and taken elsewhere. A third possibility is that crania, because of their shape, may have rolled out of the cave more often than less spherical elements. Clearly, there is an interesting taphonomic problem here involving missing heads that cannot simply be dismissed as due to small sample size.

**Fig. 3.167** Ust-Kan, bone damage. Side A. Digestive pits and the feathered end show that this was a bone fragment that a hyena swallowed. However, the piece has unquestionable cut marks. This piece, and others like it, demonstrate that hyenas were scavenging and disturbing the occupation refuse left by humans. Note also the tiny mineral flowers that overlie two of the cut marks on the right. Thus, three processes through time can be recognized: (1) butchering; (2) acid erosion; (3) mineral deposition. Actual width of image is 3.3 cm (CGT neg. IAE 6-18-03:18).

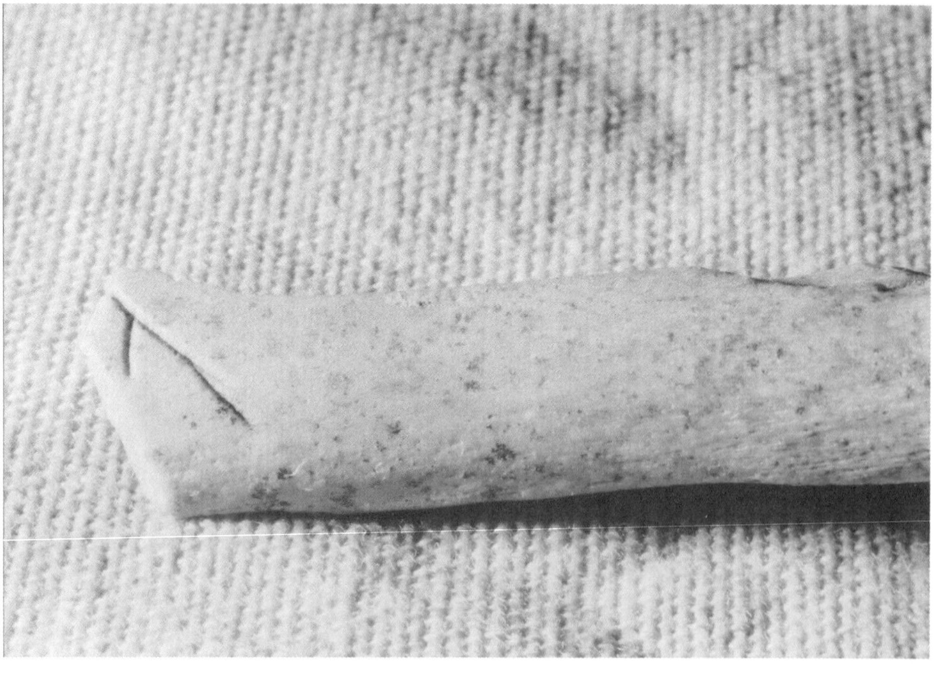

**Fig. 3.168** Ust-Kan, bone damage. Another example of a cut bone fragment that had been swallowed. The borders of the cut mark are rounded instead of sharp, but too narrow and deep to be pseudo-cuts. Actual width of image is 3.3 cm (CGT neg. IAE 9-8-00:11).

**Fig. 3.169** Ust-Kan, bone damage. This Layer-7 piece has cut marks that had to have been made before it was burned, presumably by accident. Actual width of image is 3.3 cm (CGT neg. IAE 9-8-00:12).

**Fig. 3.170** Ust-Kan, bone damage. A 12 cm long fragment of yak or bison femur found in Layer 3 has classic spiral fracturing. The two arrows point to internal spalling that seems to have been produced by tooth pressure, although there is no tooth scarring on the external surface at these locations, nor is there any sign of pounding (CGT neg. IAE 8-2-99:26).

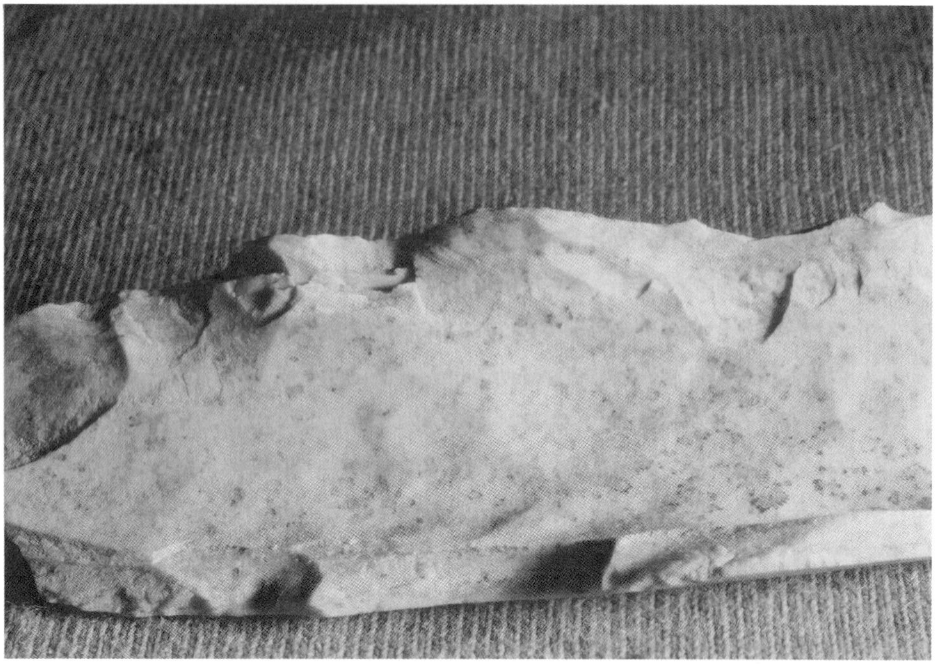

**Fig. 3.171** Ust-Kan, bone damage. Fragment with two notches on upper fracture line. Based on findings from Razboinich'ya hyena cave, small notches are associated with carnivore bone damage. Actual width of image is 4.0 cm (CGT neg. IAE 6-18-03:21).

**Fig. 3.172** Ust-Kan, cut marks. A mandible fragment of a young animal (no tooth wear, active bone remodeling) with cut marks (lower right), dints (center), and tooth scratch marks (lower left). Sequence of damage could not be determined (CGT neg. IAE 9-20-00:34).

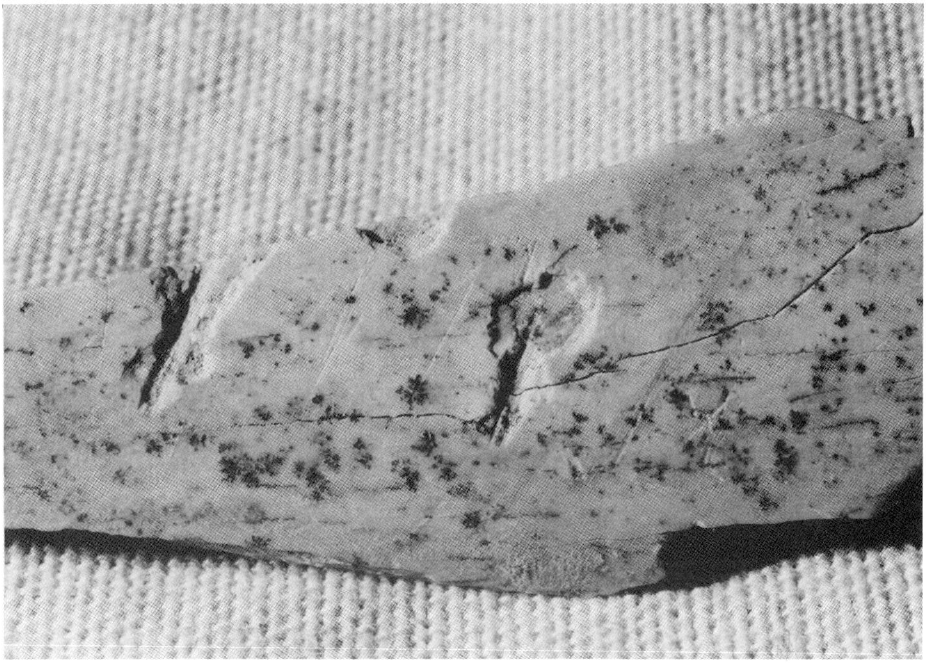

**Fig. 3.173** Ust-Kan, cut marks. The very fine cut marks that more-or-less parallel each other also parallel the two larger chop marks, so defined by their lateral spalling. This situation suggests that the same tool was used to produce both types of damage – slicing and chopping. Mineral flowers cover parts of some of the fine cut marks. Width of image is 3.3 cm (CGT neg. IAE 9-20-00:28).

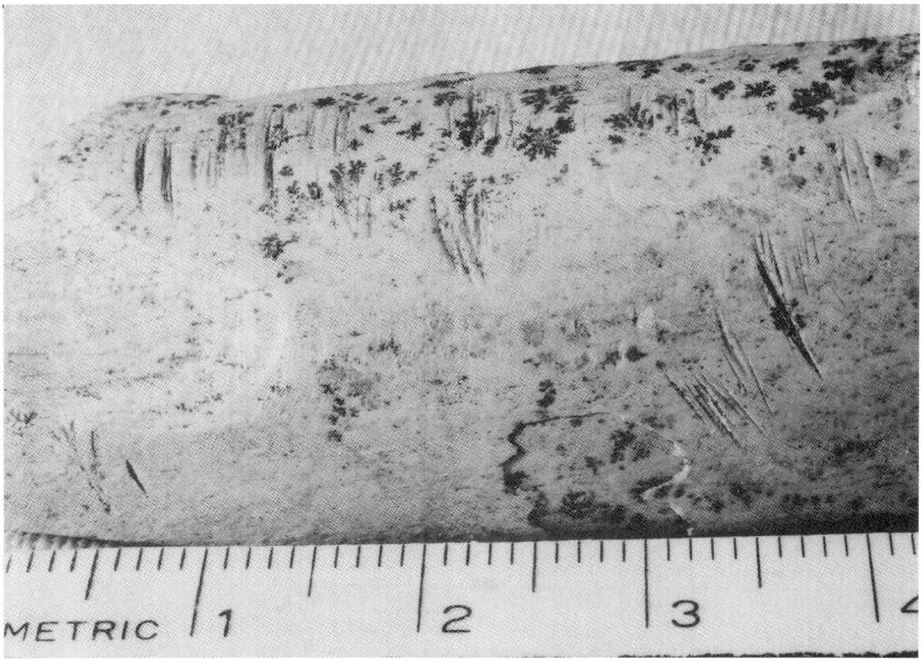


**Fig. 3.174** Ust-Kan, cut marks. This interesting piece has several swarms of fine, closely spaced cut marks. The reason for these swarms is not understood, unless the piece had served as a sort of cutting board. Note the mineral flowers in some of the cut marks (CGT neg. IAE 9-20-00:10).

**Fig. 3.175** Ust-Kan, cut marks. Another candidate for having served as a cutting board, this one from Layer 4B. There are at least 35 cut marks and 55 abrasion striations. The piece was broken some time after the cuts on the lower border were incised. Width of image is 3.3 cm (CGT neg. IAE 9-20-00:14).

**Fig. 3.176** Ust-Kan, cut marks. A third example of what looks to have been some sort of cutting board, from Layer 6–7. There are at least 29 cut marks. Width of image is 3.3 cm (CGT neg. IAE 9-20-00:16).

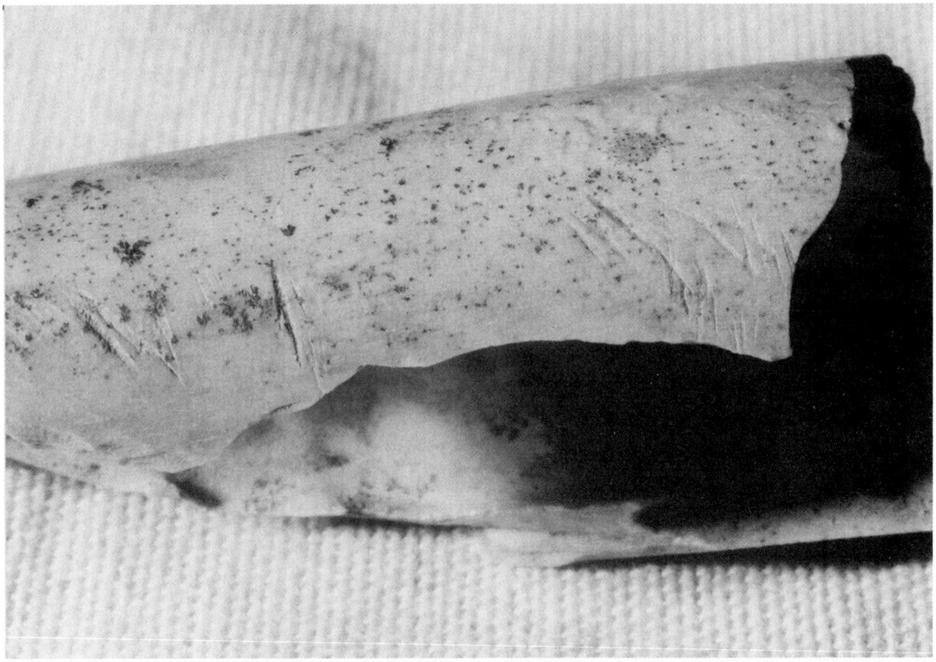

**Fig. 3.177** Ust-Kan, cut marks. This piece from Layer 7 nicely shows that defleshing occurred prior to breakage to extract marrow. Some of the 40 cut marks had crossed beyond the breakage. Width of image is 3.3 cm (CGT neg. IAE 9-20-00:20).

**Fig. 3.178** Ust-Kan, bone damage. The seven dints look suspiciously like tooth cusp marks, although no match could be made with any of the jaws in Ovodov's comparative osteology collection. The very fine scratches or cut marks on the right appear new and are presumed to have been caused by archaeological processing. Width of image is 3.3 cm (CGT neg. IAE 6-18-03:16).

**Fig. 3.179** Ust-Kan, bone damage. A hyena dental crown that had cracked in half during life. The animal lived for some time after this trauma because there is some wear at the crown–root junction. Width of image is 3.3 cm (CGT neg. IAE 6-18-03:13).

While large sample size makes Ust-Kan valuable for our perimortem taphonomy study, the stratigraphic mixing of Ust-Kan makes it less valuable for cultural replacement (microevolution) considerations. Mixing is of no real importance to our investigation since temporal divisions are more-or-less irrelevant to our objectives of identifying the perimortem damage signatures of humans and carnivores. Moreover, the faunal assemblages represented here were apparently all much the same in life. We offer a few other remarks for Ust-Kan in light of its assumed statistical reliability.

Heavy and extensive carcass processing occurred within Ust-Kan Cave. Of this we can be certain because of the large Ust-Kan sample size, and by extension to our other smaller but just as heavily processed assemblages. Large amounts of perimortem breakage characterize almost all of our assemblages with the exception of paleontological sites such as Krasny Yar and Shestakovo. The best single interpretation for such extensive damage by humans is that all possible nutrients were being extracted. To us this implies limited environmental resources that would have in turn limited human population size and demographic expansion in late Pleistocene Siberia.

Postmortem breakage is very infrequent in Ust-Kan, whereas it is generally more common in our other assemblages. While some postmortem breakage is due to archaeological recovery, and some is due to osteological characteristics of different species, especially mammoth long bones and ribs, we sense that a true taphonomic problem exists that is not simply cave versus open site preservation differences. There is also the well-known relationship between postmortem breakage and surface weathering, to which bone damage due to changes in moisture content may be enhanced by annual variation in wet–dry freezing and thawing. We suggest that the amount of postmortem breakage is a potential guide to the past environmental temperature and moisture stability wherein an assemblage is found. Our visits to Ust-Kan revealed site advantages and disadvantages for occupation and bone processing. The advantages are that it has good protection from weather conditions, it is an excellent look-out site, there is water nearby, and it is a highly defendable site should there have been inter-group conflict. On the other hand, it is hard to reach, and hauling wood for heating and cooking up the steep slope would not be very cost effective. In sum, we view the Ust-Kan bone damage as reflecting the positive and negative qualities of the cave. All things considered, we hypothesize that seasonal occupation of the cave is a more parsimonious view of its use. Seasons may have lasted for generations rather than annually. Here, as elsewhere in this study, we feel there is more evidence for discontinuity of human occupation in our Siberian study area than for continuity. End-hollowing is relatively common in Ust-Kan. It is generally less so in our other assemblages that lack direct or indirect evidence of hyenas, such as Boisman II, Bolshoi Yakor I, Kamenka, Kurla I, Mal'ta, Straschnaya, Varvarina Gora, and Yelenev. All carnivores are variously capable of end-hollowing, but we feel, as mentioned earlier, that relatively low amounts in an assemblage indicate small- to medium-sized carnivore (fox, dog, wolf) processing (and possibly also mineral-seeking herbivores), and large amounts, say around 10% or more, indicate large carnivore activity (hyena, bear, lion). Naturally, carnivore age, number of carnivores, and time involved in processing are further considerations necessary for microtaphonomy analyses.

We have generally presumed, following our Razboinich'ya studies, that notching on a fracture surface was pressure-produced by a carnivore. Ust-Kan has to some extent upset this notion. Both carnivores and humans produced notching in the Ust-Kan assemblage. Human-induced notching could best be identified when there were hammer stone or anvil abrasions close to the notch. These notches seem to have been formed by blows whose purpose was to break bones. Both stone and bone or antler tools are implicated – stone when abrasions are present, bone when they are not present.

The acid-eroded pieces fall into two fairly distinct classes: heavy erosion and very slight erosion. We presume that the former represents a complete transit of the alimentary tract, while the latter could represent vomiting or full transit encased in a wad of hair or ligament. Small numbers of lightly etched stomach bones must also be considered as products of foxes and wolves. Most of the Ust-Kan stomach bones are heavily eroded, indicating that they had been defecated.

Some stratigraphic problems with Ust-Kan are attributable to hyena digging and burrowing activity. This suggestion rests with the high frequency of stomach bones at Ust-Kan, stomach bones with cut marks, and five that had also been burned.

The hyena stomach bones with cut marks document with no question whatsoever that hyenas were scavenging the human game refuse left in Ust-Kan Cave. That scavenging activity must have involved digging, just like modern dogs jumping up into open dumpsters and pawing through the trash in search of items to eat. One wonders when the scavenging occurred at Ust-Kan. Were humans in the vicinity, or had they left days or weeks earlier in their own search for additional game? Might the hyena pack, in a state of hunger, have entered the cave while the Mousterians were still inside? Might they in their hunger have attacked the people? Whatever the scenario, we feel that the acid-eroded pieces with cut marks are especially strong evidence of the interaction between Pleistocene humans and hyenas in Siberia, and it is this evidence that makes us wonder if hyenas and other large carnivores had some effect on delaying human expansion into the New World.

In conclusion, we note that our perimortem taphonomy work has had an influence on the interpretation of cave occupancy in Siberia. Before our work became known, cave occupancy and bioturbation were not issues. Now, at Ust-Kan at least, it is being given much consideration. Postnov feels that during the time of formation of Layers 1, 9, and 10, humans rarely used the cave. During the formation of Layers 2, 3, 5, 7, and 8, humans more commonly occupied the cave. During the time of Layers 4 and 5 human presence in Ust-Kan was considerable, as assessed on artifact frequency. As in other sites in this study, such as Okladnikov Cave, human occupancy was apparently discontinuous. Discontinuity allows for the possibility of different populations entering the region, as is classically indicated for the Cro-Magnon replacement of Neanderthals in Europe and for the Russian replacement of most native Siberians in more recent times.

# 26 Ust-Kova

## Background

Ust-Kova (Ust, located; Kova, a river name) is an open site located on the left bank of the Angara River, above the mouth of the Kova River, ca. 700 km north-northwest of Irkutsk. It is situated on the 400 m wide second terrace that is 14–17 m above the Angara. Ust-Kova was discovered in 1973 by A. P. Okladnikov. Kuzmin and Orlova (1998:11) locate the site at 58°33′ N, 100°33′ E. There are four layers to the terrace, of which Layers 3 and 4 contain Paleolithic archaeological and faunal remains. The third layer is subdivided into chronologically different complexes (Drozdov *et al.* 1990). The middle layer of artifacts that occurs in carbonized clayish sediment is dated at 23 920 ± 310 BP (KRILL-381, charcoal). The early complex, related to a buried soil, is dated in the range of 28 000 to 32 000 BP (Drozdov 1981, Drozdov and Chekha 2003, Drozdov and Laukhin 1980). Vasili'ev *et al.* (2002:526) list nine carbon-14 dates based mostly on charcoal. They range in age from 34 300 BP (Lower Component) to 13 860 BP (Middle Component). Regional dating is reported by Laukhin *et al.* (1980).

The total number of stone tools and manufacturing debitage recovered from Layer 3 was 2721 (Vasilievsky *et al.* 1988). Flakes were abundant (1235), as were blade "removals" (968). Less common were blades (125), core spalls (82), cores (71), fragments of tools (63), perforators/borers (32), retouched flakes, chisel-like tools, scrapers, knives, etc. Several items of art work have been found, including items of decorative jewelry, carved figurines, and one of special interest, a mammoth figurine made of antler. Other information on cultural remains and geological considerations can be found in Akimova and Bleinis (1986), Leontiev and Stepanov (1986), Chekha (1990), Drozdov *et al.* (1990), Leontiev and Drozdov (2003), and Leontiev *et al.* (2000). Drozdov and Chekha (1990) give special attention to preservation and the processes they refer to as "paleofrosted phenomena."

The mammalian remains that Drozdov turned over to Ovodov for identification included 2906 pieces from excavations over 1978–1988, although exact provenience for some pieces was missing. Preservation, however, suggests to Ovodov that most of the unprovenienced pieces probably came from Layer 3. The species included: hare (*Lepus* sp.), six bones; beaver (*Castor fiber*), two; gray wolf (*Canis lupus*), seven; fox (*Vulpes vulpes*), 41; Arctic fox (*Alopex lagopus*), four; mammut (*Mammuthus primigenius*), 127; horse (*Equus cf. caballus*), 243; rhinoceros (*Coelodonta antiquitatis*), seven; red deer

(*Cervus elaphus*), 13; reindeer (*Rangifer tarandus*), 2394; bison (*Bison priscus*), 62. Three more species can be added to this inventory that were identified earlier by E. A. Vanheim and I. V. Foronova: brown bear (*Ursus arctos*), otter (*Lutra lutra*), and elk (*Alces alces*), although the total number of bones they examined is uncertain. The work by R. S. Vasilievsky, N. I. Drozdov and V. V. Burilov (1988:90) lists more than 10 000 bone fragments of mammoth, reindeer, bison, horse, elk, and roe deer. Among faunal remains related to the buried soil (i.e., Layer 4, the deepest cultural layer), identifications include mammoth, kulan, reindeer, and rhinoceros (total pieces: 327).

The faunal remains we examined in August 1999 came from excavations made in 1980 and 1988, under the supervision of Nicolai D. Drosdov. They are curated at IAE, Academgorodok, Novosibirsk. Further excavation is being considered.

## Findings

**1 Provenience.** Table A1.1 (site 26) shows that provenience information was largely unavailable from the faunal storage boxes, although Sub-layers 4 and 5 were most often identified. We understand that this open site was stratigraphically a limited occupation during late Pleistocene times, although the range of carbon-14 dates would not support this inference. The possibility of older and unrelated forest fires leaving charcoal debris may be involved.

**2 Species.** Table A1.2 (site 26) shows that most of the faunal remains were reindeer (58.1%), followed by horse (30.2%), and one piece each of elk and mammoth. The complete identification of the 116 pieces is due to the relatively small amount of perimortem damage in this assemblage and its intensive earlier study by Ovodov.

**3 Skeletal elements.** As with species identification, all of the pieces from Ust-Kova could be identified as to skeletal element (Table A1.3, site 26). The absence of vertebrae and ribs is due to their not being saved; however, the absence of skull elements is because we missed finding one or more Ust-Kova storage boxes.

**4 Age.** Almost the entire Ust-Kova assemblage had been adults in life (95.7%; Table A1.4, site 26). Each of the five sub-adults was a reindeer.

**5 Completeness.** There is a high frequency of whole bones (11.2%; Table A1.5, site 26), and fragments with one anatomical end (87.1%). There are only two pieces lacking both anatomical ends. These are a 24.5 cm long piece of reindeer antler and a 15.8 cm long piece of reindeer mandible. As discussed elsewhere, the high frequency of whole bones and bones with one intact anatomical end is characteristic of assemblages with minimal carnivore activity.

**6 Maximum size.** The mean of 117 Ust-Kova pieces is 8.0 cm, the range is 2.9 cm to 28.3 cm (Table A1.6, site 26). Compared with the pooled assemblage values, Ust-Kova is very similar in size. Looking at whole long bone lengths of reindeer and horse provided by Vera Gromova (1950: table 27) shows that the Ust-Kova upper range limit is about the same as the reindeer lower limits, and generally less than the lower limits of horse.

**7 Damage shape.** This category of information was not collected in 1999 as we were trying at that time to formulate a classification.

**8 Color.** Ust-Kova, being an open site, possesses almost none of the ivory colored pieces that characterize the color of bone from cave sites. Instead, open sites have many more pieces that are brown. Brown pieces make up the overwhelming majority (94.9%) of Ust-Kova, whereas ivory colored pieces (1.7%) are rare (Table A1.8, site 26). The discoloration of open site bones is largely only surficial, but the amount of discoloration or "patina" is probably a useful indicator of subsurface water movement or "Brownian movement." Contact with dry or even slightly moist soil seems not to discolor bone surfaces to the extent that soil with constant drainage does. Color alone is not necessarily a useful indicator of antiquity, but discoloration is probably a valuable characteristic for estimating bone denaturing, which if extensive makes bone of little use for carbon-14 dating or DNA analysis. Thus, one might be more suspicious of bone dating or a DNA assay from Ust-Kova, than from Razboinich'ya or Denisova caves.

**9 Preservation.** Corresponding with color, the Ust-Kova assemblage also has relatively poor bone quality. More than half (65.5%) of the 116 pieces are chalky rather than ivory in hardness (Table A1.9, site 26). This is a much higher frequency of low-quality bone than occurs in cave deposits of comparable age. On the other hand, another late Pleistocene open site assemblage examined for this project, Bolshoi Yakor, has a high frequency of hard ivory-like bone fragments. This good quality we believe is due to that site having remained frozen throughout the Holocene – up to the time of its recent excavation, which is still in progress (see Bolshoi Yakor). Another difference between cave and open sites, but one that we have not systematically recorded, is the frequent occurrence of root tracks on the surfaces of Ust-Kova bones. Root damage is very rare in cave deposits, where the lack of sunlight limits plant growth.

**10 Perimortem breakage (Figs. 3.180–3.182).** While perimortem breakage is substantial in the Ust-Kova faunal assemblage (88.8%; Table A1.10, site 26), the amount is relatively low compared to cave sites, especially our carnivore baseline standard, Razboinich'ya Cave. By comparison, Bolshoi Yakor, an open site also with a high frequency of reindeer remains, had only one of 263 pieces without evidence of perimortem breakage. Ust-Kova has 13 such pieces (11.2%), suggesting that the Ust-Kova hunters were relatively less frugal or some sampling bias exists due to post-excavation curation and disposal.

**11 Postmortem breakage.** There is a modest amount of postmortem breakage in the Ust-Kova assemblage (15.5%; Table A1.11, site 26). Most of it, if not all, seems attributable to archaeological recovery. We presume that this is due to the relatively poor preservation of these bones, and the difficulty getting them out of the ground.

**12 End-hollowing.** Perhaps one of the better means in perimortem taphonomy for hypothesizing carnivore activity is the chewing and cupping damage to bone ends,

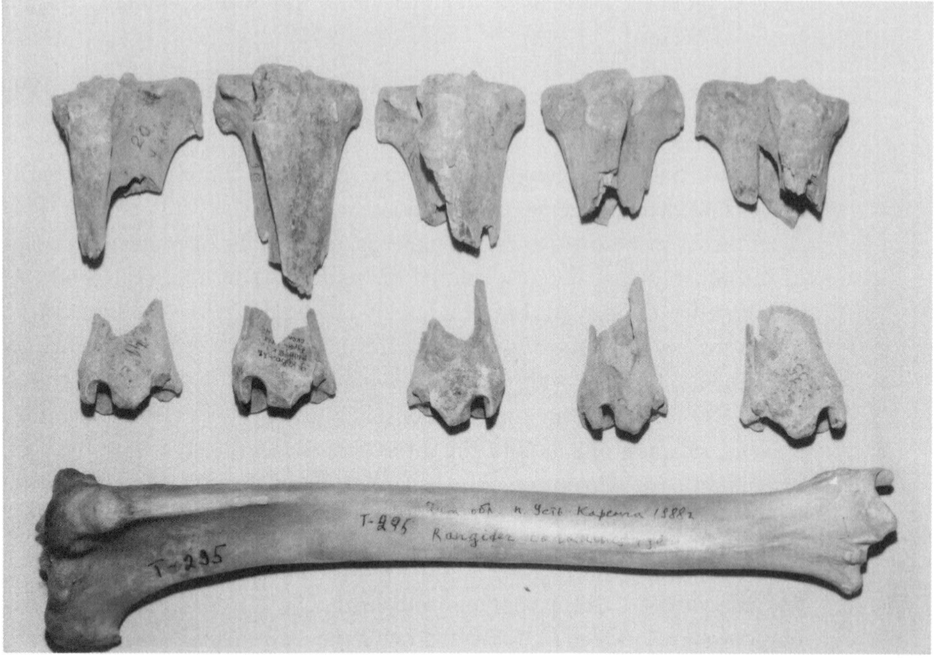

**Fig. 3.180** Ust-Kova, bone breakage. Reindeer tibiae with characteristic human damage to long bone shafts but not to ends. Carnivore damage is more often at the softer ends of bones. A whole modern reindeer tibia is at the bottom to illustrate the amount of breakage to the Pleistocene specimens. The piece in the upper right is 7.5 cm maximum diameter (CGT neg. IAE 8-6-99:2).

especially when tooth dints and scratches are also present. For what would appear to be a kill site with but brief human occupation, Ust-Kova has a very low frequency of end-hollowing (3.5%; Table A1.12, site 26). However, the four end-hollowed pieces have no associated tooth dints, tooth scratches, pseudo-cuts, or stomach acid erosion. One of the four has associated cut marks. While we feel that carnivores had indeed damaged a small amount of the Ust-Kova bone refuse, we caution that end-hollowing alone is not an absolute sign of carnivore activity, and this caution applies to all other damage types. Hence, the justification for a multi-variable damage signature.

**13 Notching.** There are two pieces (1.7%) with simple notching – that is, neither is associated with a tool-produced blunt impact on a fracture plane, as are two other notched pieces (Table A1.13, site 26). Inasmuch as there are tooth dints and tooth scratches in this assemblage, one or both of the simple notches could be due to carnivore chewing. However, none of the pieces with notching also has dints or scratches.

**14 Tooth scratches.** Four (4.3%) of the 116 pieces have tooth scratches. Of these, the number of scratches varies from two to twelve (Table A1.14, site 26). One of the pieces with tooth scratches ($n = 12$) is a complete adult reindeer astragalus. It also has tooth dints ($n = 5$). This piece was undoubtedly chewed on by a carnivore, whereas the other three were probably so damaged.

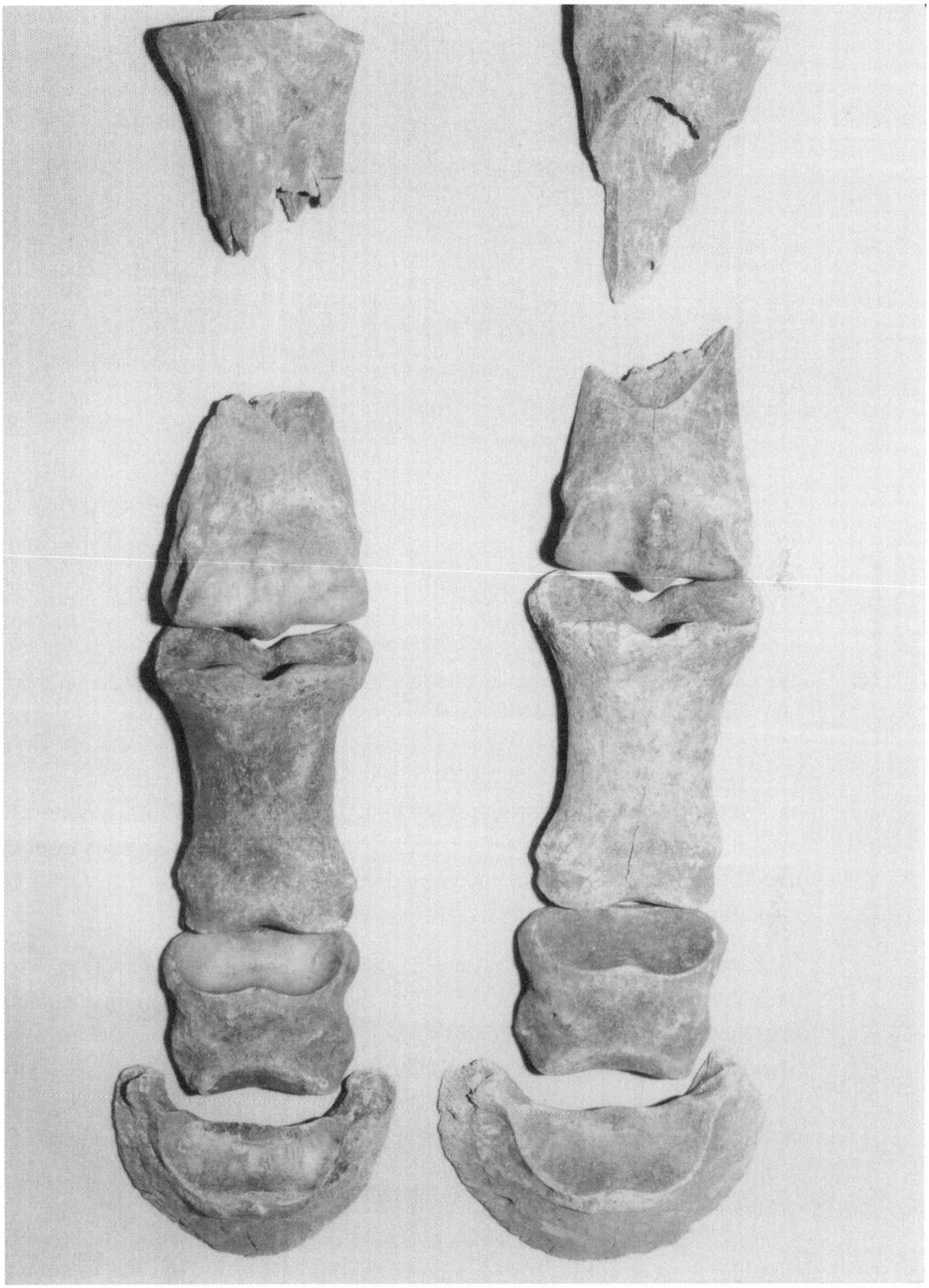

**Fig. 3.181** Ust-Kova, bone breakage. Horse metapodial shafts have been broken by humans to extract marrow as in previous figure. Lower left toe bone is 8.2 cm in diameter (CGT neg. IAE 8-3-99:26).

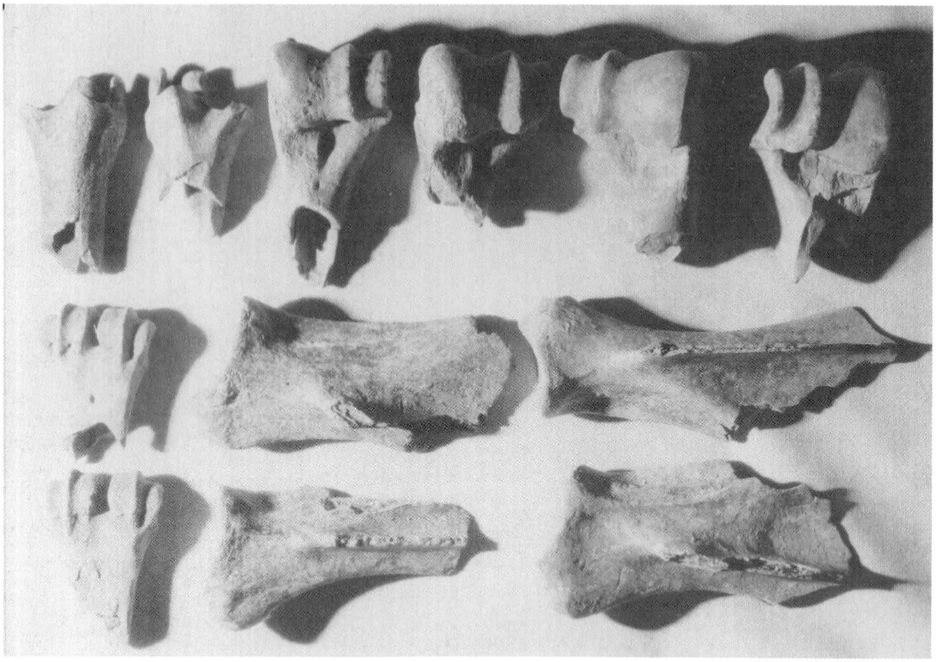

Fig. 3.182 Ust-Kova, bone breakage. Human breakage of reindeer scapulae, tibiae, and metapodial shafts. Lower left metapodial is 5.9 cm long (CGT neg. IAE 8-5-99:30).

**15 Tooth dints.** As with tooth scratching, there are a few (5.2%) of the 116 pieces that exhibit tooth dints (Table A1.15, site 26). The number of dints varies from three to eight. In addition to the astragalus mentioned above, dints occur on a distal tibia fragment, a mandible, and three metatarsals, all reindeer.

**16 Pseudo-cuts.** Two (1.7%) pieces show pseudo-cuts (Table A1.16, site 26). One of these, a metatarsal, also has tooth dints; the other, a distal humerus fragment, has tooth scratches. Taken altogether (end-hollowing, notching, tooth scratches, tooth dints, and pseudo-cuts), there is clearly identifiable carnivore damage in the Ust-Kova assemblage, although the amount is low, and the type of carnivore would appear to be relatively small, certainly not hyenas, bears, or large wolves. Such a small amount of Pleistocene carnivore damage has several possible explanations. One is that the assemblage was rapidly buried and only a brief amount of time was available for scavenger activity. Second, large scavengers such as hyenas were not present as far north as Ust-Kova, or were not present during the season(s) when this largely reindeer and horse assemblage was slaughtered. However, bears and wolves must have been in the neighborhood, so rapid burial, perhaps initially under a heavy winter snowfall, could have "hidden" the remains long enough to make their decomposition sufficient to make them no longer useful as a source of nutrients. This might especially be the case if living game animals were abundant in the region. Third, the remains that we observed were not representative of the original excavated assemblage due to post-excavation discard. Finally, the game (live and dead) in the area, with a low number of human competitors, may have been so abundant that carnivores had better opportunities to feed elsewhere.

**17 Abrasions.** Only one fragment (0.9%) has abrasions (Table A1.17, site 26). The scratch lengths of its seven closely grouped striations are about 3.0 cm. They occur on the interior aspect of the anterior border of an adult reindeer proximal scapula fragment. The border is indented and scored as if it had been chewed, but there is no corresponding damage to the external aspect, as would be expected if the specimen had been chewed by a carnivore. Hence, we infer that the abrasions are human-caused. Considering that almost 90% of the Ust-Kova sample have perimortem breakage, the low amount of abrasion stigmata suggests that the tools and anvils used to break up the skeletal elements were made of smooth-surfaced materials such as bone, antler, or wood – not grainy stone.

**18 Polishing.** There is considerable (34.5%) polishing in the Ust-Kova assemblage (Table A1.18, site 26). Given the relative "wholeness" of the assemblage it is not unexpected that polishing occurs mainly on the ends of fragments because polishing is assessed almost solely on perimortem fractures. Therefore, we feel that the percentage of end-polishing is higher than it might be if breakage had been more extensive. Only one of the five pieces with both end and middle polishing has another indicator of carnivore damage – a specimen with eight tooth dints. This piece is a 15.8 cm long adult reindeer mandible fragment with no anatomical ends. Elsewhere, White (1992) first, and subsequently agreed upon by Turner and Turner (1999), linked fragment end-polishing with abrasive action caused by stirring bone fragments and marrow broth in a ceramic vessel with a rough interior surface. But, being a pre-ceramic Upper Paleolithic site, the Ust-Kova end-polishing must be due to some other mechanism. Trampling, soil acidity, crystal matrix decomposition, and other physical and chemical activity over thousands of years are possible considerations in need of experimental simulation.

**19 Embedded fragments.** As Table A1.19 (site 26) shows, 11 of the 116 Ust-Kova pieces have one to four embedded fragments. Pieces with embedded fragments have no associated tooth dints or scratches. Only one piece with embedded fragments has an associated notch, which was caused by an impact blow that caused both the notch and the adjacent embedded fragments. Hence, embedded fragments at Ust-Kova appear mainly to be the result of human rather than carnivore activity, and the tool(s) used were smooth-surfaced, like bone, antler or wood.

**20 Tooth wear.** We did not record observations for the Ust-Kova teeth, since they had no perimortem damage.

**21 Acid erosion.** There were no examples of stomach acid erosion in the Ust-Kova assemblage. This is consistent with the low amount of severe carnivore damage that would be expected if hyenas had been present.

**22 Rodent gnawing.** One of the 116 pieces (0.9%) has rodent gnawing marks (Table A1.22, site 26). This is a well-preserved (ivory-condition) 8.7 cm long adult reindeer metatarsal with one anatomical end from Layer 4 (1988). The rarity of rodent gnawing in all of the assemblages reported herein suggests that this form of damage has very little taphonomic significance for our sampled area of Siberia. Whatever ecological or taphonomic significance this has is beyond our abilities to explain. The primary inference is that bone remained on the ground surface long enough to receive rodent attention, and was on the ground when these creatures are active. Because the Ust-Kova assemblage has relatively poor preservation, much of it likely was on the ground surface for a long time.

**23–24 Insect damage and human bone.** These variables are not present in the Ust-Kova assemblage.

**25 Cut marks.** A total of 17 Ust-Kova pieces (14.7%) exhibit 1–7 cut marks (Table A1.25, site 26). The cuts occur on three scapulae, one astragalus, three humeri, one calcaneus, one radius, two tibiae, and six butchered metapodials. There is no obvious pattern. They range in number from one to seven cuts per piece, with one cut per piece being the most common number. Only one of the cut pieces also has carnivore damage, in this case tooth scratches. It is a poorly preserved (chalky) 10.5 cm long adult reindeer scapula fragment with an intact proximal end. In this assemblage there are no pieces with both cut marks and pseudo-cut marks.

**26 Chop marks.** There are six pieces with chop marks, which vary from one to six chops each (Table A1.26, site 26). The chopped bones include the tibia, humerus, radius, and metapodial. The specimen with six chop marks is a poorly preserved (chalky) 8.7 cm long adult horse tibia with its distal end intact. The chopping is around the joint surface, suggesting that the hoof and nearly meatless metatarsal had been removed.

## Discussion

Elsewhere (Turner *et al.* 2001b) we compared the perimortem damage of Ust-Kova and Ust-Kan, open and cave archaeological sites, respectively. The differences were substantial in several regards, most notably in the expected preservational differences in protected and unprotected situations, the degree of processing, and in the bone damage attributable to hyenas at Ust-Kan. Why was there no apparent hyena damage at Ust-Kova? Was the site too far north and too cold for hyenas? N. I. Drozdov (personal communication August 2, 2002) thinks possibly not, as there are suggestions that the site was located in a micro-environment that was less harsh in the winter than other locations at the same latitude. However, the question remains – why so little, if any, hyena bone damage at Ust-Kova? One explanation might revolve around the occurrence of small objects of bone art at Ust-Kova (Fig. 3.183) and their absence at Ust-Kan and most other Upper Paleolithic sites of Siberia we have examined. Bone art may be a sign of more sophisticated human hunters, as has been frequently suggested for the European Cro-Magnon versus the nearly artless Neanderthal culture. Perhaps the Ust-Kova men were better able to keep the hyenas away from their encampment or meat cache, possibly out of fear of the dangerous humans. Obviously, more information on the paleo-climate at Ust-Kova would be useful, as well as more information on cave denning possibilities for hyenas in the Ust-Kova region.

**Fig. 3.183** Ust-Kova, artifacts. IAE museum exhibit of 22 000-year-old mammoth figurine, beads, and other items (CGT neg. IAE 8-12-03:27).

Can we not wonder about the weak carnivore signature at Ust-Kova (and Mal'ta), where bone-carving art seems to signal incoming anatomically modern Europeans capable of disconnecting the earlier hyena–Mousterian correspondence as seen in Denisova, Okladnikov, and Ust-Kan caves? Were the Ust-Kova people part of a Cro-Magnon expansion into Siberia?

# 27 Varvarina Gora

## Background

The site called Varvarina Gora (Barbara's Hill) is up-river from Kamenka, at 51°38′ N, 108°10′ E (Lbova 2000:153, Vasili'ev *et al.* 2002:528). It contains the accumulation of artifacts and faunal refuse left ca. 30 000 years ago by blade-making Upper Paleolithic hunters and their families, who camped repeatedly at the site, probably for many generations judging from its extensive refuse area. One reaches Varvarina Gora by turning off a dirt road connecting a remote and strangely out-of-place rusty factory town called Zaigraevo with a distant village by the name of Sara Bryan. At the turn off, a lightly used track winds uphill around burned stumps and new-growth forest (Fig. 3.184). Varvarina Gora is located on the lower southern slope of a conifer-covered hill overlooking the distant left bank of the northward-flowing Bryanka River.

Stone artifacts were discovered by workers constructing a power line in 1961, some poles of which had been set into the ancient site. News of the discovery of stone artifacts and bones was sent by E. A. Khamzina and D. D. Bazarov to A. P. Okladnikov, who carried out a multi-year excavation program at Varvarina Gora in the 1970s (Lbova 2000:51, Okladnikov and Kirilov 1980:31). During our site inspection on July 8, 2003, the forest was plagued by swarms of fiercely biting large black flies that instinctively attacked faces and eyes. Were these dreadful pests present when Varvarina Gora was occupied? Ovodov recalls that they were such a severe problem when he and A. Konapotski helped A. P. Okladnikov conduct the original excavations, that they necessitated smoky fires be kept burning much of the time to drive the voracious flies away.

Ludmila Lbova (2000, 2002) discusses Okladnikov's finds as well as her own subsequent additional excavations. She illustrates the deeper (Layer 2) and typically Upper Paleolithic artifacts recovered from the approximately 1100 $m^2$ area excavated at Varvarina Gora, as judged from her site maps (Lbova 2000:178–180). Stratigraphic and archaeological evidence initially indicated a single major component (Okladnikov and Kirilov 1980, Ovodov 1987a), but Lbova's re-excavations in 1986, 1992, and 1993 showed that there was also a later (Layer 1) 17 000 BP occupation by people who manufactured microblades and wedge-shaped cores. This later occupation overlay the earlier occupation dated 28 000 to 34 000 BP (Lbova, personal communication, July 8, 2003). Vasili'ev *et al.* (2002:528) list all the published dates for Varvarina Gora, which we have simplified to 30 000 BP for ease of recall. Ovodov (1987a) identified the species

Fig. 3.184 Varvarina Gora site. Ludmila Lbova (left) explains her additional work at this open site, first tested by A. P. Okladnikov. Nicolai Ovodov (right) was a member of the original dig crew (CGT neg. Varvarina Gora 7-8-03:6).

and MNI of the faunal remains from Okladnikov's excavations at Varvarina Gora. Among others, Ovodov recognized fox, wolf, bear, horse, rhinoceros, reindeer, bison, gazelle, spiral-horned antelope, and goat. Our taphonomic findings for Varvarina Gora follow, and have also been reported elsewhere (Turner 2010).

## Findings

**1 Provenience.** All of the specimens we examined were curated in the Institute of Archaeology and Ethnology, Novosibirsk. While some specimens had no provenience labels, these few were always bagged with pieces that were labeled for year of excavation and section number. The dates include 1973, 1974, 1976, and 1977. The specimens had been sorted into bags according to identifiable species, and bags of unidentifiable pieces. In addition to section numbers, some pieces were also numbered for Ovodov's specimen catalogs. Hence, our observations are based on the excavations led by Okladnikov and do not include the later work directed by Lbova. We examined 864 pieces, disregarding most unmodified loose teeth and foot bones. Our total was at least one-quarter of the collection. The excavations in the 1970s did not find the microblade horizon, so there was no reason not to pool our observations.

**2 Species.** Table A1.2 (site 27) shows that all 864 pieces were assessed for species determination (precise or general, i.e., big or small mammal, etc.). Indeterminable pieces made up 18.6% of the total, a value not too dissimilar from that of the nearby Kamenka site (21.9%). Gazelle (32.3% pieces) was the most commonly represented species, followed by horse (17.1%). Rhinoceros (6.0%), goat-sheep (4.4%), wolf (1.6%), bison (0.2%), hare (0.1%), and mammoth (0.1%) are also represented. The relatively large percentage for rhinoceros is due to there having been discovered much if not all of a single animal at Varvarina Gora. It had cut and chop marks indicating butchering.

In Ovodov's 1987 study he also identified marmot, fox, korsak, bear (a single tooth), yak, screw-horned antelope (three pieces of horn), and found a few pieces of bird bone that he did not identify. He remarked that the mammoth reported in the literature by Okladnikov was represented by ivory objects only. Ovodov felt that mammoth presence at Varvarina Gora was not the result of human hunting.

Kalmykov (2000) notes that carnivores were quite rare in the late Pleistocene Lake Baikal basin. They are rare at Varvarina Gora and Kamenka as well.

**3 Skeletal elements.** Our element sample size was 864 pieces (Table A1.3, site 27). Non-specific long bone pieces were the most common (12.4%), followed by rib pieces (11.6%). These values are nearly identical to those at nearby Kamenka (12.7% and 11.4%, respectively), although there are not quite as many cranial pieces (8.3%) as identified at Kamenka (12.5%). Still, these correspondences are increased by the presence of penis bones in both sites, and almost nowhere else in our other assemblages.

**4 Age.** Our sample size for age assessment was 864 pieces (Table A1.4, site 27). Adults were about five times more common than sub-adults. This age ratio is nearly identical to that found at Kamenka, so all the comparative comments there apply here as well.

**5 Completeness.** We had 863 usable pieces (Table A1.5, site 27). Whole pieces (3.4%) were similar in number but fewer than at Kamenka (8.7%); however, considering that we may have possibly excluded from study proportionally a few more Varvarina Gora whole bones, this difference is viewed as unimportant. Certainly, the ratio of one-ended to no-ended pieces in the two sites is similar, although there is a slightly significant statistical difference ($\chi^2 = 5.4$, 1 d.f., $0.01 < p < 0.05$). Varvarina Gora does not stand out for completeness when compared to our other assemblages.

**6 Maximum size.** The average size for 847 usable pieces is 9.0 cm (Table A1.6, site 27). This is larger than the mean for Kamenka (6.5 cm), which is entirely attributable to the several very large pieces of the rhinoceros found at Varvarina Gora. Removing those pieces reduces the Varvarina Gora mean considerably. However, with the rhinoceros pieces included, Varvarina Gora has a mean maximum diameter that is in the upper half of the size range for our pooled assemblage.

**7 Damage shape.** We evaluated this variable on the basis of 862 pieces (Table A1.7, site 27). Most frequent were long bone fragments (30.3%) and long bone flakes (13.8%). The number of long bone splinters was significantly less ($\chi^2 = 98.8$, 1 d.f., $p < 0.001$) at Varvarina Gora (3.4%) than at Kamenka (11.4%). The suggestion by Germonpré and Lbova (1996:44) that the Kamenka splinters "could be the byproducts of bone tool manufacturing" is reasonable but would have been strengthened had Varvarina Gora had a similar proportion of splinters, given the many other taphonomic similarities in the two sites.

**8 Color.** Color was assessed in 862 pieces (Table A1.8, site 27). Almost all of the pieces were ivory colored (99.2%). Pieces that were burned or possibly burned (brown or black) made up only 0.7% of the total. One was a partly burned, mostly whole horse vertebra. This vertebra is the only possible piece that might suggest roasting, although it more likely represents accidental burning. The occurrence of ivory colored pieces in Varvarina Gora and Kamenka (98.9%) is identical. Compared with our other assemblages, Varvarina Gora is at the very upper end of the ivory colored range.

**9 Preservation.** Quality could be evaluated in 863 pieces (Table A1.9, site 27). While the majority of pieces were ivory hard (65.3%), there were many pieces of chalky and intermediate quality (Fig. 3.185). There is a highly significant quality difference between Varvarina Gora and Kamenka ($\chi^2 = 1270.3$, 1 d.f., $p < 0.001$). The preservation difference is likely due to Varvarina Gora's coarser-grained sediments, southern exposure, and hillside slope – conditions the combination of which must have led to more humic acid percolation, more and deeper ground freezing and thawing, and a longer period of surface weathering before final burial. In addition, the many highly weathered rhinoceros bones indicate that they had been exposed to surface weathering for a relatively long period of time. Varvarina Gora quality is much like that of other open sites. It is Kamenka that has the unusually high quality of preservation for an open site.

**10 Perimortem breakage.** We could assess 862 pieces for perimortem damage (Table A1.10, site 27). Like nearly all of our sites, Varvarina Gora has a very high frequency of

**Fig. 3.185** Varvarina Gora, rhinoceros. Preservation was poor for a complete animal found at the site. Like desert conditions, those at this site caused bone to become largely chalk, which crumbles and breaks easily, unlike in the dry Altai caves, where the organic component of bone remains for millennia (CGT color IAE 7-30-03:35).

perimortem breakage (97.0%). This amount of breakage is only slightly more than that found at nearby Kamenka (94.7%). While the difference is statistically significant ($\chi^2 = 4.95$, 1 d.f., $0.01 < p < 0.05$), it is just barely so. Given the previous discussion of the very significant bone quality differences (i.e., arising from environmental and depositional history) between Varvarina Gora and Kamenka, we feel that the manner in which these ancient humans extracted nutrients from game animals and their use of bone for fabricational purposes was effectively the same in both sites. This view is backed up by the lack of color difference in these two sites, i.e., little or no evidence for roasting that would scorch or blacken bone.

Compared with our other assemblages, Varvarina Gora is not unusual with respect to perimortem damage.

**11 Postmortem breakage.** We evaluated 862 pieces for postmortem damage (Table A1.11, site 27). Of these, 9.6% showed postmortem breakage. Much of this happened in the excavation of the adult rhinoceros, most of whose skeleton was crumbly and chalky due to weathering and poor preservation. Even so, there is much more postmortem breakage at Varvarina Gora than at nearby Kamenka (1.1%). The difference is so large that there is no need for a formal statistical test. Relative to our other assemblages, the postmortem breakage is not so much unusual for Varvarina Gora than it is due to the exceptionally good preservation at Kamenka. Varvarina Gora is near the middle of the postmortem breakage range for all our assemblages.

**12 End-hollowing.** Due to the large number of pieces without anatomical ends, we could assess only 418 pieces for end-hollowing (Table A1.12, site 27). End-hollowing occurs in 1.2% of the Varvarina Gora sample, a value not significantly different from that of Kamenka (0.8%) ($\chi^2 = 0.19$, 1 d.f., $p > 0.5$). Hence, discussion of Kamenka regarding our objective of developing multiple criteria for defining carnivore damage applies here also. By itself, end-hollowing does not suggest much carnivore activity at these two sites.

Compared with our other assemblages, especially those where there had been an unquestionable presence of hyenas, Varvarina Gora end-hollowing is decidedly at the very low end of the range of occurrence. The presence of wolf bones at Varvarina Gora may not point to their having caused the end-hollowing, because Ovodov (1987a) suggested that the occurrence and types of several wolf bones, as well as their breakage, indicates that these animals may have been hunted for food as well as fur. On the other hand, carnivores that caused the end-hollowing are likely not represented in the Varvarina Gora assemblage, so scavenging wolves and foxes could still have produced the small amount of end-hollowing.

**13 Notching (Fig. 3.186).** Our sample for notching consisted of 856 pieces (Table A1.13, site 27). Of these, 5.7% had one or more notches. The most frequent number of notches per piece was one (4.2%). Occurrence and intensity of notching is less than at

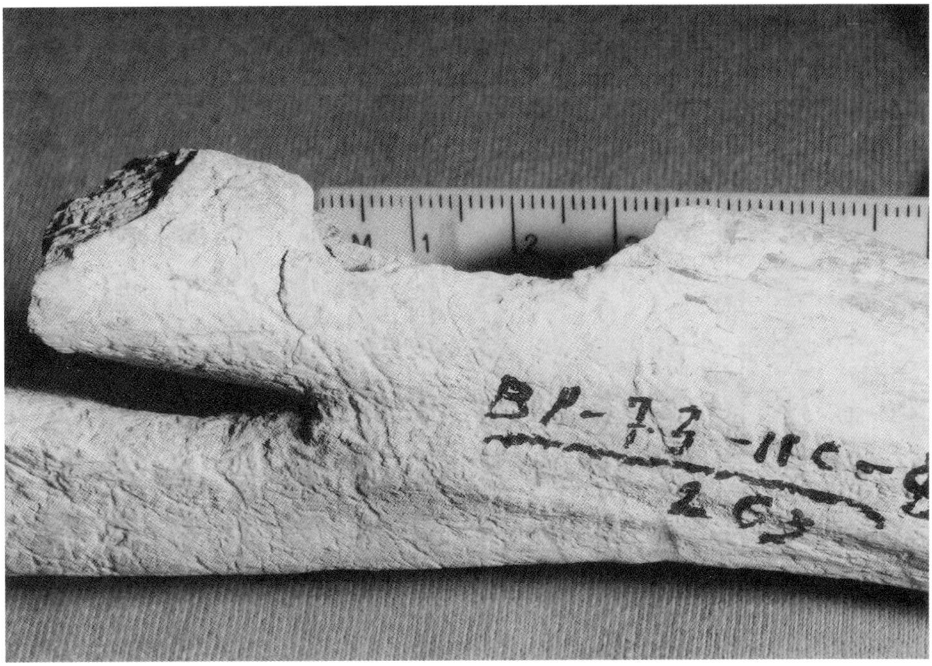

**Fig. 3.186** Varvarina Gora, bone damage. Horse radius with large impact notch, external view. Root damage and soil erosion are extensive (CGT neg. IAE 7-30-03:20).

Kamenka (9.8% occurrence). Although the difference is statistically significant ($\chi^2 = 8.32$, 1 d.f., $0.001 < p < 0.01$), we do not believe that it signals any special manner of cultural or taphonomic importance because chalky pieces with crumbly edges would not be scored as notched in the face of uncertainty caused by possible postmortem damage. Compared with our other assemblages, notching at Varvarina Gora is unexceptional for an open site. It clearly had less notching than the amount associated with hyenas and other carnivores.

**14 Tooth scratches.** We were able to assess the occurrence of tooth scratches in 703 pieces (Table A1.14, site 27). Only 1.9% had one or more scratches. The most frequent number of scratches per piece was five (0.4%), but two pieces had more than seven scratches. Occurrence and intensity of tooth scratches is slightly less than at Kamenka (2.8% occurrence), but the difference is not statistically significant ($\chi^2 = 1.1$, 1. d.f., $0.2 < p < 0.5$). As far as tooth scratches alone indicate carnivore activity, bone refuse at the two sites was processed similarly. Compared with our other assemblages, Varvarina Gora tooth scratching is unexceptional for an open site, and has much less than in the carnivore cave sites.

**15 Tooth dints (Fig. 3.187).** A total of 682 pieces could be evaluated for tooth dints (Table A1.15, site 27). Only 1.9% had one or more dints. The most frequent number of dints per piece was three (0.6%), and only one piece had more than seven dints. Occurrence and intensity of tooth dints is less than at Kamenka (3.3%), but the difference

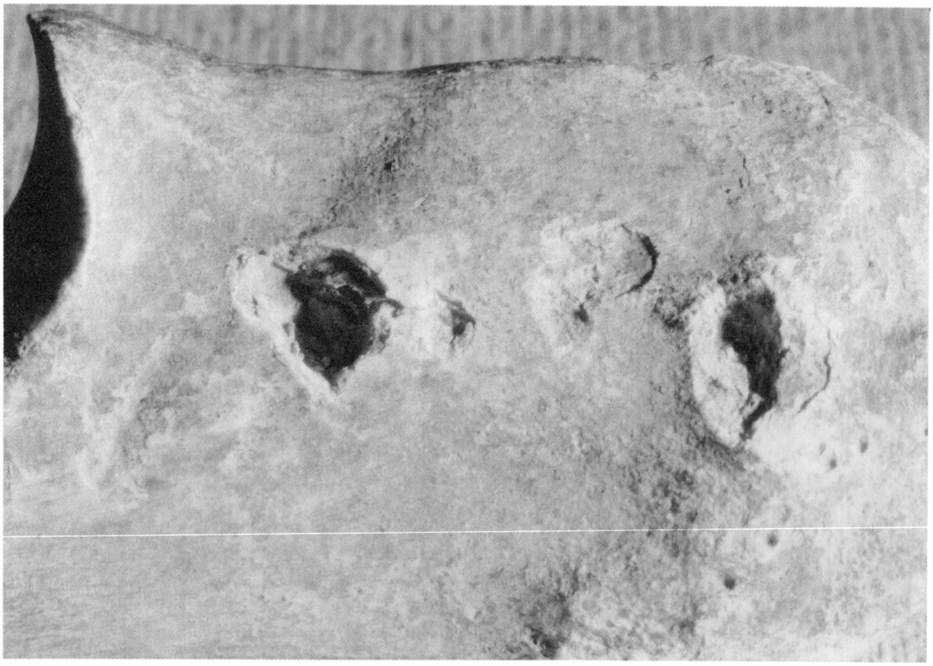

**Fig. 3.187** Varvarina Gora, bone damage. Gazelle ulna (1976) with tooth dints. Tiny embedded fragments are in the right dint. Width of image is 3.3 cm (CGT neg. IAE 7-30-03:1).

is not statistically significant ($\chi^2 = 2.41$, 1 d.f., $0.05 < p < 0.2$). Our comments for tooth scratching apply equally to tooth dinting – namely, there is only a small amount of carnivore damage that can be inferred for Varvarina Gora.

**16 Pseudo-cuts.** We could identify no examples of pseudo-cuts in all 689 usable pieces. This again reflects a minimum of carnivore processing of Varvarina Gora bone refuse.

**17 Abrasions.** In 686 pieces, only 0.4% had one or more abrasion grooves per piece (Table A1.17, site 27). This is practically the same as seen at Kamenka (0.2%), for which we earlier proposed that bone breakage for marrow extraction was accomplished mainly with the use of non-abrasive hammers and anvils such as bone, antler, or wood instead of gritty abrasion-producing hammer stones and anvil stones. Stone could well have been used for bone breakage, but it would have had to have been fine-grained so as not to leave abrasion grooves.

A low frequency of abraded pieces is characteristic of all the assemblages in our study. Given the generally coarse- to medium-grained stone sources that we have noticed in all the site areas visited, there seems to be a reasonable basis for proposing that stone was not the major material used in the perimortem bone breakage.

**18 Polishing.** A total of 834 pieces could be assessed for polishing (Table A1.18, site 27). Polishing was very common in Varvarina Gora (96.0%). At Kamenka there was less (90.0%), the difference in which is statistically significant ($\chi^2 = 20.19$, 1 d.f., $p < 0.001$). The difference is attributable to slope, where some polishing was possibly caused by bone movement downhill, since the amount of slope at the time of deposition seems to have been a few degrees greater at Varvarina Gora than at Kamenka. Also, because the proportion of chalky pieces is greater at Varvarina Gora than at Kamenka, bone refuse at Kamenka was better protected from weathering, which implies less time passed before burial, and once buried the perimortem and postmortem polishing processes were less energetic. Compared with our other assemblages, Varvarina Gora is decidedly at the upper end of the range for polishing. This range does not break down as neatly for open versus cave sites as do some other variables.

**19 Embedded fragments.** We were able to assess embedding in 858 pieces (Table A1.19, site 27). There were very few pieces that had embedded fragments (0.9%). Of these, having two embedded fragments was most common (0.3%). There were no pieces with more than four embedded fragments. Kamenka had a similarly low occurrence of pieces with embedded fragments (1.5%). Compared with our other assemblages, Varvarina Gora is at the low end of the range for embedded pieces.

**20 Tooth wear.** We evaluated tooth wear on 36 pieces of tooth-bearing bone (Table A1.20, site 27). Individuals we considered to be young made up 13.9% of this total. Kamenka had almost half this amount, but both assemblages are too small for meaningful chi-square comparisons. Compared with our other samples, Varvarina Gora would appear to be at the low end of the range for the presence of young individuals.

**21 Acid erosion.** We could assess 864 pieces for acid erosion (Table A1.21, site 27). Only 0.5% of this total had acid erosion. Of the four eroded pieces, three were highly

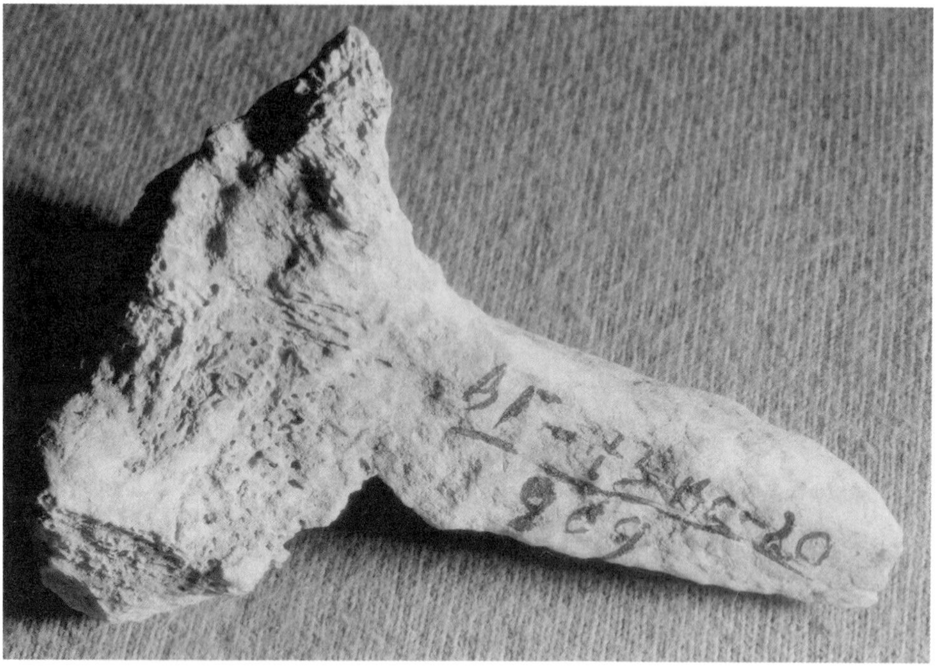

**Fig. 3.188** Varvarina Gora, bone damage. Gazelle proximal tibia fragment, with soil damage that has partly obscured condition, that was likely a hyena stomach bone. Maximum length is 6.1 cm (CGT neg. IAE 7-30-03:27).

rounded, like water-worn pebbles. The fourth was highly eroded, but gritty rather than greasy feeling (Fig. 3.188; see 3.193, 3.194). None is completely like the digestive damage seen in the hyena cave assemblages. In addition, there were two other pieces that were corroded on one surface, as if they had been in the process of dissolving by some agency other than stomach acid. These two were not considered as having digestive damage. The 0.5% of acid-eroded pieces is somewhat less than occurred at Kamenka (1.3%), but not significantly so ($\chi^2=2.85$, 1 d.f., $p>0.08$). Compared with our other assemblages, Varvarina Gora is at the low end of the range for acid erosion.

**22–24 Rodent gnawing, insect damage, and human bone.** Varvarina Gora had no examples of these considerations.

**25 Cut marks (Figs. 3.189–3.192).** We could assess 714 pieces for cut marks (Table A1.25, site 27). Fully 8.8% of all pieces had one or more cut marks. The number of cuts per cut piece range more-or-less equally from one to more than seven. The largest number of cut marks on a given Varvarina Gora piece is 15. There is one 15.8 cm long piece of a distal end of an adult horse humerus with five ultra-fine cut marks varying from 2.0 to 5.0 cm in length. The piece is extraordinarily well preserved, equal to that found in cave sites, suggesting that other very fine cut marks may have been erased from less well-preserved pieces. This in turn suggests that we have to some degree underestimated cutting in Varvarina Gora and perhaps other open sites where preservation is not as good

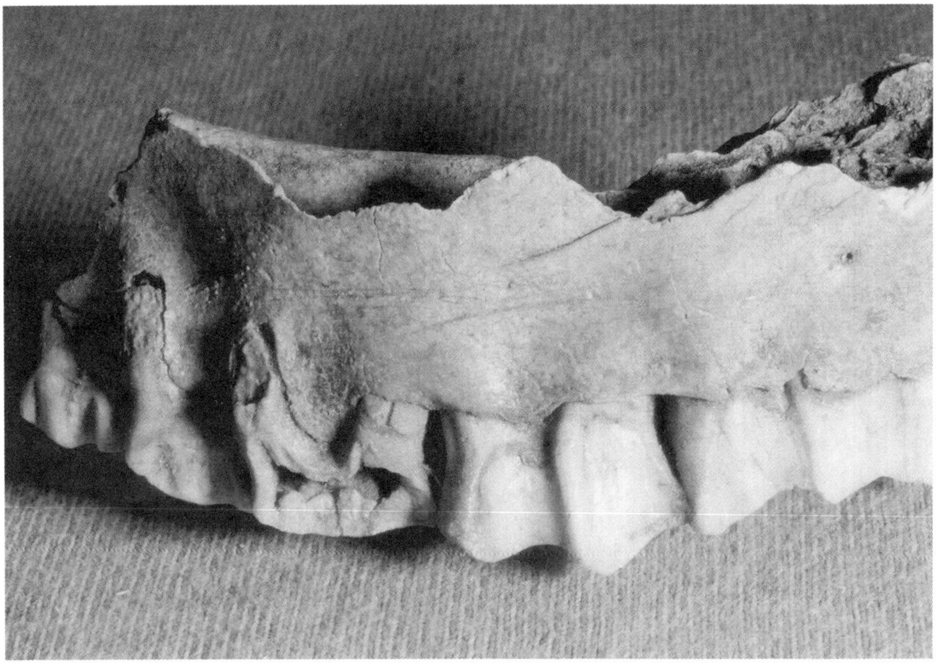

**Fig. 3.189** Varvarina Gora, bone damage. Cut marks on gazelle maxilla. The third tooth from the right was broken in life and shows subsequent wear on the fracture (CGT neg. IAE 7-30-03:24).

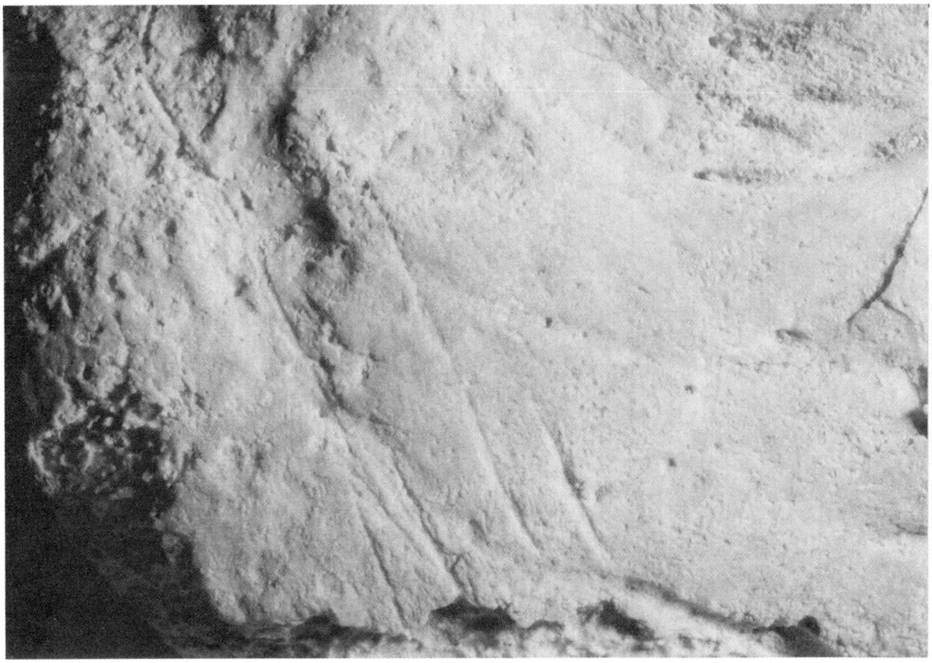

**Fig. 3.190** Varvarina Gora. Horse distal tibia with growth lines that resemble cut marks. Width of image is 3.3 cm (CGT neg. IAE 7-30-03:28).

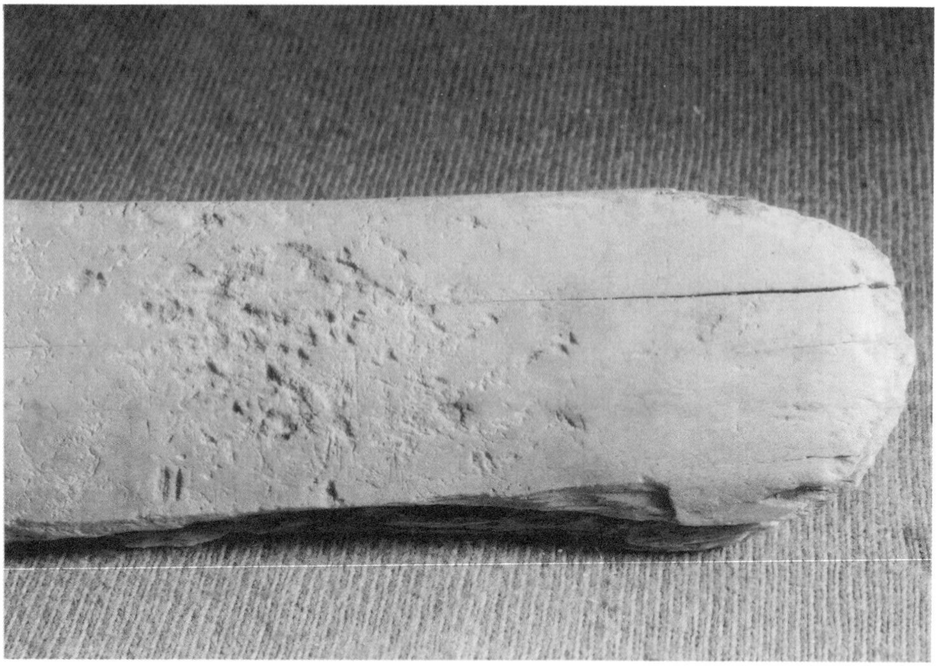

**Fig. 3.191** Varvarina Gora, bone damage. Horse (?) rib fragment with chop and cut marks, abrasions and rounded end. A sort of anvil or cutting board. Possibly used in stone tool production. Width of image is 6.8 cm (CGT neg. IAE 7-30-03:30).

**Fig. 3.192** Varvarina Gora, bone damage. An immature wolf tibia with a cut mark on the distal joint. In all of the food refuse recovered at Varvarina Gora and all of the other sites presented in this volume, there is no suggestion that wolves were used as a food source. Ovodov suggests that they were hunted for their fur. Width of image is 3.3 cm (CGT neg. IAE 7-30-03:26).

**Fig. 3.193** Varvarina Gora, bone damage, stomach bone. Insofar as can be determined, there is no direct evidence of hyenas in the Varvarina Gora assemblage. Hence, a wolf may have defecated or regurgitated this stomach bone. Width of image is 3.3 cm (CGT neg. IAE 7-30-03:12).

**Fig. 3.194** Varvarina Gora, bone damage. The surface texture of this piece lacks the “greasy” feel of nearly all stomach bones. However, the appearance matches. It is possible that the piece was originally a stomach bone whose surface polish was eroded by soil conditions. Width of image is 3.3 cm (CGT neg. IAE 7-30-03:7).

as in most cave sites. Nearby Kamenka has a slightly higher frequency of cut marks (9.5%), but the difference is not significant ($\chi^2 = 0.18$, 1 d.f., $p > 0.3$). As related earlier, Kamenka bone was excellently preserved, so if we are under-counting cut marks it does not show up in these two neighboring sites with notable preservation differences.

Compared with our other assemblages, Varvarina Gora is more-or-less in the middle of the range for cut mark frequency.

**26 Chop marks.** There were 747 pieces that could be evaluated for chop marks (Table A1.26, site 27). Of these, 13.8% of all pieces had one or more chop marks. Of chopped pieces, the majority (8.6%) had only one chop mark. Two chop marks were less common (2.9%), and only a few other pieces had 3–6 grooves due to chopping. The occurrence of chop marks at Kamenka was less (8.6%), and most of these were pieces with only one chop mark. The frequency difference in the occurrence of chop marks at Varvarina Gora and Kamenka is significant ($\chi^2 = 8.94$, 1 d.f., $p < 0.005$). We suggest that this difference is due to Varvarina Gora having proportionally more pieces of large mammals (horse, rhinoceros, "big") than Kamenka. Certainly, the large mammal pieces at Varvarina Gora have more chop marks than do pieces of smaller mammals. Moreover, Varvarina Gora shows a large number of pieces with associated chop marks and notching. It would seem that gaining access to the marrow cavity of large mammals like horse and rhinoceros was more commonly accomplished by the use of large, heavy, sharp-edged stone artifacts like the Varvarina and Kamenka cores illustrated by Lbova (2002:65–66) than by hammering with a non-lithic piece of bone, horn, or wood. Conceivably, the choice of material illustrates ancient awareness of the "stiletto heel effect" in which a great deal of kinetic energy can be focused on a small area, causing great damage that might not occur when the impacting surface has more area.

Compared with our other assemblages, Varvarina Gora is in the upper part of the range for chop mark frequency.

As mentioned in "background," there were two cultural levels in Varvarina Gora. Figs. 3.195–3.198 illustrate some stone artifacts from both levels.

## Discussion

Varvarina Gora is a regional neighbor to, and in the same river valley as, Kamenka. Both sites are located on the lower slopes of low hills adjacent to the valley. Both were repeatedly used, open steppe sites with early dates around 30 000 years ago, a time of late Pleistocene relative warmth. Ovodov (1987a:138) feels the faunal assemblage at Varvarina Gora indicates a pronounced steppe-like condition with some forest-steppe and forest-tundra. He also views the site as having been occupied for a long time because of the multiple remains of hearths, dwellings, and numerous "grinding" stones. Both faunal and cultural remains indicate forest vegetation in the steppe and on the northern slopes of hills. Wood would have been needed to warm the dwellings in the winter.

Both Varvarina Gora and Kamenka show similar tool types that the occupants used to hunt and process the same sorts of game animals. These were mainly gazelles and horses.

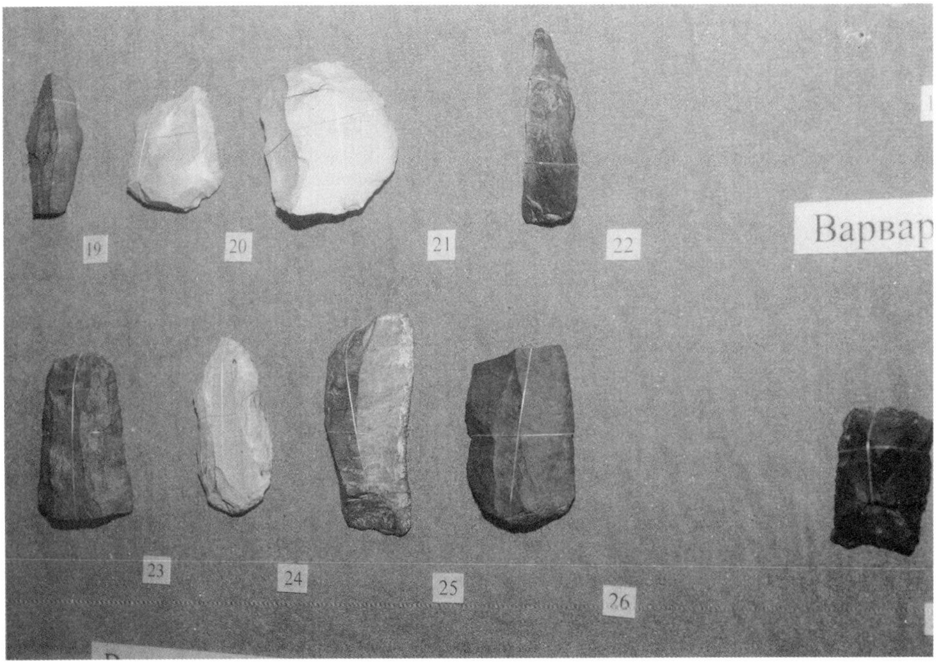


**Fig. 3.195** Varvarina Gora. Stone tools, excavated in Layer 1 (ca. 17 000 BP) by Ludmila Lbova and associates, on exhibit in the Museum of Buryatia Scientific Center, Ulan-Ude (CGT neg. MBSC 7-4-03:35).

**Fig. 3.196** Varvarina Gora. Stone tools excavated by A. P. Okladnikov and associates on exhibit in the former IHPP museum. Text reads: “Paleolithic site Varvarina Gora 34,000 BP.” All the faunal remains discussed and illustrated here are from Okladnikov’s excavations (CGT color IHPP 2-10-84:34).

**Fig. 3.197** Varvarina Gora. Additional stone artifacts on exhibit with those shown in the previous figure (CGT color IHPP 2-10-84:35).

**Fig. 3.198** Varvarina Gora. Additional stone tools on exhibit with those in the previous two figures (CGT color IHPP 2-10-84:36).

The natural- and human-caused perimortem taphonomy of both sites is remarkably similar, with the exception of Varvarina Gora having gone through less favorable conditions for bone preservation than the exceptionally well-preserved assemblage Lbova recovered from Kamenka. The occurrence of chopping at Varvarina Gora is greater than at Kamenka, which reflects the larger proportion of pieces of big animal bone at Varvarina Gora.

The perimortem taphonomic characteristics of Varvarina Gora are broadly like those of our other open sites. Carnivore damage was slight, and there is no direct or indirect evidence of hyenas. If meat had been cooked, it must have been by hide bag- or wooden bowl-boiling, because the evidence for roasting is weak to non-existent. As elsewhere in this study, perimortem processing was intensive – that is, most bones were broken, and broken into many small pieces whose average maximum diameter is only 9.0 cm, a value that would be much smaller had our protocol not excluded from study pieces smaller than 2.5 cm in diameter.

# 28 Volchiya Griva

## Background

Volchiya Griva (Wolf's Mane) is an open *beast solonetz* (saline soil) 20 000 $m^2$ in area, similar in zoogeological features to Shestakovo, discussed previously. It is located on a ridge in the Barabinsk (Baraba) steppe, near Mamontovoe village, west-southwest of Novosibirsk. Like Shestakovo, it is north of the late Pleistocene southern boundary of the West Siberian Plain. Kuzmin and Orlova (1998:7) give its location as 54°63′ N, 80°25′ E. The site has been known since 1957 (Alexeeva and Vereschagin 1970, Maschenko and Leshchinskiy 2001), and the findings we report herein are based on the recovery from excavations conducted in 1991. Here, as at Shestakovo, Sergei V. Leshchinskiy (1998, 1999, 2001a, 2001b) and Vasily N. Zenin directed the excavations that in five field seasons produced more than 5000 pieces of bones and teeth from sediments they calculate to be about 2.5% of the total bone-bearing area. Of the 5000 pieces, 98% are mammoth, and 2% are bison and wolf. The MNI numbers in the hundreds (Leshchinskiy 2001b). Three layers were established, with weathering in the upper layer being greater than in the middle or lower layers. Judging from the photographic illustrations in Maschenko and Leshchinskiy (2001: figures 1–7), most of the very large bones such as ribs, long bones, scapulae, and crania have undergone extensive postmortem breakage. Some of these large bones would be exceptionally difficult, expensive, and time-consuming to stabilize and transport to the University of Tomsk for reconstruction and study. Mammoths make up about 98% of their total. An MNI of 50 individuals could be identified. Mammoth calves are at least 26% of the total MNI. Considering all immature individuals, they total 42%, a value not unlike that of Shestakovo, and similar to the 50% death rate for modern young herbivorous livestock in similar pauperized environments (Leshchinskiy 2001a:297, after Kovalskiy 1974). Other identified species found at Volchiya Griva include horse (MNI = 3), bison (MNI = 3), and one wolf. Rodent chewing was identified on some mammoth bones (Maschenko and Leshchinskiy 2001).

Leshchinskiy (2001a:295) reports that carbon-14 dates for Volchiya Griva range from approximately 15 000 to 10 500 years ago, a period of time that corresponds to the second half of the Sartan (see also Firsov and Orlova [1971] for additional dating information). The geochemical and geomorphological qualities of Volchiya Griva are similar to those of Shestakovo (discussed earlier, where the reader can find the descriptive and theoretical considerations). Upper Paleolithic human presence was found at Volchiya Griva, but it has yet to be described. Orlova *et al.* (2000b) provide further radiocarbon and stratigraphic information. Another mammoth site in saline soil in the region is named Kochegir and dates to about the same time (Leshchinskiy *et al.* 2003).

In sum, the geological findings at Volchiya Griva parallel those of Shestakovo. Our expectation for perimortem bone damage should be that very little evidence will be found indicating human butchering or other carcass processing and body part utilization. We examined the Volchiya Griva assemblage in the Department of Paleontology and Historical Geology at Tomsk University in June 2002.

## Findings

**1 Provenience.** Table A1.1 (site 28) lists the assigned provenience designations. As the excavators themselves treat this site in their publications as a single temporal unit, we do the same.

**2 Species.** Our sample of 113 pieces contains only mammoth (74.3%; Table A1.2, site 28), and bison (25.7%). We did not see the horse or wolf pieces reported by Maschenko and Leshchinskiy (2001).

**3 Skeletal elements.** Feet (15.0%) and ribs (15.0%) are the most frequent items (Table A1.3, site 28). However, when pieces identified as "feet" are combined with other skeletal elements located inferior to the leg bones, then nearly half (46.9%) of our sample is in actuality bones of the feet. Maschenko and Leshchinskiy (2001) report a similar finding. Long bones are conspicuous by their under-representation, some of which is due to salvage and conservation considerations.

**4 Age.** Our Volchiya Griva sample has mostly adult bones (69.9%; Table A1.4, site 28). However, sub-adults are well represented (20.3%). As with skeletal elements, these percentages are similar to those reported by the excavators.

**5 Completeness.** Our sample has a large number of whole bones (69.0%; Table A1.5, site 28), and bones with one anatomical end (12.4%). Bones that have no anatomical ends are much fewer (18.6%). The relatively good preservation situation and the very small amount of identifiable carnivore damage are responsible for this aspect of preservation.

**6 Maximum size.** As expected for a predominately mammoth and bison paleontological site, the Volchiya Griva mean for 117 pieces is large (17.0 cm), as is its range (4.2 cm to 69.1 cm). Compared with the pooled assemblage values, Volchiya Griva is significantly larger. Looking at the whole long bone lengths of mammoth and bison provided by Vera Gromova (1950: table 27), Volchiya Griva upper range limit is larger than that of bison, but generally smaller than her mammoth elements.

**7 Damage shape.** By far the most numerous of our 113 pieces was classified as undamaged (69.9%) (Table A1.7, site 28). Much rarer are pieces we call medial rib fragments (9.7%). All other types are quite infrequent. As noted above, this sample has very little perimortem damage.

**8 Color.** As an indication of preservation, the color of the Volchiya Griva sample suggests a relatively good preservation context, because almost half of the pieces are ivory colored (48.7%) (Table A1.8, site 28). The brown coloration that we associate with soil-staining by mineral-bearing sub-surface water movement is understandable here in this "mineral oasis" context.

**9 Preservation.** Seemingly contradictory to the above remarks about good preservation at Volchiya Griva, most of the bone pieces in our sample have a chalky quality (91.1%) (Table A1.9, site 28). We feel that the contradiction is due largely to the nature of mammoth bone and its relatively thin cortex. Another possibility is that the salts and other concentrated chemicals in Volchiya Griva had erosive effects on bone.

**10 Perimortem breakage.** Here again, perimortem damage is minimal, at least in comparison to the faunal remains from other sites such as Okladnikov Cave. Only 8.0% (Table A1.10, site 28) of our sample has identifiable perimortem breakage. Given that this site existed for about 4000 years, this suggests that the bones we studied were sealed in the mud and mud-cracks soon after the death of the animals, and that carnivores, humans, and other scavengers who might have exploited their remains were infrequent visitors to the site.

**11 Postmortem breakage.** Despite what appear to be good conditions for preservation, at least for an open site, there is a large amount of postmortem breakage (46.0%) (Table A1.11, site 28). Some of the 52 pieces with postmortem breakage may have resulted from trampling; however, we suspect that most happened during excavation, although we saw no trowel or shovel marks.

**12 End-hollowing.** Only one piece has end-hollowing (0.9%) (Table A1.12, site 28). It is a chalky 7.5 cm long adult bison carpal bone that also has two tooth dints.

**13–14 Notching and tooth scratches.** No examples of these variables were found in the Volchiya Griva sample.

**15 Tooth dints.** One example of tooth dinting was found (Table A1.15, site 28). It is the specimen mentioned under end-hollowing.

**16–19 Pseudo-cuts, abrasions, polishing, and embedded fragments.** No examples of these variables were found.

**20 Tooth wear.** No teeth were studied.

**21–22 Acid erosion and rodent gnawing.** No examples of these variables were found, although Maschenko and Leshchinskiy (2001) report that some mammoth bones had rodent gnawing.

**23–26 Insect damage, human bone, cut marks, and chop marks.** No examples of these variables were found.

## Discussion

Considering all the forms of perimortem damage that help distinguish that done by humans and non-human animals, we wholly agree with the excavators that there is very little evidence of carnivore and/or human scavenging of the animal remains at the Volchiya Griva *beast solonetz*. But why is this so? In the taphonomic and forensic literature, and in our personal experience with various-sized animals, large and small, which we found dead upon the ground, all were, within months, even days, markedly scavenged and scattered. Some animal examples that the senior author and his oldest daughter have made records and time-sequential photographs of, commencing from the known day the animals died, all but completely disappeared within a week (a small beef calf) to a year (a fully grown beef bull, and an old horse named Lucky). The scavengers in these cases were dogs, coyotes, buzzards, ravens, rodents, and first-on-the-scene scavengers, namely flies and other zoophagous insects. But we personally have experience with one significant exception – a non-scavenged adult steer that died on its ranch. The ever-present carrion birds were never seen circling over its carcass. This carcass was never scavenged, although it eventually decayed after a few years, whereas all other dead animals found on the ranch over a ten-year period had been scavenged and scattered almost entirely. Is it possible that the salt-tolerant bacterial and chemical concentrations at Volchiya Griva caused some deaths that scavengers naturally avoided?

# 29 Yelenev Cave

## Background

Yelenev Cave was named by Nicolai Ovodov in recognition of Alexei S. Yelenev's early-twentieth-century explorations of more than 50 caves in the vicinity of Krasnoyarsk. Yelenev was a mathematics teacher before the Russian Revolution of 1919. Sometime afterwards he moved east to Irkutsk, where his notes, papers, and journals were lost. Yelenev was not alone in exploring Yenisei River caves. A man named Proskuryakov (1889, 1890, 1893) was also doing this in the late 1800s. Following Ovodov's initial testing, the cave was extensively excavated for 12 seasons in the 1970s and 1980s under the directorship of archaeologist Nicolai P. Makarov. Ovodov assisted him, as did various staff members of the Krasnoyarsk Regional Museum, and numerous highschool children on summer vacation. V. P. Cheka *et al.* conducted studies at Yelenev Cave to help reconstruct the regional environment in the late Pleistocene and early Holocene. O. V. Andrenko has prepared an article on the paleoecology of Yelenev Cave, and has also identified the avian remains. Further environmental reconstruction was developed by the principle investigators (Ovodov and Martynovich 1994, 1999). In August 2000 a day-trip was arranged by Ovodov to visit the cave. He was aided by Valery Gorchakovsky, who owned a dascha 300 m or so from the cave (Figs. 3.199–3.201). The party consisted of the present three authors and Elaina Popkova. Yelenev Cave is an 8 m high, 4 m wide, and 23 m deep remnant of an ancient underground water-formed grotto, now located 17 m above the present level of the Yenisei River, near the summit of a steep limestone cliff that rises nearly vertically from the river's edge. The river here runs through a wide canyon draped with taiga forest that is dramatically broken by stretches of bare pillar-shaped left-bank vertical canyon walls. The south-facing cave is situated 20 km up-river from Krasnoyarsk and 30 km down-river from the Krasnoyarsk hydro-electric dam (Figs. 3.202–3.205). The cave's coordinates are 55°58′ N, 92°29′ E. Due to the near-vertical cliff face, Yelenev Cave is most safely reached by hiking up from the river at a less abrupt down-stream boat-landing, and on reaching the summit above the cave, lowering oneself by rope down to the entrance. The cave is one of at least five that Ovodov has explored in this stretch of the river between the city and the dam. Before the construction of the dam, the cold, swift-flowing Yenisei River would freeze over in the winter. Now it forms an aquatic year-round north–south barrier to all except power boat operators. Like many other limestone grottos in southern Siberia, the low level of light,

**Fig. 3.199** Odyssey. Yelenev Cave. Our day trip to this cave began in Krasnoyarsk, where we boarded a 60-passenger hydroplane river boat that provides travel service up and down the Yenisei River. We disembarked 35 km upriver at a small left-bank community of dacha cabins and vegetable gardens, and walked up a steep road to the dacha belonging to Valery Gorchakovsky, a friend of Nicolai Ovodov, whom he got to know when excavating Yelenev Cave. There, we had lunch before hiking on to the cave. Son Dmitry is shown starting a fire in the outdoor stove to heat water for tea. Next to Dmitry is a large tub full of Neolithic pottery sherds that were found during several seasons of dacha use. Ovodov and his dog Charlie look on (CGT neg. Yelenev 7-14-00:12).

**Fig. 3.200** Odyssey. Yelenev Cave. Sherd types suggest this riverside locality has been occupied since earliest Neolithic times. The sherds and stone artifacts were picked up over the years while building the cabin, fences, and working the garden plot (CGT neg. Yelenev 7-14-00:11).

**Fig. 3.201** Odyssey. Yelenev Cave. Valery Gorchakovsky picks some fresh onions and lettuce from his garden for our lunch while Ovodov looks on (CGT neg. Yelenev 7-14-00:13).

**Fig. 3.202** Odyssey. Yelenev Cave, trail to the cave. After lunch Ovodov leads the way to the summit of the high bluff overlooking the Yenisei River. It is easier to reach the cave by descending from the cliff above it than trying to reach it from below (CGT neg. Yelenev 7-14-00:14).

**Fig. 3.203** Odyssey. Yelenev Cave, view of the Yenisei River right bank from the cliff summit above the cave. Power lines across the river come from the hydroelectric dam up-river to the right. Left to right: Dmitry Gorchakovsky, Elaine Popkova, Olga Pavlova, Nicolai Ovodov (standing) (CGT neg. Yelenev 7-14-00:15).

**Fig. 3.204** Odyssey. Yelenev Cave. Looking down at the hydroplane river boat that maintains a frequent schedule. In this view from the cliff top the vessel is proceeding further up-river. On the flatter riverside opposite Yelenev Cave, numerous dachas have been erected above the high-water level of the river that varies depending on electricity demand in and around Krasnoyarsk, including the regional section of the Trans-Siberian electrified railroad (CGT neg. Yelenev 7-14-00:19).

**Fig. 3.205** Odyssey. Yelenev Cave. The arrow points to the location of the cave. The view is up-river through haze that is due to a combination of high humidity and distant forest fire smoke (CGT neg. Yelenev 7-14-00:16).

except near the entrance, is unfavorable for plant growth, hence there is very little root damage to the osteological remains found in the cave deposits. During our visit the senior author was impressed with how few mosquitoes there were around and at the cave site. Ticks infected with an encephalitis virus are common in the region, presumably carried with the introduction of cattle. Finally, the condition of the cave midden and its contents suggests that Yelenev was generally a dry cave, although it may have been damp in the rainy season.

Twenty-two layers were identified in a total depth of 6.5 m, 2.5 m of which contained human refuse, artifacts, and a small number of disarticulated human bones (Ovodov, personal recollections, visit to Yelenev Cave, July 14, 2000). Not all of the faunal remains were saved due to museum storage space limitations. Ribs, vertebrae, and small unidentifiable fragments were discarded and unrecorded. Ovodov feels that the Neolithic hunters began their butchering of game at the kill site, which along with the excavation discards presumably helps explain most of the missing skeletal elements. Because some fetal bone was recovered, Ovodov further feels that Yelenev Cave was occupied by humans mainly during the summer, so some amount of seasonally induced bias needs also to be taken into account. Based on the recovered species of small mammal and bird remains, Ovodov and Martynovich (1994, 1999, 2000) proposed that forest formation started in the late Pleistocene and played an increasingly important ecological role as the tundra-steppe habitat was replaced with taiga due to climatic warming. The early and middle Holocene game animals that the Yelenev residents hunted included hares (39.2%), roe deer (46.23%), maral deer (6.3%), bear (0.8%), elk (0.6%), fox (3.8%), sable (1.6%), and beaver (0.2%). Badger, gray wolf, marmot and musk deer were also recognized, as were 15 species of waterfowl and grouse. Of the 66 500 fish elements recovered, 6500 were of Neolithic age, and 54 500 were of Mesolithic age. Most were eelpouts, followed by grayling, carp, pike, sturgeon, and others.

Artifacts recovered were typical Neolithic objects ranging from bone harpoons and fishing hooks to stone arrow points and knives, some pottery, and a few other classes of objects. Fires had been prepared within the cave as hearths were encountered. Layer 10 was especially rich in ash. However, very few burned bones were found, and those that were more likely than not were burned accidentally. As will be illustrated, roasting of game was certainly rare at Yelenev Cave, if it was practiced at all.

Carbon-14 dates indicate that the main human occupation was Mesolithic and Neolithic – that is, commencing about 9000 years ago. Some 49 carbon-14 dates have been obtained, including a 13 600 BP date related to pre-human paleontological materials. However, no examples of extinct fauna were recovered (Ovodov, personal communication; Vasili'ev *et al.* 2002:524). Kuzmin and Orlova (1998:18) list ten carbon-14 dates, all based on charcoal, that range from 10 485 BP to 8205 BP.

## Findings

**1 Provenience.** At least 14 layers were archaeologically recognized, as well as several horizontal squares. However, as with other sites described in this volume, we could see no major temporal or area differences in the types, treatment, or occurrence of the faunal remains, so we have pooled the various archaeological units. Table A1.1 (site 29) lists the provenience units for the 241 studied pieces of bone.

**2 Species.** Three-quarters of the 241 pieces we examined (excluding 65 pieces of human remains) belonged to roe deer (74.7%), with goat-sheep making up the remainder except for one fox bone (Table A1.2, site 29). This absolutely high amount of species identification results from the intensive analysis by Ovodov, coupled with the relative completeness of many of the sampled fragments. About 100 additional pieces of unidentifiable long bone fragments were only measured for their maximum length.

**3 Skeletal elements.** Metapodials (24.9%), non-specific long bones (11.6%), mandibles (11.2%), and scapulae (10.4%) make up more than half of the skeletal elements (Table A1.3, site 23). Since skull fragments are usually quite easy to identify, an absolutely low frequency (skull, 3.3%; maxilla, 2.5%) confirms Ovodov's view that initial butchering occurred elsewhere. They would have been saved after excavation for identification purposes, whereas the very low occurrence of bones of the trunk (vertebrae, 0.4%; ribs, 4.6%; and pelvis, 2.9%) is due in large part to the post-excavation discarding of vertebrae and ribs.

**4 Age.** Only 19 (8.6%) of the 221 pieces whose age could be determined were identified as sub-adult on the basis of tooth eruption, tooth wear, and epiphysial union (Table A1.4, site 29). Whether this suggests a fall–winter hunting schedule, or possibly some form of aboriginal conservation or culling practice cannot be determined at this time.

**5 Completeness.** As Table A1.5 (site 29) shows, there is a relatively low frequency of complete bones (7.0%), and high frequencies of bones with one (45.2%) or no anatomical ends (47.7%). In contrast with another site where similarly sized ungulates were food items (Bolshoi Yakor, late Pleistocene reindeer), the Yelenev Cave bones were much less broken and smashed. More than 90% of the Bolshoi Yakor bones were missing both anatomical ends. This suggests that the Yelenev Cave occupants did not need to be as frugal as their Ice Age counterparts, and reflects the discarding of unidentifiable fragments mentioned previously.

**6 Maximum size.** The mean for 340 Yelenev pieces is 6.0 cm, and the range is 1.5 cm to 17.5 cm. Compared with the pooled assemblage, the dimensions of the Yelenev assemblage are significantly smaller. Looking at the lengths of whole long bones for roe deer and goat-sheep provided by Vera Gromova (1950: table 27) shows that the upper range limit of Yelenev is generally smaller for both species.

**7 Damage form.** Table A1.7 (site 29) shows each of the damage types we identified for the 241 pieces. Flakes (21.6%), butts (18.3%), cracked-open pieces (11.6%), and segments (10.8%) make up more than 50% of the total number of types. Given that most of the bones recovered from the cave excavation represent limb bones, we find it somewhat surprising that the frequency of splinters (0.4%) is so low. While major long bones when smashed tend to wind up as flakes, damaged metapodials not infrequently are represented by splinters. On the whole, there is nothing especially distinctive about the kinds of damage or the frequencies of each type. We see no sure sign of any particular damage signature that might have resulted from specialized local carcass processing practices, economic necessity, special processing tools, or idiosyncratic butchering behaviors.

**8 Color.** With the exception of one piece that is brown colored, a 6.3 cm long medial rib fragment with end-polishing, all the other 240 pieces (99.6%) are ivory colored (Table A1.8, site 29). The singular brown piece might represent roasting, although at the time of our examination we saw no sure sign of thermal modification in any piece as large as 2.5 cm. However, there were three burned pieces smaller than 2.0 cm. These could easily have burned accidentally in or under the fire hearths. Said another way, none of the Yelenev bone, whole or fragmentary, shows any sure sign of having been cooked, let alone roasted.

**9 Preservation.** All 241 pieces are in excellent condition – we have characterized each as having an ivory rather than a chalky quality (Table A1.9, site 29). High bone quality correlates strongly with deposition in a limestone cave away from the entrance drip line.

**10 Perimortem breakage.** All but eight (3.3%) of the 241 pieces exhibit some degree of perimortem breakage (Table A1.10, site 29). All but one of the pieces without perimortem breakage is a whole bone. The single exception is a fragment of a sub-adult tibial epiphysis. Recall that whole bones are defined as having both anatomical ends and that there can be minor perimortem damage somewhere on the element, i.e., there are 17 whole bones in this sample. Hence, the processing of this assemblage was extensive, as much as that seen in late Pleistocene Siberian archaeological sites, but the degree of completeness is greater.

**11 Postmortem breakage.** Only two (0.8%) of the 241 pieces has identifiable postmortem breakage. In both cases, excavation or laboratory processing probably caused the damage (Table A1.11, site 29). Thus, the nearly total absence of postmortem breakage suggests that Yelenev Cave received little or no post-depositional disturbance of an identifiable sort.

**12 End-hollowing.** Carnivores seem to have played only a small part in the perimortem modification of the Yelenev faunal remains. Only one (0.4%) of the 241 pieces exhibits end-hollowing (Table A1.12, site 29). It is a complete, 15.2 cm long, roe deer sub-adult mandible that also has eight tooth dints and two cut marks. We suspect the end-hollowing was caused by a relatively small carnivore such as a dog, but it is possible that an even smaller fox was responsible.

**13 Notching.** Another presumed (large) carnivore source of damage is tooth-sized semicircular notching on a fracture plane, although some notching must have resulted from impacts by stone or bone tools. Thirteen (5.4%) pieces exhibit notching (Table A1.13, site 29). Twelve have only one notch; one has two notches. Their occurrence seems random. They were found on a scapula, on five long bone flakes, three cracked-open metapodials, a metapodial flake, a mandible, and on two ulna segments. Because only two of the notched pieces also have tooth dints, and none of the notched pieces have tooth scratches, we are of the opinion that most if not all of the notching resulted from bone processing by humans.

**14 Tooth scratches.** Four (1.6%) of the 241 pieces exhibit tooth scratches (Table A1.14, site 29). Each piece also has tooth dints. The number of scratches and dints in these four pieces is: two scratches and ten dints; three and six; two and thirteen; and one and four. The co-occurrence of scratches and dints is more strongly associated than is the co-occurrence of notching and dints, or as seen below, the co-occurrence of dints and notching.

**15 Dints.** Table A1.15 (site 29) shows that there are 19 (7.9%) pieces with tooth dints. The range in dint number per piece is 1–13; the mean is 4.2 for pieces with dints. Only two pieces with dints also have notching. This low association leads us to even more strongly feel that notching in this assemblage is the result of human rather than carnivore activity.

**16 Pseudo-cuts.** There is a surprisingly large number of pieces with damage that we judged as representing pseudo-cuts. Twenty-five pieces (10.4%) have from one to four pseudo-cuts (Table A1.16, site 29). In this assemblage, with an absolutely low number of tooth scratches, and a relatively low number of tooth dints, it would appear that we have misidentified several stone tool cut marks as being the very similar sorts of pseudo-cut marks first recognized in the Razboinich'ya hyena cave deposits. In fact, only two pieces with pseudo-cuts have associated tooth dints, and none is associated with tooth scratches. Hence, true stone tool cut marks and pseudo-cut marks cannot always be differentiated – at least, we cannot do so. However, several of the pseudo-cuts are very fine and visible only with the aid of a ×20 hand lens. They may well be cut marks, but their shallow depth, extreme fineness, and lack of a sharp V-shaped cross-section make them questionable as such.

**17 Abrasions.** Nine pieces (3.7%) of the 241 have abrasions, 2–36 each (Table A1.17, site 29). Given the great amount of perimortem breakage, and the gritty nature of the limestone cave, whose rubble would have served well as anvils and hammer stones, we wonder if perhaps most of the perimortem breakage was done with some sort of smooth bone or antler mallet. Smooth-surfaced pounding tools would not be expected to leave abrasion striations.

**18 Polishing.** Polishing has been found on the ends and middle of bone fragments, as well as on both areas. On the basis of the Razboinich'ya hyena cave findings, mid-section polishing is much more common in carnivore bone accumulations than in those deposited by humans. Yelenev Cave is no exception. Only six pieces (2.5%) have polishing solely in the middle, although 89 (37.1%) have polishing on one or both ends and in the middle of these fragments (Table A1.18, site 29). Perhaps most importantly, 71 fragments (29.6%) have no identifiable polishing. This suggests that the occurrence of polishing all over a fragment is not due solely to abrasive actions by or within the midden itself, otherwise there should be no specimens without polishing. One splinter-shaped metapodial butt had its pointed end polished for at least 4.0 cm, suggesting that it had been used as an expedient tool, perhaps in an awl-like function.

**19 Embedded fragments.** Fragments still adhering to a blunt impact area or at a notch are defined as being embedded. Twenty-five pieces (10.4%) of the 241 have 1–8 embedded fragments (Table A1.19, site 29). Embedding occurs most often at locations where the cortex is thin and is underlain by abundant spongy bone – for example, the proximal head of the humerus. In the case of human bone, embedded fragments are a useful indication of processing by other humans that involved smashing and pounding. An obvious exception is when embedded fragments occur in what is undoubtedly the tooth penetration area of a large carnivore. Although five (20.0%) pieces with embedded fragments also had tooth dints or scratches, none occurred in a tooth dint. Unquestionable cases of embedding associated with human processing include one set of three embedded fragments in a chop mark, one embedded fragment in another chop mark, and three sets of 2–4 embedded fragments in impact notches. Thus, most if not all of the fragment embedding in the Yelenev sample is attributable to human butchering activity.

**20 Tooth wear.** Twenty-one pieces (8.7%) of the 241 were tooth-bearing fragments of mandibles or maxillae (Table A1.20, site 29). Wear in all instances except one ranged from "present but dentine not exposed" to "dentine exposed but cusps still present." These would represent mainly sub-adults to younger adults. The exception was a roe deer adult maxilla from Square 9-GB, Level 2, where the degree of crown loss was between that where cusps are almost worn off and the pulp chamber is exposed. In deer of normal dental development this is presumably an old adult. There is nothing else distinctive about this specimen. Three individuals with teeth and associated occlusal wear possessed deciduous teeth. These specimens were classified as sub-adults. The presence of worn deciduous teeth suggests several postnatal and weaned months of life. Hence, these animals could have been fall kills, assuming normal seasonal birthing time for this species.

**21 Acid erosion.** Elsewhere, in hyena caves such as Razboinich'ya, there are numerous bone fragments evidencing the destructive surface modification of their having passed through the stomach and gut of hyenas or other large carnivores, or regurgitated before reaching the intestinal tract. Only one (0.4%) of the 241 Yelenev Cave pieces evidenced such erosion (Table A1.21, site 29). This is a complete 2.9 cm diameter adult roe deer astragalus from Layer 9A. There are four other fragments, each with one remaining anatomical end, that are highly polished but not eroded. These are a 4.4 cm adult calcaneus, a 3.3 cm adult toe, a 1.6 cm adult mandible fragment, and a 2.1 cm adult toe. The two undersized specimens were fully entered into the database because of their exceptional polishing, which might represent some form of carnivore chewing or digestion. Even with the four exceptional specimens, the very low frequency of stomach-acid-damaged bone seems convincing evidence that large carnivores were not normal occupants of Yelenev Cave.

**22–23 Rodent gnawing and insect damage.** There are no cases of identifiable rodent gnawing or insect damage in our study sample, nor did Ovodov see any sign of gnawing in the many other pieces that were not studied herein.

**24 Human bone.** Yelenev Cave produced 65 pieces of human bone, representing the very fragmentary and incomplete remains of five individuals, three adults and two sub-adults (Figs. 3.206–3.212). The possibility of cannibalism was evaluated, but no

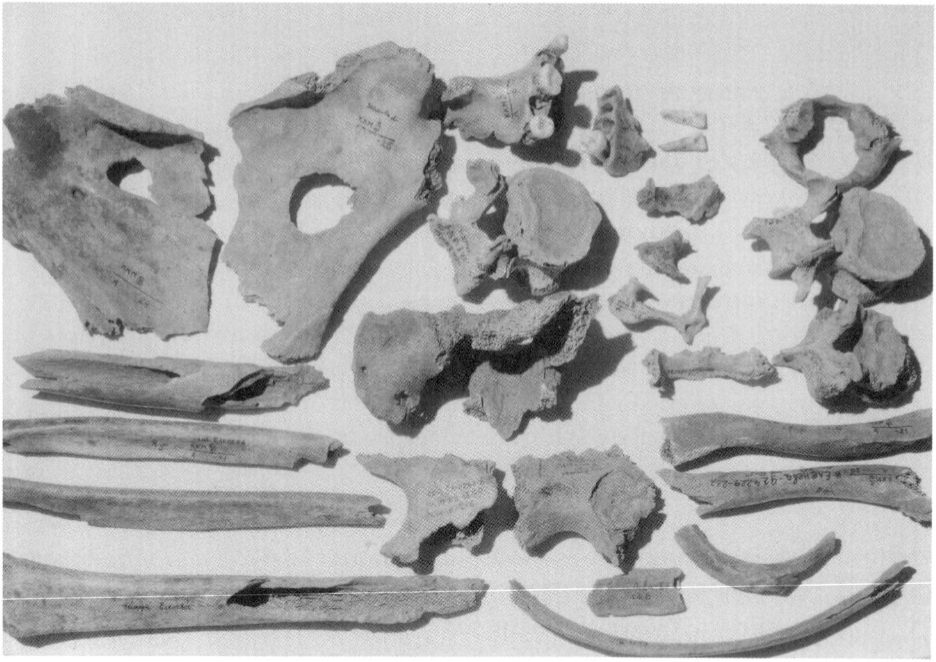

**Fig. 3.206** Yelenev Cave, early Neolithic human remains. Shown are the pieces of skeletal elements that are thought to have possibly belonged to a young adult female – light to moderate tooth wear, no pathology, gracility of bone size and muscle markings, and wide sciatic notch. There is chewing damage and polishing. The cause of the holes in the scapulae is unknown. The humerus in the lower left is 19.5 cm in length (CGT neg. IAE 7-23-98:3).

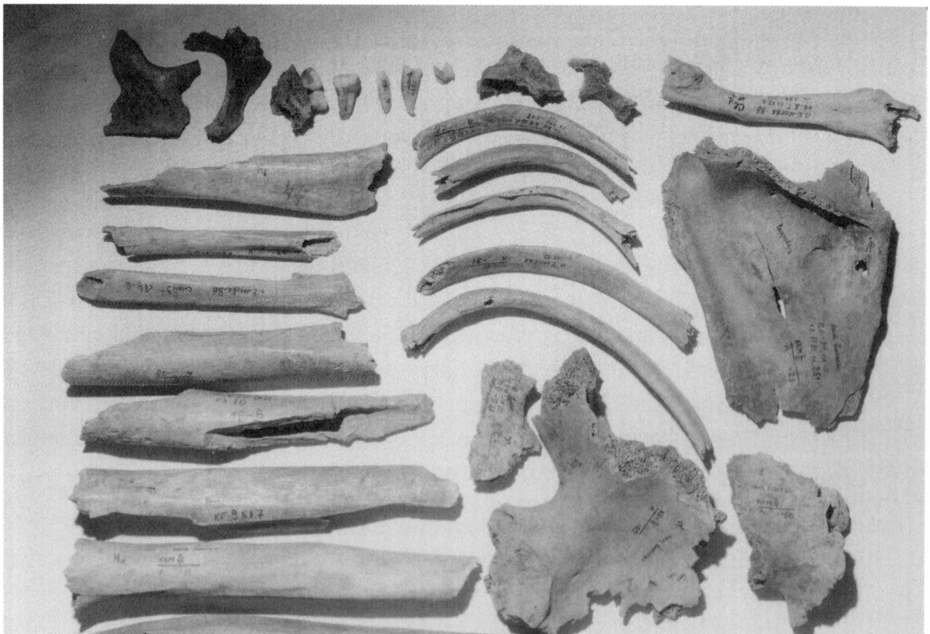

**Fig. 3.207** Yelenev Cave, early Neolithic human remains. These bones seem to belong to a middle-aged adult female. There is chewing damage. The ulna fragment in the lower left is 21.0 cm in length (CGT neg. IAE 7-23-98:23).

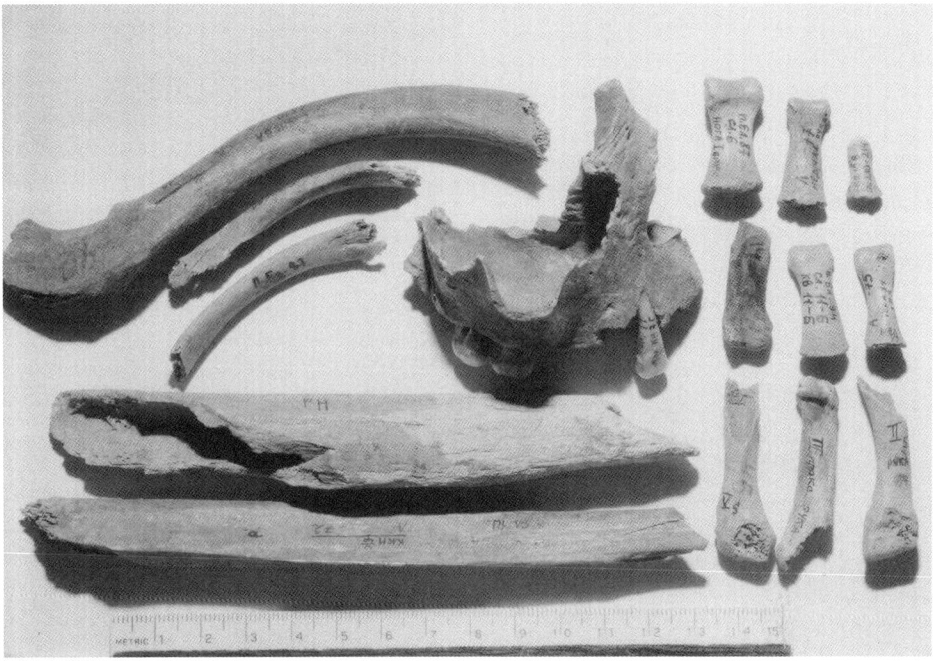

**Fig. 3.208** Yelenev Cave, early Neolithic human remains. This set of bones seems to represent another middle-aged adult, but whose sex cannot be even estimated. Scale is 15 cm (CGT neg. IAE 7-23-98:28).

**Fig. 3.209** Yelenev Cave, early Neolithic human remains. Mandible fragment and teeth are those of a 5–6-year-old child. The long bone fragments could not be matched with any individual in this assemblage, and the curved long bone piece in the middle may not be human. Scale is in centimeters (CGT neg. IAE 7-23-98:34).

**Fig. 3.210** Yelenev Cave, early Neolithic human remains. The pieces probably belonged to one individual, a 12–15-year-old sub-adult (CGT neg. IAE 7-23-98:29).

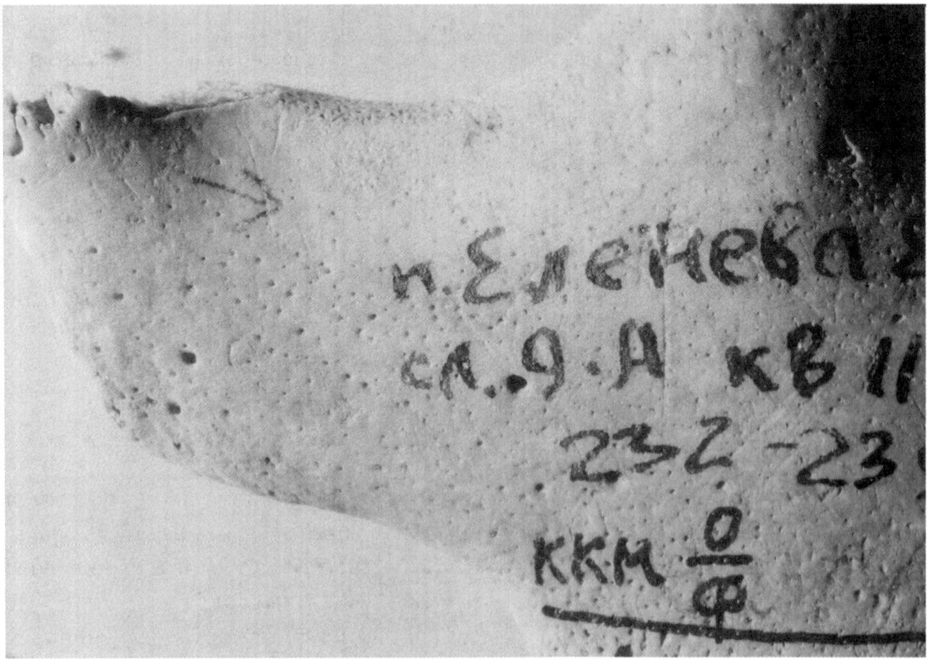


**Fig. 3.211** Yelenev Cave, early Neolithic human remains. Cut marks on a zygomatic bone. This piece is shown also in the upper left corner of Fig. 3.207. Width of image is 3.6 cm (CGT neg. IAE 7-23-98:21).

**Fig. 3.212** Yelenev Cave, early Neolithic human remains. Abrasions on pelvis near sciatic notch. Actual width of image is 3.3 cm (CGT neg. IAE 7-23-98:6).

definite conclusion could be reached due to the extensive carnivore damage (78.8% of all pieces had tooth scratches, dints, and extensive polishing). Nevertheless, these human remains come closest to being considered as cannibalized as any human remains the senior author has studied in Russia since 1984. On the basis of very limited dental morphological information, the Yelenev people seemed to have been about equally related to Northeast Asians and Europeans, as has been noted for other Neolithic Siberians. The results of the human perimortem analysis and epigenetic affinity assessment are reported in Turner *et al.* (2001b).

**25 Cut marks.** Bones with cut marks are relatively common (17.8%) in Yelenev Cave. Of the total, cut pieces most often have only one incision, although a few have more (Table A1.25, site 29). Cuts seem to occur mostly at joints and muscle and ligament attachment sites, or at random. As discussed above regarding pseudo-cuts, it is likely that the actual number of cuts in this assemblage is greater than 17.8%, which should be viewed as a minimal estimate.

**26 Chop marks.** Chopping occurs less frequently than cutting. Only 9.1% of the Yelenev assemblage have chop marks, ranging from one to four per piece. One chop mark is the most frequent number (Table A1.26, site 29). There are two examples of chop marks occurring on fractures, which suggests stone tools were used to break these bones. Six chop marks have tiny adhering bone fragments.

## Discussion

Yelenev Cave is one of our two post-Pleistocene assemblages, the other being Boisman II, an open site on the Sea of Japan coast. Both provide different contexts – cave and open, forest and coast – against which to see the sorts of differences that might exist between Pleistocene and post-Pleistocene perimortem taphonomy in Siberia.

# 30 Zhemchuzhnaya Cave

## Background

Zhemchuzhnaya Cave (Pearl Cave) is a natural animal trap located on the left bank of the Biryusa River, a left-bank tributary of the Yenisei River, 50 km up-river from Krasnoyarsk at approximately 55°58′ N, 91°58′ E. All of the recovered bones were coated with a hard, gray, limy deposit that is or is like travertine (Fig. 3.213). This coating prohibited our study of the faunal remains without damaging the natural coating. The coating had to have accumulated while the bones were immersed in mineral-saturated water or mud. Hence, in addition to its difficult access and limited living space, the cave was wet and unsuitable for occupation by human and non-human animals alike. The faunal remains were collected by Nicolai Ovodov in the 1980s from below the entrance on the cave "floor" before it sloped down to a lower grotto (Ovodov and Martynovich 2000). Because of its unique type of preservation we have retained Zhemchuzhnaya in our set of studied sites despite having no quantitative data to characterize the perimortem taphonomy, and because one piece may have had human modification.

## Findings

Due to the limy coating we were unable to obtain quantitative data for our standard set of variables. Ovodov recovered about 100 bones from Zhemchuzhnaya Cave; all but one rodent-like occiput and an adult bear femur belonged to sub-adult bears. On the basis of the right half of the lower jaw there were five bear cubs of very similar ages. Each had its first lower molar about to erupt at the time of death. A sixth bear with all of its permanent teeth erupted, but lost postmortem, was identified, also on the basis of a right mandible half. It was a sub-adult with bone remodeling of the ascending ramus in progress at its time of death. The maximum length of the older bear mandible is 16.5 cm, smaller than a comparative adult lab specimen that is 27.4 cm long. In a few places where the coating had broken off in excavation, the actual bone surface could be seen. Preservation was good, with ivory colored and relatively hard bone evident. One sub-adult bear tibia has a series of actual or pseudo-cut marks. Fig. 3.213 illustrates the nature of the mineral coating.

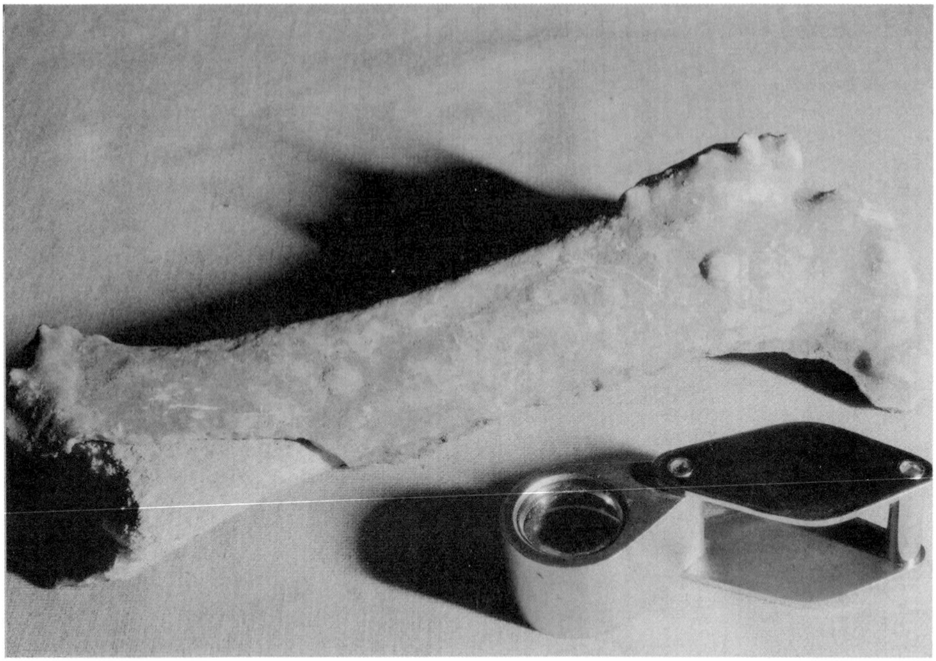

**Fig. 3.213** Zhemchuzhnaya Cave, travertine coating on bone. Bone length is 14.5 cm (CGT neg. IAE 7-16-01:8).

## Discussion

Ovodov and Martynovich (2000) have commented on the bears found in Zhemchuzhnaya Cave as part of their study of climate and environment reconstruction of the Biryusa karst region. In addition to the possible cut marks, our interest lies in the difference between bones preserved in wet and dry limestone caves. In the former, there seems to be considerable leaching of soluble bone protein (the dry bone sticks strongly to the tongue), whereas in the latter most of the protein matrix seems to have been preserved (the bone does not stick to the tongue), even though both appear surficially to be identically preserved.

# 4 Discussion: analyses, comparisons, inferences, and hypotheses

If you're not taking flack, you're not over the target.

(US Air Force maxim)

## Summary of our descriptive perimortem taphonomic findings

Chapter 3 provided site-by-site reviews of the perimortem taphonomy of all 30 assemblages in this study. Those reviews showed some degree of carnivore presence in every site. Their presence was in some instances identified by the recovery of actual skeletal or dental remains of bear, *Canis* sp.?, carnivore sp.?, fox, hyena, lion, and wolf. In the absence of actual carnivore skeletal and dental parts, their use of and activity in each site was inferred on the basis of tooth marks, which have similarities to bone damage caused by modern domesticated dogs – tooth dints, tooth scratches, end-hollowing, notched fractures, fragment end- and middle-polishing, and embedded fragments. Also, pseudocuts and acid erosion (stomach bones) are attributed to carnivores even though there is a slight chance both could have been produced occasionally by humans. Hence, this chapter discusses carnivore recognition largely in the absence of actual carnivore remains in an assemblage. Thus, uniformitarianism is the tool for believing the effects of past activity, where similar, were caused by similar agencies known in the present.

It is interesting to note that not one of the thousands of pieces of bone examined in this study had a stone or bone weapon fragment embedded in it. This contrasts markedly with what B. Bratlund (1991) found in late Pleistocene reindeer bones recovered at Stellmoor, Schieswig-Holstein. Hence, we have to ask: How much of the bone in our sites represents actual human hunting, and how much represents human scavenging or introductions by carnivores? Like our scarcity of evidence for roasting of game, we similarly have a scarcity of direct evidence for human hunting.

Bone-damaging behavior is not restricted to hyenas. They simply have the capacity to damage bone to complete destruction. As of this writing we have not found a comparative study of bone damage done by various known animals. Carnivores in general damage bone, with larger predators with larger teeth and more powerful jaws (lions, tigers, wolves, hyenas, bears) doing more damage than smaller ones (leopards, foxes, wolverines, martens, others). Non-carnivores can also damage bone. Rodent gnawing is well known, but damage by vultures and deer and other herbivores is less well known. The latter chew horns and other elements to extract minerals and salt. Aside from chewing, other sources of damage by animals include elephant trampling, scattering by birds and carnivores, burrowing by insects, and scratching by crabs. In fact, most large and small land animals die on the ground surface, which exposes their remains to severe and often complete destruction. It is almost miraculous that there is any record of ancient terrestrial life. Even humans, with our unusual behavior of burying the dead, leave many dead on the ground. And humans are well known for their bone-damaging activities.

In an extensive literature review by Gary Haynes (1982:266) dealing with prey carcass utilization, the author focused on the behavior of timber wolves. Haynes allows that prey carcasses are not as fully utilized when hunting is easy, but more so when hunting is difficult. When there is an abundance of sub-adult or older prey, carcass utilization by wolves is minimal. As Table A1.10 shows, perimortem breakage (marrow extraction) is extensive in our assemblages. From this feature alone, we infer that hunting was difficult for humans and carnivores alike in late Pleistocene Siberia.

Richard Klein (1975) proposed a novel approach for distinguishing whether a bone assemblage was the result of human or carnivore activity. This he suggested on the basis of a carnivore–herbivore MNI ratio. His main assumption was that carnivores eat other carnivores more than do humans. We tried this approach, but quite early in our study found that neither it nor MNI used in other ways was going to be helpful for our particular assemblages because they are so severely broken up. Moreover, identifying the number of individuals is a goal different from that which we are focused on; namely, defining the bone damage signatures of carnivores and humans based on the perimortem damage differences in carnivore paleontological sites and archaeological sites. As discussed in Chapter 3, there is good evidence for some carnivore activity in most of our assemblages. However, if we use only identified species, then Klein's carnivore–ungulate ratio would be useless in many cases, i.e., zero actual carnivores / $\Sigma n$? identified herbivores. On the other hand, if we generate a ratio of carnivore-damaged pieces out of the total number of pieces, then we can get a statistically adequate way of estimating the intensity of carnivore activity in any given assemblage.

Approaching the problem in this manner takes away the certainty that a carnivore is in an assemblage, and substitutes an inferred presence. However, as will be shown below, the inference is very strong. Our goal also differs from Klein's in that we want to identify the degree of carnivore presence in an assemblage of bone that we already know from independent cultural evidence had to some degree resulted from human hunting and butchering activity. On the other hand, as we must admit, it is possible some of what would appear to be carnivore damage may have been due to human bone processing. This being the case, then perhaps the best overall estimate of carnivore activity would require multiple damage features on every fragment. One combination might be that carnivore determination requires that some number of the presumed carnivore damage features be present – dints, acid damage, end-hollowing, for example, or scratches, dints, polishing, or some other combination of two or more damage types believed to be due primarily to carnivores. If we use fragments that have one or more of the damage types generally regarded as indications of carnivore utilization (tooth dints, scratches, end-hollowing, etc.), we find that in Ust-Kova, for example, there was carnivore damage to 11.2% (13 / 116) of the assemblage. When mid-fragment polishing and overall polishing are added to this set, then carnivore representation rises to 18.1% (21 / 116). That is, about one-fifth of the total assemblage possesses one or more damage types usually attributed to carnivore chewing. We suspect that this estimate is low, in fact, because carnivores probably can crack bones open without always leaving their diagnostic tooth marks, or, with digestion, not leaving any bone whatsoever (Payne and Munson 1985; discussed in S. J. M. Davis [1987]). In the above percentages, only one damage feature was needed to identify a piece as having been processed by a carnivore. If we require that two damage types be present on a given piece to establish carnivore presence, then there is only a 2.6% (3 / 116) incidence in the Ust-Kova example. This low frequency strikes us as being too rigorous, but it probably ensures that a true carnivore presence has been identified.

## Analytical findings

### Bone damage, site definitions, literature comparisons, food ways

At the beginning of this study it was assumed that at least seven perimortem damage stigmata result from human butchering and cooking practices. The combination or signature was first identified in cannibalized prehistoric Native American skeletal remains. They include: cut marks, perimortem breakage, anvil abrasions, burning, missing vertebrae, fragment end-polishing (pot polishing), and chop marks. The first five were identified by the senior author and associates (Turner and Turner 1999). Tim White (1992) proposed the end-polishing following a cannibalized assemblage of prehistoric Anasazi Indians found in Mancos Canyon, southwestern Colorado. The seventh trait was added following the work of Carmen Pijoan (1997) who, along with her co-workers, differentiated chop marks from cut marks in cannibalized prehistoric Mexican Indian assemblages. The combination of damage types has been found repeatedly as a multivariant signature of cannibalistic body processing of prehistoric Indians of the southwestern United States and Mesoamerica. Several of those assemblages were discovered in archaeological contexts that largely ruled out the possibility that the damage was caused by carnivores or any other identifiable non-human agency. In the prehistoric American Southwest and Mesoamerica, humans alone were responsible for this distinctive multi-feature processing signature. Comparisons between the human butchering and that of large and small game animals (Dice 1993, Turner and Turner 1999, White 1992) showed remarkably similar kinds and amounts of damage. It was on the basis of this overall similarity of butchering that cannibalism was cautiously proposed.

Operating on the principle of uniformitarianism, we assumed that these same stigmata would also identify butchering and carcass processing of non-human animals carried out by late Pleistocene Siberian hunters. However, since this assumption had never been tested in Siberia, we added to our multi-site survey paleontological assemblages whose context and content indicated perimortem bone damage caused only or mainly by carnivores, in particular, hyenas. It was quickly learned that missing vertebrae would not serve to identify late Pleistocene human activity since vertebrae were rare in both the archaeological and paleontological contexts (Table A1.3; on average <5.0%).

Since nearly all of the sites in this study are pre-ceramic, there is no chance that polishing could have occurred by stirring bone bits in a pottery cooking vessel. If polishing were caused mainly by human cooking practices, then the vessels must have been made of wood, leather, or tightly woven basketry. Boiling crushed bone fragments and adhering soft tissue in wood, hide, or basket containers to extract calorie-rich marrow grease would have required the addition of heated stones, whose abrasive surfaces would have caused much the same sort of polishing as the abrasive interior walls of pottery cooking jars. However, there would be a difference because hot stones in bag-, wood bowl-, or basket-boiling would also polish the middle portions of bone fragments, not just the ends, as would occur when stirred in a geometrically spherical pottery cooking vessel. Hence, polishing was adjusted from end-polishing to an end-middle polished condition. We continued to assume that polishing, in this modified condition, was solely or mainly the product of humans, even though we had a few examples of bones chewed and polished all over by domestic urban and ranch dogs. Analysis at the end of our study showed end-middle polishing would have to be abandoned, along with missing vertebrae, as a marker trait of human presence. As Table A1.27 shows, end-middle fragment polishing is more frequent in the hyena sites than in the archaeological sites without hyenas, and in the archaeological sites with mixed human and hyena presence. While these results do not rule out boiling in non-ceramic cooking vessels, they do show that identical end-middle polishing had happened as a result of carnivore chewing, and probably some other agency as well, possibly geochemical.

As for abrasions (caused in the American Southwest by slippage during the cracking open of marrow bones with the use of a hammer stone and stone anvil), the findings herein proved disappointing as an indicator of human presence. The mainly hyena sites have an unweighted mean of 1.7% abraded pieces (weighted mean, 0.6%). The mixed archaeology and hyena sites' unweighted mean is 2.0% (weighted, 2.3%). The archaeology–no hyena sites' unweighted mean is 0.6% (weighted, 0.5%). In other words, the stigmata that we define as abrasions, and attribute to humans alone, occur almost with the same frequency in the three types of contextual settings. The difference between mainly hyena and mixed is statistically significant ($\chi^2 = 18.9$; 1 d.f.; $p < 0.001$), whereas it is not in the mainly hyena/no hyena comparison ($\chi^2 = 0.16$; 1 d.f.; $p > 0.05$). By themselves, abrasions do not distinguish between carcass and bone processing by carnivores and humans. In the above-mentioned cannibalism studies, the kinds of perimortem damage that might possibly be found in fully paleontological faunal assemblages were not systematically studied due to the lack of appropriate comparative paleontological collections with perimortem damage. While there are a few well-known Pleistocene animal traps in North America, such as the La Brea tar pits in southern California, these are inappropriate for identifying a carnivore damage signature. Indeed, much of the reason for the present study was to take advantage of the ideal kind of comparative faunal assemblages that the Siberian hyena cave sites provide for taphonomy research. This is especially true for perimortem breakage.

Perimortem breakage is very common in both archaeological and paleontological assemblages, averaging about 85%. Statistical comparisons provided in Table A1.28 show that the amount of perimortem breakage in mainly hyena sites and non-hyena archaeology sites is about the same as in hyena and mixed hyena–archaeology sites.

Hence, by itself, perimortem breakage is of no value as a perimortem damage indicator of human activity, at least not quantitatively. While we kept our eyes open for potential distinctive qualitative features of perimortem breakage, none was recognized. At this point, by elimination, we are left with only burning, chopping, and cutting as indicators of butchering and cooking activity by humans.

Table A1.29 shows the frequencies of pieces with burning, cutting, and chopping. Burning is rare to absent in most of our assemblages. Only one burned piece was found in 1638 pieces (0.08%) that came from the mainly hyena sites. This is significantly less than the occurrence of burning in the mixed archaeology–hyena type of sites (1.6%; $\chi^2 = 41.9$; 1 d.f.; $p < 0.001$) and in the archaeology, no hyena sites (1.3%; $\chi^2 = 9.4$; 1 d.f.; $p < 0.01$). Burned bone pieces occurred around five times more often in the two classes of archaeological sites than in the mainly hyena sites. Burning remains an indicator of human presence if natural burning such as forest or steppe fires can be excluded as the cause. However, based on our study, its frequency is so low that sample size needs to be substantial (100 or more pieces) to expect to find a 1.0–2.0% occurrence of burned fragments.

Whatever ways late Pleistocene Siberians cooked their game, our assemblages (5170 pieces from 13 sites) show without any question that roasting was not the usual or preferred on-site practice. Had it been, many more burned pieces would be expected. Moreover, we would also expect to have some pieces with both burned and unburned areas, dependent on the thickness of overlying protective soft tissue. No such incompletely burned piece was found in any assemblage, not even in the thousands of pieces excluded from study because they were smaller than our minimal size selection criterion of 2.5 cm or because of excessive root damage.

Chop and cut marks are the best indicators in our study of carcass butchering by late Pleistocene Siberians. Chop marks were found on 9 / 1618 pieces (0.5% unweighted mean) from the mainly hyena sites, the same percentage as cut pieces. In the two types of archaeological sites, pieces with chop marks are more numerous but fewer than those with cuts – 9.2% and 8.7% unweighted means, respectively. In both cases the difference between them and the mainly hyena sites is statistically significant ($\chi^2 = 124.3$; 1 d.f.; $p < 0.001$; $\chi^2 = 144.6$; 1 d.f.; $p < 0.001$). There are roughly 20 times more chopped pieces in the two types of archaeological sites than in the mainly hyena group. Statistically speaking, chopping is definitely a perimortem taphonomic indicator of human presence in late Pleistocene Siberia.

Cut pieces are very rare in the mainly hyena group. Eight pieces had cut marks (8 / 1608). Four came from Maly Yaloman, and four from Straschnaya. Both sites had small amounts of archaeological materials, so cut pieces here are not totally unexpected. Both unweighted and weighted means indicate less than a 1.0% average occurrence of pieces with cut marks in our mainly hyena group. This frequency is significantly less that the 12.0% ($\chi^2 = 247.7$; 1 d.f.; $p < 0.001$) and 12.6% ($\chi^2 = 150.1$; 1.d.f.; $p < 0.001$) unweighted averages (15.1%, 10.1% weighted) for our two classes of archaeological sites. Since some of the mainly hyena sites did have small amounts of archaeological refuse, we can say that their extent or degree of occupation, as measured by cutting of bone, was something like 20–30 times less than in the mixed and no-hyena archaeological groups. Cut marks on bone are the preeminent indicator of perimortem carcass processing by humans of late Pleistocene Siberia, and reasonable evidence of human occupation, at least in our study area.

## What is an archaeological site?

The very low occurrence of burning, chopping, and cutting that defines the mainly hyena sites compared with the relatively much greater frequencies in the two types of archaeological sites leads us to question the definition of a late Pleistocene Siberian archaeological site. Some of our mainly hyena sites have been presented as archaeological (Derev'anko *et al.* 1998). Deciding on what constitutes an archaeological site is an old and incompletely resolved question in American archaeology, especially for legally important archaeological surveys.

In the American Southwest, the phrase "one pot sherd does not an archaeological site make" sums up the problem. In late Pleistocene Siberia, the analogy could be: a few stone flakes do not constitute an archaeological site. A standardized threshold or minimal content has not been defined (Fish 1993) for distinguishing between discoveries of two or three potsherds or stone chips or a few lines scratched into a rock surface (i.e., cultural "float"), and the more abundant amounts of pottery, worked stone debris, rock art, or other material culture remains that everyone would agree is at minimum an archaeological site. Patently, the further back in human evolution we go, the more the definition must be adjusted to the level of presumed cultural evolution.

Is chimpanzee "ant fishing" any more evidence of tool using than stick selection by desert doves in constructing their nests? Here, we could take the position that any sign of ancient human activity is important and should be preserved by some sort of classification and recording standard. On the other hand, the sites herein are largely conserved and concentrated within protective stone walls and roofs of caves, whereas open sites are many times less likely to be preserved due to lack of protection from decay and scatter. In this sense it could be said that humans have been everywhere in our study area, so finding a few stone chips or cut bones here and there is to be expected almost anywhere.

Humans, as well as animals, seek out shelter in times of bad weather, so we should expect to find tiny amounts of human refuse wherever shelter existed. To kill time, those waiting out a passing summer rainstorm or a winter blizzard may chip away at a stone core, whittle on a bone, build a fire when possible, or do any number of other things. Does the refuse from this sort of relief from boredom constitute an archaeological site? Does it tell us anything other than one or more people sat around waiting for a storm to pass or for a pack of hyenas or threatening cave lion to go away?

Try as we did to find some sort of cultural information in our human perimortem damage stigmata, we noticed nothing. There is no identifiable cultural information within a cut itself. Cut variations such as being wide or narrow, deep or shallow, long or short are idiosyncratic and random, not cultural. Location says something about butchering practices, but we feel that butchering techniques had become universal around the world by the end of the Pleistocene. So really, we have no specific cultural information in either a cut or its location, in a chop mark or its location, or in a burned bit of bone. Said another way, while our mainly hyena sites have a trace of human-generated perimortem damaged bone debris, that debris tells us nothing other than that it was left by stone-tool-using, meat-eating humans. Is that enough information to designate a low-yielding location as an archaeological site? This question is, of course, critically important from a statistical viewpoint. Small samples often are considerably different from larger ones, even when drawn from the same statistical universe. Some researchers think the differences are due to cultural differences, ignoring or being unaware that small samples simply cannot contain all the variation that a larger sample may possess. Kara-Bom might be a good example of too much significance being given to a relatively small archaeological collection.

On the other hand, the mixed archaeological–hyena sites and the archaeological–no hyena sites have on average yielded nearly identical amounts of human-processed bone. The presence of hyenas does not seem to have any overall effect on the amount of human-processed bone. However, there is some interesting inter-site variation in the frequencies of burning, cutting, and chopping. Okladnikov Cave and Kara-Bom stand out because of their low frequencies of these stigmata. The Kara-Bom situation could be explained on the grounds that it is an open site, and as such was perhaps not a location where butchering was practiced to the extent that it was in the more permanent shelter and safety of a cave. But this does not help explain the small stigmata frequencies of the sheltered Okladnikov cave. Perhaps here there may have been some manner of hyena effect. One effect that we can imagine is that hyenas could have carried many more bones of their own kills and scavenging back to Okladnikov Cave than were carried to the other mixed hyena–archaeology sites. The hyena effect would be to change the ratio of cut to uncut bones, etc. Finally, Shestakovo has almost no identifiable human-caused perimortem damage, because despite the presence of stone artifacts, it is considered to be a mammoth natural death site, not a kill site resulting from human hunting. The very small amount of cutting at Shestakovo probably indicates bone or ivory extraction more than butchering of meat, although both could have occurred. Do those few cuts make it an archaeological site?

By way of comparison, Michael C. Wilson (1983) analyzed bone damage in a 3000-year-old bison bone bed in Alberta, Canada, for the purpose of determining whether it was a kill site or the result of a natural catastrophic die-off. Even though a small collection of stone artifacts were found with the bison bones, Wilson concluded the remains represented a die-off. The bone damage he attributed mainly to carnivore scavenging on the basis of tooth marks, end-hollowing, and other damage types caused by carnivores. The stone artifacts were not hunting weapons, and there were no cut marks. He allowed that the stone artifacts represented some scavenging by humans also. In our terms, the bison bone bed was a paleontological site, not an archaeological one. As in our earlier discussion, sites may contain a small number of stone artifacts, but they are only marginally at best an archaeological site.

Following is one final example of assessing whether faunal remains in a site that also had stone artifacts were damaged by carnivores or humans. This example comes from the Yukon Territory of Canada. The site is called Blue Fish Cave I. Bone damage was analyzed by B. F. Beebe (1983). She found much carnivore damage, mainly tooth scratches. There was bone breakage also, but Beebe could not tell if it was done by humans or carnivores. The absence of cut marks, chop marks, and burning would favor the view that the bone damage was done by wolf-sized carnivores. In other words, there is no link between the 15 000-year-old Pleistocene fauna and the stone artifacts. These and other such assemblages lead us to think that claims for bone damage and bone accumulations as having been done by humans are not acceptable in the absence of cut marks, chop marks, and burning. The assumption that bone accumulations are evidence of human presence should be carefully considered at all times. For example, Kanowski (1987:52–53) years ago warned:

> Among the non-human agencies responsible for producing bone collections, archaeologists have to take into account such predators as leopards in Africa and hyenas in European caves. But probably strangest of all bone gatherers are African porcupines – compulsive collectors of bone and similar material. These large, mainly vegetarian creatures, are like rats and other rodents in that they spend a lot of time gnawing at suitable objects to wear down their teeth, their incisors would otherwise grow out of control to unusable lengths.

Although porcupine presence has been recognized by Ovodov in our study, the very low frequency of gnawing (Table A1.22) pretty much eliminates them as significant players in our perimortem taphonomy investigation.

In conclusion, the Siberian late Pleistocene perimortem taphonomic signature of human butchering and cooking can be defined as accidental bone burning, and intentional chopping and cutting. Assuming our samples are representative, we propose that a ratio of burning, chopping, and cutting of approximately 1:9:12, respectively, could be used in future excavations to distinguish between primary site usage by humans and non-humans. Non-human usage would be indicated if the actual numbers of these three traits were very low and the ratio lower, perhaps on the order of 1:5:5 or less. We suggest that an archaeological site such as our assemblages be defined by the 1:9:12 ratio.

## Some other comparisons

Here we make three more comparisons with published bone damage information, starting in the present and working back to the Upper Paleolithic. There are many such studies and our choices were purely random. We will have a brief discussion for each assemblage that is compared at this point. We have organized our comparisons on the basis of Old World and New World assemblages on the assumption that bone damage will be less in the New World because late Pleistocene cave hyenas seemingly did not expand into North America on account of the frigid conditions of Beringia. Our study found no evidence of hyenas in Siberia north of 55° latitude.

### A modern assemblage

Kevin T. Jones (1983) studied the way of life of the South American Aché foraging in eastern Paraguay. One aspect of his interesting work was an analysis of the frequency of damage due to marrow extraction in long bones he found in various camp sites. Based on

231 bones, he did not find marrow extraction damage, or only butchering marks of bones with less than one milliliter volume. All the bones that had 3 ml to more than 5 ml volume had damage caused by marrow extraction. In other words, bone breakage correlated highly with bone size. In our study, which considered breakage in all types of bone (Table A1.10), breakage was very common, even in the sites such as Neolithic Boisman II, where there was no evidence of hyenas. Thus, it would appear that bone breaking for marrow extraction was more commonly practiced by the Siberians, along with additional breakage by large carnivores. We can add that the Siberian breakage was not also done to make bone tools, since almost no bone tools have turned up in the Siberian sites. All this hints at a greater scarcity and difficulty of obtaining food for the Siberians than for the tropical forest-dwelling Aché. The absence of abundant tropical plant foods, including nuts, fruits, and tubers, probably forced the Siberians to waste very little of their largely animal-based nutrients.

## A Neolithic assemblage

Thomas F. Kehoe (1983) studied *Bos* (cattle) bone damage in a German Neolithic site called Schreckensee, which dates about 3000 BCE. The cattle bones were recovered from lake deposits. A total of 519 pieces were studied and analyzed for cut and gnaw marks. Presumably dogs were the principle gnawers. A total of 14% showed gnaw marks. There were about three times more bones with cut marks (40%).

Kehoe did not break down gnawing into the various categories that we did, but if we use only our tooth scratches (Table A1.14) for comparison it is obvious that sites we have identified as having a hyena presence have much more gnawing – for example, Denisova (20.9%), Dvuglaska (38.5%), Okladnikov (57.0%). Some of our sites with hyena presence have less gnawing than the German Neolithic site – Kara Bom (10.9%), Ust-Kan (9.0%). However, if we add in our other carnivore damage types such as dints, notching, and hollowing, all our sites with hyena presence exceed the German Neolithic gnawing frequency. Our sites that have no clear-cut evidence of a hyena presence but do exhibit some carnivore damage, such as Bolshoi Yakor I, Kamenka, and Varvarina Gora, were probably damaged by wolves. In this comparison, the gnawing by presumed dogs and presumed wolves is considerably less than that attributed to the jaws of the more powerful hyenas. Because of the low occurrence of bear and wolf bones in our assemblages, we attribute most of the Siberian damage to hyenas. This would have to be the case for the Neolithic Boisman II (1.6% tooth scratches), where dog bones were recovered by Alexander Popov (personal communication, June 9, 2000).

## An Upper Paleolithic assemblage

Jean Bouchud (1975) studied the faunal remains from the famous French site called Abri Pataud, excavated by Hallam L. Movius, Jr. Abri Pataud carbon-14 dates between roughly 22 000 and 33 000 BP, with various cultural manifestations (Movius 1975, site profile, no page number). Compared to our sites of comparable antiquity such as Denisova, Kaminnaya, and Okladnikov Caves, the difference in animal representation is dramatic. The Abri Pataud Perigonian IV horizon had 21 119 (97.4%) pieces of reindeer, and only three pieces of cave hyena (<1.0%). Two things stand out: (1) the Abri Pataud hunters were concentrating on reindeer, whereas the Siberians were hunting on a much more opportunistic basis; (2) there is almost no hyena presence at Abri Pataud, whereas hyena presence was common in the Siberian sites. This suggests much more discontinuous occupation of the Siberian archaeological sites because humans would not have been present during hyena utilization of the archaeological sites.

The number of Abri Pataud pieces of human bone and teeth studied by Ginette Billy (1975) is greater than those found in any of the Siberian sites. This difference adds to our view that the Siberia hyenas were likely responsible for the rarity of human remains in late Pleistocene Siberia.

One way to assess the reliability of the intensity of bone damage indicated in Table A1.27 is to compare our findings with assemblages, as just related. However, there may be a substantial difference in methodology and scoring standards between observers, so at this time we limit our inter-observer comparisons and emphasize intra-observer data (Table A1.31). Larger carnivores such as wolves and bears are rare in the American Southwest sampling area, and there are no species with abilities comparable to the powerful bone-crushing ability of hyenas. Therefore, we would expect the North American carnivore damage to be most like that of the Siberian assemblages that lack evidence of hyenas.

As can be seen in Table A1.29, bone accumulations believed to be the result primarily of hyena activity have the greatest frequency (27.5%) of perimortem bone damage, even when only end-hollowing, tooth dints, tooth scratches, and acid erosion are considered. The frequency would be greater if notching and polishing were included. We have excluded them for this inter-regional comparison because both may have been caused to a small degree by humans as well as carnivores. The assemblages that are known to have a mixture of human and hyena presence have about half the amount of damaged pieces (15.2%). The least amount of Siberian carnivore damage (2.0%) occurs in those assemblages where there is no secure evidence of hyenas, only human and small carnivore activity, the latter identified mainly by the small size of dints and shallow, weak scratches. As expected, the amount of North American carnivore damage (0.9%) is much like that of the Siberian material that is considered to have no evidence of hyenas. Although very similar, nevertheless the difference is statistically significant. While the significance could be due to the more concentrated effort to identify carnivore damage in the present study, it seems better to assume that site context and fewer and probably less aggressive North American carnivores are responsible. Thus, the carnivore signature is easily recognized and, importantly, it varies in intensity. This variation in intensity or severity of damage, as these simple comparisons demonstrate, provides additional confidence in the value and usefulness of perimortem bone damage for certain bioarchaeological problems. Problems that immediately come to mind include considerations of site abandonment, variable discontinuity of site use, predator pressure, distinguishing bone damage caused by carnivores and humans, and other narrowly and broadly defined topics. Because our work focuses on late Pleistocene Siberian perimortem conditions, one topic that we are unable to avoid has to do with the possible stratigraphic blurring and mixing of Siberian and possibly many Eurasian archaeological sites because of the use of the same sites by hyenas at different times.

Comparisons between our Siberian assemblages show that archaeological and paleontological sites with no actual hyena remains had the lowest amount of these eight kinds of bone damage. Quite the opposite condition was found in those sites with carnivore bones and no or very little archaeological refuse. These two groups were more-or-less expected, which we initially (and still do) referred to as archaeological assemblages and paleontological assemblages. What was unexpected was the intermediate condition of certain archaeological sites where, as it turned out, a substantial carnivore presence was identified. Tables A1.28–A1.30 show this threefold contextual grouping. The intermediate group is referred to as "mixed" in Tables A1.27–A1.30. Not all of our sites are shown in these tables. Those left out are either Holocene in age, their faunal assemblage is numerically small, or they are not especially relevant to the concern here of defining a late Pleistocene carnivore processing signature.

Table A1.27 shows the six most distinctive carnivore processing stigmata and their average frequencies in the mainly hyena, mixed, and no-hyena groups. These are end-hollowing, tooth dints, tooth scratches, notches, stomach bones, and end-middle polishing. Tooth dints, scratches, and stomach bones are the three most distinctive of the six. It is noteworthy that a few stomach bones turned up in the no-hyena sites. Although few in number, these acid-eroded pieces from Bolshoi Yakor, Kamenka, etc., must represent bone fragments swallowed by wolves, foxes, and possibly even the human inhabitants of these sites.

Generally speaking, archaeological sites lacking evidence of hyenas have an unweighted average of 8.1% bone damage attributable to carnivores (the weighted average is only slightly higher at 11.2%). These carnivores, we assume, were wolves, small bears, wolverines, and foxes – that is, medium- to small-sized carnivores. Between four and five times (unweighted average, 38.2%; weighted, 30.8%) more damage occurred in the largely non-archaeological, mainly hyena sites. The mixed archaeological–hyena sites have, on average, three times (unweighted 24.4%; weighted 23.2%) more damage than the archaeological no-hyena sites.

Table A1.28 provides comparisons of the three groups using chi-square goodness-of-fit tests. Most of the comparisons are significant at the $p > 0.05$ and higher confidence levels. The number and size of the differences provides a parallel pattern to the above remarks. There are more (8 / 12 comparisons) smaller differences between mainly hyena and mixed than between mainly hyena and no-hyena sites. Simply adding up the chi-square values shows clearly the closer relationship between the hyena and mixed groups ($\Sigma_{\chi}^{2} = 918.2$) than between the hyena and no-hyena groups ($\Sigma_{\chi}^{2} = 2656.2$). As would be expected, the mixed and no-hyena groups add up to an intermediate value ($\Sigma_{\chi}^{2} = 2151$). Table A1.28 also shows the summed chi-square values for the human perimortem damage stigmata. The mixed and no hyena groups have a low value ($\Sigma_{\chi}^{2} = 160.3$) befitting two sorts of archaeological assemblages. The cultural damage is reversed. The hyena and mixed groups have a greater summed chi-square value ($\Sigma_{\chi}^{2} = 446.8$) than the hyena and no-hyena groups ($\Sigma_{\chi}^{2} = 320.8$), which may reflect differences in butchering and cooking practices that have little or nothing to do with the way we have combined sites for the three-group analysis.

One outstanding characteristic of our archaeological faunal assemblages is the enormous amount of bone breakage associated with humans. This is especially noticeable when contrasted with bone deposits from paleontological sites. Examples of such include the frozen mammoths of far northern Siberia, the frozen Alaskan muck deposits near Fairbanks, the La Brea tar pits in Los Angeles, the multiple-mammoth deposit near Waco, Texas (and other such natural die-offs), or natural trap assemblages. In this study we have bone breakage in paleontological contexts that parallels the archaeological situation. Some sense of the degree of breakage can quickly be obtained by comparing the means and ranges in long bone length given in Tables A1.6 and A1.30. While the measurements combine species and skeletal elements, and do not include the few whole bones we excluded from our study because they lacked perimortem damage, nevertheless nearly all of the species in Table A1.2 are medium to large mammals, most of which have been identified in one or another of our assemblages. By referring back to Table A1.2, a crude idea of species content for each assemblage can be gained, and with that in mind, compared with the relevant species provided in Table A1.27. The point of this simple comparison is to emphasize that late Pleistocene carnivores of Siberia were like humans in the great amount of bone breakage and size reduction they were capable of doing. Where it is possible to identify the species from which a given skeletal element was derived, comparing its maximal dimension shows it to be generally less than 15.0% of the lower end of the range provided in Table A1.30.

Comparisons of the maximal size measurements in Table A1.30 show that the mainly hyena and the mixed groups are metrically very similar. Both have mean dimensions that are less than that of the no-hyena group, indicating slightly less human processing reduction in the latter group. However, some of this lesser processing is attributable to the larger means and ranges of the two no-hyena sites with much mammoth bone – Mal'ta and Shestakovo. This particular inter-group difference would be even larger had it been possible for the archaeologists to recover intact all the big mammoth bone pieces we saw in published site photographs and maps. Both size reduction and type of damage are useful for establishing a carnivore damage signature. However, the kinds of damage are considerably more specific to carnivores than is maximal size, which was produced similarly by both humans and carnivores. Examination of Table A1.31 shows that there is a general pattern of similarity in the frequencies of skeletal elements. A chi-square comparison with 18 d.f. of most of the elements from Razboinich'ya and Okladnikov (excluding "teeth," long bone fragments, and coprolites because of their very large numbers and/or not being a skeletal element), while producing a statistically significant difference ($\chi^2 > 40$; $p < 0.01$), is not as large as might be expected given all the possible sources of variation in these two sites. By itself, this comparison is not especially supportive of our view that Okladnikov was more of a paleontological than an archaeological site. However, a chi-square comparison of hyena coprolite occurrence in Razboinich'ya, Logovo Gieny, and Okladnikov shows no statistically significant difference ($\chi^2 = 3.61$, 2 d.f., $0.05 < p < 0.20$). This comparison suggests that because the archaeological recovery of coprolites was very similar in the two caves, the actual frequency of coprolites must have been about the same in each of the three caves. This implies that the amount of time spent by hyenas in the three caves must have been about the same, assuming a similar number of hyena occupants for each cave, a not unreasonable assumption judging from hunting pack size in Kalahari spotted hyenas (Mills 1990:77). So, the general pattern of preserved hyena skeletal elements and the frequency of their coprolites, in addition to the kinds of perimortem damage reviewed elsewhere, leave little room for doubting that Okladnikov Cave was used as much if not more by hyenas than by humans. The implications for stratigraphic disturbance are rather obvious.

The lengths of undamaged long bones published by Vera Gromova (1950) are provided in the site descriptions in Chapter 3. Her measurements are included to help the reader appreciate the marked degree to which perimortem bone breakage had reduced whole long bones to very small fractions of their original size. Similar fragmentation and size reduction occurred to all other parts of the thousands of skeletons represented in this study. At the end of each assemblage's description there is a brief discussion of our main inferences about the perimortem taphonomy of the site. Much of our description may be of interest only to the small number of specialists in Siberian taphonomy. For this reason, the reader can skim or even ignore most of the descriptions, and concentrate on our more general discussions of each site. However, for the reader who focuses on the discussions and is uncomfortable with our inferences, we urge him or her to go back and read the descriptions of the relevant variables.

We propose that the Siberian late Pleistocene no-hyena baseline signature of carnivore activity is an average of about 10% damage, as recorded by the six traits listed in Table A1.27. This baseline signature would represent small- to medium-sized carnivores. When larger carnivores, especially hyenas, were present in a given region, the damage signature increases fourfold to 38.2% (weighted, 30.8%) in paleontological sites, and threefold (24.4%; weighted, 23.2%) in archaeological sites. Said another way, the amount of presumed hyena bone-processing activity in the mixed archaeology and hyena sites is more like that found in hyena sites than in non-hyena archaeological sites. We view this amount of hyena presence in these mixed archaeology sites to be potentially of considerable interpretative importance for three main reasons.

First, as discussed preliminarily in Chapter 3, the presence of hyena remains in an archaeological site should sound an alarm about the likelihood of substantial stratigraphic disturbances caused by the digging behavior of these animals (see Brain's 1981 discussion of hyena dens, pp. 57–58). The nature of the disturbance will to a large degree be dependent on how the archaeological site was utilized by humans. If cave occupation was seasonal, say, in the summer when there was less need for warmth-producing fires, then hyenas could prowl, scavenge, and dig in the human midden for maybe half the year. This would certainly be enough time for even a few animals to annually churn the human refuse into a random mixture, bringing to the surface artifacts and refuse of earlier times and falsely associating older remains with "recent" objects. The stratigraphic effect of carnivore digging would be to blur the boundary separating earlier autochthonous cultural remains and artifacts left later by incoming Upper Paleolithic Europeans, as exemplified by Mal'ta. Movements into Siberia by external alien bands most probably did occur, and probably repeatedly over the millennia due to fluctuations in the more easterly Eurasian population density, pressure, and response to natural catastrophes such as large-scale earthquakes, droughts, forest fires, epidemics, conflict, and so forth. Site formation processes are not limited to just within a site itself. Instead of inferring "replacement" or "discontinuity," undetected hyena disturbance of midden would more likely be interpreted as "gradual local evolution." Indeed, this seems to be the prevailing Siberian archaeological view (flake industry evolving into blade industry: Derevianko 2001, 2005) based on the same sites we have identified as having a substantial mixture of human and hyena occupation. Without an evaluation of the extent of bioturbation, the argument is potentially unsupportable for local tool type evolution, and cultural continuity between the Middle and Upper Paleolithic.

Second, the senior author has had six years of excavation experience with highly stratified prehistoric village middens in the Aleutian Islands. Upon visiting the Siberian mixed archaeology–hyena cave sites he could distinguish no Pleistocene stratigraphic levels comparable to the hundreds of easily discerned layers in Aleutian archaeological sites, or, for example, in the very distinct Holocene (Neolithic) layers of Denisova Cave. Hyenas were extinct by the end of the Pleistocene, so the Holocene Denisova strata were unaffected by their digging behavior. In the Aleutians, there were no terrestrial carnivores larger than foxes, and Aleut settlements seem to have been continuously occupied for thousands of years. Hence, foxes did not den in the Aleutian village middens, and even minutely thin strata remained distinct, pristine, and readily recognizable. The excavation of Okladnikov Cave was nearly completed in 1987 when Sergei Markin led the senior author to see the last bit of deposit remaining deep in the cave. Even that small section seemed odd, as mentioned in Chapter 3. At Kaminnaya Cave, he again could see no distinct Pleistocene strata, even with the detailed explanation that Markin provided about the stratigraphy. Ust-Kan was the same story. While Alexander Postnov pointed out where he thought his levels were, clear and distinct banding was not apparent. At the time of these visits, the senior author's inability to recognize distinct strata was attributed to inexperience with Pleistocene archaeology. Now, there may be a better explanation, which is bioturbation caused by carnivore disturbance and consequential stratigraphic blurring.

Third, the amount of carnivore perimortem damage should correspond positively with the amount of site usage by these animals in the same manner as the amount of human refuse reflects occupation duration and intensity in much of the world. In this respect, then, Denisova, Kaminnaya, Kara-Bom, Okladnikov, and Ust-Kan are more like the six hyena caves than like the eight no-hyena sites. Okladnikov Cave stands out as having as much of a carnivore perimortem bone damage signature (37.3%) as the unweighted average for the six hyena sites combined (38.2%) (Table A1.27). Had Okladnikov Cave produced only a handful of stone or bone artifacts, we would have no reluctance to classify it as a paleontological site on the grounds of perimortem damage. If we choose to ignore the possibility of midden disturbance by hyenas, we cannot ignore the fact that these creatures and probably other species of digging animals occupied these caves in the absence of competing humans. In other words, human use of the cave and open sites was almost certainly seasonal, sporadic, or decades apart. We are not alone in this view. Wrinn (2010) concluded that there was so much "carnivore ravaging of bone assemblages, [which] suggests a competitive situation for human foragers, and the overall pattern is consistent with ephemeral occupations and low population density."

There had to have been discontinuities in the human occupation of these Siberian sites, the possible reasons for which are numerous – local reduction of food supplies, epidemics in game animals and/or humans, carnivore predation on the humans, competition with other human groups, exceptionally severe winters, forest and steppe wildfires, human infertility, and so forth. These breaks could have been seasonal, decades, centuries, or millennia in duration. Said another way, the longer the periods of absence, the greater the opportunity for short- or long-distance introduction of new cultural and human genetic variation. We suspect that there were many introductions of "novelty" caused by fluctuating human population pressure in southern and eastern Eurasia. We envision late Pleistocene Siberian winter conditions, especially the Late Glacial Maximum, as a significant obstacle to continuous local human occupation. The marked need for winter warmth most likely drove humans into the fuel-rich and wind-protected forests, and out of the frigid, barren caves, and possibly even out of the region by making north to south fall migratory treks of hundreds of kilometers, leaving the caves to the over-wintering hyenas and other animals. One needs look no further for evidence of long-distance migratory treks than ethnographic accounts of annual migrations by Alaskan and Canadian Eskimo bands across vast tracts of land.

In sum, our observations and analyses of perimortem bone damage have the potential to make significant contributions to the study of archaeological site formation and content. Based on our analyses we proposed the following. (1) Archaeological sites without a hyena presence have a much lower level of carnivore damage. (2) In sites with mixed human and hyena content, there were periods of human site disuse. (3) Disturbance of stratigraphy and probable blurring of strata by hyena and other animal digging can lead to false inferences about site and regional cultural evolution. (4) Comparison of paleontological and archaeological faunal assemblages shows that when the latter indicate the presence of hyenas, these sites have perimortem damage signatures closer to those of paleontological than archaeological sites. (5) By implication, the presence of some animal remains in mixed human–hyena sites may have little to do with human procurement.

We make some of the above remarks in consideration of ethological information in the following literature review of modern carnivores.

## Damage signatures

The bone damage signature of carnivores can be distinguished from that of humans in late Pleistocene Siberia and elsewhere. However, in most of our Siberian assemblages there is a small fraction of the perimortem damage that cannot be attributed to either carnivore or human uses. Gary Haynes (2002a; see especially his figures 4–6) illustrates some strikingly similar bone damage done by humans and carnivores. Pickering and Wallis (1997) found similar bone damage done by chimpanzee bone gnawing.

We identify one such ambiguity as "pseudo-cuts." The greatest number of pseudo-cuts was found in the Denisova Cave assemblage (29.6% of 115 pieces). For the vast majority of bone damage, cut marks and burning readily identify human processing. Unlike our sites evidencing both carnivore and human presence, a site in the Crimea with

Mousterian artifacts and carnivore remains yielded only four pieces out of 3500 with cut marks (Enloe *et al.* 2000). Some of our assemblages (Table A1.25) had as many as 20% of all pieces with cut marks. This significant difference could reflect a greater degree of game processing in Siberia than in the milder climate of the Crimea, or it could be due as well to differences in identification methods. Tooth dints, scratches, end-hollowing, and polishing characterize carnivore damage. The most common form of perimortem damage – breakage – is not diagnostic. Pickering and Wallis (1997) found that captive chimpanzees left perimortem damage to bone similar to that left by feeding carnivores. We have not found any bone damage that we could attribute with certainty to human chewing, but the Pickering and Wallis experiment suggests there could have been some human chewing of bone, but we had no way to empirically distinguish human from carnivore chewing. We believe that humans did, in fact, chew on bones, but our assemblages provide no clue to it having actually happened. Cut marks and burning are our diagnostic human bone damage characteristics. If our Siberian folk had chewed on bones without cut marks and burning, we are unable to distinguish whether humans or carnivores left the "chewing" marks. In other words, we question the assumption that all bones in an archaeological site represent the hunting and collecting activity of those ancient humans.

## Cooking

As is well known, eating raw, unprocessed food takes a lot of chewing, sometimes several minutes per mouthful, and even then not all the nutrients are released. Some processing such as crushing plants and meat improves the nutrient yield, but cooking with heat provides the highest yield because plant cell walls burst and animal connective tissue breaks. A medium rare BBQ T-bone steak with some salad, such as the acidified stomach contents of a reindeer, says it all.

Our assemblages suggest that cooking of meat was either rarely practiced within the living area of archaeological sites, or it was cooked by hot stone and bag-basket boiling, or in earth ovens. Because there is so little evidence of thermally altered stones and no acceptable evidence of earth ovens, at least as reported by Russian excavators, we hypothesize that cooking was rarely done. There is effectively no evidence whatsoever of roasting in any of our Pleistocene sites.

An inference of minimal cooking in southern Siberia is largely counter-intuitive if one is unfamiliar with cooking practices in the New World Arctic. Many accounts of Eskimo cooking practices report that meat was essentially only warmed in slightly heated water. Whale and seal blubber was eaten rare, as were the stomach contents of herbivores.

Susan Keates (2003:48) discusses Chinese hominid diet, suggesting cut marks, burned bones, and limited species are indications of diet, but she does not discuss cooking *per se*. She also notes that there are Australian and Southeast Asian ethnographic examples of animals having been cooked in their own skins. Unbutchered cooking should leave heat damage to bones with little protective muscle tissue, such as tail bones, metapodials, phalanges, and horns. We have no examples of such burning, so we infer that whole-body cooking was either not done, or it took place outside of the excavated areas of all our sites. However, since we did not systematically examine rodent bones (they usually did not meet our minimal-size criterion), these small animals would have been most likely the sorts that might have been cooked without butchering. (For a review of small-animal cooking in the American Southwest, see Turner and Turner 1999:32.) With respect to identifying diet, we have cut marks on a large number of species, as would be expected when food resources were limited; that is, everything edible was consumed. In sum, we have no evidence of any cooking practices.

## Meat caches

Storage pits are well documented for Upper Paleolithic sites of the central Russian plain (Soffer 1985:253–258, 1987). They are associated with structures built with mammoth bone. Pit sizes vary, but their depths are limited to 1.0 m, the depth assumed to represent the limit of the summer–fall thaw depth. This further suggests that Paleolithic storage pits were excavated during the warm times of the year. Were they dug in anticipation of fall hunting success? Their contents included mammoth, horse, reindeer, and also some carnivore bones.

Storage of food for winter consumption must have been as important, if not more so, for the late Pleistocene Siberians discussed herein, just as it was for all terrestrial, riverine, and marine coast hunters and fishermen of northern Eurasia, northern North America, and Greenland in historic times. Caches of meat, fish, eggs, birds, and other edible items are well described in historic accounts and recent ethnographies of Arctic and sub-Arctic peoples. In the senior author's experience, Aleuts dried summer-caught spawning fish – salmon for the most part – and stored them underground in oiled sealion stomachs. By chewing, their oral enzymes helped digest the raw fish. They also prepared a well-preserving oil called "stinky seal oil," which was kept in containers made of driftwood, placed in pits dug into the cold but not always frozen winter ground of the Aleutian Islands. Eskimos buried in rock-covered pits meat, fish, birds, and blubber, any and all of which preserved well in the frozen Arctic winter earth. Forest-dwelling Athabaskan Indians stored dried meat and fish high in trees, where it remained frozen all winter and out of reach of most carnivores. Meat and berries were crushed together by Canadian Indians to make pemmican, which kept very well above ground or below, frozen or unfrozen. All sorts of foods were cached by northern peoples to feed themselves when winter hunting expeditions failed to provide the necessary protein and fat to survive in the endlessly frigid far north. Of course, caching was practiced in many other parts of the world, but it is a major survival feature of northern peoples.

The problem with caching is that carnivores can be just as hungry in the winter when their own hunting fails to turn up a meal. Meat caches can be a windfall for them if discovered, and if the cache can be dug up or pulled down. Excluding bears, which retire in various sorts of shelter in the winter, all other carnivores are active, and even bears in late fall are well known for their tearing open of caches laid down from human summer and fall fishing and hunting. Fall and early winter cabin break-ins by bears, even with humans inside, are not unusual in the far northern New World.

All of the middle-sized to big modern Alaskan and Canadian carnivores have their late Pleistocene Siberian counterparts, except for the cave hyenas, cave bears, and cave lions. Based on the modern problems of cache destruction in Alaska, and bears and wolves coming into camps to get food they can smell from far away, we need to think about what this addition could have meant for human winter survival. Was the meat of game animals killed by late Pleistocene human hunters ever cached with hyenas living in the vicinity? Bears and wolverines are able to get at tree and pit meat caches in Alaska because both species can climb, as well as dig and push aside protective piles of rock.

Without some storage of food for the winter, Siberian humans would have faced periods of starvation when hunting was impossible or difficult due to stormy weather. Native names for the winter months are sometimes poignant in this respect. It is our speculation that the addition of hyenas to the list of Siberian winter hazards was a serious contributing problem for human survival. Not only might the humans themselves be killed by these powerful, bone-crunching, social hunters, but so also would their meat caches be at risk. Meat caches function best when established some distance apart so that they can be drawn upon in emergencies during migratory winter hunting. The problem with this strategy is that they remain largely unguarded and vulnerable to carnivore thievery. We suggest that losses of meat caches to various winter-hungry carnivores contributed to the limiting of Siberian human population size and related northward expansion in late Pleistocene times.

Although we feel that the Siberian folk must have used storage pits, the evidence for them is not good. The index of the premier English-language synthesis of the Siberian Paleolithic (Derev'anko *et al.* 1998), has no reference to meat caches or storage pits, nor do the book's site maps show any indications of storage pits. Examination of other Siberian site reports turns up no mention of storage pits. We suspect this absence is an excavation sampling error rather than a cultural practice of not storing food for winter needs.

## Site disturbance

Siberian archaeological cave sites had a significant carnivore presence, and as such probably underwent stratigraphic disturbances as well as bone introductions unrelated to human occupants. Removal of bone is also possible. Ovodov recalls a case in point that occurred during the 1973 excavation of Varvarina Gora. A one-year-old dog (Laika), wandering in the excavated area, grabbed a neck vertebra of the rhinoceros that had been buried in the earth for 34 000 years and tried to escape with it into the surrounding forest.

While all sorts of large carnivore remains have been found in Siberian cave and open sites, both archaeological and paleontological, we emphasize the hyena contribution to site disturbance because their remains were much more commonly found than those of other carnivores. Brown bear remains have been found in a number of caves; these mostly appear to have been chosen as denning sites or natural vertical karst "traps"

(Ovodov 1977a). In addition to the quantity of hyena remains, our emphasis on hyenas as a major potential source of site bioturbation does not exclude garbage-seeking or denning bears or wolves, but also includes the bone-damage signature as defined by the characteristics identified in the prolific presence of hyenas in Razboinich'ya Cave.

Burrowing activity by hyenas likely disturbed stratigraphy. The "natural" taphocenosis (assemblage of bones and teeth found buried together) was likely altered to various degrees within a site. Understanding the complicated stratigraphic situations in the Pleistocene Altai cave sites has been recognized as a serious problem (Derevianko *et al.* 1997a, 2001d). However, no previous workers have considered whether hyenas contributed to the complexity. We feel that hyenas may have been a very significant source of stratigraphic disturbance and bioturbation, based on comparisons of sites where diagnostic tool types are out of stratigraphic expectations (Turner *et al.* 2001b), as well as judging from the burrowing and denning activities of modern African hyenas (Brain 1981, Mills 1990). Although climatic conditions differed between the African and Siberian habitats, we propose that the open-country dens of the African hyenas were less adaptive in the frozen tracts of open Siberia. The burrowing discussed below may have been used in the Siberian sites along with simple dug-out basins in cave fill as an added means to ward off the cold of winter wind.

Sticking with thinking about intra-site disturbance lets us make a comparison of Siberian cave stratigraphy, especially Denisova Cave, and Aleutian shell middens. Both types of site have a striking similarity. In Denisova Cave the Holocene strata are strikingly apparent and horizontal (Fig. 3.10). The underlying Pleistocene deposits have almost no recognizable stratigraphy. We propose that this dichotomy is due to the bioturbation of the Pleistocene deposits by hyenas. Aleut shell middens have almost no recognizable stratigraphy in the prehistoric levels, but historic strata are clear and easy to recognize. This was most probably due to repeated prehistoric pithouse construction that more-or-less ceased in historic times due to the majority of Aleuts having been taken away in the summer and fall to hunt sea otters for Russian fur hunters. Without men in the villages when house construction would have most likely occurred, there would be very little disturbance. Hence, we propose that in Siberia the stratigraphic mixing was caused by burrowing hyenas. In the Aleutians it was caused by men constructing pithouses. The bioturbation in Siberia ceased with the extinction of the hyena. It ceased in the Aleutians with the Aleut males being taken away by Russians to hunt for sea otters in the distant coastal waters of British Columbia and California.

## Review of studies of modern carnivores with emphasis on hyenas

Before we begin this section, we want to briefly comment on our attitude toward animals in general. Hal Herzog's (2010) cleverly titled book on this subject, *Some We Love, Some We Hate, Some We Eat: Why It's So Hard to Think Straight About Animals*, might well have added "Some We Fear." Those we do fear are often big and powerful. But some very small animals are often feared as well – for example, scorpions, snakes of all sizes, bats. Fear arises from perceived threats of danger and harm. As the reader will learn shortly, we accept that fear of hyenas, probably as much as any large carnivore, has existed throughout all of human prehistory and history. They are big, nocturnal, hunt in packs, make scary vocalizations, and eat almost anything – including people. They have been vilified in almost everything we have read in our learning about them.

Studies of modern carnivores, and their relationships with humans outside of Siberia, are a source for inferring how humans and hyenas might have interacted in late Pleistocene Siberia (Aigusta 1963). We focus on hyenas because their remains (bones, teeth, scats, stomach bones) are much more common than those of bears, lions, or wolves. It is for this quantitative reason that we attribute much of the carnivore bone damage signature reviewed in Chapter 3 to hyenas.

The literature on living and extinct carnivores is extraordinary – too much to review here. In the English language alone, there are books devoted exclusively to one or more species or classes of carnivore (e.g., Busch 2000, Ewer 1973, Gittleman 1996, 2001; Kruuk 1972, 2002, Lynch 2005, Werdelin and Solounias 1991), or that have lengthy sections on carnivores (e.g., Benecke 1999). Books are often based on proceedings and articles, of which there are hundreds dealing with carnivores. This academic attention is not surprising when so much information points to the fact that humans and terrestrial carnivores are and have been in direct competition for food and territory. An article by Geist *et al.* (2009) that deals primarily with bears and wolves is an outstanding summary. Moreover, it reviews the various anti-predator mechanisms humans have devised as far back in time as *Homo erectus*. Our concern with modern carnivores is to help us better understand the damage they do to bones, and to learn how much of a danger they are and were to modern humans. We explicitly assume that their danger today was about the same or greater in the late Pleistocene of Siberia. Recent news accounts illustrate this danger:

> Chignik, Alaska – An autopsy has concluded that rural teacher Candice Berner, 32, was killed by animals, and the head of the Alaska State Troopers said Thursday [February 26, 2010] that wolves are the likely suspect. However, the autopsy could not say for sure what animals are to blame. Berner's body was found Monday about a mile outside Chignik Bay. The body had been dragged off the road to the village's lagoon and was surrounded by wolf tracks. (Anonymous 2010a)
>
> Cooke City, Mont. Bear rampage kills 1, injures 2 at camp ground. At least one bear rampaged through a heavily occupied camp ground Wednesday [July 21, 2010] near Yellowstone National Park in the middle of the night, killing one person and injuring two others during a terrifying attack that forced people to hide in their vehicles as the victims were torn from their tents . . . [one survivor] suffered several lacerations and crushed bones from bites on her arms. (Anonymous 2010b:A2 )

While wolves and bears are today common in Alaska and other western US states, and Canadian provinces, in modern Africa the spotted hyena is the most common carnivore, outnumbering lions many-fold, wrote J. Pickrell (2002) when reporting on the observations of several field workers. Given the wealth of hyena remains in our late Pleistocene Siberian sites, it may well be true that they were as common then as today in Africa (Figs. 4.1–4.2). If these living hyenas had snarled at the photographer, he might have seen a frightening mouth full of large teeth, as shown in Fig. 4.3. Hyenas have long conflicted with human populations: "African legends and folklore associate the hyena with witchcraft and the supernatural" (African Wildlife Foundation 2008).

**Fig. 4.1** African spotted hyena (*Crocuta crocuta*), Tanzania. This animal is closely related to the larger Siberian cave hyenas discussed in this book (Matthew J. Betz photo, color, August, 1999).

**Fig. 4.2** African spotted hyena, Tanzania. M. G. L. Mills (1990:36) reports observing as many as nine Kalahari spotted hyenas feeding on an animal they had killed. H. Kruuk (2002:64) reports weight as being more than 70 kg (>150 lb). The Siberian hyenas are believed to have been larger and heavier, making them even more dangerous than modern spotted hyenas (Matthew J. Betz photo, color, August, 1999).

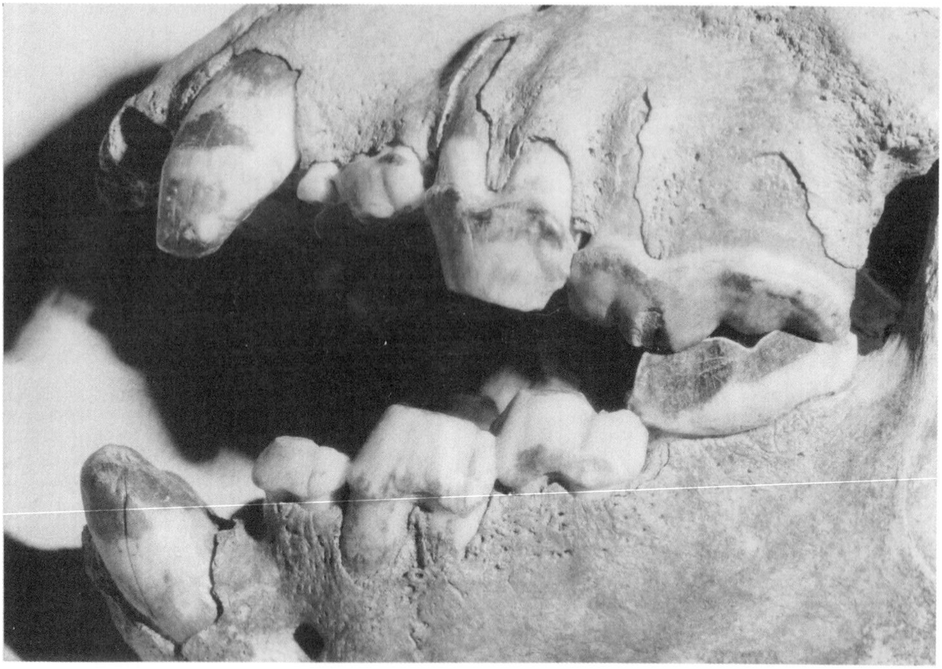

Fig. 4.3 Fifteen-year-old hyena maxillary and mandibular teeth. Specimen found in Fanats (Fanatic's Cave), Khakasia (CGT neg. Krasnoyarsk Regional Museum 8-7-98:2).

## Modern hyenas

### Hunting

There are many accounts and observations of hyena hunting in the ethological literature. One of the most respected observers is Hans Kruuk (2002:104), who described a kill scene he witnessed in the Serengeti:

> A pack of spotted hyenas chased a zebra family at night, and by biting at her legs and flanks they manage to slow down one of the mares. The stallion attacks the hyenas but there is little he can do against a dozen of the tormentors, and within minutes of the first bite the mare is down, while the rest of her family runs on. Hyenas tear away the flesh, and more of them join. A total of 34 hyenas eat from the victim, and 40 minutes after being pulled down there is nothing left of the zebra, just a large, dark stain on the grass and a steaming heap of stomach contents. Somewhere a hyena will be chewing on a jaw, reducing it to no more than a set of teeth. But the rest of the prey, including all the large bones, will be digested totally. . . .

Such observations demonstrate that the chance of any prey item landing itself in the fossil record (e.g. by getting neatly piled up in a cave) is extremely slim.

It was Hans Kruuk who first recognized that hyenas live in social groups that he called "clans." These were composed of up to 100 individuals. They recognized and scent-marked the boundary of their own territory, guarding it from incursions by members of other clans in the vicinity. How might late Pleistocene Siberian humans have recognized these territorial borders? If a clan were very large, it likely was important to do so. We speculate that it was done by recognition of clan members, the same way we recognize neighborhood dogs or by the experienced eyes of ranchers who can recognize individual beeves even from a great distance. Careful attention to animals and their characteristics, including behavior, is unquestionably a very old human talent as evidenced in the cave art of the European Upper Paleolithic.

## Dens

African hyenas make dens by burrowing and enlarging tunnels of other burrowing creatures. When there are natural caves or overhangs for shelter, like their European relatives, they use these also (Brain 1981). These underground dens are spacious enough for even large adult females. Van Lawick and van Lawick-Goodall (1970:108) noted:

> Once Demon [a hyena pup] followed Black Angel [an adult female] down one of the dens only to emerge hastily, half hidden in a cloud of dust, as the female started digging.... Black Angel grabbed one pup by the skin of its back and, followed by two other pups, vanished into the den.

Owens and Owens (1984:254–256) provide a brief description of an African brown hyena den:

> Star [a female hyena] had enlarged an existing springhare hole for her den. Three deep trenches in the sand led to separate tunnels underground, each concealed by a thicket of acacia bush. During the day she slept in the patchy shade about fifteen yards away ... [When this den was abandoned and the cubs moved to a new den far away, the Owenses had a chance to explore the den] ... I crawled head first into the open trench and then into a tunnel about two and a half feet high. By lowering my head and shoulders I could just squeeze inside ... the tunnel ran straight for about twelve feet and then made a turn to left ... [wriggling to the end, Della Owens came to a chamber] ... about five feet in diameter and three feet high ... three small tunnels and two larger ones led from the chamber.

Her husband, Mark, was crawling and exploring a similar and connecting tunnel. The den was clean and free of dung. Only a few bones were present. The den was full of fleas, offering the Owenses an explanation, at least in part, why hyena mothers move their cubs to new dens.

## Scavenging

The African Wildlife Foundation (2008) comments that species of the African Kalahari hyena possess several adaptations for scavenging. The adaptations are: (1) large and powerful bone-crushing teeth and jaws that enable these creatures to break bones and get to the marrow; (2) they can digest bone – what they cannot fully digest are hair, hooves, and horn; (3) they can travel long distances in search of carrion (their "laughing" vocalization is used to alert other clan members up to three miles away of a food source).

S. M. Cooper (1990:136), in his discussion of hyena scavenging, reports that half of the meals found by hyenas were carrion, and most of these remains were bones. He also reports that during his two-year study in Botswana, herd animals were hunted by the spotted hyena, most frequently zebra foals. Three-quarters of their food came from hunting animals that weighed less than 150 kg of consumable mass. J. D. Skinner *et al.* (1998) report on the number of bones found around two brown hyena maternity dens in the Namib desert. Between 1982 and 1996, 14 385 bones had accumulated. This suggests that a fairly large number of the bones excavated in our assemblages were carried in by non-human cave occupants, presumably hyenas. Hyena bone accumulation was the subject of a study by Pickering *et al.* (2004). These, and other studies, importantly Brain (1981), make us markedly skeptical about who or what deposited bone refuse in archaeological sites where there is evidence of hyena presence.

### Growth

L. G. Frank, S. E. Glickman, and P. Licht (1991) report on the unusual prenatal and neonatal behavior of infant hyenas that results from the large titration of testosterone in the female womb. This hormone-rich prenatal environment leads to aggressive behavior in the newborn, of which there are two infants at a given birth. The stronger of the neonates inflicts wounds and even death upon the weaker neonate. This aggression is aided by the fact that newborn hyenas have a full set of teeth at birth. This high titer of testosterone in developing and pregnant females leads to strong masculinization in females, which contributes to their larger body size compared to males, as well as other anatomical features.

## Siberian humans and hyenas

At this point we offer some thoughts about the possible relationships between late Pleistocene Siberian humans and carnivores, especially hyenas. To begin, it is important to note that there are some living Siberian carnivores whose ancestors survived the megafaunal extinction event of the deeply cold climatic shift at the Late Glacial Maximum (LGM) temporally near the terminal Pleistocene. Among the terrestrial survivors there remain the tiger, bear, wolf, snow leopard, wolverine, foxes, and smaller species. Gone are the large cave lions and cave hyenas. Their extinction along with the megafaunal forms such as the mammoth, wooly rhinoceros, bison, horse, and other large species suggests a rather close predator–prey relationship. As the large LGM prey died off, so did their larger predators. Cycles in prey population size could have swung so low near the end of the Pleistocene that the considerably smaller population size of predators must have dipped below some minimal recovery point, bringing about their extinction, possibly even before that of the herbivorous megafauna. One conjecture for why these now extinct carnivores failed to shift their hunting to smaller species suggests an epigenetic behavioral tie with the big herbivores. As in Africa today, Siberian hyenas may have specialized in hunting only a few species, and their hallmark scavenging ability would have been reduced due to competition with several other species of meat-eaters. Another scenario for their extinction could be that they simply were unable to deal with the LGM cold despite their widespread Old World ecological and altitudinal distribution. Yet another possibility is that Upper Paleolithic people more-or-less commandeered their den sites, leaving the large female cave carnivores with fewer and more vulnerable locations to give birth to and raise their newborn. One more possibility, and most interesting to us, is that the big carnivores were unable to compete as well with the technologically progressive Upper Paleolithic hunters as they did with the late Middle Paleolithic Mousterian people. We will return to this idea later when we review the question: Who were the Upper Paleolithic humans?

Today, the surviving descendant predators make their living from kills of medium- to small-sized creatures like pigs, deer, reindeer, sheep, goats, hares, marmots, mice, and so forth, whose Siberian ancestors also passed through the extinction filter. The larger of these carnivores can, of course, bring down a large herbivore of moose or elk size, but it is risky business for even predator packs such as wolves. The odds of a carnivore such as a member of a wolf pack being killed or maimed are surely less than if it was hunting a healthy moose or elk by itself. Despite escaping terminal Pleistocene extinction, modern Siberian carnivores, both terrestrial and aquatic, are much more at risk for their lives and numbers today because of human hunters and competition with humans for game animals and living space than ever in the late Pleistocene. What must have been a late Pleistocene carnivore dominance over humans, or at least a standoff, is now completely reversed, all because of numerous forms of technology and knowledge that enhance human reproduction and survival.

Behavioral studies of modern Siberian carnivores provide valuable insight into the probable life ways of their Pleistocene ancestors. The Siberian tiger has been and continues to be intensively studied both in the wild and in captivity (see Quammen 2003 for references). Its behavior and habits of today must mirror those of its late Pleistocene progenitors. The fact that the modern Siberian tiger hunts humans as well as deer and other creatures should not escape our thinking as we attempt to reconstruct the relationship between hyenas, other carnivores, and Upper Paleolithic humans. Ethnographers have compiled a wealth of information about animal behavior obtained from native Siberian hunters and fishermen such as the Koryaks Kets, Nivkys, and Nenets, whose lives were until very recently closely linked to the terrestrial, riverine, and marine animals of their territories. Native stories about, and fear of, wolf pack and bear attacks, especially polar bears, help enrich our reconstruction. There are historic accounts and descriptions of animal behavior by all kinds of observers, ranging from the most sensationalistic newspaper reporters to carefully made observations by lay persons with good common sense, and professionally trained scientists of animal behavior. Bottom line? Large wild adult carnivores are not to be trifled with, despite their initially cute and cuddly infant forms and behaviors. Comparisons of Siberian natural history and animal behavior records can be made with others recorded for the same or closely related species elsewhere in the world. Such comparisons help identify variation that can be useful for reconstructing Pleistocene behavior. For example, behavioral and ecological studies of Alaskan wolves, bears, wolverines, and other carnivore species are ethological classics (for a summary, see Guthrie 1990), rich in detail, and markedly valuable for their Arctic environmental relationships that would apply to the Pleistocene climate of Siberia. Most significant are the behavioral studies of living carnivores whose close relatives went extinct in Siberia. Of these, the observations of African lions and hyenas, especially the spotted hyena, are of utmost importance to our thinking about the relationships between late Pleistocene humans and hyenas of Siberia. These will be reviewed shortly. In the meantime, let us try to pose a few questions that seem relevant to this discussion.

Behavioral studies of captive hyenas from Kenya by psychologists Laurence Frank, Christine Drea, and Stephen E. Glickman at the University of California, Berkeley, are providing powerful insights into questions revolving around some of the evolutionary uniqueness of the spotted hyena – female aggression, newborn cub aggression, female dominance, etc. (www.abc.net 1999). They have learned, for example, that testosterone is manufactured by the placenta in pregnant females, flooding the developing fetuses with the male sex hormone. This leads to anatomical masculinization of female external genitalia and behavioral female aggression, but also neonatal sibling aggression. As noted above, born with a full set of sharp deciduous teeth, newborn hyena siblings promptly engage in combat and play that can result in serious wounds and often death of the weaker siblings. Among hyenas, "tooth and claw" natural selection begins soon after birth. Not only are hyenas born fighters, they are also born cannibals, as our perimortem taphonomic findings suggested time and time again. For hyenas, nutrients come from hunting, cannibalism, and scavenging. The hyena is not alone in its scavenging behavior. Kruuk found that lions commonly scavenge from hyena kills. Both the cave lion and cave hyena of late Pleistocene Siberia probably scavenged from human kills of large animals, which would have decreased the efficiency of human hunting in terms of nutritional yield. This, in turn, must have put a damper on human population growth.

After a successful hunt by some or most members of a hyena clan, the high-ranking dominant female(s) often takes over the downed carcass, and if still caring for cubs, ensures that – like her – they too get a good share of the kill during the ensuing feeding frenzy by clan members. The California psychologists also note that among other carnivores, unlike hyenas, lion cubs are low in the social hierarchy, so that when food is scarce, they are subject to starvation. Patently, as hunting and food-processing machines, hyenas have several evolutionary advantages over other carnivorous predators, including even tool-using humans, whose slow-maturing infants not only are unwieldy baggage for any nomadic human group, but are also toothless, weak prey for prowling hyenas. And attack humans hyenas do.

## Modern attitudes about hyenas

Accounts of hyenas in the modern popular press and folk culture are often peppered with negative connotations: cowardly, ugly, fearsome, sneaky, foul, thieving, dirty, repulsive, hermaphroditic, cannibalistic, night-stalkers, and so on. There can be little doubt that the modern human attitude about hyenas leans toward its possessing a repulsive, monstrous, and evil character. It should be noted, however, that not all field workers would completely agree. Owens and Owens (1984) discuss and illustrate friendly interactions they had with some Kalahari hyenas.

Among other carnivores, the lion is viewed as noble, proud, distinguished. Wolves share these subjective qualities, as do most other large carnivores. Nevertheless, as Hans Kruuk (2002:64) notes:

> The spotted hyaena is, despite its reputation, a large, wolf-like predator, often hunting the African plains and even the forests in packs . . . they have a considerable crime record . . . hyenas are also killers of people. . . . In Malawi . . . they killed and ate 27 people over 5 years. Many . . . of the victims were children.

L. van der Post (1961:119) noted that Bushmen have an expression that is used when a person is undergoing some manner of deep personal disaster. It places hyenas in a negative light: "The time of the hyaena is upon her [or him]."

What might have been the attitude about hyenas among late Pleistocene Siberian peoples? Since there seems to be no Siberian archaeological carved bone or stone depictions or rock art of hyenas, one might infer that not creating images of them – as was done for prehistoric birds, mammoths, bison, rhinoceros, fish, and various herbivores in Upper Paleolithic Europe – means they were unimportant in the lives of Pleistocene Siberians. Jacobson (2005) has found rock art depictions of large late Pleistocene animals in the Altai Mountains of Siberia and Mongolia, so rock art existed in our study area. On the other hand, hyenas might have been as feared and reviled then as they are today in Africa. Depicting them might have been considered foolhardy and dangerous in the animistic world view of the past. Accepting for the moment that ethnographic analogy is the best way to estimate the Paleolithic attitude about hyenas, a few accounts from Africa, where humans and hyenas still co-exist, provide insight into what might have been an uncomfortable psychological human–hyena relationship in the past.

In van der Post's (1961:219) recounting of Bushman myths about animals, most creatures are looked upon quite favorably. However, not so for carrion eaters such as hyenas, jackals, and crows:

> The choice of the hyaena as villain-in-chief is another example of this inborn regard for exact truth which characterized the first man of Africa. The hyena emerges out of its hole only in the hours of darkness. Although an animal of great strength and powerful jaws, it kills only the weak and prefers to live on the courage, initiative and labour of others, scrounging what it can of the remains of the lion's or leopard's banquet. For the Bushman it was the most clearly accredited representative of the power of darkness and principality of evil. Many a time, as I have listened to its wail alone in the night, miles from shelter and the sound of other men, I have thought of it as the cry of the damned and been troubled with emotions so far out of the range of my awareness that I cannot shape or name them.

Hans Kruuk (2002:186–189) discusses carnivores in a cultural context. He proposes that hyenas are more often associated with witchcraft than any other animal in Africa:

> The animal involved in witchcraft more than any other is the spotted hyena, a species which generally is utterly loathed throughout the continent. . . . This loathing goes beyond feelings based on mere ecological competition: it may well be that a primitive fear is involved, arising from the knowledge that hyenas are the living "mausoleum of the dead", as someone described them. Aren't the animals' weird laughing noises and its slinking nocturnal movements around one's house (often followed by some disaster to the occupants) almost proof that in some devilish way it is under control of supernatural powers? . . . Everybody is aware that people known to be witches ride hyenas at night (that is why hyenas' backs are sloping), laughing madly, while casting their spells on other people.

In a brilliant book-length review of human attitudes about big predators extracted from scientific, literary, and mythological sources, David Quammen (2003:133) sums up humanity's universal reaction to predator corpse-eating:

> Bury the corpse, cremate it, put it up on a platform to be picked clean by birds, pile rocks over it . . . even cook it and eat it yourselves . . . but by all means don't leave it to be gnawed at by leopards or hyenas.

Long before fire, cave art, and the two-million-year-old manufactured protective culture that distinguishes the hominid lineage, there was real danger – one common danger being large predatory carnivores. As Quammen (2003:329) poignantly remarks:

> We'll never know whether [our early ancestors'] fear of big-toothed felines was acute, reverentially muted, or dulled into routine amid a welter of other dangers ... death by predation must have seemed ordinary. No one had escaped the awareness of being eaten.

Later in human evolution that biological awareness must have been reinforced each generation by story-telling elders, who themselves were living proof of the value of their words, however phrased. Group membership provides protection to the individual. Children must have been educated in the basic survival law dealing with large carnivores – stay close to the group, do not wander off alone, stay away from the meat-eaters. Today, for most children, the predators are human, and the edict is: *stay away from strangers*.

Abhorrence must have been heightened for hyenas if in ancient times, as today, they entered camps at night and dragged away helpless young or elderly victims. The entry of predatory hyenas into late Pleistocene camps of sleeping Siberian people likely meant an easy meal, until the domestication of the dog, whose keen senses of smell and sound would have made them superb sentinals. The evolution of weapons and counter-weapons took a quantum leap with the beginning of domestication of the wolf more than 30 000 years ago, the directly carbon-14 dated age of the Razboinich'ya dog.

In a stunningly simple discussion of ecological tropic levels, Quammen (2003:75) describes the essence of that concept, with its energy transfer in efficiencies and losses from the ultimate sunlight energy capture by plant photosynthesis through plant-eating rodents and herbivores, and eventually to the precariously perched carnivores at the top of the pyramid. Quammen concludes: "Big fierce animals are inherently rare." This is certainly the case for living Siberian carnivores, whose lives are lived out mainly in a solitary fashion. Block the sunlight, and the tropic pyramid collapses from the bottom up. Add another large carnivore (efficient modern humans) and the pyramid collapses from the top down. Even the Pleistocene hyena and wolf packs must have been widely spaced across the Siberian landscape. However, they, like humans, hunted in groups. The more eyes, ears, and noses there are surveying the landscape for prey, the greater are the chances for discovery. While descriptions of African hyena dining seem frenzied, even hazardous to ears and snouts, the benefits of the cooperative hunting pack surely reduced the risks of being a luckless solitary predator. But adding modern humans to the mix of predators at a time of severe climate change must have resulted in deadly competition. While we do not know the precise sequence of events, we do know the outcome – humans survived into the Siberian Holocene, whereas some of the four-footed carnivores did not. Because there are few species of large terrestrial carnivores left in the world, inferences about their behavior, particularly that of hyenas, in late Pleistocene Siberia, will have to depend largely on analogies from living hyenas. Very large carnivores such as cave lions, cave bears, and cave hyenas became extinct before the end of the Pleistocene. These terminal Pleistocene extinctions apparently occurred all over Siberia in conjunction with megafaunal extinctions. Tatiana Krakhmalnaya (1999) has inventoried these creatures' similar extinction in the Ukraine, whereas in Siberia carnivores such as wolves, bears (*Ursus arctos*), and others did not go extinct. At first glance this would seem to be a simple case of extinction related to a disappearing food supply – that is, the megafauna. However, judging from the species found in Razboinich'ya Cave, the hyenas were subsisting on a wide range of game animals, so while the megafaunal extinction may well have reduced their prospective food supply, it seems unlikely that conditions reached the point of widespread starvation or infertility. However, the problem may not have been quantitative; rather, it might have been qualitative. Live prey with little fat and stressed by nutritional, epidemiological, and physiological factors may have been like an interior Native Alaskan winter problem: eating fat-poor hares and muskrats often led to starvation.

The large predator–large prey co-extinctions are not the only co-occurrences in late Pleistocene Siberia. There is the appearance of the Upper Paleolithic tool tradition and associated artistic expressions, both linked with anatomically modern humans. While the Mousterian folk may have had a long but shaky co-existence with the large Siberian carnivores, the latter may have been no match for the better-equipped modern humans, both the European Cro-Magnons and their Far Eastern counterparts. Is it just a coincidence that hyenas, which had occupied Eurasia since at least the Miocene, should cease to exist throughout most of their northern range following the appearance of anatomically modern humans in Eurasia? Moreover, there is an inverse relationship between hyena presence and the occurrence of human remains. Later in this chapter we explore some taphonomic and taxonomic issues involving late Pleistocene Siberian human remains.

## Human predation by carnivores

The living large carnivores of Siberia include bears, wolves, leopards, and tigers. With the exception of wolves, the others are largely solitary hunters. The tiger has the worst reputation as a killer of humans. However, one Primorsky biologist who has studied tiger behavior for more than 20 years, Victor Georgivich Yudin, said in an interview with Natalia Boyarkina (2000:20):

> I can assure you [that] if tigers hunted people, they [people] would never go into the taiga. Yet, there are thousands of people calmly collecting mushrooms, berries, and nuts.

He added:

> The tiger cannot make eye contact with humans very long. He possibly will not notice your trembling knees and wet pants if you look at him in the same manner as a wife looks at her husband who came home drunk without any salary in his pockets.

Nevertheless, they *are* dangerous. Yudin emphasized that the person who encounters a tiger in the tiaga, and tries to run away, will be chased down instinctively.

Among the largest of living Siberian carnivores, the Amur tigers of Primorsky Territory are seemingly a relic population of a species that once was distributed over much of Asia. They are brilliantly illustrated in an article by M. Hornocker (1997). Today, this large solitary forest-dwelling animal feeds on boar, red and roe deer, occasional dogs, and other similarly sized species. Weighing 250–300 kg (550–660 lb), tigers need 4–5 kg (9–11 lb) of fresh meat daily. In prehistoric times they may have also fed on humans, as they even now are known to enter and prowl around small isolated native villages. However, guns have given humans an overwhelming advantage, and modern cases of man-eating are very rare. Quammen (2003:383) notes that there were no recent cases until February 1976,

> when a tractor driver was killed by a tiger along a road near the village of Lazo [northeast of the port city of Nakhodka]. The tractor driver hadn't committed any notable provocation, and the tiger lingered to feed on the corpse.

Quammen cites other subsequent killings, including one in which a tiger killed and then carried a man up a tree, where he was completely eaten. During Quammen's winter trip to the Russian Far East to learn about the Siberian tiger, he was loaned a translation of a 1925 work on Russian and Manchurian tigers, prepared by Nikolai Baikov (1925), a self-trained naturalist and ethnographer. Baikov's writings were an important source for Quammen, who quoted him frequently. One quote is especially graphic (Quammen 2003:389):

> When a tiger openly attacks a person, it executes one or several jumps, beating the human along the side of the head or shoulders with blows of its paws. . . . The tiger digs its claws as deeply as possible into the head or body trying to rip off the clothing. It can open up the spine or the chest with a single whack.

The causes of such killings vary. We are unconcerned about placing blame; instead we want to emphasize that human life in the late Pleistocene was surely as threatened by large carnivores as it is in their habitat today, probably more so, since some of the large Siberian carnivores no longer exist.

Even small carnivores in many parts of the world can pose a danger to humans. For example, in late June 2010, two separate coyote attacks on little girls occurred in a New York City suburb. There was no sign that the animals were rabid; nevertheless, the girls were treated for rabies as a precaution (Anonymous 2010c). In late October 2009, a young woman was hiking alone on a national park trail in eastern Canada, where she was attacked by two coyotes. She died the next day of injuries inflicted by the two animals (Anonymous 2009).

## Other attacks on humans

A mountain lion attacked a seven-year-old boy in a recreation area near Boulder, Colorado, on April 15, 2006. The animal was a 22.5 kg (80 lb) female that pounced on the boy, who was last in a seven-member family group hiking in the recreational area. The cat bit the boy's head and made other puncture wounds. The boy lived and the cat was eventually shot and killed (Anonymous 2006).

Attacks by bears are well known, usually in the wild when a sow is protecting her cub(s). At other times there is no explanation for an attack. Recently a geologist was exploring alone in a remote valley of the Alaska Range. He was attacked by a grizzly twice, leaving him with a gnawed and clawed right arm. "The [first] attack lasted 10 to 15 seconds, then the animal lumbered away. The bear shortly returned and continued his attack" (Pemberton 2010, p. A10).

The following account is a direct quote (Anonymous 2008):

> Bears devour 2 Russian miners. Moscow – A pack of enormous bears searching for food killed and ate two men at mines in Russia's Pacific Kamchatka region and have kept hundreds from reaching the mine.
>
> A pack of up to 30 Kamchatka bears, which are similar to grizzlies, prowled around two mines of a platinum-mining company where they killed the two guards last Thursday, local officials were quoted by the Russian ITAR Tass news agency as saying.
>
> About 400 workers have refused to return to the mines for fear of the bears, which stand nearly 10 feet tall on their hind legs and weigh up to 1500 pounds, Interfax reported.
>
> About 10 bears have also been seen near the village of Khalino sniffing fish remains and other garbage.
>
> Fish poaching forces the bears to seek other sources of food. Bears frequently attack humans in the scarcely populated region.

A day later, another northern bear attack was reported, this one on Alaska's Kenai Peninsula, 50 miles southwest of Anchorage (Thiessen 2008: p. A8):

> A 21-year-old woman was attacked just 25 yards away from a tourist lodge: A lodge guest heard her screams. Looking out a window, the guest saw that a bear . . . had her head in its jaws and it dragged her a few feet. Rushing outside, the guest scared away the 500–800 pound bear. The woman suffered major cuts on her face and head.

The most highly publicized recent attack was by an Alaskan Peninsula grizzly that attacked, killed, and ate Timothy Treadwell and his girlfriend, Amie Huguenard (Herzog 2005). While Treadwell flirted with disaster every summer that he spent being with the Alaskan bears, and was regarded by many as a "crazy," the fact remains that the two people's voices were recorded by a running camera's audio microphone during their attack, being killed trying to defend themselves, and then eaten (Herzog 2005).

Few other attacks by wolves, mountain lions, or other large carnivores have been as graphically documented.

Another recent bear attack occurred in Montana, when 44-year-old Johan Otter and his 18-year-old daughter, Jenna, were hiking on August 25, 2005, in Glacier National Park. They came upon a grizzly who savagely attacked and mauled them for five minutes to protect her cubs. Johan received the worst of the attack. His scalp was torn off, five vertebrate were fractured as was one of his eye sockets, three ribs and nose were broken, and he suffered five major bites throughout his body. Both survived the attack by following posted warnings that should they be attacked they should lie down and tuck into a defensive fetal position (Kim 2005).

Numerous historic accounts of North American bear attacks on humans are reviewed by Valerius Geist (1989), who suggests that while historically bears were occasionally a lethal problem for Native Americans and European colonists, the super bear, *Arctodus*

(Alaskan giant short-faced carnivorous bear), was, in our words, the atomic bomb of the North American late Pleistocene carnivores. In sum, bear and tiger attacks have occurred recently in Siberia and even small carnivores such as coyotes have attacked humans (Anonymous 2007).

While hyenas and humans can get along together, there are more contemporary reports of hyena attacks on humans than there are accounts of attacks on humans by other terrestrial carnivores, including wolverines, wolves, bears, lions, mountain lions, and tigers. Large numbers of African hyenas certainly contribute to this differential ratio, just as does the encroachment on hyena territory by human groups in Africa and Eurasia. We suspect similar but smaller imbalances and disturbances also existed in the late Pleistocene of Siberia, where human groups must have been very small but detrimental to hyena and other predator control of the climatically unstable landscape. To help imagine the Siberian situation, we recount a few modern hyena attacks on humans: (1) On July 19, 2000, an 11-year-old Baltimore boy named Mark Garrat Shea was attacked and killed by a hyena while sleeping in his tent. The boy had been on a safari for more than two weeks in Zimbabwe and Botswana's Moremi Game Reserve, south-central Africa, where the attack occurred. It was believed that the tent flaps had not been secured. The number of people in the safari was not reported, but judging from other accounts, the number of humans seems almost irrelevant (*The Namibian*, August 11, 2000). (2) In 1999, three men were attacked and eaten in three different locations within the Fedis administrative area of eastern Ethiopia, about 30 km south of Hadar, northeastern Africa. Three others were seriously injured by hyena attacks in the same district. In southern Ethiopia's El Kere and Bare regions, 50 people, 35 of them children, were victims of hyena attacks between October 1998 and January 1999 (www.igorilla.com 1999). (3) In June 2005, a rabid hyena attacked and killed nine people and injured 15 others in the rural Dedza district of Malawi, about 43 miles south of the capital, Lilongwe, east-central Africa. The victims of the ferocious hyena ranged in age from 5 to 65. Hundreds of villagers were reported to have sought refuge in a school during the day-long rampage (www.Herald-Sun.com, 2005). A nearly identical report appeared in the *Arizona Republic* newspaper on June 14, 2005 (p. A2). Some victims thought the large animal that attacked them was a hyena, while others thought it might have been a lion.

An attack in June 2005 by two hyenas in eastern India was thwarted by a family dog. In the village Raghunathpur, Bokaro district, a woman and her seven-year-old boy were severely injured in the hyena attack, but lived through it thanks to the family's dog, which engaged the hyenas. One of the hyenas tried to drag the boy away but the family dog stopped it. When police arrived, the villagers were running for their lives. One of the hyenas was shot to death, the other disappeared into the forest. A third victim was injured (India News 2005).

John Clarke (1970), in his book *Man is the Prey*, tells of an African hyena attack on a man who received just one shearing bite that tore away most of his face below his eyes and most of his lower jaw. Other accounts suggest that hyenas that attack humans aim for the head.

Gordon Grice (2010:50) came across a 2005 medical report that said "an American doctor working in Nigeria mentioned the case of a woman who lost consciousness while giving birth and awoke to find a hyena eating her child."

Modern first-hand accounts of hyena attacks are most numerous in Africa, but there are also reports for other areas within the remaining hyena territory, including India. Encroachment by humans with associated destruction of wildlife habitat undoubtedly underlies many of the attacks. Some of these attacks involve rabid animals, such as one by a hyena that reportedly killed at least three people and injured 16 others in Blantyre, Malawi, in March 2003. The dead included two mutilated women and a three-year-old girl, who was dragged away and not found at the time of the report (Associated Press 2003). One of several recent attacks by presumably healthy hyenas is reported to have occurred in Botswana's Moremi Game Reserve, as mentioned above. Despite the fact that over-populating humans are in some sense responsible for hyena attacks on human settlements and livestock, nevertheless, it is patently obvious that these nocturnal hunters are much less fearful of humans than are many other more solitary predators, and even social carnivores such as wolves.

## Did late Pleistocene Siberian hyenas hunt humans?

Basically, this question cannot be answered. However, it should be considered. There is a small but significant number of modern first- and second-hand accounts of autochthonous terrestrial carnivores attacking and sometimes consuming humans almost everywhere in the world (Kruuk 2002:64–65, 75), although not all such encounters end in violence. This is as true for hyenas as it is for other large carnivores (van Lawick-Goodall and van Lawick-Goodall 1970, Owens and Owens 1984). Nevertheless, there are enough tragic accounts to assume that *ethnographic analogy* would be useful for developing preliminary inferences about Pleistocene human and carnivore interactions. Of course, there would have to be allowances made for differences due to modern human population size and territorial encroachment. Today, humans numerically dominate carnivores, but the opposite was probably the case throughout most of the Pleistocene. Technological changes through time would also have to be weighed in the balance. Because our perimortem bone damage signature so much involves hyena activity, we will focus on them, even though in the back of our minds we are fully aware that predation by other carnivores must have been an ever-present danger to Pleistocene Siberians, especially children. The danger level would have correlated with human and predator population sizes, values that we have no precise way to determine. We can estimate the number of individual predators from cave sites, but such estimates would have to be adjusted for the amount of time deposition occurred in a cave. If we assume that the amount of deposition time is roughly the same for all caves, then a relative danger level can be proposed. For example, Baryshnikov and Vereschagin (1996) found that in 12 West European cave sites 135 hyena first lower molars were found, the average per cave being 11.3. By way of comparison, Ovodov and Martynovich (n.d.) recovered 246 lower first molars in Razboinich'ya Cave. They proposed that at least 137 animals are represented. These values suggest that the danger to humans by hyenas was at least as great in the Altai Mountains as it was in West Europe.

Assuming that hyenas were a danger to late Pleistocene Siberians, is there any evidence they *hunted* humans? We have evidence that they ate humans. The remains could represent hunting events, but could also represent scavenging. We propose that scavenging might have been practiced more during the warmer months, when the ground had thawed. Hunting of humans might have been practiced more frequently during the colder times of the year, when finding anything to eat was difficult, especially digging in the frozen ground.

## Did humans eat hyenas?

If we accept that late Pleistocene Siberian hyenas sometimes consumed living and/or dead humans (acid-eroded human teeth, human bone in animal caves) as do present-day African hyenas, is there any evidence that humans anywhere ate hyenas? Speaking on a worldwide scale, humans characteristically do not eat terrestrial carnivores. The dog is the chief exception, primarily in prehistoric Mexico, where small dogs (chihuahuas) were raised as a specialty food (Aguilar-Moreno 2006:329). There are some accounts of dogs having been eaten in Arctic starvation situations. Bear meat was eaten at certain feasts held by Siberian tribes and Ainus (Sokolova 2000:121). Parts of powerful carnivores such as tigers were made into medicinal potions by the Chinese. Still, these practices are so rare that it can be generalized that humans do not eat terrestrial carnivores as part of our normal diet, nor do we eat hyenas under any circumstances.

What about Paleolithic humans having eaten hyenas? Our assemblages show no evidence of humans butchering hyenas, as were the findings of G. A. Bonch-Osmolovsky (1931). Writing about his study of cut marks on bones found in Crimean Paleolithic sites, such as Kirk-Koba, he notes that about 80% of his sample (size not specified) had cut marks. The cut carnivore bones were identified as bear, lynx, wolf, and fox, but no cut hyena bones. Also, the cut marks on the bones of these fur-bearing animals do not necessarily indicate that they were eaten.

O'Hanlon (2010) reports that a bone assemblage found in an Iberian cave included hyena remains. The finder, Antonio Rodríguez-Hidalgo, dates the assemblage to 117 000–183 000 years ago. At least one of the hyena bones had cut marks that he proposes suggest consumption. Since the deposit suggests a hyena den, the cut marks could be pseudo-cuts. We have no cut marks on bones that could be identified as those of hyenas. Hence, our assemblages have no potential evidence of ancient Siberians having eaten hyenas. It would appear that the negative views of hyenas held by modern Africans were also held by Paleolithic Siberians. Hyenas were not only dangerous, they likely also dug up human burials and did other damage to human living areas.

## Hyenas and archaeological stratigraphy

While digging and burrowing activity are well-documented behavior for modern African and Israeli hyenas, we know of no references to their digging behavior in Siberian late Pleistocene archaeological sites. However, judging from the heterogeneous mixed condition of the Razboinich'ya hyena cave deposits (Ovodov, personal observations), it would appear that these ancient bone-crushing scavengers and predators were as much inclined to dig, scratch, burrow, disturb, and generally mix up their denning areas and scavenging localities as their modern relatives. Pristine stratigraphy should not be expected where hyena presence can be demonstrated. It is especially important to take into account bioturbation when making inferences about cultural continuity or discontinuity when hyena presence can be recognized. Inasmuch as it is generally considered that Siberian cave hyenas went extinct about 13 000 to 14 000 years ago, stratigraphic conditions should differ in pre- and post-extinction levels. Ovodov and Martynovich (2005) suggest some indirect evidence for Siberian hyena burrowing, namely that during reproductive periods of the year, female hyenas burrowed as well as sought out caves, because there are caves that were quite suitable for occupation, but lack all signs of hyena or human use. Two such empty horizontal caves with much Pleistocene soil, named Erkina and Bezymyannaya, are located near the Razboinich'ya hyena cave.

## Who were the late Pleistocene humans of Siberia?

We know what did the animal damage in our assemblages, but who were the human butchers? This question arises because there are almost no Paleolithic human remains in Siberia. The senior author was the first worker to describe the dental morphology of Siberian Paleolithic human teeth (Mal'ta, Denisova, and Okladnikov Caves). The examination of the few teeth found in the two cave sites was carried out in 1987 under hurried and minimally acceptable examination conditions (no chance to re-examine the teeth, only 30 minutes allowed, poor lighting, and no comparative teeth at hand except the ASU-DAS standard reference plaques). Later, the cave teeth were re-studied under better conditions and at her leisure by Moscow State University biology student E. G. Shpakova (Shpakova and Derevianko 2000, Shpakova 2001), for the most part using a system developed by A. A. Zoubov (1977). In her later discussion of Siberian Pleistocene teeth (Shpakova 2005:421), she acknowledges that both Turner (1990c) and Valery Alexeev (1998) thought the teeth to have: "distinctly archaic characteristics." She also proposed that the Denisova teeth were perhaps more archaic than those from Okladnikov Cave. Despite this interpretation of archaism, she goes on to conclude that both Denisova and Okladnikov cave teeth have "anatomically modern affinities" (Shpakova 2005:422) without any reference as to what are the characteristics and variation of modern human teeth (e.g., Turner 1991, Scott and Turner 1997).

Despite our having sorted through >1 000 000 whole and fragments of bone and loose teeth, we found not a single example of human bone or teeth in the faunal collections. The few bits of human remains had been recognized either in the field or later in the lab, when Ovodov was making his preliminary species identifications over the years prior to our project. These were curated separately from the non-human materials. Hence, the following discussion about the people is based on separate earlier examinations by the senior author in the 1980s. These examinations strikingly revealed two facts: first, there is very, very little in the way of late Pleistocene Siberian human remains, and what does exist is extremely fragmentary and incomplete; second, most of the human remains had belonged to children, so osteological comparisons are not possible within or outside late Pleistocene Siberia because of incomplete growth. Only the morphology of unerupted or erupted permanent teeth can provide suggestions about evolutionary grade (modern, archaic) and racial affiliation (European, East Asian). Of the late Pleistocene Eurasian teeth the senior author has examined, few showed the characterizing size differences that exist between modern human groups; for example, European upper lateral incisors tend to be small, whereas those of East Asians tend to be large. The point here is that Shpakova really does not know much about modern dental variation, let alone Middle Paleolithic variation. Her criticism of Turner is unwarranted, even at the novice level of tooth identification (Table A1.32).

The most famous Paleolithic Siberians are the two sets of fragmentary remains of Upper Paleolithic (ca. 15 000 BP) children found by Gerasimov at Mal'ta, south of Irkutsk. He believed them to have been pathological, so no study was conducted. The senior author found that the unerupted permanent incisors and molar teeth of the three-year-old child belonged to a dentally modern human. Moreover, the Mal'ta teeth were more like those of modern Europeans than like modern East Asians (Turner 1990a, Haeussler 1996, Haeussler and Turner 2000). Because of its very young age (9–12 months), the second child had not developed enough permanent crown to be useful for description or comparison. Deciduous teeth are not used in the Arizona State University dental anthropology system (Turner *et al.* 1991), mainly because of very few inter-group differences.

Less well known are the fragmentary remains of two anatomically modern individuals discovered at Afontova Gora, the multi-component site within the city limits of Krasnoyarsk (Gryaznov 1932). Much has been made of a sub-adult frontal bone recovered in Afontova Gora because of the attached flat nasal bones, which suggested to Alexeev that the individual was Mongoloid (Alexeev 1998, Alexeev and Gokhman 1983, Turner 1983a). This single frontal bone became the basis for Alexeev's belief that Siberia was occupied only by Mongoloids until Europeans arrived in Neolithic times. However, all infants and young children have about the same degree of flatness to their nasal bones, regardless of race, so the racial affinity of the Afontova Gora child is ambiguous. A better case for Asian affiliation is present in a mandibular first molar of an anatomically modern child found by N. I. Drozdov at Listvenka (15 000–16 000 BP), a site located up-river from Krasnoyarsk and down-river from the Krasnoyarsk hydroelectric dam. This fragment of a mandible had an unerupted first molar, the morphology of which is much more often found in Northeast Asians than in Europeans (Haeussler 1996, Haeussler and Turner 2000).

The Mal'ta and Listvenka Upper Paleolithic teeth have no hint of archaic morphology, but some teeth from two Altai Middle to late Paleolithic sites do have less modern and more Neanderthal-like characteristics. A handful of sub-adult bones and four teeth were found in Okladnikov Cave. Two teeth of comparable antiquity were found in Denisova Cave. The senior author reported on these few teeth from both caves in 1990 as having more of an overall Neanderthal morphology than that of the European Cro-Magnons or East Asian Sinodonts, teeth of three groups he had personally studied. Shpakova and Derevianko (2000) disagreed, despite neither having ever actually examined Neanderthal,

Cro-Magnon, or Pleistocene East Asian teeth. The senior author disregards their opinion on the grounds of inexperience. Antiquity-uncertain Siberian human remains include a very Eskimo-like old adult male skull found at a site called Kordyupov, Yakutia, who is thought to date to the late Paleolithic. Very little is known about the provenience of this skull. Its relatively good preservation casts doubt on its Paleolithic antiquity. Whatever its antiquity, dentally it was a modern Northeast Asian–American Sinodont. A site called Kirkalinskaya Cave had some human bone whose antiquity may be late Pleistocene. Again, antiquity is uncertain. We have recognized Neanderthal or Neanderthal-like teeth from Denisova and Okladnikov caves with hyena digestive polishing and erosion (Figs. 3.24, 3.25, 3.133). We can recognize this damage with certainty based on our examinations of scores of non-human stomach bones and teeth.

## Siberian Mousterians replaced?

The tiny bits of middle–late Paleolithic Siberian bone and teeth suggest that the Mousterian culture-bearers of Siberia were more closely related to European Neanderthals than to anatomically modern humans such as Cro-Magnons or early East Asian Sundadonts and Sinodonts (Turner 1983a). Neanderthal teeth are distinctly different from those of Cro-Magnons (Bailey and Turner 1999). Admittedly, this proposition is based on very few teeth and a single DNA study based on a bone fragment found in Denisova Cave. Given this dental and genetic evidence, there is a stronger case for early Siberians to have been replaced by anatomically modern humans, such as those found at Mal'ta and Listvenka, than having been their ancestors. We base this opinion on the following considerations. To begin with, stone tool specialists Okladnikov and H. Marie Wormington thought that the Mal'ta stone tools much resembled those found in European Upper Paleolithic sites (Turner 1983c). Mal'ta and Upper Paleolithc sites of Europe such as Kostienki, Mezin, and others, are all situated at ca. 50° N, similar to the latitude of Vanciyverm, and demonstrably liveable Winnipeg, Canada (see Anikovich *et al.* 2007 for related considerations). A population drift of Central Russian Plain folk to Mal'ta would have followed the environmentally tolerable 50° N latitude.

Teeth, stone tools, and mobile art objects suggest a *site intrusion* for Mal'ta. These independent lines of evidence, which Irving Rouse (1986) so strongly advocated as being required for proposing an actual migration, are offered as being supportive of Okladnikov's view of a cultural link between Mal'ta and European Cro-Magnons. Dentally, all known European Upper Paleolithic teeth are similar to each other and those of modern Europeans. The best-known northern Upper Paleolithic example of Eurodonts are the Sunghir teeth (Zoubov 2000) found at a location east of Moscow. Compare the Mal'ta teeth (Figs. 3.82–3.83) with the Sunghir teeth (see Figs. 4.10–4.12). The Mal'ta teeth are very much like those of Sunghir and all European Cro-Magnons (see especially Zoubov in Alexeeva and Bader 2000). Mal'ta, Sunghir, other Cro-Magnons, and modern European teeth are characterized by having little or no maxillary incisor shoveling, reduced lateral maxillary incisor breadth, four-cusped lower second molars, and several other polymorphisms whose expressions are usually in the weaker range. If further study shows these relationships to be the case, then the notion of local evolution of Siberians, as believed by Derevianko, should be rejected. For a detailed discussion of the near-universal dismissal of the local or multi-regional human evolution concept, see Brauer (2007). Genetic evidence likewise favors a single origin for anatomically modern humans (Cavalli-Sforza *et al.* 1994, Chen *et al.* 1995). On the other hand a few workers still favor the multi-regional theory for modern human origins (Clark and Willermet 1997), and the possibility of Neanderthal–anatomically modern human hybridization (Kozintsev 2003). Local evolution, at least in stone tool technology, was championed also by Larichev *et al.* (1992). More recently, Derevianko (2005:11) has embraced the idea of local evolution. We believe that a theory of local evolution is less parsimonious than one of biocultural replacement, as currently believed for Europe (Stringer and Gamble 1995, Tattersall and Schwartz 2000, Trinkhaus and Svoboda 2005, Turner 1995). This view is not a return to "migrationism"; rather, it is based on the evidence that Rouse requires be present before proposing a migration event. In western Siberia, Bagashev (1998) found craniometric data that suggested prehistoric populations to be intermediate between Europeans and Northeast Asians. In Turner's 1984 study of the IAE skeletal collections, one western Siberian site in particular, Sopka, showed dental crown and root trait frequencies between those found in Europe and Northeast Asia (Turner forthcoming). Intermediate forms often result from hybridization, although not always if clinal factors are at play. Clines in northern Eurasia are more closely linked to migrations, such as that of the Golden Horde, than to single-factor gradients of natural selection (Turner 1984). More likely clines result from genetic drift/founder's effect, at least in eastern Asia (Turner 1992b).

Recently, Johannes Krause *et al.* (2007) proposed that a separate Neanderthal-like DNA species they call Denisavans existed in the Altai (see also Bower 2010:6). We agree with these authors that more osteological, dental, and DNA evidence is needed to draw such an unexpected inference. Moreover, and to repeat ourselves, the strong possibility for hyena disturbance in the strata of Denisova Cave makes dating iffy at best. Since there is an off-chance that *Homo erectus* could have been inhabiting the area near Denisova Cave (the deep Kamara open site a few kilometers down-river), a genetic study is needed to reconstruct the mitochondrial DNA pattern of Northeast Asian *Homo erectus* if technically possible.

Admittedly, we have drifted from our topic of perimortem taphonomy, but it was necessary to help identify who the late Pleistocene Siberian tool-users were, and to explore the complexity of population history of Ice Age Siberia.

## Why are there so few late Pleistocene human skeletal remains in Siberia?

Despite the many hardships, a lot of archaeological work has nevertheless taken place in late Pleistocene Siberian sites. The first discovery of a Paleolithic presence occurred in 1871 with the recovery of stone artifacts in a deeply buried Irkutsk site at what is now called the Military Hospital site. Since then the recovery of Paleolithic relics from cave and open sites easily totals in the tens of thousands of stone artifacts, cores, and debitage. In some Upper Paleolithic sites, well-preserved bone artifacts have been found, including human figurines, carvings of water fowl, scratched drawings of mammoths, needles, and unique dog and fox burials, etc. In contrast, the number of human bones and teeth that have been recovered is disproportionately small.

We realize that there could have been multiple causes for the missing human remains; however, one may have been the major reason for the rarity of Paleolithic human remains in Siberia. Possible causes include:

**1** *Small population size.* Of the possibilities, small population size must have been a contributing if not primary cause for the low number of Siberian late Pleistocene human remains. We should not expect to find many burials if there were very few people living in the vastness of Siberia – a vastness that is only beginning to be significantly sampled. But, if the dead had been buried in cave middens off and on during periods of time involving thousands of years and hundreds of generations, should we not expect at least one or two complete human skeletons by now? Let's do the numbers.

If the Siberian population size was small, then there would have been fewer deaths per generation than if the population size had been large. The fewer the deaths per generation, the lower the chances of finding human remains. For example, if we assume that the population size of late Pleistocene Siberia consisted minimally of 100 bands containing 20 people each, then each generation there would have been 2000 deaths. The late Pleistocene occupation occurred minimally 30 000 years. Assume a generation lived for 20 years on average, then there would have been 1500 generations, or 3 000 000 human deaths. Assume a larger population size and a longer occupation, and there will be a larger number of hypothetical deaths. If we further assume that taphonomic destruction will allow the preservation of only 1.0% of all deaths, then there would be 30 000 sets of human remains to be found. If only 1.0% of all late Pleistocene sites have been excavated, then only 300 individuals could be expected to be discovered.

With the exception of anatomically modern Mal'ta, there are no late Pleistocene "burials" in Siberia. The foremost expert on Mal'ta, German Medvedev, questions whether the two children that Gerasimov found at this site were actually buried (personal communication, August 4, 2001, Mal'ta site visit). There are a number of inferences that could be drawn from the above demographic exercise – the population was actually smaller, the taphonomic destruction was greater, the occupation was shorter, there were more sites, and, as discussed above, places of excavation may not match the places of burial.

**2** *The dead were not buried.* Perhaps the dead were not buried, but left on frozen ground during winter, or, regardless of season, cremated, placed in isolated nooks and crannies, or left in trees away from Paleolithic camps and habitation sites. A mortuary act but not a burial was practiced by Crow Indians (scaffold burial) in the northern US Plains. An example is shown in Ubelaker (1989:5). The nearly complete absence of Paleolithic human remains in Siberian caves fits this scenario, but it is contrary to Neanderthal and Cro-Magnon burial practices throughout Europe (Mellers 1996). Moreover, the Mal'ta children were found within the habitation and activity area of their camp. What little bone exists for the Siberian Upper Paleolithic favors burial within a habitation area, although there certainly must also have been deaths far away from encampments.

**3** *Dead intentionally set out for scavengers.* Similar to abandonment, but with a purposeful goal, might have been a mortuary practice in which the deceased were placed in the open on purpose, away from or on the edge of camp areas so as to be consumed by scavengers. Although rare, disposal of the dead in this manner is not unknown in modern times. There is such a practice in India, where the deceased are left to be eaten by vultures. The Crow Indian scaffold "burial" cited above could also be assigned to this non-burial class. A recent article on Serengeti spotted hyenas, illustrated with marvelous photographs by A. Shah and M. Shah, notes that in Africa "the Masai leave corpses in the bush for hyenas to dispose of" (Carroll 2005:55).

**4** *Poor preservation.* While this possibility certainly would apply to open sites, it is less a concern for dry cave midden wherein faunal bone preservation is generally excellent (Table A1.9). Even so, two of the richest Siberian Upper Paleolithic sites in terms of bone artifacts are open (Mal'ta and Ust-Kova). Preservation does not seem to be the main problem, although the quality of the faunal remains in these two sites is not as good as that found in cave remains.

**5** *The deceased were buried in locations away from the cultural deposits that Russian archaeologists have focused on.* Because excavation costs are considerable in Siberia, and the excavation season is short, it is not surprising that random testing, rarely productive outside of identifiable cultural areas, is seldom done. Hence, the possibility of late Pleistocene off-site burial practice is not now a testable hypothesis. Other possible burial practices with low probability of ever being found are illustrated by Ubelaker (1989).

**6** *The deceased were cannibalized by their companions or families.* Their remains were scattered in the habitation areas like any other bones discarded after eating. This possibility has a fair chance of being tested if the remains were left lying about the habitation area, and if the dead had been processed and damaged the same as game animals. However, to date the tiny amount of human skeletal and dental remains recovered from late Pleistocene Siberian sites shows but few signs of butchering, nor have we found acceptable various claims for human cannibalism reported in the literature or rumored in meetings or campfire talk.

For a century there has been controversy about possible Neanderthal cannibalism, especially in the large but very fragmentary Krapina assemblage from Croatia. Breakage and marks on the bones were responsible for the cannibalism claim. The claim was dispelled by Mary D. Russell (1987), who conducted a microscopic study of the damage. She concluded that about one-quarter of the damage was caused by or related to excavation, and the rest was caused by sedimentary pressure and/or roof fall. About ten years later, A. Defleur *et al.* (1999) proposed Neanderthal cannibalism based on perimortem taphonomic analysis in both the human and faunal remains at Moula-Guercy, Ardeche, France. There are other reasonably certain cases of cannibalism in European Middle Paleolithic to Bronze Age times, most with tentative explanations leaning toward starvation cannibalism (for example, Boulestin 1998). Where these European cases differ from the Siberian situation is in the occurrence of abundant Neanderthal bone in Europe and the near total absence of Pleistocene human remains in Siberia. The Krapina bone assemblage represents at least 43 adults and sub-adults, and almost twice that number based on 282 teeth (Russell 1987:373). The simplest inference for the difference in MNI for Europe and Siberia is that there were not as many Paleolithic people in Siberia as there were in Europe. Such a condition would, as previously discussed, provide fewer bodies per generation.

Accounts of cannibalism abound in the ethnographic literature (Travis-Henikoff 2008). In the Arctic, when cannibalism is vividly recounted it mostly is said to result from starvation crises. However, Boas relates an Arctic non-starvation tale that involved a psychopathic cannibal. The story is credible insofar as there are well-researched modern accounts of psychopathic cannibals outside the Arctic. This story Boas (1888) collected during his fieldwork with Central Canadian Eskimos. In essence it involves a huge bad man who killed, chopped up, and ate many people. In time, he was killed and chopped up by neighboring Inuit. Our point is that had cannibalism occurred in the late Pleistocene of Siberia, it might not have been caused solely by starvation.

Renowned Russian physical anthropologist Valery Alexeev (1998), after reviewing some of the possible explanations for the low yield of Paleolithic human remains in Siberia, half-heartedly concluded, as did V. I. Gromov earlier (1924), that cannibalism could have been the reason. Alexeev also wondered whether the small pieces of human bone found in Okladnikov Cave might have been cannibalized (personal communication, July 29, 1990). While cannibalism may have occurred in the Neolithic Yelenev Cave site (Turner *et al.* 2003, Turner 2003), we have not found an assemblage that meets the standards established in the American Southwest for hypothesizing cannibalism (Turner 1983b, Turner and Turner 1999). There was one piece of human skull found in Eckina Cave, probably parietal. It was found 4.12 m deep. It had no cut marks or signs of carnivore damage, but it did have perimortem breakage and ivory hardness and color.

On the basis of M. P. Gryaznov's (1932) published photographs, Ovodov and Martynovich (n.d.) note the perimortem breakage of a left ulna from Afontova Gora that they feel might represent cannibalism. If it is still possible, the entire Afontova Gora human assemblage needs re-study.

**7** *Human bone was used to make artifacts.* Bones of deceased humans might have been used as fabricational materials for bone tools or ornaments. There are hundreds of examples of human bone having been used to make a wide variety of artifacts in prehistoric Mexico (Turner and Turner 1999). Of all the possibilities for explaining the rare occurrence of late Pleistocene Siberian human skeletal remains, this mechanism has the best chance of being tested because of the recovery of late Pleistocene Siberian bone artifacts and bone refuse that might be identified as showing one or another stage in the manufacture of artifacts. Paleolithic Siberian bone artifacts are illustrated in several reports, starting with the finds made by Gerasimov at Mal'ta, and most recently with Altai finds reported by Derevianko and Shunkov (2005), and Derevianko and Rybin (2005). None of these reports indicates that the bone source was human, nor do the illustrations suggest a human source for the bone. It is safe to say that in our faunal examinations of at least 1 000 000 pieces of bone we have come across fewer than five pieces that we feel had been slightly modified for use as a tool of some sort. None of these few "worked" pieces was identifiable as human. The rarity of human bones in late Pleistocene Siberian sites was *not* due to human bone having been used for fabricational purposes.

**8** *Hyena disturbance*. The deceased were considerately buried. Later, corpses were dug up by scavenging individuals or packs of hyenas or wolves. Referring to his own observations and those of others, Hans Kruuk (2002:204) wrote about hyenas robbing human graves:

> Stripped hyaenas hang around human habitation wherever they occur, throughout the Middle East and the northern parts of Africa. Even in medieval bestiaries they are shown as graverobbers – Hyenas are not alone in this gruesome specialization, as wolves are also known to dig up human corpses around towns and villages. The two species often occur jointly around desert settlements in the Middle East and I have watched them scavenging together near kibbutzes in Israel.

Perhaps hyenas were digging up the dead Siberian humans wherever they found them. In Europe, caves were used for burials, as were open encampments such as Sunghir. Alan Mann (personal communication, September 13, 2007) found in his excavations at Les Pradelles, a Middle Paleolithic Mousterian site in Charente, France, a number of Neanderthal teeth "that were clearly eaten by hyenas and then regurgitated." Les Pradelles also produced numerous hyena bones and scats.

In the thinly soiled high Arctic, modern carnivores attack rock-covered surface cairn burials, the burial rituals of which were carefully followed by the Central Canadian Eskimo (Boas 1888:205), but the post-funerary attention to the deceased was minimal:

> It is strange that, though the ceremonies of burying are very strictly attended to and though they take care to give the dead their belongings, they do not heed the opening of the graves by dogs or wolves and the devouring of the bodies and do not attempt to recover them when the graves are invaded by animals.

## Extinction of megafauna

The issue of megafauna extinction at the end of the Pleistocene has been much discussed. Paul S. Martin (1984) maintains that human hunters were a principal cause. Other workers see both humans and climate change involved (Haynes 2002b), while others believe that climate alone was responsible (Guthrie 1990). N. K. Vereschagin (1971) reviewed this topic and related issues in considerable detail based on Russian Pleistocene paleontology and archaeology. He seems to favor the human and environmental combination as responsible for at least the demise of mammoths. On the other hand, Orlova *et al.* (2000a) do not see humans as having played an important role in the extinction of the Siberian mammoth population. Several years earlier, W. E. Garutt (1964) set the stage for Vereschagin's work by inventorying all of the reported paleontological and archaeological mammoth finds in Eurasia, mainly Russia, and Alaska as of 1964. Orlova *et al.* (2000a) identified more than 120 sites in Siberia and the Russian Far East, with about 530 carbon-14 dates.

Whatever the cause of the megafaunal extinctions – temperature, vegetative changes, habitat reductions, disease, etc. – we feel human hunting added to the predation pressure by large carnivores. Today, in the Siberian, Primorsky, Mongolian, north China, and Korean regions, five species of large animals are in danger of going extinct: the tiger, Przewalski's horse, Asiatic wild ass, wild Bactrian camel, and the sika (Fisher *et al.*

1969). Modern human hunting and habitat encroachment certainly hint that humans had a hand in the late Pleistocene megafaunal extinctions.

Given the many thousands of years of co-existence between the Mousterians and Arctic Steppe fauna, the extinction of both may be a consequence of the arrival of modern humans in Siberia, coincidentally with climate change that replaced the Arctic steppe habitat with taiga and tundra. Our study of perimortem bone damage has led us to the topic of megafaunal extinction, although not in a fully satisfactory way. We envision humans and carnivores consuming every species they encountered, which suggests climate change alone was not responsible for the demise of megafauna. Why? Because would not some species be expected to retain their niches or rapidly adapt to similar ones?

## Northern limit of cave hyena distribution

As our site inventory and discussion elsewhere have shown, late Pleistocene cave hyena remains have been found across most of Eurasia, from England to the Sea of Japan. However, not all Pleistocene cave excavations have produced remains of hyenas; Mezmaiskaya Cave in the northwestern Caucasus, for example, had no hyena remains (Baryshnikov *et al.* 1996). In European Russia, hyena remains have been found in the upper level of the "recent" loess of Wurm times. They have been found in Aurignacian sites in the Crimea, but not afterwards (Golomshtok 1938:221). While today these creatures have a more southerly Old World distribution, in late Pleistocene times they roamed further north. Ageeva *et al.* (1978) proposed that at the time of the expansion and eventual reduction of the northern Eurasian mammoth fauna, cave hyenas had lived as far as 56° N latitude (Table A1.33). Vereschagin and Baryshinkov (1984:496) agree. Law (2007: n.p.) stated that "*Crocuta crocuta* has been recorded from as high as 4000 meters in east Africa and Ethiopia." Such height indicates that despite their southerly presence they have a tolerance of cold. It would be roughly freezing at that elevation when the evening temperature on the plains below was 70°F. Today, the northern limit for hyenas is about 35° N (roughly the same latitude as Cyprus and Tehran). Some 100–170 striped hyenas are thought to live in Israel (Horwitz and Smith 1988:472), and at least one of their dens has been the subject of actualistic perimortem taphonomic study (Horwitz and Kerbis 1991). Becker and Reed (1993) studied bone damage in Nubia caused by hyenas. Their interpretations paralleled those of Horwitz and Kerbis. In Russia, the cities of Moscow, Krasnoyarsk, and Bratsk are located on the fifty-sixth parallel, the northern limit of Pleistocene hyena distribution. We add for comparative purposes that also located at about 56° N is the north shore of Lake Baikal, Bering Island, Port Moller (on the Alaska Peninsula), the Belcher islands in the southern part of Hudson Bay, and Glasgow, Scotland. At this latitude continental winters are long and cold, although less so where oceanic currents introduce relatively warm water. Ageeva *et al.* (1978) additionally note that paleontological and archaeological finds of Siberian hyenas and other associated faunal remains suggest that the preferred hyena habitat was a sort of middle-mountainous relief covered with both steppe and forest vegetation. Such a combination of climatic latitudinal limit, vegetation–game animal associations, and foothill and low mountain cave utilization sites could have been the conditions that created the boundary across which further northward human dispersal was also challenged in late Pleistocene times. Our site backgrounds (Chapter 3) confirm the fifty-sixth parallel hyena boundary hypothesis, although we have one site (Ust-Kova) north of that latitude.

Since 1978, reports have appeared dealing with archaeological sites and their faunal contents on or near the fifty-sixth parallel. In addition to our own identifications listed in Table A1.2, we can add what S. N. Astakhov (1999) has reviewed and synthesized from the various excavations in the cluster of Afontova Gora sites located along and near the left bank of the Yenisei River in Krasnoyarsk – research that began in 1884 and continues to the present day. The late Pleistocene Afontova Gora large mammal inventory is familiar (mammoth, rhinoceros, horse, bison, reindeer, bear, wolf, etc.), but no hyenas. While the absence of hyena remains supports the fifty-sixth parallel hypothesis, the fact that Afontova Gora is a riverside setting could mean that its environmental conditions did not encourage hyena visitations. There are stone caves not far up-river, Yelenev, for example, so cave absence is not an issue. The finding of what is believed to be a dog in Afontova Gora may explain the absence of hyena remains as dogs could have served as sentinals and counter-hyena guards.

In addition to the sites with hyenas in our 30-site inventory, we would like to relate what Ageeva *et al.* (1978) had to say concerning a 1977 discovery by speleologists from Abakan, in the territory of Khakasia – namely, some details about a nearly complete, large (height at shoulder estimated to be 90 cm) and old (teeth very worn; Fig. 4.3) hyena skeleton and other animal remains at the bottom of a vertical cave trap called "Fanatic's Cave," the fauna of which fits the steppe-forest concept. Unlike the massively damaged hyenas of our inventoried sites, the Fanatic's Cave animal indicates a solitary death due to its fall into the uninhabited animal trap. Other Pleistocene species from Fanatic's Cave identified by these authors included sable, wolverine, wolf, bear, fox, and bison – steppe and forest dwellers whose presence in the trap was not fully understood. The location of Fanatic's Cave in the Serengeti-like Minusinsk Plain of southeastern Siberia would have been a favorable area for large herds of seasonally migrating ungulates such as horse, kulan, yak, and bison, whose numbers would have provided resident hyenas and other large carnivores many opportunities for hunting, scavenging, and demographic stability, if not population size. The numerous late Pleistocene archaeological finds from the upper Yenisei River basin corroborate this inference of abundance (Vasili'ev 1996).

At the fifty-sixth parallel both hyenas and humans of late Pleistocene times can be hypothesized as having been intensively competitive due to limited resources. We propose that the fifty-sixth parallel was the rubicon that humans had to overcome to reach the New World, and at this latitude hyenas possessed advantages that humans lacked, most notably having the ability to hunt at night, and being climatically adapted in contrast to the "night-blind" and "tropical" humans burdened with their helpless and slow-to-mature infants. We propose that the fifty-sixth parallel marked the location of a significant barrier to human dispersal northward, as much as was the Arctic Circle at 66.5° N latitude. There, most large carnivores had dropped out of the barrier mix and only cold, long periods of mentally depressing (Arctic hysteria) darkness, and patchy environment remained as major inhibitors to human population expansion into Beringia.

Hoffecker (2005) reviews the far northern settlement by humans, which he finds to be a late human expansion.

According to Kurtén (1976), no fossil remains of cave hyenas have been discovered in the New World. Still, some late Pleistocene inhabitants of Siberia did reach Alaska, notably saiga antelopes, humans and their dogs. We feel that it was the fully domesticated dog that helped break through the hyena barrier and made possible the transit to eastern Beringia. One example of such aid is helping transport all the camping, personal, and hunting gear needed for human survival in the high Arctic, equipment the likes of which the Siberian Mousterians did not possess. Dogs could have been used to pull sophisticated sleds with runners across more-or-less level ice and snow (Fig. 4.4), or pull simple two-pole travois sleds, sufficient for use in grasslands, or used to carry "saddle bags" in forests, rocky terrain, and other situations. Modern Na-Dene Indians of Alaska used saddle-bag-carrying dogs, as shown in a late-nineteenth-century postcard (Fig. 4.5). Oddly, dogs seem not to have filled the imaginations of artists who have depicted the migration of Siberians to Alaska, or as aids in hunting. As with modern Eskimo use of dogs to help hunt polar bear, late Pleistocene dogs could have aided in the hunt for game animals; after all, their behavioral evolution involved hunting in social packs. Until the introduction of modern snowmobiles, dogs were indispensible for human existence in the Old and New World Arctic.

**Fig. 4.4** One of several valuable uses that dogs serve across the Arctic. Malamute dog team and sled belonging to Patricia Bradley, driving sled. Jacqueline A. Turner and Korri Dee Turner, passengers. Temperature was ca. 0°F. Patricia Bradley is daughter of N. P. S. archaeologist Zorro Bradley and wife Natalie, Fairbanks, Alaska. These dogs are about the size of the Razboinich'ya dog (CGT color Fairbanks 12-30-75:34).

**Fig. 4.5** Modern Na-Dine Indians of Alaska used saddle-bag-carrying dogs, as shown in this late-nineteenth-century postcard. Photographer unknown.

Similarly, tailored clothing was mandatory in the sub-zero Arctic winter weather (Fig. 4.6). Tailored clothing was being manufactured at least 20 000 years ago in Siberia, as evidenced by fine bone needles, such as the one we found at Dvuglaska Cave and those from Afontova Gora. We also feel that shamans (Fig. 4.7) were necessary for the Beringian trek. Their visions of where to find game and other needed resources, based on environmental cues learned early in their practices, would have enhanced the probability of group survival. Their animism was not without a pragmatic basis. In addition, their awareness of community life and individual behaviors would have served their people to cope with personal problems and the depressing low light level of winter months. Indirectly, we propose that the occurrence of a dog skull in Razboinich'ya Cave is suggestive evidence of Siberian shamanism at least 30 000 years ago.

The Razboinich'ya bear skull (Figs. 3.135–3.136) is even more suggestive of shamanistic presence because of the widespread bear cult practiced all across the far north of Europe, Siberia, and North America (Fig. 3.137). Sokolova (2000:129) believes that the bear cult is an ancient ritual:

> The similarity of rites observed among all the peoples inhabiting the vast territories of Northern Europe and North America testifies to the extremely ancient origin of these ceremonies which stem from a respectful attitude to the bear and date back to the Upper Paleolithic.

We propose that the Razboinich'ya bear skull, which has no evidence whatsoever of hyena damage, and hence must have been placed in the cave by Paleolithic humans, is direct physical evidence for Sokolova's view. Having been found in close association with a directly dated dog skull, it means the cult could be as old as 30 000 BP.

Fig. 4.6 Finely tailored wind-tight leather clothing assembled with the aid of delicate bone needles must have been as important for survival in Siberian winters as it is for modern tribes living in the far north. Shown are beautifully crafted Evenk gloves, Anadyr River, Chukchi Peninsula. Made in the 1890s, these items belong to the Museum of Anthropology and Ethnography, St. Petersburg, Russia. Photographed in Smithsonian Institution's Natural History Museum (CGT color 1-10-89:24).

## A hyena barrier to Beringia?

We feel that demography lies at the heart of the late arrival of humans to the New World. Simply said, there could be little or no population expansion if there was little or no population growth. Instead of a possible expansion rate of, say, 30 km (20 miles) per generation of hunters, there would be no need for expansion if the number of hunters did not exceed the holding capacity of a given region. To expand, carnivore predation of humans or their food supply had to be reduced. A 1918 comment referred to by David Quammen (2003:38) reinforces one restricting factor in a "barrier" hypothesis, namely, child-napping, which would have contributed to the limiting of human population size in late Pleistocene Siberia:

> Careless mothers leave children unattended [remarked R. G. Burton in 1918] and a panther [leopard], skulking around the village outskirts after dark, even though not a confirmed man-eater, will always be ready to carry off a child if no one is watching.

Such a viewpoint fits well with Quammen's (2003:70) discussion of carnivore food size preferences and restrictions. Prey determination is based on what size animal a predator can kill and swallow. These restrictions apply to the solitary predators, but upper size

**Fig. 4.7** Inasmuch as shamanism was present throughout the New World, the practice must have been part of the late Pleistocene Paleo-Indian culture brought from Siberia. Some of the bone objects found at Mal'ta and Ust-Kova may have been used by shamans. Shown is a shaman doll, ca. 60 cm high, on sale at the Irkutsk Regional Museum gift shop (CGT color 10-1-00:29).

limit can easily be extended for the social predators – hyenas and wolves – whose effective mass is multiplied by pack size. The hyena's bone-crushing jaws and teeth circumvent the swallowing restriction of other carnivores.

While Pleistocene cave lions and bears must have been terrifying – for example, the menacing 30 000-year-old paintings of them in Chauvet Cave, France (Clottes 2003) – we wonder whether carnivore horror may not have been even greater for the killing packs of night-prowling cave hyenas and wolves. The unattended toddler would make a nice, easily snatched evening meal. Many animals are depicted in the cave and mobile art of the European Upper Paleolithic mammoth or Arctic steppe fauna: lion, feline forms, bear, rhinoceros, red deer, hind, chamois, horse, muskox, ibex, goat-sheep, reindeer, wolf, hare, megaloceros, mammoth, bison, auroch, and many others, including salmon, sole, other fish, seals, and birds, such as owls, and even a grasshopper (Leroi-Gourhan 1967). While we have failed to find any hyena portrayal in our literature search, Guthrie (2005:236–267) has done so. According to this eminent zoologist, whose book on Paleolithic art shows scores of mammals he identifies, he found depictions of hyenas to be very rare. He proposes only two possible hyena depictions. Keep in mind that Guthrie is a zoologist, not an art historian. Hence, his identifications carry considerable credibility. One image of a hyena was found at La Marche, France. However, there are no spots indicated as is the case of the second depiction, found at Parpallo, Spain. The Spanish scene shows a smaller animal attacking a larger and horned mammal. The attacking creature seems to be clenched to the throat of the prey. Guthrie suggests that the predator is a hyena because of the erect short tail and the spotted coat. With the bones and teeth of hyenas so abundant in the Paleolithic, one wonders what was the reason for the rarity of hyena depictions. Other Upper Paleolithic predators are rather commonly depicted. Fear and revulsion come quickly to mind, as discussed in our section on living hyenas. Anderson (1984:60) also finds "that there are few representations of the spotted hyena in Paleolithic art, [but there is] an ivory sculpture from La Madeleine."

Does this rarity of hyena images mean that cave hyenas were so unimportant that they did not merit depicting? Or were they so horrifying that it was too dangerous to portray them, even in the deep, dark sanctuaries of European limestone caves? Vereschagin and Baryshinkov (1984:496) noted that large amounts of cave hyena remains have been found in numerous European cave sites. Gary J. Galbreath (personal communication, September 14, 2007) also related that large numbers of hyenas have been found all over Europe, adding that "the individuals were quite large by modern hyena standards." Thus, there are a variety of reasons to be suspicious about the role hyenas might have played in Siberian late Pleistocene human demography. After all, carnivores and humans were competing for the same food and shelter resources.

There is another possible limiting factor for human population size in Siberia – human sacrifice to persistent and troublesome predators. Citing the 1925 edition of the 1825 writings of the self-educated Russian naturalist of Primorsky Territory and Manchuria, N. A. Baikov, Quammen (2003:356) related that an infant might be tied to a tree as a sacrifice to an aggressive marauding Amur tiger.

While hyenas and humans seem to have been strongly competitive associates in southern Siberia, one wonders if that association was maintained as Upper Paleolithic

Siberian people pushed further north toward Beringia in the late Pleistocene. Ovodov's finding of cave hyena remains and stone artifacts in Geographic Society Cave demonstrates that the association existed even as far east as the Sea of Japan. If eastern Siberian Upper Paleolithic people used the Amurian–Pacific continental shelf as a route to Beringia, one might expect the hyena–human relationship to continue beyond the 55° N limit because of the warming effect the Pacific Ocean would have had on the continental shelf. However, the association seems to have ceased before humans reached Alaska. Kurtén and Anderson (1980:198) report that bone-eating hyenids had reached North America, although their fossils are rare, and none has been identified as a cave hyena (*Crocuta spelaea*).

There are no known depictions of hyenas in Siberian rock or mobile art, even though images have been identified which are attributable to late Pleistocene–early Holocene fauna (Jacobson 2005:121). Jacobson notes that one Altai image can be confidently identified as having a Pleistocene age, namely that of a wooly rhinoceros.

In sum, we envision a diminishing food supply and shelter availability as one proceeds northward in Siberia. At some point the competition between humans and hyenas (and other carnivores) must have so intensified that the latter preyed on humans with increased frequency. The human groups were reduced in number to a level of bare survival. As the food resources diminished or became harder to come by with increased latitude, there must have been a zone in which we suggest the competition was so intense that humans were unable to proceed northward. At the same time, hyena groups were unable to expand due to the same limited food supply that both species competed for. We propose that the competitive struggle climaxed at 55–56° N, the northern limit of known hyena remains, and about the northern limit of dated late Pleistocene archaeological sites.

While climate may be the sole or main controlling factor for the northern limit of human occupation, as revealed by their archaeological residue in Siberia, we feel otherwise. The northern limit of Pleistocene hyenas seems to have been about 55–56° N. With one exception, Kuzmin (1994; see also Kuzmin and Tankersley 1996) lists no early Upper Paleolithic (39 000–24 000 BP) sites above 55° N. Kuzmin does list a number of archaeological sites north of 55° in latest Upper Paleolithic times (24 000–10 000 BP). One example is Ust-Kova at 58° N. The deepest cultural layer produced a radiocarbon date, based on charcoal, of 14 000 BP. Despite the recovery of a large faunal assemblage, there was very little carnivore damage and no direct or indirect evidence of hyenas at Ust-Kova.

It would be foolish on our part to over-emphasize the potential contribution of hyenas to the late human colonization of Beringia, but we want the reader to consider the possibility of such an additional inhibitor to the late peopling of the Americas relative to the spread of modern humans everywhere else in the world, except deep pelagic Oceania.

The reader may also be wondering how a study of perimortem bone damage could wind up with discussions of dogs, clothing, cults, shamans, missing burials, and human teeth. To our way of thinking the bone damage can be readily described, but for fuller understanding it has to be put into some sort of cultural, temporal, populational, and environmental context. This context involves issues that came up in the course of our data collection.

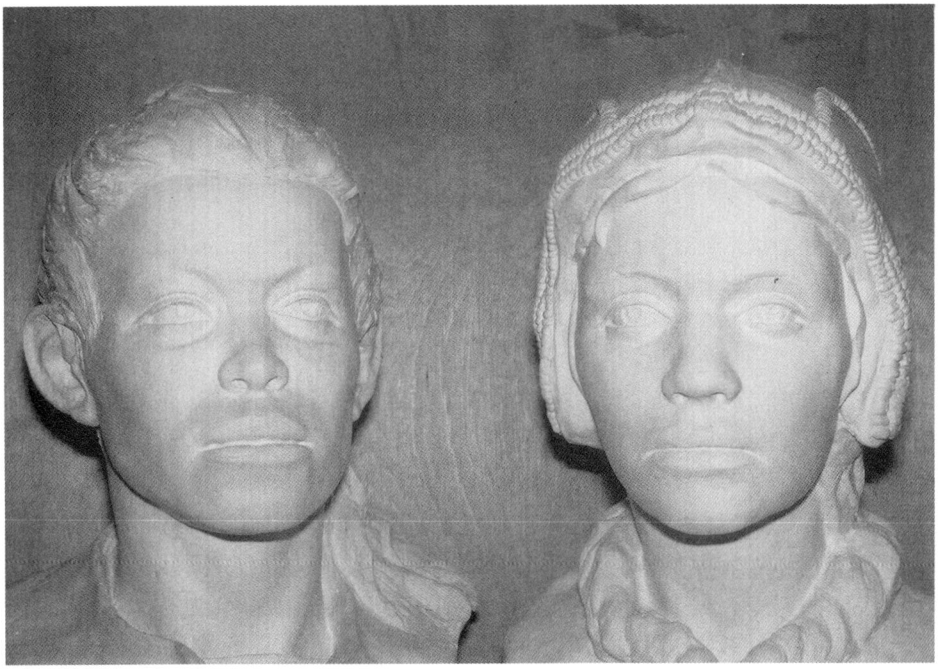

**Fig. 4.8** Facial reconstructions by M. M. Gerasimov of the Sunghir sub-adults. Their prognathism follows from the fact that they had large teeth. Although large, they were simplified and morphologically like teeth of modern Europeans. These reconstructions are the best ever made, showing what Cro-Magnons looked like. Given their dental similarity to that of the Mal'ta child, had he or she lived to a comparable age, they would likely have looked similar to the Sunghir pair (CGT color LPR 1-6-81:19).

This context showed that hyenas were a significant part of the southern Siberian landscape, perhaps so much so that they contributed to the limiting of the Mousterian hold on Siberia, allowing the entry and expansion of bands of European Cro-Magnon people to as far east as Lake Baikal (Mal'ta). Upper Paleolithic burials in European Russia, such as Sunghir (Figs. 4.8–4.12), are dentally very much like modern Europeans and Mal'ta. A second prong of modern humans entered far eastern Siberia. These East Asians were dentally very unlike the Europeans. Called Sinodonts, these anatomically modern humans reached the New World, whereas the Cro-Magnons and the Siberian Mousterians did not.

The human and animal figurines found at Mal'ta and Ust-Kova, both open sites evidencing butchering of many hunted animals, proclaim much more symbolic activity than the inhabitants at earlier sites such as Denisova and Okladnikov caves.

However, the marked differences in carnivore bone damage between the earlier and later assemblages strongly suggest that carnivores were less of a threat in later than in earlier sites. Perhaps 20 wolves prowling at night about the Mal'ta and Ust-Kova camp fires would have been less of a terror than 20 larger cave hyenas (Fig. 4.13).

A great deal of government-funded archaeological excavation has taken place in Siberian late Pleistocene sites. The first discovery of Siberian Paleolithic stone tools

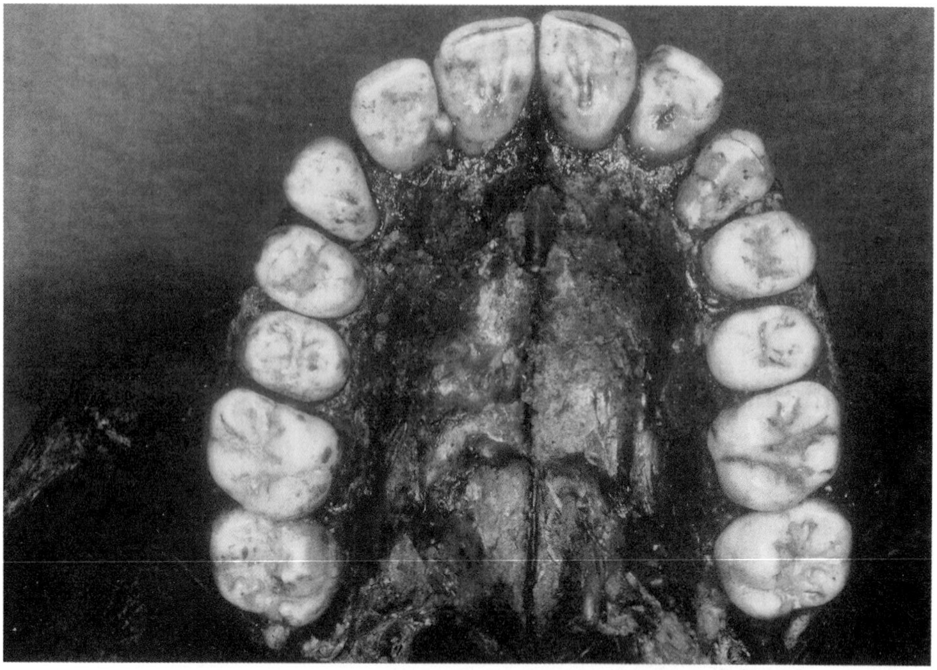

**Fig. 4.9** Maxillary teeth, Sunghir A (12–15-year-old male). Like modern Europeans, Sunghir A has weak central incisor shoveling, no double-shoveling, a reduced breadth of the lateral incisors, strong expression of Carabelili's trait on the first molar, no cusp 5 on the first molars, no first molar enamel extension, and other similarities. These features are also found in the Mal'ta child (Fig. 3.82) (CGT color LPR 12-25-80:1).

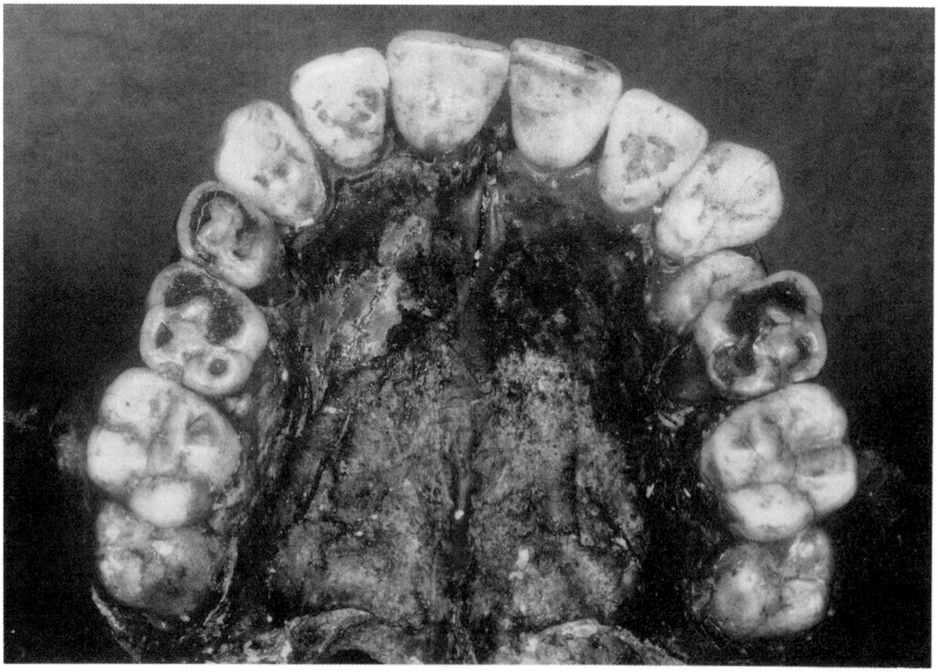

**Fig. 4.10** Maxillary teeth, Sunghir B (9–10-year-old female?). The same comments on the dental morphology of Sunghir A also apply here (CGT color LPR 12-25-80:5).

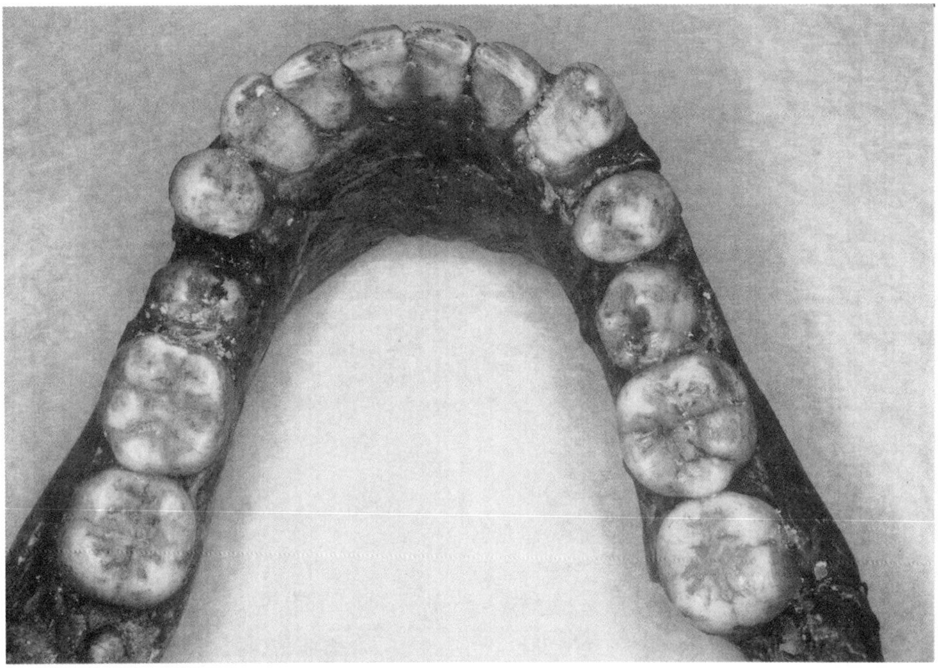

**Fig. 4.11** Mandibular teeth of Sunghir A. Most notable, and like modern Europeans, is the absence of the protostylid and cusps 6 and 7 on the first molars, and lack of cusp 5 on the second molars. These features are also found in the Mal'ta child (Fig. 3.83) (CGT color LPR 12-25-80:4).

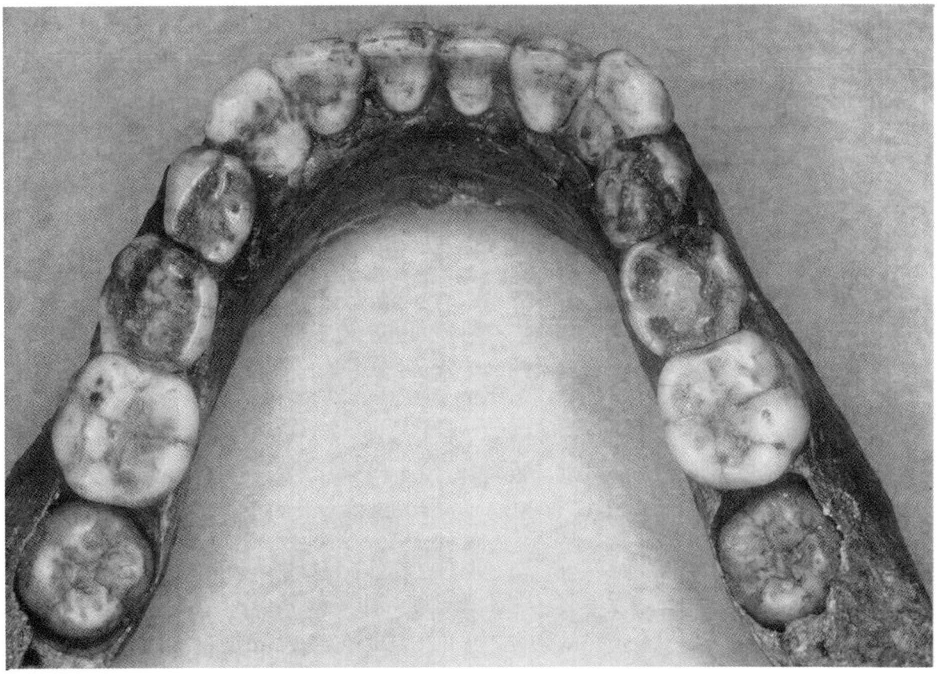

**Fig. 4.12** Mandibular teeth of Sunghir B. The same comments apply as on the dental morphology of Sunghir A (CGT color LPR 12-25-80:7).

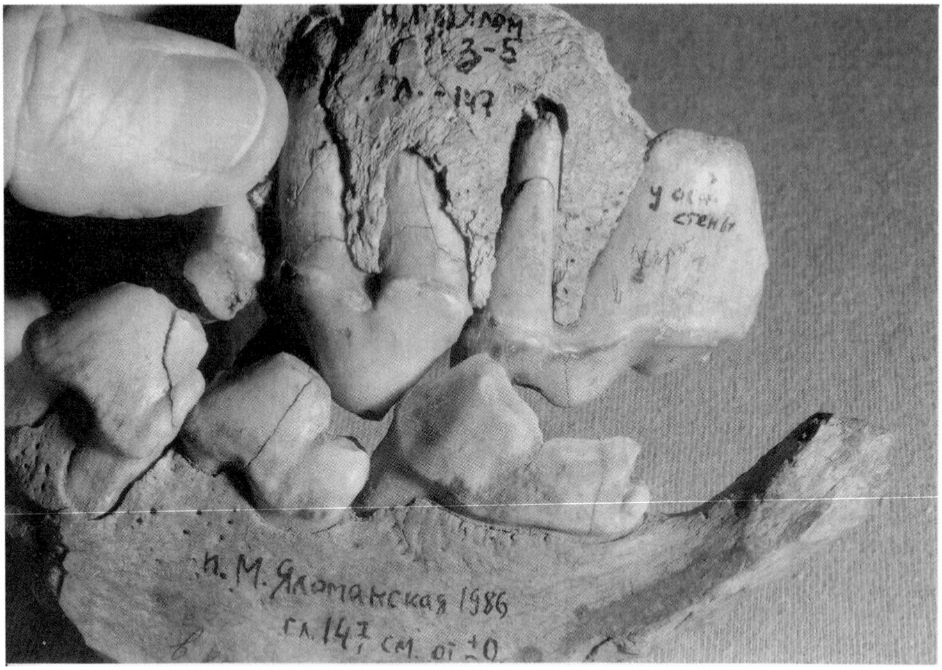

**Fig. 4.13** Hyena jaw fragments from Maly Yaloman Cave (1986). We conclude our illustrations with this photograph to make the point that it was the large quantity of perimortem bone damage caused by carnivores that enabled us to propose that this Altai cave site, as well as Okladnikov Cave, despite their stone artifacts, were used as much by carnivores as by humans. For example, 48% of the Maly Yaloman bone assemblage had hyena stomach-acid erosion, and the amount of carnivore bone damage at Maly Yaloman was equal to that found in the Razboinich'ya hyena cave. As we came to realize, hyenas must have been a major feature of the Siberian landscape, so much so that we envision them as having contributed significantly to the late human colonization of the New World, and contributing to the discontinuous occupation of Siberian regions, which allowed nomadic Upper Paleolithic Europeans to replace the Siberian Mousterian Neanderthals (CGT neg. IAE 8-5-03:35).

occurred in 1871 in Irkutsk, at what is now called the Military Hospital site. Since then the recovery of Paleolithic remains from cave and open sites easily totals in the tens of thousands of stone artifacts, cores, debitage, and several bone artifacts, including human figurines, carvings of waterfowl, scratched drawings of mammoth, unique dog and fox burials, etc. In contrast, the number of human bones and teeth that have been recovered is disproportionately (and disappointingly) small.

A combination of technology, dog domestication, climate and vegetation change, megafaunal and hyena extinctions, and the arrival of anatomically modern human Sinodonts from Northeast Asia were all necessary elements for explaining the late colonization of the Americas. We emphasize the competitive role that hyenas may have played in the lives of late Pleistocene Siberians.

At a drizzly late night vodka-fueled field campfire discussion, one of our 20 or so Russian colleagues replied, after being asked why the New World was colonized so late:

"Why would anyone want to leave the paradise of southern Siberia?" Our bone damage study cannot answer that question, but it does offer a scenario of why the colonization was so late in human migratory history. Natural selection molded our species to explore. We are now robotically searching out our distant neighboring planet, Mars. If we should discover bones, Nicolai, Olga, and I will stand ready to help.

# 5 Conclusions for seven questions

This is a detective story, but a rather odd one. The clues are bones.

(C. K. Brain 1981:3)

Our selection of >9000 pieces of bone for this study, out of at least 1 000 000 pieces in 30 assemblages, was based on each having a 2.5 cm minimum diameter, and minimal plant root damage. Each selected piece was scored for 26 variables. This study is based on >230 000 observations.

*Question 1* What was the perimortem bone damage signature of late Pleistocene Siberian carnivores?

*Conclusion 1* The diagnostic damage done by big carnivores, especially hyenas, includes tooth dints, tooth scratches, end- and mid-shaft polishing, end-hollowing, notching, and stomach bones. While we think of hyenas as being bone-breakers, so also were late Pleistocene humans. Statistically speaking, bone breakage is not a reliable feature for identifying the breaker. Tooth dints, tooth scratches, and stomach bones are, in combination, the best indicators of hyena and other large carnivore presence.

*Question 2* What was the perimortem bone damage signature of late Pleistocene Siberian humans?

*Conclusion 2* Human causation of bone damage is best recognized by the occurrence of cut marks, chop marks, and burning. However, burning is very rare in our assemblages. But if archaeological context can rule out natural burning, such as lightning-ignited steppe or forest fires, burning is a prime indicator of human presence.

*Question 3* How was meat prepared for eating?

*Conclusion 3* It follows from conclusion 2 that roasting was almost never involved in the perimortem damage of our assemblages, at least on-site. Thus, we must rule out roasting as a regular means of cooking meat. We feel that if meat had been cooked at all, it must have been done with stone-boiling in perishable containers. But given the absence of fire-altered stew-soup heating stones, bowl-shaped stone pots, earth ovens, or in-skin roasting of small animals, meat was generally not cooked by late Pleistocene Siberians. Pollen analyses of our sites do not suggest that flowering plant materials contributed much to the average diet.

*Question 4* What happened to the remains of >1000 generations of Siberians (at least 1 000 000 individuals)?

*Conclusion 4* Given the excellent preservation of non-human faunal remains in our assemblages, a mystery quickly arose. Why are there almost no human remains in our assemblages? Possibilities include: off-site burial locations, cannibalism, no burial, on-site burials followed by hyena scavenging. Because we have one human tooth from Denisova Cave that had been swallowed and acid-eroded like our hyena stomach bones, we favor the latter inference. In a shallow animal cave ca. 100 m from Okladnikov Cave, Ovodov found some human bone that probably came from Okladnikov Cave. The bone must have been carried to the animal cave by some carnivore, presumably a hyena, who dug it up from its burial site.

*Question 5* Who were the late Pleistocene inhabitants of Siberia?

*Conclusion 5* The human remains in our Middle Paleolithic assemblages include a very small number of fragmentary bones and teeth, mainly from Okladnikov Cave. The bone has almost no morphological information for affinity assessment, but a few of the teeth can be identified as Neanderthal or Neanderthal-like. These teeth came from Denisova and Okladnikov caves. An Upper Paleolithic child's dentition from the more recent Mal'ta site looks decidedly like those of Cro-Magnons and modern Europeans. The molar teeth from the Listvenka site look like those of anatomically modern and living Northeast Asians and Native Americans. We suggest that the population history of Siberia appears to be like that of late Pleistocene Europe – namely, replacement, not local evolution. We expect skeptics will say this inference is based on much too small a sample. Our response is that, statistically speaking, these teeth make up the known universe of late Pleistocene Siberian human teeth. They are not a sample. DNA analysis from a single Denisova finger bone proved to be Neanderthal.

*Question 6* Did hyenas by their burrowing activity blur the stratigraphy of cave and open sites, hence misleading researchers into believing in cultural continuity and local evolution?

*Conclusion 6* The competition between humans and hyenas in Africa is today quite intense, which must have been the case also in the Siberian Paleolithic. Cave and open archaeological sites show considerable use and disturbance by hyenas. We propose that what may look like local cultural evolution based on gradual changed ways of manufacturing stone artifacts from Middle to Upper Paleolithic times was caused by hyena activity mixing sharp, distinct strata into a blur where non-evolutionary stratigraphic borders have been erased. The few Pleistocene human teeth suggest there had been population replacement, not local evolution. Do a few teeth trump thousands of stone artifacts? Yes, if the archaeology does not take into account the possibility of bioturbation.

*Question 7* Was there a hyena barrier to Beringia?

*Conclusion 7* While controversy swirls about the date when Paleo-Siberians first crossed Beringia and entered the New World, the most convincing evidence is that their arrival into Alaska began about 13 500 years ago. This is the time that the Pleistocene was ending, and northern environmental conditions were shifting from Arctic (mammoth) steppe to tundra and taiga. The Bering Sea was on the rise, flooding coastlines in Siberia and Alaska. This is about the same time that hyenas are believed to have gone extinct. The northern limit where hyena remains have been found in Siberia is ca. 55–56° N. With a few inconclusive exceptions this is also the northern limit where late Pleistocene Siberian archaeological sites have been discovered. This concordance leads us to believe that a sort of hyena barrier to Beringia existed. We envision the barrier to have been more like hit-and-run guerilla warfare than like the sharply drawn borders of trench warfare during World War I in Europe. As Paleo-Siberians and hyenas competed at 55–56° N latitude for the same limited resources, their mutual struggle for survival must have intensified. Food scarcity perhaps drove hyenas to kill and eat more often unprotected children and weak elderly individuals, thus slowing down northern expansion by humans because of insufficient population growth. The rate of northward human population expansion was without question also influenced by Arctic cold and patchy resources, but importantly, by the heightened competition with carnivores, especially pack-hunting hyenas. Reaching western Beringia had to wait until the hyena population thinned and they eventually went extinct along with mammoths and other large animals that hyenas opportunistically preyed upon.

So, this is the account of our Siberian odyssey that subconsciously began on the Amur River in 1979, and certainly began gestating by 1984 when the senior author first met Ovodov and Pavlova. By 1987 our odyssey led along a muddy mountain road into the Altai. In 1998 we were formally collecting perimortem damage information. Data collection ceased in 2006 because of the senior author's deteriorating vision. We hope the reader will find our account interesting, useful, and sufficiently inspiring to venture forth in a similar way elsewhere in the world.

## Another tale of winter death

Late one snowy winter afternoon, before the collapse of the Soviet Union, three speleology hobbyists were out searching for new caves. They were in the cavern-pocked limestone karst terrain near the middle Yenisei River. With night approaching, they were about to call it quits as the pale sunlight turned watery evening gray, when the dog with them, named Volchik (little wolf), barked where it had come upon a shallow depression near the summit of a low limestone hill. It suggested that beneath there might be a cave, access to which was blocked. After a few minutes of digging, the depression suddenly collapsed. Experience had taught them to be careful in situations like this, so the digger was tethered to a rope held by the other men, who stood away from the depression that was now a gaping, smooth-walled, circular black hole about half a meter wide, down into which the digger might have disappeared.

The digger descended into the near-vertical shaft aided by the rope the other two held. The descent was some 15 m to the bottom. Eons of earthquake movement and subsequent erosion had transformed this portion of an ancient subterranean water channel into a vertical animal trap.

At the base of the shaft was a grotto with a small, cone-shaped mound of loose sediments that had fallen from the surface. Shining his light around, the digger saw he was in a large grotto from which exited other tubes. The grotto floor was littered with bones of small animals he recognized as field mice, shrews, hares, bats, and remains of small carnivores – weasel, fox, and sable. What caught his eye next was the complete skeleton of a large wolf-sized animal. The speleologists had found animal remains in other caves, and knew how to excavate them in careful archaeological ways. The speleologist also knew that a complete skeleton was an important paleontological find. He carefully dug up and bagged the remains, which he carried to the surface, now almost as dark and cold as in the grotto. Eventually a vertebrate paleontologist many kilometers down-river learned of the find and acquired the bones for study.

His scientific examination determined that it was a big animal (at least 90 cm at the shoulder), old (teeth much worn, osteophytic growths on some joints), probably female (relative size), and cave hyena (dental morphology, cranial shape). The skeleton had to be at least 13 000 years old because cave hyenas in northern Eurasia went extinct by that time. It was evident that the animal had accidentally fallen into the trap and was unable to climb back out, or the fall had killed it. Aside from the anatomical facts he had gleaned from his examination, there was little else that could be inferred about its life history.

He had no way of knowing that this hyena had led a winter attack on humans in Two-Eyed Cave more than 20 000 years ago. Nor could he tell that she had been the alpha female of her clan, whose vast territory included the attack cave and the animal trap. He presumed that her clan was fairly well off, because their steppe-forest habitat would have supported large herds of horses, bison, and other large grazing animals. Nor could he know that she was alone for many seasons, the only remaining member of her clan. The others had all been poisoned to death by eating the many small fish they easily caught near the shores of shallow coves along the great river. These fish were infected with parasites that secreted a neurotoxin that destroyed their ability to swim much below the water's surface, and then only in a floppingly backward circular manner for a few days before they died. It was much easier for the clan to wade into the shallow waters and lap up mouthfuls of these tasty dying fish than to chase down the big game animals out on the steppe grasslands. Old Long Clitoris never ate creatures that lived in water or air, so she escaped the convulsive poisoning that killed all the others of her clan.

Alone, she had to scavenge more than hunt. Her pristine teeth began to wear and chip from cracking open scavenged bones for marrow grease, something she never had to do as alpha female. Her search for food slowed because her joints were stiff and painful.

Eventually, she could no longer hunt; instead, she wandered aimlessly, scavenging whatever animal remains she came across or occasionally catching some small rodent. Even field mice became too difficult to catch, so her general health deteriorated with malnutrition. Alone, slow, weak, she had become easy prey for wolves.

It was snowing hard the evening a pack of wolves actually caught her scent. She smelled them also, and turned away in an attempt to reach the forest edge, where she had a better chance of a stand-off than being out on the open rocky hillside. She hobbled as fast as she could, with fear as her stimulant. Blinded in part by the whipping snow, the growing darkness, and her failing vision, she did not see the black hole until her body bounced crazily back and forth against the sides of the lightless animal trap. She landed hard, with near-instant pain shooting throughout her crumpled body. She lay and slowly decayed in that position undisturbed for thousands of years.

When discovered and examined there were no clues to tell that these bones were the remains of Old Long Clitoris.

Like most things archaeological, very little remained to piece together a life history for her. Nevertheless, the animal-loving paleontologist imagined that she had in life occupied an important social position. Believing this, he respectfully placed her skull in his little lab on a ceiling-high shelf next to a shiny silver samovar and a time-stained wooly rhinoceros skull. This high shelf was a special place, almost shrine-like. She was once again an alpha female.

Figs. 2.10 and 2.12 show Old Long Clitoris.

# Appendix 1

## Tables

**Table A1.1** The faunal assemblages

| Site name | Site location | No. pieces studied | Pieces studied % | Nature of site | Year of perimortem taphonomy study |
|---|---|---|---|---|---|
| 1 Afontova Gora | Krasnoyarsk | (5000*) | – | Open with dwelling structures | 2006 |
| 2 Boisman II | South of Vladivostok | 1247 | 14.1 | Shell mound | 1999, 2000 |
| 3 Bolshoi Yakor I | Lake Baikal | 263 | 3.0 | Open kill camp | 2001 |
| 4 Borabashevskaya | South of Vladivostok | 30 | 0.3 | Temporary rock shelter? | 1999 |
| 5 Denisova | Altai Mountains | 116 | 1.2 | Human and hyena cave | 1999 |
| 6 Dvuglaska | Khakasia | 205 | 2.3 | Human and hyena cave | 2000 |
| 7 Gosudarev Log I | Krasnoyarsk | 14 | 0.1 | Open kill site? | 2000 |
| 8 Kamenka | East of Ulan-Ude | 552 | 6.3 | Human open camp | 2003 |
| 9 Kaminnaya | Altai Mountains | 771 | 8.7 | Human and hyena cave | 2000 |
| 10 Kara-Bom | Altai Mountains | 108 | 1.2 | Human open camp | 2001, 2003 |
| 11 Kirkalinskaya | Altai Mountains | 4 | 0.04 | Human and animal cave | 1999, 2000 |
| 12 Krasny Yar | Ob River, Novosibirsk | 203 | 2.3 | River deposit | 2000 |
| 13 Kurla I | Northwest coast of Lake Baikal | 72 | 0.8 | Human open camp | 2000 |
| 14 Malaya Seeya | Middle Yenisei | 144 | 1.6 | Human open camp | 1999 |
| 15 Mal'ta | South of Irkutsk | 186 | 2.1 | Human open lakeside camp | 2000 |
| 16 Maly Yaloman | Altai Mountains | 178 | 2.0 | Human and hyena cave | 2003 |
| 17 Mokhovo Mine | Kemerovo, Tom River | 12 | 0.1 | Open bison and human site | 2000 |
| 18 Nizhneudinskaya | Irkutsk | 36 | 0.4 | Animal trap, cave | 2001 |
| 19 Okladnikov*** | Altai Mountains | 168 | 1.9 | Human and hyena cave | 1999, 2000, 2003 |
| 20 Proskuryakova | Middle Yenisei | 47 | 0.5 | Hyena cave | 1999 |
| 21 Razboinich'ya | Altai Mountains | 739 | 8.4 | Hyena cave | 1998, 1999, 2000, 2003 |
| 22 Sarala | Khakasia | 44 | 0.5 | Mostly animal cave | 2001 |
| 23 Shestakovo | Southeast of Tomsk | 159 | 1.8 | Mammoth death site | 2001, 2002 |
| 24 Straschnaya | Altai Mountains | 527 | 6.0 | Hyena and human cave | 2001 |
| 25 Ust-Kan | Altai Mountains | 1654 | 18.5 | Human and hyena cave | 1999, 2000, 2003 |
| 26 Ust-Kova | North of Irkutsk | 116 | 1.3 | Open kill camp | 1999 |
| 27 Varvarina Gora | East of Ulan-Ude | 864 | 9.8 | Human open camp | 2003 |
| 28 Volchiya Griva | West of Novosibirsk | 113 | 1.3 | Mammoth death site | 2002 |
| 29 Yelenev | Yenisei, south of Krasnoyarsk | 241 | 2.7 | Human and fox cave | 2000 |
| 30 Zhemchuzhnaya** | Krasnoyarsk | 0 | 0.0 | Animal trap, cave | 2001 |
| Total pieces studied | – | 8813 (total) | – | – | – |

* 1 Afontova Gora: incompletely studied; not included in project total.
** 30 Zhemchuzhnaya: all bones were covered with hard mineral deposit and could not be studied without destroying the mineral coating.
*** Okladnikov damage = 655 pieces.

**Table A1.2a** Species (piece count, %)

| Site | No. | Artiodactyl | Badger | Bear | Beaver | Big mammal | Bird | Bison | Camel | *Canis* | *Capra* |
|---|---|---|---|---|---|---|---|---|---|---|---|
| 1 Afontova Gora | (5000) | – | – | – | – | – | – | – | – | – | – |
| 2 Boisman II | 840 | 0.0 | 0.0 | 0.0 | 0.0 | 10.4 | 5.6 | 0.0 | 0.0 | 8.2 | 0.0 |
| 3 Bolshoi Yakor I | 263 | 0.0 | 0.0 | 0.0 | 0.0 | 4.2 | 0.0 | 0.0 | 0.0 | 0.0 | 0.0 |
| 4 Borabashevskaya | 30 | 0.0 | 0.0 | 0.0 | 0.0 | 0.0 | 0.0 | 0.0 | 0.0 | 0.0 | 0.0 |
| 5 Denisova | 116 | 0.7 | 0.0 | 2.6 | 0.0 | 0.9 | 0.0 | 3.4 | 0.0 | 0.0 | 0.0 |
| 6 Dvuglaska | 205 | 0.0 | 0.0 | 0.0 | 0.0 | 4.9 | 0.0 | 2.4 | 0.0 | 0.0 | 0.0 |
| 7 Gosudarev Log I | 14 | 0.0 | 0.0 | 0.0 | 0.0 | 92.9 | 0.0 | 0.0 | 0.0 | 0.0 | 0.0 |
| 8 Kamenka | 552 | 0.0 | 0.0 | 0.0 | 0.0 | 7.2 | 0.0 | 2.0 | 0.4 | 0.0 | 0.0 |
| 9 Kaminnaya | 770 | 0.0 | 0.0 | 0.1 | 0.0 | 8.1 | 0.3 | 0.9 | 0.0 | 0.1 | 0.0 |
| 10 Kara-Bom | 108 | 0.0 | 0.0 | 0.0 | 0.0 | 23.1 | 0.0 | 1.8 | 0.0 | 0.9 | 0.0 |
| 11 Kirkalinskaya | 4 | 0.0 | 0.0 | 0.0 | 0.0 | 0.0 | 0.0 | 0.0 | 0.0 | 0.0 | 0.0 |
| 12 Krasny Yar | 203 | 0.0 | 0.0 | 7.4 | 5.4 | 0.0 | 0.0 | 5.9 | 0.0 | 0.0 | 0.0 |
| 13 Kurla I | 72 | 0.0 | 0.0 | 0.0 | 0.0 | 0.0 | 0.0 | 0.0 | 0.0 | 0.0 | 0.0 |
| 14 Malaya Seeya | 144 | 0.0 | 0.0 | 0.0 | 0.0 | 0.0 | 0.0 | 0.0 | 0.0 | 0.0 | 0.0 |
| 15 Mal'ta | 186 | 0.0 | 0.0 | 0.0 | 0.0 | 10.7 | 0.0 | 4.3 | 0.0 | 0.0 | 0.0 |
| 16 Maly Yaloman | 178 | 0.0 | 0.0 | 0.0 | 0.0 | 1.7 | 0.0 | 0.0 | 0.0 | 0.0 | 55.1 |
| 17 Mokhovo Mine | 12 | 0.0 | 0.0 | 0.0 | 0.0 | 0.0 | 0.0 | 50.0 | 0.0 | 0.0 | 0.0 |
| 18 Nizhneudinskaya | 36 | 0.0 | 0.0 | 94.4 | 0.0 | 0.0 | 0.0 | 0.0 | 0.0 | 0.0 | 0.0 |
| 19 Okladnikov | 168 | 3.4 | 0.6 | 0.0 | 0.0 | 10.1 | 0.0 | 1.2 | 0.0 | 0.0 | 0.0 |
| 20 Proskuryakova | 47 | 0.0 | 0.0 | 0.0 | 0.0 | 0.0 | 0.0 | 0.0 | 0.0 | 0.0 | 0.0 |
| 21 Razboinich'ya | 638 | 0.0 | 0.2 | 1.2 | 0.0 | 1.2 | 0.8 | 0.0 | 0.0 | 0.0 | 0.6 |
| 22 Sarala | 44 | 0.0 | 0.0 | 0.0 | 0.0 | 9.1 | 0.0 | 0.0 | 0.0 | 0.0 | 0.0 |
| 23 Shestakova | 159 | 0.0 | 0.0 | 0.0 | 0.0 | 0.0 | 0.0 | 3.8 | 0.0 | 0.0 | 0.0 |
| 24 Straschnaya | 529 | 0.0 | 0.0 | 0.0 | 0.0 | 6.4 | 0.0 | 0.0 | 0.0 | 0.0 | 0.0 |
| 25 Ust-Kan | 1654 | 0.0 | 0.0 | 0.0 | 0.0 | 3.6 | 0.0 | 0.06 | 0.0 | 0.0 | 0.0 |
| 26 Ust-Kova | 116 | 0.0 | 0.0 | 0.0 | 0.0 | 0.0 | 0.0 | 0.0 | 0.0 | 0.0 | 0.0 |
| 27 Varvarina Gora | 864 | 0.0 | 0.0 | 0.0 | 0.0 | 17.1 | 0.0 | 0.2 | 0.0 | 0.0 | 0.0 |
| 28 Volchiya Griva | 113 | 0.0 | 0.0 | 0.0 | 0.0 | 0.0 | 0.0 | 25.7 | 0.0 | 0.0 | 0.0 |
| 29 Yelenev | 241 | 0.0 | 0.0 | 0.0 | 0.0 | 0.0 | 0.0 | 0.0 | 0.0 | 0.0 | 0.0 |
| 30 Zhemchuzhnaya | 0 | – | – | – | – | – | – | – | – | – | – |

**Table A1.2b** Species (piece count, %)

| Site | No. | Carnivore | Big deer | Small deer | Elk | Fox | Gazelle | Goat-sheep | Hare |
|---|---|---|---|---|---|---|---|---|---|
| 1 Afontova Gora | (5000) | – | – | – | – | – | – | – | – |
| 2 Boisman II | 840 | 0.7 | 0.0 | 1.3 | 0.0 | 0.2 | 0.0 | 0.0 | 0.0 |
| 3 Bolshoi Yakor I | 263 | 0.0 | 0.0 | 0.0 | 0.0 | 0.8 | 0.0 | 0.0 | 2.7 |
| 4 Borabashevskaya | 30 | 0.0 | 0.0 | 0.0 | 0.0 | 0.0 | 0.0 | 0.0 | 0.0 |
| 5 Denisova | 116 | 0.9 | 0.0 | 1.7 | 0.0 | 0.0 | 0.0 | 7.8 | 0.0 |
| 6 Dvuglaska | 205 | 0.0 | 0.0 | 0.0 | 0.0 | 0.0 | 0.0 | 0.0 | 0.0 |
| 7 Gosudarev Log I | 14 | 0.0 | 0.0 | 0.0 | 0.0 | 0.0 | 0.0 | 0.0 | 0.0 |
| 8 Kamenka | 552 | 0.0 | 0.0 | 1.3 | 0.0 | 0.0 | 57.1 | 0.7 | 0.0 |
| 9 Kaminnaya | 770 | 0.0 | 0.4 | 0.8 | 0.0 | 0.1 | 0.0 | 2.8 | 0.0 |
| 10 Kara-Bom | 108 | 0.0 | 0.0 | 0.9 | 0.0 | 0.0 | 0.9 | 0.0 | 3.7 |
| 11 Kirkalinskaya | 4 | 0.0 | 0.0 | 0.0 | 0.0 | 0.0 | 0.0 | 100.0 | 0.0 |
| 12 Krasny Yar | 203 | 0.0 | 3.9 | 7.9 | 26.6 | 0.0 | 0.0 | 0.0 | 0.0 |
| 13 Kurla I | 72 | 0.0 | 0.0 | 0.0 | 0.0 | 0.0 | 0.0 | 0.0 | 0.0 |
| 14 Malaya Seeya | 144 | 0.0 | 0.0 | 0.0 | 0.0 | 0.0 | 0.0 | 66.7 | 0.0 |
| 15 Mal'ta | 186 | 0.0 | 3.2 | 0.0 | 0.0 | 0.0 | 0.0 | 0.0 | 0.0 |
| 16 Maly Yaloman | 178 | 0.6 | 1.1 | 1.1 | 0.0 | 7.9 | 0.0 | 8.4 | 0.6 |
| 17 Mokhovo Mine | 12 | 0.0 | 0.0 | 0.0 | 0.0 | 0.0 | 0.0 | 0.0 | 0.0 |
| 18 Nizhneudinskaya | 36 | 0.0 | 0.0 | 0.0 | 0.0 | 0.0 | 0.0 | 0.0 | 0.0 |
| 19 Okladnikov | 168 | 0.0 | 0.0 | 0.0 | 0.0 | 0.0 | 0.0 | 16.7 | 0.0 |
| 20 Proskuryakova | 47 | 0.0 | 0.0 | 0.0 | 0.0 | 0.0 | 0.0 | 0.0 | 0.0 |
| 21 Razboinich'ya | 638 | 0.0 | 0.0 | 0.2 | 0.0 | 31.0 | 0.0 | 2.5 | 0.0 |
| 22 Sarala | 44 | 0.0 | 0.0 | 0.0 | 0.0 | 0.0 | 0.0 | 18.2 | 0.0 |
| 23 Shestakova | 159 | 0.0 | 0.0 | 0.0 | 0.0 | 0.0 | 0.0 | 0.0 | 0.0 |
| 24 Straschnaya | 529 | 0.0 | 0.0 | 0.0 | 0.0 | 0.4 | 0.0 | 0.0 | 0.0 |
| 25 Ust-Kan | 1654 | 0.0 | 0.0 | 0.0 | 0.0 | 0.6 | 0.0 | 0.1 | 0.0 |
| 26 Ust-Kova | 116 | 0.0 | 0.0 | 0.0 | 0.9 | 0.0 | 0.0 | 0.0 | 0.0 |
| 27 Varvarina Gora | 864 | 0.0 | 0.0 | 0.0 | 0.0 | 0.0 | 32.3 | 4.4 | 0.1 |
| 28 Volchiya Griva | 113 | 0.0 | 0.0 | 0.0 | 0.0 | 0.0 | 0.0 | 0.0 | 0.0 |
| 29 Yelenev | 241 | 0.0 | 0.0 | 0.0 | 0.0 | 0.4 | 0.0 | 24.9 | 0.0 |
| 30 Zhemchuzhnaya | 0 | – | – | – | – | – | – | – | – |

**Table A1.2c** Species (piece count, %)

| Site | No. | Herbivore | Horse | Hyena | Lion | Mammoth | Marmot | Medium mammal | Pig | Reindeer |
|---|---|---|---|---|---|---|---|---|---|---|
| 1 Afontova Gora | (5000) | – | – | – | – | – | – | – | – | – |
| 2 Boisman II | 840 | 0.0 | 0.0 | 0.0 | 0.0 | 0.0 | 0.0 | 7.9 | 0.7 | 0.0 |
| 3 Bolshoi Yakor I | 263 | 0.0 | 0.0 | 0.0 | 0.0 | 0.0 | 0.0 | 0.0 | 0.0 | 0.4 |
| 4 Borabashevskaya | 30 | 0.0 | 0.0 | 0.0 | 0.0 | 0.0 | 0.0 | 0.0 | 0.0 | 0.0 |
| 5 Denisova | 116 | 0.0 | 5.2 | 0.0 | 0.0 | 0.4 | 0.0 | 0.0 | 0.0 | 0.0 |
| 6 Dvuglaska | 205 | 0.0 | 0.5 | 0.0 | 0.0 | 0.0 | 0.0 | 0.0 | 0.0 | 0.0 |
| 7 Gosudarev Log I | 14 | 0.0 | 7.1 | 0.0 | 0.0 | 0.0 | 0.0 | 0.0 | 0.0 | 0.0 |
| 8 Kamenka | 552 | 0.2 | 4.7 | 0.0 | 0.0 | 0.0 | 0.0 | 3.8 | 0.0 | 0.0 |
| 9 Kaminnaya | 770 | 0.0 | 3.6 | 1.3 | 0.1 | 0.1 | 2.5 | 3.1 | 0.0 | 0.1 |
| 10 Kara-Bom | 108 | 0.9 | 0.9 | 0.9 | 0.0 | 1.8 | 11.1 | 6.5 | 0.0 | 0.0 |
| 11 Kirkalinskaya | 4 | 0.0 | 0.0 | 0.0 | 0.0 | 0.0 | 0.0 | 0.0 | 0.0 | 0.0 |
| 12 Krasny Yar | 203 | 0.0 | 18.2 | 0.0 | 7.9 | 1[a] | 0.0 | 0.0 | 0.0 | 3.0 |
| 13 Kurla I | 72 | 0.0 | 0.0 | 0.0 | 0.0 | 0.0 | 0.0 | 0.0 | 0.0 | 100.0 |
| 14 Malaya Seeya | 144 | 0.0 | 8.3 | 0.0 | 0.0 | 0.0 | 0.0 | 0.0 | 0.0 | 25.0 |
| 15 Mal'ta | 186 | 0.0 | 1.1 | 0.0 | 0.0 | 49.5 | 0.0 | 0.0 | 0.0 | 9.1 |
| 16 Maly Yaloman | 178 | 0.0 | 4.5 | 9.5 | 0.0 | 0.0 | 1.1 | 0.0 | 0.0 | 0.0 |
| 17 Mokhovo Mine | 12 | 0.0 | 0.0 | 0.0 | 0.0 | 0.0 | 0.0 | 0.0 | 0.0 | 0.0 |
| 18 Nizhneudinskaya | 36 | 0.0 | 0.0 | 0.0 | 0.0 | 0.0 | 0.0 | 0.0 | 0.0 | 0.0 |
| 19 Okladnikov | 168 | 0.0 | 4.2 | 14.3 | 0.0 | 5.4 | 0.0 | 0.0 | 0.0 | 0.0 |
| 20 Proskuryakova | 47 | 0.0 | 0.0 | 51.1 | 0.0 | 0.0 | 0.0 | 0.0 | 0.0 | 0.0 |
| 21 Razboinich'ya | 638 | 0.0 | 0.2 | 11.7 | 0.0 | 0.0 | 0.2 | 0.0 | 0.0 | 0.0 |
| 22 Sarala | 44 | 0.0 | 4.5 | 0.0 | 2.3 | 0.0 | 0.0 | 0.0 | 0.0 | 0.0 |
| 23 Shestakova | 159 | 0.0 | 1.3 | 0.0 | 0.0 | 87.4 | 0.0 | 0.0 | 0.0 | 3.1 |
| 24 Straschnaya | 529 | 0.0 | 0.0 | 0.0 | 0.0 | 1.9 | 0.2 | 0.0 | 0.0 | 0.0 |
| 25 Ust-Kan | 1654 | 0.5 | 0.5 | 0.4 | 0.0 | 0.0 | 0.0 | 2.3 | 0.0 | 0.0 |
| 26 Ust-Kova | 116 | 0.0 | 30.2 | 0.0 | 0.0 | 0.9 | 0.0 | 0.0 | 0.0 | 68.1 |
| 27 Varvarina Gora | 864 | 0.0 | 17.1 | 0.0 | 0.0 | 0.1 | 0.0 | 2.3 | 0.0 | 0.0 |
| 28 Volchiya Griva | 113 | 0.0 | 0.0 | 0.0 | 0.0 | 74.3 | 0.0 | 0.0 | 0.0 | 0.0 |
| 29 Yelenev | 241 | 0.0 | 0.0 | 0.0 | 0.0 | 0.0 | 0.0 | 0.0 | 0.0 | 0.0 |
| 30 Zhemchuzhnaya | 0 | – | – | – | – | – | – | – | – | – |

[a]: A nearly complete mammoth skull was recovered in 2000 by Sergei Vasiliev.

**Table A1.2d** Species (piece count, %)

| Site | No. | Rhinoceros | Roe deer | Saiga | Seal | Sea mammal | Sheep | Small mammal | Tiger | Wolf |
|---|---|---|---|---|---|---|---|---|---|---|
| 1 Afontova Gora | (5000) | – | – | – | – | – | – | – | – | – |
| 2 Boisman II | 840 | 0.0 | 0.0 | 0.0 | 0.4 | 0.4 | 0.0 | 4.4 | 0.0 | 0.0 |
| 3 Bolshoi Yakor I | 263 | 0.0 | 0.0 | 0.0 | 0.0 | 0.0 | 0.0 | 0.0 | 0.0 | 0.0 |
| 4 Borabashevskaya | 30 | 0.0 | 0.0 | 0.0 | 0.0 | 0.0 | 0.0 | 0.0 | 0.0 | 0.0 |
| 5 Denisova | 116 | 0.0 | 0.0 | 0.0 | 0.0 | 0.0 | 0.0 | 0.0 | 0.0 | 0.9 |
| 6 Dvuglaska | 205 | 0.1 | 0.0 | 0.0 | 0.0 | 0.0 | 0.0 | 0.0 | 0.0 | 0.0 |
| 7 Gosudarev Log I | 14 | 0.0 | 0.0 | 0.0 | 0.0 | 0.0 | 0.0 | 0.0 | 0.0 | 0.0 |
| 8 Kamenka | 552 | 0.4 | 0.0 | 0.0 | 0.0 | 0.0 | 0.0 | 0.0 | 0.0 | 0.4 |
| 9 Kaminnaya | 770 | 0.4 | 0.4 | 0.0 | 0.0 | 0.0 | 1.0 | 1.6 | 0.0 | 0.0 |
| 10 Kara-Bom | 108 | 0.9 | 0.0 | 0.0 | 0.0 | 0.0 | 0.0 | 0.0 | 0.0 | 0.0 |
| 11 Kirkalinskaya | 4 | 0.0 | 0.0 | 0.0 | 0.0 | 0.0 | 0.0 | 0.0 | 0.0 | 0.0 |
| 12 Krasny Yar | 203 | 3.0 | 0.0 | 4.4 | 0.0 | 0.0 | 0.0 | 0.0 | 0.0 | 6.4 |
| 13 Kurla I | 72 | 0.0 | 0.0 | 0.0 | 0.0 | 0.0 | 0.0 | 0.0 | 0.0 | 0.0 |
| 14 Malaya Seeya | 144 | 0.0 | 0.0 | 0.0 | 0.0 | 0.0 | 0.0 | 0.0 | 0.0 | 0.0 |
| 15 Mal'ta | 186 | 2.7 | 0.0 | 0.0 | 0.0 | 0.0 | 0.0 | 1.1 | 0.0 | 0.0 |
| 16 Maly Yaloman | 178 | 0.0 | 0.0 | 0.0 | 0.0 | 0.0 | 0.0 | 0.0 | 0.0 | 3.9 |
| 17 Mokhovo Mine | 12 | 0.0 | 0.0 | 0.0 | 0.0 | 0.0 | 0.0 | 0.0 | 0.0 | 0.0 |
| 18 Nizhneudinskaya | 36 | 0.0 | 0.0 | 0.0 | 0.0 | 0.0 | 0.0 | 0.0 | 0.0 | 0.0 |
| 19 Okladnikov | 168 | 6.5 | 0.0 | 0.0 | 0.0 | 0.0 | 1.2 | 0.0 | 0.0 | 0.0 |
| 20 Proskuryakova | 47 | 0.0 | 0.0 | 48.9 | 0.0 | 0.0 | 0.0 | 0.0 | 0.0 | 0.0 |
| 21 Razboinich'ya | 638 | 0.0 | 0.0 | 0.0 | 0.0 | 0.0 | 0.0 | 0.3 | 0.0 | 4.7 |
| 22 Sarala | 44 | 0.0 | 0.0 | 0.0 | 0.0 | 0.0 | 2.3 | 0.0 | 0.0 | 0.0 |
| 23 Shestakova | 159 | 0.0 | 0.0 | 0.0 | 0.0 | 0.0 | 0.0 | 0.0 | 0.0 | 2.5 |
| 24 Straschnaya | 529 | 0.0 | 0.0 | 0.0 | 0.0 | 0.0 | 0.0 | 0.0 | 0.0 | 0.0 |
| 25 Ust-Kan | 1654 | 0.0 | 0.06 | 0.06 | 0.0 | 0.0 | 0.0 | 1.0 | 0.0 | 5.5 |
| 26 Ust-Kova | 116 | 0.0 | 0.0 | 0.0 | 0.0 | 0.0 | 0.0 | 0.0 | 0.0 | 0.0 |
| 27 Varvarina Gora | 864 | 6.0 | 0.0 | 0.0 | 0.0 | 0.0 | 0.0 | 0.1 | 0.0 | 1.6 |
| 28 Volchiya Griva | 113 | 0.0 | 0.0 | 0.0 | 0.0 | 0.0 | 0.0 | 0.0 | 0.0 | 0.0 |
| 29 Yelenev | 241 | 0.0 | 74.7 | 0.0 | 0.0 | 0.0 | 0.0 | 0.0 | 0.0 | 0.0 |
| 30 Zhemchuzhnaya | 0 | – | – | – | – | – | – | – | – | – |

**Table A1.2e** Species (piece count, %)

| Site | No. | Yak | Indeterminable |
|---|---|---|---|
| 1 Afontova Gora | (5000) | – | – |
| 2 Boisman II | 840 | 0.0 | 59.9 |
| 3 Bolshoi Yakor I | 263 | 0.0 | 89.7 |
| 4 Borabashevskaya | 30 | 0.0 | 100.0 |
| 5 Denisova | 116 | 0.0 | 74.1 |
| 6 Dvuglaska | 205 | 0.0 | 91.2 |
| 7 Gosudarev Log I | 14 | 0.0 | 0.0 |
| 8 Kamenka | 552 | 0.0 | 21.9 |
| 9 Kaminnaya | 770 | 0.3 | 71.4 |
| 10 Kara-Bom | 108 | 0.0 | 45.4 |
| 11 Kirkalinskaya | 4 | 0.0 | 0.0 |
| 12 Krasny Yar | 203 | 0.0 | 0.0 |
| 13 Kurla I | 72 | 0.0 | 0.0 |
| 14 Malaya Seeya | 144 | 0.0 | 0.0 |
| 15 Mal'ta | 186 | 0.0 | 18.3 |
| 16 Maly Yaloman | 178 | 2.2 | 2.2 |
| 17 Mokhovo Mine | 12 | 0.0 | 50.0 |
| 18 Nizhneudinskaya | 36 | 0.0 | 5.6 |
| 19 Okladnikov | 168 | 0.0 | 36.3 |
| 20 Proskuryakova | 47 | 0.0 | 0.0 |
| 21 Razboinich'ya | 638 | 0.0 | 45.1 |
| 22 Sarala | 44 | 0.0 | 63.6 |
| 23 Shestakova | 159 | 0.0 | 1.9 |
| 24 Straschnaya | 529 | 0.0 | 91.1 |
| 25 Ust-Kan | 1654 | 0.0 | 86.1 |
| 26 Ust-Kova | 116 | 0.0 | 0.0 |
| 27 Varvarina Gora | 864 | 0.0 | 18.6 |
| 28 Volchiya Griva | 113 | 0.0 | 0.0 |
| 29 Yelenev | 241 | 0.0 | 0.0 |
| 30 Zhemchuzhnaya | 0 | – | – |

**Table A1.3a** Skeletal elements (piece count, %)

| Site | No. | Vault | Base | Maxilla | Mandible | Tooth | Antler-horn | Hyoid | Vertebra | Rib |
|---|---|---|---|---|---|---|---|---|---|---|
| 1 Afontova Gora | (5000) | – | – | – | – | – | – | – | – | – |
| 2 Boisman II | 1242 | 1.3 | 0.1 | 0.5 | 4.5 | 0.1 | 1.2 | 0.0 | 6.7 | 1.8 |
| 3 Bolshoi Yakor I | 263 | 1.9 | 0.0 | 0.8 | 1.1 | 0.0 | 1.1 | 0.0 | 1.1 | 9.9 |
| 4 Borabashevskaya | 30 | 0.0 | 0.0 | 0.0 | 0.0 | 0.0 | 0.0 | 0.0 | 0.0 | 0.0 |
| 5 Denisova | 116 | 0.9 | 0.0 | 0.9 | 2.6 | 0.9 | 0.0 | 0.0 | 2.6 | 7.8 |
| 6 Dvuglaska | 205 | 1.0 | 0.0 | 0.5 | 1.5 | 0.5 | 0.5 | 0.0 | 2.0 | 7.3 |
| 7 Gosudarev Log I | 14 | 0.0 | 0.0 | 0.0 | 14.3 | 0.0 | 0.0 | 0.0 | 14.3 | 0.0 |
| 8 Kamenka | 552 | 3.3 | 0.2 | 1.3 | 7.2 | 0.0 | 0.5 | 0.0 | 5.8[a] | 11.4 |
| 9 Kaminnaya | 771 | 1.1 | 0.0 | 0.5 | 3.9 | 1.4 | 0.3 | 0.0 | 1.5 | 3.9 |
| 10 Kara-Bom | 108 | 0.0 | 0.0 | 0.9 | 4.6 | 0.0 | 1.8 | 0.0 | 3.7 | 11.1 |
| 11 Kirkalinskaya | 4 | 0.0 | 0.0 | 0.0 | 0.0 | 0.0 | 25.0 | 0.0 | 0.0 | 0.0 |
| 12 Krasny Yar | 203 | 4.4 | 0.0 | 2.5 | 30.0 | 0.0 | 2.0 | 0.0 | 1.5 | 0.0 |
| 13 Kurla I | 72 | 2.8 | 0.0 | 0.0 | 16.7 | 0.0 | 4.2 | 0.0 | 5.5 | 4.2 |
| 14 Malaya Seeya | 144 | 2.8 | 0.0 | 0.0 | 4.2 | 0.0 | 0.7 | 0.0 | 2.8 | 9.7 |
| 15 Mal'ta | 186 | 2.7 | 0.0 | 0.0 | 0.5 | 0.5 | 2.1 | 0.0 | 22.6 | 29.6 |
| 16 Maly Yaloman | 178 | 0.6 | 0.0 | 2.8 | 15.7 | 16.8 | 0.6 | 0.0 | 1.7 | 2.2 |
| 17 Mokhovo Mine | 12 | 0.0 | 0.0 | 0.0 | 0.0 | 8.3 | 0.0 | 0.0 | 0.0 | 16.7 |
| 18 Nizhneudinskaya | 36 | 2.8 | 0.0 | 0.0 | 13.9 | 2.8 | 2.8 | 0.0 | 2.8 | 19.4 |
| 19 Okladnikov | 168 | 0.0 | 0.6 | 0.6 | 8.9 | 3.0 | 0.6 | 0.0 | 6.5 | 0.6 |
| 20 Proskuryakova | 47 | 29.8 | 0.0 | 0.0 | 4.3 | 0.0 | 25.5 | 0.0 | 0.0 | 0.0 |
| 21 Razboinich'ya | 638 | 0.5 | 0.3 | 0.9 | 4.7 | 3.0 | 0.0 | 0.0 | 8.3 | 10.2 |
| 22 Sarala | 44 | 0.0 | 2.3 | 2.3 | 2.3 | 0.0 | 0.0 | 0.0 | 2.3 | 9.1 |
| 23 Shestakova | 159 | 0.6 | 0.6 | 0.0 | 1.3 | 0.0 | 0.0 | 1.3 | 18.3 | 20.1 |
| 24 Straschnaya | 527 | 0.2 | 0.0 | 0.2 | 1.5 | 0.6 | 1.1 | 0.0 | 1.5 | 5.3 |
| 25 Ust-Kan | 1654 | 1.0 | 0.0 | 0.06 | 1.0 | 1.2 | 0.1 | 0.0 | 1.2 | 4.7 |
| 26 Ust-Kova | 116 | 0.0 | 0.0 | 0.0 | 0.9 | 0.0 | 0.9 | 0.0 | 0.0 | 0.0 |
| 27 Varvarina Gora | 864 | 0.0 | 0.0 | 2.1 | 6.2 | 0.0 | 0.1 | 0.0 | 4.6 | 11.6 |
| 28 Volchiya Griva | 113 | 0.0 | 0.0 | 0.0 | 0.0 | 0.0 | 0.0 | 0.0 | 14.2 | 15.0 |
| 29 Yelenev | 241 | 3.3 | 0.0 | 2.5 | 11.2 | 0.0 | 0.0 | 0.0 | 0.4 | 4.6 |
| 30 Zhemchuzhnaya | 0 | – | – | – | – | – | – | – | – | – |

[a]: One fragment of a sacrum is included with Kamenka vertebrae.

**Table A1.3b** Skeletal elements (piece count, %)

| Site | No. | Scapula | Humerus | Radius | Ulna | Pelvis | Femur | Patella |
|---|---|---|---|---|---|---|---|---|
| 1 Afontova Gora | (5000) | – | – | – | – | – | – | – |
| 2 Boisman II | 1242 | 2.1 | 2.9 | 1.1 | 1.4 | 1.8 | 1.8 | 0.0 |
| 3 Bolshoi Yakor I | 263 | 3.4 | 0.8 | 1.5 | 0.0 | 0.0 | 1.1 | 0.0 |
| 4 Borabashevskaya | 30 | 0.0 | 0.0 | 0.0 | 0.0 | 0.0 | 0.0 | 0.0 |
| 5 Denisova | 116 | 1.7 | 1.7 | 0.9 | 0.0 | 0.9 | 0.9 | 0.0 |
| 6 Dvuglaska | 205 | 2.0 | 1.0 | 2.0 | 0.5 | 1.0 | 0.5 | 0.0 |
| 7 Gosudarev Log I | 14 | 21.4 | 0.0 | 0.0 | 0.0 | 21.4 | 7.1 | 0.0 |
| 8 Kamenka | 552 | 6.3 | 4.9 | 4.9 | 1.7[b] | 3.1 | 5.1 | 0.2 |
| 9 Kaminnaya | 771 | 1.2 | 0.9 | 0.9 | 0.8 | 1.2 | 0.8 | 0.1 |
| 10 Kara-Bom | 108 | 0.0 | 1.8 | 2.8 | 0.9 | 1.8 | 2.8 | 0.0 |
| 11 Kirkalinskaya | 4 | 0.0 | 25.0 | 25.0 | 0.0 | 0.0 | 25.0 | 0.0 |
| 12 Krasny Yar | 203 | 2.5 | 12.8 | 5.9 | 3.4 | 2.5 | 4.4 | 0.0 |
| 13 Kurla I | 72 | 12.5 | 0.0 | 1.4 | 0.0 | 1.4 | 1.4 | 0.0 |
| 14 Malaya Seeya | 144 | 2.1 | 2.1 | 0.0 | 0.0 | 0.7 | 2.1 | 0.0 |
| 15 Mal'ta | 186 | 0.5 | 2.1 | 1.6 | 0.5 | 2.7 | 1.1 | 0.0 |
| 16 Maly Yaloman | 178 | 1.1 | 2.8 | 1.1 | 1.7 | 1.7 | 0.0 | 2.2 |
| 17 Mokhovo Mine | 12 | 0.0 | 8.3 | 0.0 | 0.0 | 0.0 | 8.3 | 0.0 |
| 18 Nizhneudinskaya | 36 | 5.6 | 19.4 | 2.8 | 0.0 | 0.0 | 2.8 | 0.0 |
| 19 Okladnikov | 168 | 2.4 | 5.4 | 0.6 | 1.8 | 3.0 | 1.2 | 0.0 |
| 20 Proskuryakova | 47 | 0.0 | 8.5 | 8.5 | 0.0 | 0.0 | 0.0 | 0.0 |
| 21 Razboinich'ya | 638 | 2.5 | 4.7 | 2.8 | 3.1 | 3.9 | 3.4 | 0.2 |
| 22 Sarala | 44 | 0.0 | 4.5 | 2.3 | 4.5 | 0.0 | 0.0 | 0.0 |
| 23 Shestakova | 159 | 0.6 | 1.3 | 1.3 | 0.6 | 0.6 | 1.3 | 1.9 |
| 24 Straschnaya | 527 | 0.2 | 0.0 | 0.0 | 0.4 | 0.6 | 0.2 | 0.0 |
| 25 Ust-Kan | 1654 | 0.4 | 0.3 | 0.2 | 0.2 | 0.06 | 0.3 | 0.0 |
| 26 Ust-Kova | 116 | 6.9 | 11.2 | 6.9 | 0.0 | 0.0 | 0.0 | 0.9 |
| 27 Varvarina Gora | 864 | 6.6 | 6.0 | 6.4 | 3.7 | 3.4 | 5.0 | 0.0 |
| 28 Volchiya Griva | 113 | 0.0 | 0.9 | 1.8 | 0.9 | 0.9 | 0.0 | 1.8 |
| 29 Yelenev | 241 | 10.4 | 5.4 | 1.2 | 7.1 | 2.9 | 1.2 | 0.0 |
| 30 Zhemchuzhnaya | 0 | – | – | – | – | – | – | – |

[b]: Two fragments of joined radius and ulna are included with Kamenka ulnae.

**Table A1.3c** Skeletal elements (piece count, %)

| Site | No. | Tibia | Fibula | Long bone | Foot | Calcaneus | Metapodial | Toe | Epiphysis |
|---|---|---|---|---|---|---|---|---|---|
| 1 Afontova Gora | (5000) | – | – | – | – | – | – | – | – |
| 2 Boisman II | 1242 | 1.4 | 0.0 | 8.9 | 5.4 | 0.6 | 4.0 | 1.6 | 0.0 |
| 3 Bolshoi Yakor I | 263 | 0.4 | 0.0 | 41.8 | 4.2 | 0.0 | 0.0 | 0.0 | 0.0 |
| 4 Borabashevskaya | 30 | 0.0 | 0.0 | 0.0 | 0.0 | 0.0 | 0.0 | 0.0 | 0.0 |
| 5 Denisova | 116 | 1.7 | 0.0 | 37.9 | 3.4 | 1.7 | 8.6 | 2.6 | 0.0 |
| 6 Dvuglaska | 205 | 0.0 | 0.0 | 47.8 | 0.0 | 0.5 | 5.9 | 0.5 | 0.0 |
| 7 Gosudarev Log I | 14 | 0.0 | 0.0 | 7.1 | 7.1 | 0.0 | 0.0 | 0.0 | 0.0 |
| 8 Kamenka | 552 | 4.9 | 0.0 | 12.7 | 0.7 | 1.4 | 7.1 | 2.0 | 0.0 |
| 9 Kaminnaya | 771 | 1.7 | 0.0 | 37.0 | 1.9 | 1.0 | 5.8 | 6.5 | 0.0 |
| 10 Kara-Bom | 108 | 1.8 | 0.0 | 25.9 | 3.7 | 0.0 | 3.7 | 6.5 | 0.0 |
| 11 Kirkalinskaya | 4 | 0.0 | 0.0 | 0.0 | 0.0 | 0.0 | 0.0 | 0.0 | 0.0 |
| 12 Krasny Yar | 203 | 7.9 | 0.0 | 0.0 | 0.0 | 1.5 | 15.3 | 3.0 | 0.0 |
| 13 Kurla I | 72 | 1.4 | 0.0 | 1.4 | 2.8 | 1.4 | 11.1 | 30.5 | 0.0 |
| 14 Malaya Seeya | 144 | 0.7 | 0.0 | 50.0 | 0.0 | 0.7 | 17.4 | 0.0 | 0.0 |
| 15 Mal'ta | 186 | 2.1 | 0.0 | 9.7 | 5.4 | 0.0 | 4.8 | 4.8 | 0.0 |
| 16 Maly Yaloman | 178 | 2.8 | 0.0 | 3.9 | 9.0 | 1.1 | 7.9 | 19.7 | 0.0 |
| 17 Mokhovo Mine | 12 | 0.0 | 0.0 | 8.3 | 0.0 | 0.0 | 16.7 | 0.0 | 0.0 |
| 18 Nizhneudinskaya | 36 | 2.8 | 5.6 | 2.8 | 0.0 | 2.8 | 0.0 | 11.1 | 0.0 |
| 19 Okladnikov | 168 | 1.2 | 0.0 | 32.1 | 1.8 | 1.2 | 5.9 | 6.5 | 0.0 |
| 20 Proskuryakova | 47 | 0.0 | 0.0 | 21.3 | 0.0 | 0.0 | 2.1 | 0.0 | 0.0 |
| 21 Razboinich'ya | 638 | 3.9 | 0.9 | 26.0 | 1.1 | 0.8 | 3.0 | 3.1 | 0.0 |
| 22 Sarala | 44 | 2.3 | 0.0 | 40.9 | 0.0 | 0.0 | 4.5 | 2.3 | 0.0 |
| 23 Shestakova | 159 | 1.3 | 3.1 | 3.1 | 25.8 | 1.9 | 3.1 | 8.2 | 0.6 |
| 24 Straschnaya | 527 | 0.2 | 0.0 | 31.7 | 0.6 | 0.0 | 4.2 | 1.9 | 0.0 |
| 25 Ust-Kan | 1654 | 0.2 | 0.06 | 42.2 | 1.3 | 0.06 | 2.6 | 2.6 | 1.1 |
| 26 Ust-Kova | 116 | 27.6 | 0.0 | 0.0 | 5.2 | 5.2 | 34.5 | 0.0 | 0.0 |
| 27 Varvarina Gora | 864 | 8.9 | 0.0 | 12.4 | 0.1 | 0.7 | 7.5 | 0.2 | 0.0 |
| 28 Volchiya Griva | 113 | 0.9 | 4.4 | 0.9 | 26.5 | 0.0 | 8.0 | 13.3 | 8.8 |
| 29 Yelenev | 241 | 5.0 | 0.0 | 11.6 | 0.0 | 0.0 | 0.0 | 24.9 | 0.0 |
| 30 Zhemchuzhnaya | 0 | – | – | – | – | – | – | – | – |

**Table A1.3d** Skeletal elements (piece count, %)

| Site | No. | Penis | Unknown |
|---|---|---|---|
| 1 Afontova Gora | (5000) | – | – |
| 2 Boisman II | 1242 | 0.0 | 50.7 |
| 3 Bolshoi Yakor I | 263 | 0.0 | 29.3 |
| 4 Borabashevskaya | 30 | 0.0 | 100.0 |
| 5 Denisova | 116 | 0.0 | 22.4 |
| 6 Dvuglaska | 205 | 0.0 | 25.4 |
| 7 Gosudarev Log I | 14 | 0.0 | 7.1 |
| 8 Kamenka | 552 | 0.7 | 14.7 |
| 9 Kaminnaya | 771 | 0.0 | 27.5 |
| 10 Kara-Bom | 108 | 0.0 | 25.9 |
| 11 Kirkalinskaya | 4 | 0.0 | 0.0 |
| 12 Krasny Yar | 203 | 0.0 | 0.0 |
| 13 Kurla I | 72 | 0.0 | 1.4 |
| 14 Malaya Seeya | 144 | 0.0 | 2.8 |
| 15 Mal'ta | 186 | 0.0 | 6.4 |
| 16 Maly Yaloman | 178 | 0.0 | 3.9 |
| 17 Mokhovo Mine | 12 | 0.0 | 33.3 |
| 18 Nizhneudinskaya | 36 | 0.0 | 0.0 |
| 19 Okladnikov | 168 | 0.0 | 16.1 |
| 20 Proskuryakova | 47 | 0.0 | 0.0 |
| 21 Razboinich'ya | 638 | 5.4 | 5.6 |
| 22 Sarala | 44 | 0.0 | 65.9 |
| 23 Shestakova | 159 | 0.0 | 3.1 |
| 24 Straschnaya | 527 | 0.0 | 49.7 |
| 25 Ust-Kan | 1654 | 0.1 | 38.8 |
| 26 Ust-Kova | 116 | 0.0 | 0.0 |
| 27 Varvarina Gora | 864 | 0.7 | 11.8 |
| 28 Volchiya Griva | 113 | 0.0 | 2.7 |
| 29 Yelenev | 241 | 0.0 | 1.7 |
| 30 Zhemchuzhnaya | 0 | – | – |

**Table A1.4** Age (piece count, %)

| Site | No. | Adult | Sub-adult | Adult? | Unknown |
|---|---|---|---|---|---|
| 1 Afontova Gora | (5000) | – | – | – | – |
| 2 Boisman II | 513 | 7.6 | 80.5 | 1.4 | 10.5 |
| 3 Bolshoi Yakor I | 263 | 0.8 | 47.5 | 0.0 | 51.7 |
| 4 Borabashevskaya | 30 | 0.0 | 0.0 | 100.0 | 0.0 |
| 5 Denisova | 116 | 0.9 | 43.1 | 31.9 | 24.1 |
| 6 Dvuglaska | 205 | 0.5 | 56.1 | 0.0 | 43.4 |
| 7 Gosudarev Log I | 14 | 0.0 | 92.9 | 0.0 | 7.1 |
| 8 Kamenka | 552 | 13.2 | 65.0 | 0.0 | 21.7 |
| 9 Kaminnaya | 771 | 2.2 | 41.1 | 0.0 | 56.7 |
| 10 Kara-Bom | 108 | 1.8 | 45.4 | 7.4 | 45.4 |
| 11 Kirkalinskaya | 4 | 25.0 | 75.0 | 0.0 | 0.0 |
| 12 Krasny Yar | 203 | 6.4 | 92.6 | 0.0 | 1.0 |
| 13 Kurla I | 72 | 0.0 | 88.9 | 0.0 | 11.1 |
| 14 Malaya Seeya | 144 | 0.7 | 87.5 | 11.8 | 0.0 |
| 15 Mal'ta | 186 | 15.0 | 45.2 | 0.0 | 39.8 |
| 16 Maly Yaloman | 178 | 6.2 | 75.3 | 0.0 | 18.5 |
| 17 Mokhovo Mine | 12 | 0.0 | 66.7 | 0.0 | 33.3 |
| 18 Nizhneudinskaya | 36 | 36.1 | 50.0 | 0.0 | 13.9 |
| 19 Okladnikov | 168 | 8.3 | 42.9 | 32.7 | 16.1 |
| 20 Proskuryakova | 47 | 4.3 | 87.2 | 8.5 | 0.0 |
| 21 Razboinich'ya | 638 | 7.4 | 52.5 | 0.0 | 40.1 |
| 22 Sarala | 44 | 0.0 | 77.3 | 0.0 | 22.7 |
| 23 Shestakova | 159 | 30.2 | 59.1 | 0.0 | 10.7 |
| 24 Straschnaya | 527 | 0.6 | 41.4 | 0.0 | 58.1 |
| 25 Ust-Kan | 1654 | 1.2 | 42.6 | 2.7 | 53.4 |
| 26 Ust-Kova | 116 | 4.3 | 95.7 | 0.0 | 0.0 |
| 27 Varvarina Gora | 864 | 12.4 | 58.8 | 0.0 | 28.8 |
| 28 Volchiya Griva | 113 | 20.3 | 69.9 | 2.6 | 7.1 |
| 29 Yelenev | 241 | 7.9 | 83.8 | 0.0 | 8.3 |
| 30 Zhemchuzhnaya | 0 | – | – | – | – |

**Table A1.5** Completeness (piece count, anatomical ends, %)

| Site | No. | Whole | One end | No ends |
|---|---|---|---|---|
| 1 Afontova Gora | (5000) | – | – | – |
| 2 Boisman II | 501 | 13.8 | 37.5 | 48.7 |
| 3 Bolshoi Yakor I | 262 | 0.0 | 9.2 | 90.8 |
| 4 Borabashevskaya | 30 | 0.0 | 0.0 | 100.0 |
| 5 Denisova | 116 | 6.0 | 7.8 | 86.2 |
| 6 Dvuglaska | 205 | 3.4 | 4.9 | 91.7 |
| 7 Gosudarev Log I | 14 | 7.1 | 42.9 | 50.0 |
| 8 Kamenka | 552 | 8.7 | 37.5 | 53.8 |
| 9 Kaminnaya | 714 | 6.4 | 13.6 | 80.0 |
| 10 Kara-Bom | 108 | 8.3 | 25.9 | 65.7 |
| 11 Kirkalinskaya | 4 | 0.0 | 50.0 | 50.0 |
| 12 Krasny Yar | 203 | 23.2 | 58.6 | 18.2 |
| 13 Kurla I | 72 | 1.4 | 62.5 | 36.1 |
| 14 Malaya Seeya | 144 | 0.0 | 9.0 | 91.0 |
| 15 Mal'ta | 186 | 17.2 | 31.7 | 51.1 |
| 16 Maly Yaloman | 178 | 32.6 | 28.1 | 39.3 |
| 17 Mokhovo Mine | 12 | 0.0 | 41.7 | 58.3 |
| 18 Nizhneudinskaya | 36 | 8.3 | 30.6 | 61.1 |
| 19 Okladnikov | 168 | 5.4 | 23.8 | 70.8 |
| 20 Proskuryakova | 47 | 0.0 | 55.3 | 44.7 |
| 21 Razboinich'ya | 638 | 8.6 | 28.5 | 62.8 |
| 22 Sarala | 44 | 4.5 | 9.1 | 86.4 |
| 23 Shestakova | 158 | 56.3 | 27.2 | 16.5 |
| 24 Straschnaya | 527 | 0.9 | 8.0 | 91.1 |
| 25 Ust-Kan | 1647 | 2.5 | 6.1 | 91.4 |
| 26 Ust-Kova | 116 | 11.2 | 87.1 | 1.7 |
| 27 Varvarina Gora | 863 | 3.4 | 46.0 | 50.6 |
| 28 Volchiya Griva | 113 | 69.0 | 12.4 | 18.6 |
| 29 Yelenev | 241 | 7.0 | 45.2 | 47.7 |
| 30 Zhemchuzhnaya | 0 | – | – | – |

**Table A1.6** Maximum size (piece count, including whole pieces, in centimeters)

| Site | No. | Mean | S.D. | Range | S.E. |
|---|---|---|---|---|---|
| 1 Afontova Gora | (5000) | – | – | – | – |
| 2 Boisman II | 513 | 7.4 | 2.9 | 2.5–22.5 | 0.128 |
| 3 Bolshoi Yakor I | 264 | 6.0 | 4.0 | 2.5–50.5 | 0.246 |
| 4 Borabashevskaya | 30 | – | – | (2.5–5.0) | – |
| 5 Denisova | 116 | 5.2 | 2.2 | 2.3–13.8 | 0.204 |
| 6 Dvuglaska | 429 | 5.2 | 2.8 | 1.7–28.2 | 0.135 |
| 7 Gosudarev Log I | 14 | 11.5 | 7.0 | 4.7–30.0 | 1.871 |
| 8 Kamenka | 550 | 6.5 | 2.9 | 2.9–20.7 | 0.126 |
| 9 Kaminnaya | 774 | 6.2 | 3.0 | 2.4–25.0 | 0.108 |
| 10 Kara-Bom | 24 | 6.6 | 4.2 | 3.0–24.8 | 0.857 |
| 11 Kirkalinskaya | 4 | 12.6 | 6.8 | 5.2–19.6 | 3.400 |
| 12 Krasny Yar | 204 | 19.7 | 10.3 | 3.1–43.5 | 0.721 |
| 13 Kurla I | 72 | 4.3 | 1.6 | 2.1–9.5 | 0.189 |
| 14 Malaya Seeya | 144 | 8.3 | 3.4 | 3.2–30.8 | 0.283 |
| 15 Mal'ta | 186 | 16.5 | 12.7 | 3.0–82.5 | 0.931 |
| 16 Maly Yaloman | 177 | 4.9 | 3.5 | 2.3–28.1 | 0.267 |
| 17 Mokhovo Mine | 12 | 10.2 | 4.2 | 6.2–21.0 | 1.212 |
| 18 Nizhneudinskaya | 36 | 10.9 | 6.1 | 4.7–28.5 | 1.017 |
| 19 Okladnikov | 180 | 7.9 | 4.7 | 2.1–26.9 | 0.350 |
| 20 Proskuryakova | 47 | 11.9 | 3.4 | 7.1–18.8 | 0.496 |
| 21 Razboinich'ya | 599 | 6.5 | 3.2 | 2.4–24.4 | 0.13 |
| 22 Sarala | 44 | 8.8 | 4.6 | 3.7–21.2 | 0.693 |
| 23 Shestakova | 155 | 16.1 | 11.6 | 2.4–68.0 | 0.932 |
| 24 Straschnaya | 527 | 5.4 | 2.2 | 2.5–16.2 | 0.096 |
| 25 Ust-Kan | 1656 | 4.8 | 1.6 | 2.2–19.9 | 0.056 |
| 26 Ust-Kova | 117 | 8.0 | 3.8 | 2.9–28.3 | 0.351 |
| 27 Varvarina Gora | 847 | 9.0 | 5.3 | 2.5–48.5 | 0.184 |
| 28 Volchiya Griva | 113 | 17.0 | 10.3 | 4.2–69.1 | 0.969 |
| 29 Yelenev | 340 | 6.0 | 2.8 | 1.5–17.5 | 0.152 |
| 30 Zhemchuzhnaya | 0 | – | – | – | – |

**Table A1.7a** Damage shape (piece count, %)

| Site | No. | Skull and mandible | | | | | | Tooth | |
|---|---|---|---|---|---|---|---|---|---|
| – | – | Base | Vault | Tooth-bearing | Socket | Symphysis | Condyle | Root | Whole crown |
| 1 Afontova Gora | (5000) | – | – | – | – | – | – | – | – |
| 2 Boisman II | 0 | – | – | – | – | – | – | – | – |
| 3 Bolshoi Yakor I | 263 | 0.0 | 0.0 | 1.1 | 0.0 | 0.0 | 0.4 | 0.0 | 0.0 |
| 4 Borabashevskaya | 30 | 0.0 | 0.0 | 0.0 | 0.0 | 0.0 | 0.0 | 0.0 | 0.0 |
| 5 Denisova | 0 | – | – | – | – | – | – | – | – |
| 6 Dvuglaska | 205 | 0.0 | 0.0 | 0.0 | 0.5 | 0.0 | 0.0 | 0.0 | 0.5 |
| 7 Gosudarev Log I | 14 | 0.0 | 0.0 | 0.0 | 0.0 | 14.3 | 0.0 | 0.0 | 0.0 |
| 8 Kamenka | 552 | 0.0 | 0.8 | 3.3 | 1.3 | 0.5 | 1.6 | 0.0 | 0.0 |
| 9 Kaminnaya | 557 | 0.2 | 0.4 | 1.8 | 0.0 | 0.2 | 0.0 | 0.0 | 1.0 |
| 10 Kara-Bom | 108 | 0.0 | 0.0 | 0.0 | 0.9 | 0.0 | 0.0 | 0.0 | 0.0 |
| 11 Kirkalinskaya | 4 | 0.0 | 0.0 | 0.0 | 0.0 | 0.0 | 0.0 | 0.0 | 0.0 |
| 12 Krasny Yar | 0* | – | – | – | – | – | – | – | – |
| 13 Kurla I | 72 | 0.0 | 2.8 | 4.2 | 2.8 | 1.4 | 6.9 | 0.0 | 0.0 |
| 14 Malaya Seeya | 0* | – | – | – | – | – | – | – | – |
| 15 Mal'ta | 185 | 0.5 | 2.7 | 0.0 | 0.0 | 0.0 | 0.5 | 0.5 | 0.0 |
| 16 Maly Yaloman | 178 | 0.0 | 0.0 | 10.7 | 5.1 | 0.0 | 0.6 | 0.0 | 8.4 |
| 17 Mokhovo Mine | 12 | 0.0 | 0.0 | 0.0 | 0.0 | 0.0 | 0.0 | 0.0 | 8.3 |
| 18 Nizhneudinskaya | 36 | 0.0 | 2.8 | 13.9 | 0.0 | 0.0 | 0.0 | 0.0 | 2.8 |
| 19 Okladnikov | 169 | 4.2 | 0.0 | 5.1 | 4.2 | 0.0 | 0.0 | 4.8 | 12.5 |
| 20 Proskuryakova | 0* | – | – | – | – | – | – | – | – |
| 21 Razboinich'ya | 739 | 0.0 | 2.3 | 0.0 | 1.6 | 0.1 | 0.1 | 0.3 | 1.6 |
| 22 Sarala | 44 | 0.0 | 0.0 | 4.5 | 0.0 | 0.0 | 0.0 | 0.0 | 0.0 |
| 23 Shestakova | 159 | 0.0 | 0.6 | 1.3 | 0.0 | 0.0 | 0.0 | 0.0 | 0.0 |
| 24 Straschnaya | 527 | 0.0 | 0.0 | 0.0 | 0.0 | 0.0 | 0.0 | 0.0 | 0.0 |
| 25 Ust-Kan | 1558 | 0.1 | 0.2 | 0.06 | 0.3 | 0.2 | 0.2 | 0.1 | 0.5 |
| 26 Ust-Kova | 0 | – | – | – | – | – | – | – | – |
| 27 Varvarina Gora | 862 | 0.0 | 1.5 | 3.4 | 0.8 | 0.7 | 1.5 | 0.0 | 0.0 |
| 28 Volchiya Griva | 113 | 0.0 | 0.0 | 0.0 | 0.0 | 0.0 | 0.0 | 0.0 | 0.0 |
| 29 Yelenev | 241 | 1.2 | 0.4 | 7.5 | 0.4 | 1.2 | 2.9 | 0.0 | 0.0 |
| 30 Zhemchuzhnaya | 0 | – | – | – | – | – | – | – | – |

* Information on damage form not collected for Krasny Yar since this bone locality in a bluff overlooking the Ob River was a water-deposited assemblage and much of the damage would likely have been done by ice, water, tumbling, and scouring. Similarly, no information was collected on damage form from Malaya Seeya and Proskuryakova Grotto since these were the first assemblages studied and the damage form variants had not yet been defined.

**Table A1.7b** Damage shape (piece count, %*)

| Site | No. | Vertebra | | Long bone and uncertain* | | | | | Phalanx | |
|---|---|---|---|---|---|---|---|---|---|---|
| – | – | Body | Spine | Flake | Fragment | Splinter | Epiphysis | Cracked-open | Segment | Butt |
| 1 Afontova Gora | (5000) | – | – | – | – | – | – | – | – | – |
| 2 Boisman II | 0 | – | – | – | – | – | – | – | – | – |
| 3 Bolshoi Yakor I | 263 | 0.0 | 0.4 | 59.3 | 6.6 | 25.1 | 0.0 | 0.0 | 0.0 | 1.9 |
| 4 Borabashevskaya | 30 | 0.0 | 0.0 | 0.0 | 100.0 | 0.0 | 0.0 | 0.0 | 0.0 | 0.0 |
| 5 Denisova | 0 | – | – | – | – | – | – | – | – | – |
| 6 Dvuglaska | 205 | 0.5 | 1.0 | 62.9 | 6.4 | 15.1 | 0.0 | 0.0 | 0.0 | 3.4 |
| 7 Gosudarev Log I | 14 | 0.0 | 14.3 | 0.0 | 35.7 | 0.0 | 0.0 | 0.0 | 0.0 | 0.0 |
| 8 Kamenka | 552 | 1.6 | 1.6 | 11.2 | 36.9* | 11.4 | 0.4 | 0.5 | 0.0 | 3.6 |
| 9 Kaminnaya | 557 | 0.2 | 1.1 | 38.4 | 25.7 | 9.7 | 0.0 | 2.2 | 1.8 | 5.2 |
| 10 Kara-Bom | 108 | 2.8 | 0.0 | 37.0 | 17.5 | 9.3 | 0.0 | 0.0 | 0.0 | 3.7 |
| 11 Kirkalinskaya | 4 | 0.0 | 0.0 | 25.0 | 0.0 | 0.0 | 0.0 | 0.0 | 25.0 | 25.0 |
| 12 Krasny Yar | 0 | – | – | – | – | – | – | – | – | – |
| 13 Kurla I | 72 | 0.0 | 5.5 | 2.8 | 25.0 | 0.0 | 0.0 | 12.5 | 0.0 | 29.2 |
| 14 Malaya Seeya | 0 | – | – | – | – | – | – | – | – | – |
| 15 Mal'ta | 185 | 4.3 | 11.3 | 9.7 | 17.8 | 1.1 | 0.5 | 0.5 | 1.1 | 1.6 |
| 16 Maly Yaloman | 178 | 1.1 | 0.6 | 2.2 | 13.5 | 0.0 | 1.7 | 4.5 | 0.0 | 6.2 |
| 17 Mokhovo Mine | 12 | 0.0 | 0.0 | 8.3 | 75.0 | 0.0 | 0.0 | 0.0 | 0.0 | 0.0 |
| 18 Nizhneudinskaya | 36 | 0.0 | 0.0 | 8.3 | 33.4 | 0.0 | 0.0 | 0.0 | 0.0 | 0.0 |
| 19 Okladnikov | 168 | 0.0 | 0.4 | 21.7 | 52.4 | 1.7 | 0.0 | 0.1 | 0.5 | 0.0 |
| 20 Proskuryakova | 0 | – | – | – | – | – | – | – | – | – |
| 21 Razboinich'ya | 739 | 0.7 | 5.5 | 13.4 | 31.1 | 6.1 | 0.0 | 0.1 | 0.8 | 0.1 |
| 22 Sarala | 44 | 0.0 | 0.0 | 36.4 | 11.4 | 27.3 | 0.0 | 0.0 | 9.1 | 4.5 |
| 23 Shestakova | 159 | 5.0 | 5.7 | 5.0 | 7.5 | 0.0 | 0.0 | 0.0 | 8.8 | 1.9 |
| 24 Straschnaya | 527 | 0.0 | 0.2 | 71.8 | 0.2 | 17.6 | 0.0 | 0.0 | 0.0 | 1.5 |
| 25 Ust-Kan | 1558 | 0.3 | 0.7 | 53.5 | 18.6 | 15.3 | 0.0 | 0.3 | 0.3 | 1.3 |
| 26 Ust-Kova | 0 | – | – | – | – | – | – | – | – | – |
| 27 Varvarina Gora | 862 | 1.6 | 1.0 | 13.8 | 30.3 | 3.4 | 0.0 | 0.1 | 0.0 | 0.0 |
| 28 Volchiya Griva | 113 | 0.9 | 4.4 | 3.5 | 0.0 | 1.8 | 0.0 | 0.0 | 0.0 | 0.0 |
| 29 Yelenev | 241 | 1.7 | 0.0 | 21.6 | 8.3 | 0.4 | 0.4 | 11.6 | 10.8 | 18.3 |
| 30 Zhemchuzhnaya | 0 | – | – | – | – | – | – | – | – | – |

* Difficult to identify small pieces lacking diagnostic anatomical landmarks were classified as "fragments." The solid foot bones were only rarely damaged, so they were generally not studied, the main exception being the calcaneus, which is why it is inventoried separately from the other foot bones (carpals and tarsals, astragalus). The few that have been included were regarded as mostly whole. The specific damage to foot bones was most often tooth dints or tooth scratches and was inventoried as such.

**Table A1.7c** Damage shape (piece count, %)

| Site | No. | Ribs | | | Pelvis | | Scapula | | All bones |
|---|---|---|---|---|---|---|---|---|---|
| – | – | Medial | Proximal | Distal | Center | Fragment | M–W (medium to whole) | Fragment | Undamaged |
| 1 Afontova Gora | (5000) | – | – | – | – | – | – | – | – |
| 2 Boisman II | 0 | – | – | – | – | – | – | – | – |
| 3 Bolshoi Yakor I | 263 | 0.0 | 0.0 | 0.0 | 0.0 | 0.0 | 0.0 | 0.8 | 0.0 |
| 4 Borabashevskaya | 30 | 0.0 | 0.0 | 0.0 | 0.0 | 0.0 | 0.0 | 0.0 | 0.0 |
| 5 Denisova | 0 | – | – | – | – | – | – | – | – |
| 6 Dvuglaska | 205 | 5.9 | 0.5 | 0.0 | 0.5 | 0.0 | 0.0 | 0.0 | 0.0 |
| 7 Gosudarev Log I | 14 | 0.0 | 0.0 | 0.0 | 14.3 | 0.0 | 0.0 | 0.0 | 7.1 |
| 8 Kamenka | 552 | 6.2 | 3.3 | 0.5 | 2.7 | 0.0 | 0.0 | 0.0 | 7.6 |
| 9 Kaminnaya | 557 | 0.2 | 0.5 | 0.2 | 0.4 | 0.0 | 0.0 | 0.0 | 3.8 |
| 10 Kara-Bom | 108 | 5.5 | 1.8 | 0.0 | 1.8 | 0.0 | 0.0 | 0.0 | 7.4 |
| 11 Kirkalinskaya | 4 | 0.0 | 0.0 | 0.0 | 0.0 | 0.0 | 0.0 | 0.0 | 0.0 |
| 12 Krasny Yar | 0 | – | – | – | – | – | – | – | – |
| 13 Kurla I | 72 | 0.0 | 2.8 | 0.0 | 1.4 | 0.0 | 0.0 | 0.0 | 1.4 |
| 14 Malaya Seeya | 0 | – | – | – | – | – | – | – | – |
| 15 Mal'ta | 185 | 3.8 | 13.0 | 8.6 | 0.5 | 1.1 | 0.0 | 0.5 | 11.9 |
| 16 Maly Yaloman | 178 | 1.1 | 1.1 | 0.0 | 1.7 | 0.0 | 0.0 | 0.0 | 14.6 |
| 17 Mokhovo Mine | 12 | 0.0 | 8.3 | 0.0 | 0.0 | 0.0 | 0.0 | 0.0 | 0.0 |
| 18 Nizhneudinskaya | 35 | 22.2 | 0.0 | 0.0 | 0.0 | 0.0 | 0.0 | 0.0 | 8.3 |
| 19 Okladnikov | 168 | 0.1 | 0.0 | 0.0 | 0.0 | 0.0 | 0.3 | 0.3 | 0.0 |
| 20 Proskuryakova | 0 | – | – | – | – | – | – | – | – |
| 21 Razboinich'ya | 739 | 15.4 | 8.2 | 1.5 | 0.9 | 0.3 | 0.0 | 1.5 | 1.6 |
| 22 Sarala | 44 | 0.0 | 0.0 | 0.0 | 0.0 | 0.0 | 0.0 | 0.0 | 4.5 |
| 23 Shestakova | 159 | 1.3 | 0.6 | 0.0 | 0.6 | 0.0 | 0.0 | 0.0 | 57.7 |
| 24 Straschnaya | 527 | 0.6 | 1.3 | 1.1 | 0.2 | 0.0 | 0.0 | 0.0 | 0.8 |
| 25 Ust-Kan | 1558 | 0.5 | 0.7 | 0.0 | 0.06 | 0.0 | 0.2 | 0.0 | 2.0 |
| 26 Ust-Kova | 0 | – | – | – | – | – | – | – | – |
| 27 Varvarina Gora | 862 | 5.6 | 4.2 | 0.9 | 2.8 | 0.0 | 0.0 | 0.0 | 3.5 |
| 28 Volchiya Griva | 113 | 9.7 | 2.6 | 4.4 | 0.9 | 0.0 | 0.0 | 0.0 | 69.9 |
| 29 Yelenev | 241 | 4.1 | 2.1 | 0.0 | 0.0 | 0.0 | 0.0 | 0.0 | 5.8 |
| 30 Zhemchuzhnaya | 0 | – | – | – | – | – | – | – | – |

**Table A1.7d** Damage shape (piece count, %)

| Site | No. | | All bones |
|---|---|---|---|
| – | – | Irregular | Mostly whole |
| 1 Afontova Gora | (5000) | – | – |
| 2 Boisman II | 0 | – | – |
| 3 Bolshoi Yakor I | 263 | 3.8 | 0.8 |
| 4 Borabashevskaya | 30 | 0.0 | 0.0 |
| 5 Denisova | 0 | – | – |
| 6 Dvuglaska | 205 | 2.4 | 0.5 |
| 7 Gosudarev Log I | 14 | 7.1 | 7.1 |
| 8 Kamenka | 552 | 0.9 | 4.0 |
| 9 Kaminnaya | 557 | 0.0 | 4.5 |
| 10 Kara-Bom | 108 | 2.8 | 9.2 |
| 11 Kirkalinskaya | 4 | 0.0 | 25.0 |
| 12 Krasny Yar | 0 | – | – |
| 13 Kurla I | 72 | 0.0 | 1.4 |
| 14 Malaya Seeya | 0 | – | – |
| 15 Mal'ta | 185 | 0.0 | 8.1 |
| 16 Maly Yaloman | 178 | 5.6 | 14.6 |
| 17 Mokhovo Mine | 12 | 0.0 | 0.0 |
| 18 Nizhneudinskaya | 36 | 0.0 | 5.6 |
| 19 Okladnikov | 168 | 0.6 | 0.2 |
| 20 Proskuryakova | 0 | – | – |
| 21 Razboinich'ya | 739 | 0.7 | 1.3 |
| 22 Sarala | 44 | 2.3 | 0.0 |
| 23 Shestakova | 159 | 0.6 | 3.1 |
| 24 Straschnaya | 527 | 4.2 | 1.3 |
| 25 Ust-Kan | 1558 | 3.4 | 0.5 |
| 26 Ust-Kova | 116 | 0.0 | 0.0 |
| 27 Varvarina Gora | 862 | 0.6 | 3.6 |
| 28 Volchiya Griva | 113 | 0.0 | 1.8 |
| 29 Yelenev | 241 | 0.0 | 1.2 |
| 30 Zhemchuzhnaya | 0 | – | – |

**Table A1.8** Color (piece count, %)

| Site | No. | Ivory | Brown | White | Black | Black-brown |
|---|---|---|---|---|---|---|
| 1 Afontova Gora | (5000) | – | – | – | – | – |
| 2 Boisman II | 1247 | 99.1 | 0.2 | 0.0 | 0.6 | 0.0 |
| 3 Bolshoi Yakor I | 262 | 77.5 | 18.7 | 1.5 | 2.3 | 0.0 |
| 4 Borabashevskaya | 30 | 100.0 | 0.0 | 0.0 | 0.0 | 0.0 |
| 5 Denisova | 116 | 94.8 | 5.2 | 0.0 | 0.0 | 0.0 |
| 6 Dvuglaska | 205 | 98.0 | 0.5 | 1.0 | 0.5 | 0.0 |
| 7 Gosudarev Log I | 14 | 92.9 | 7.1 | 0.0 | 0.0 | 0.0 |
| 8 Kamenka | 552 | 98.9 | 1.1 | 0.0 | 0.0 | 0.0 |
| 9 Kaminnaya | 771 | 96.6 | 0.5 | 0.0 | 3.0 | 0.0 |
| 10 Kara-Bom | 108 | 94.4 | 5.6 | 0.0 | 0.0 | 0.0 |
| 11 Kirkalinskaya | 4 | 100.0 | 0.0 | 0.0 | 0.0 | 0.0 |
| 12 Krasny Yar | 202 | 36.1 | 53.5 | 6.9 | 3.5 | 0.0 |
| 13 Kurla I | 72 | 98.6 | 0.0 | 0.0 | 1.4 | 0.0 |
| 14 Malaya Seeya | 144 | 90.3 | 4.9 | 2.1 | 2.8 | 0.0 |
| 15 Mal'ta | 186 | 99.0 | 0.0 | 0.5 | 0.5 | 0.0 |
| 16 Maly Yaloman | 178 | 96.1 | 3.9 | 0.0 | 0.0 | 0.0 |
| 17 Mokhovo Mine | 12 | 100.0 | 0.0 | 0.0 | 0.0 | 0.0 |
| 18 Nizhneudinskaya | 36 | 100.0 | 0.0 | 0.0 | 0.0 | 0.0 |
| 19 Okladnikov | 168 | 90.5 | 6.5 | 0.0 | 2.4 | 0.6 |
| 20 Proskuryakova | 47 | 100.0 | 0.0 | 0.0 | 0.0 | 0.0 |
| 21 Razboinich'ya | 637 | 92.9* | 6.4 | 0.6 | 0.0 | 0.0 |
| 22 Sarala | 44 | 100.0 | 0.0 | 0.0 | 0.0 | 0.0 |
| 23 Shestakova | 159 | 97.5 | 2.5 | 0.0 | 0.0 | 0.0 |
| 24 Straschnaya | 527 | 100.0 | 0.0 | 0.0 | 0.0 | 0.0 |
| 25 Ust-Kan | 1654 | 94.6 | 2.7 | 0.1 | 2.6 | 0.0 |
| 26 Ust-Kova | 116 | 1.7 | 94.9 | 0.0 | 3.4 | 0.0 |
| 27 Varvarina Gora | 862 | 99.2 | 0.6 | 0.1 | 0.1 | 0.0 |
| 28 Volchiya Griva | 113 | 48.7 | 51.3 | 0.0 | 0.0 | 0.0 |
| 29 Yelenev | 241 | 99.6 | 0.4 | 0.0 | 0.0 | 0.0 |
| 30 Zhemchuzhnaya | 0 | – | – | – | – | – |

* Includes 35 Razboinich'ya pieces that have large amounts of black mineral "flowers" on ivory colored surfaces. Most of the cave assemblages have some to many of these flowers.

**Table A1.9** Preservation (piece count, %)

| Site | No. | Ivory | Chalky | Intermediate |
|---|---|---|---|---|
| 1 Afontova Gora | (5000) | – | – | – |
| 2 Boisman II | 513 | 98.0 | 1.9 | 0.0 |
| 3 Bolshoi Yakor I | 263 | 90.1 | 9.9 | 0.0 |
| 4 Borabashevskaya | 30 | 100.0 | 0.0 | 0.0 |
| 5 Denisova | 116 | 100.0 | 0.0 | 0.0 |
| 6 Dvuglaska | 205 | 94.1 | 5.9 | 0.0 |
| 7 Gosudarev Log I | 14 | 14.3 | 85.7 | 0.0 |
| 8 Kamenka | 552 | 94.4 | 4.0 | 1.6 |
| 9 Kaminnaya | 770 | 99.6 | 0.4 | 0.0 |
| 10 Kara-Bom | 108 | 67.6 | 32.4 | 0.0 |
| 11 Kirkalinskaya | 4 | 100.0 | 0.0 | 0.0 |
| 12 Krasny Yar | 203 | 95.6 | 4.4 | 0.0 |
| 13 Kurla I | 72 | 95.8 | 4.2 | 0.0 |
| 14 Malaya Seeya | 144 | 84.7 | 15.3 | 0.0 |
| 15 Mal'ta | 186 | 49.5 | 50.0 | 0.5 |
| 16 Maly Yaloman | 178 | 96.6 | 1.7 | 1.7 |
| 17 Mokhovo Mine | 12 | 100.0 | 0.0 | 0.0 |
| 18 Nizhneudinskaya | 36 | 100.0 | 0.0 | 0.0 |
| 19 Okladnikov | 168 | 73.2 | 25.0 | 1.8 |
| 20 Proskuryakova | 47 | 91.4 | 8.5 | 0.0 |
| 21 Razboinich'ya | 625 | 93.8 | 6.2 | 0.0 |
| 22 Sarala | 44 | 100.0 | 0.0 | 0.0 |
| 23 Shestakova | 159 | 11.9 | 88.0 | 0.0 |
| 24 Straschnaya | 527 | 96.8 | 3.2 | 0.0 |
| 25 Ust-Kan | 1654 | 99.4 | 0.6 | 0.0 |
| 26 Ust-Kova | 116 | 34.5 | 65.5 | 0.0 |
| 27 Varvarina Gora | 863 | 65.3 | 18.4 | 16.2 |
| 28 Volchiya Griva | 113 | 8.8 | 91.1 | 0.0 |
| 29 Yelenev | 241 | 100.0 | 0.0 | 0.0 |
| 30 Zhemchuzhnaya | 0 | – | – | – |

**Table A1.10** Perimortem breakage (piece count, %)

| Site | No. | Present | Absent |
|---|---|---|---|
| 1 Afontova Gora | (5000) | – | – |
| 2 Boisman II | 1022 | 81.8 | 18.2 |
| 3 Bolshoi Yakor I | 263 | 99.6 | 0.4 |
| 4 Borabashevskaya | 30 | 100.0 | 0.0 |
| 5 Denisova | 116 | 96.5 | 3.5 |
| 6 Dvuglaska | 204 | 99.5 | 0.5 |
| 7 Gosudarev Log I | 14 | 28.6 | 71.4 |
| 8 Kamenka | 552 | 94.7 | 5.2 |
| 9 Kaminnaya | 770 | 94.3 | 5.7 |
| 10 Kara-Bom | 103 | 88.3 | 11.6 |
| 11 Kirkalinskaya | 4 | 100.0 | 0.0 |
| 12 Krasny Yar | 196 | 60.2 | 39.8 |
| 13 Kurla I | 72 | 100.0 | 0.0 |
| 14 Malaya Seeya | 144 | 99.3 | 0.7 |
| 15 Mal'ta | 162 | 75.9 | 24.1 |
| 16 Maly Yaloman | 178 | 74.7 | 25.3 |
| 17 Mokhovo Mine | 8 | 87.5 | 12.5 |
| 18 Nizhneudinskaya | 36 | 94.4 | 5.6 |
| 19 Okladnikov | 168 | 94.0 | 6.0 |
| 20 Proskuryakova | 47 | 100.0 | 0.0 |
| 21 Razboinich'ya | 607 | 92.4 | 7.6 |
| 22 Sarala | 44 | 97.7 | 2.3 |
| 23 Shestakova | 155 | 18.7 | 81.3 |
| 24 Straschnaya | 526 | 98.1 | 1.9 |
| 25 Ust-Kan | 1654 | 97.6 | 2.3 |
| 26 Ust-Kova | 116 | 88.8 | 11.2 |
| 27 Varvarina Gora | 862 | 97.0 | 3.0 |
| 28 Volchiya Griva | 113 | 8.0 | 92.0 |
| 29 Yelenev | 241 | 96.7 | 3.3 |
| 30 Zhemchuzhnaya | 0 | – | – |

**Table A1.11** Postmortem breakage (piece count, %)

| Site | No. | Present | Absent |
|---|---|---|---|
| 1 Afontova Gora | (5000) | – | – |
| 2 Boisman II | 1099 | 8.2 | 91.8 |
| 3 Bolshoi Yakor I | 263 | 2.7 | 97.3 |
| 4 Borabashevskaya | 30 | 0.0 | 100.0 |
| 5 Denisova | 115 | 18.3 | 81.7 |
| 6 Dvuglaska | 204 | 10.3 | 89.7 |
| 7 Gosudarev Log I | 14 | 85.7 | 14.3 |
| 8 Kamenka | 552 | 1.1 | 98.9 |
| 9 Kaminnaya | 767 | 4.6 | 95.4 |
| 10 Kara-Bom | 106 | 26.4 | 73.6 |
| 11 Kirkalinskaya | 4 | 25.0 | 75.0 |
| 12 Krasny Yar | 200 | 38.5 | 61.5 |
| 13 Kurla I | 72 | 0.0 | 100.0 |
| 14 Malaya Seeya | 144 | 11.1 | 88.9 |
| 15 Mal'ta | 179 | 28.5 | 71.5 |
| 16 Maly Yaloman | 178 | 4.5 | 95.5 |
| 17 Mokhovo Mine | 11 | 63.6 | 36.4 |
| 18 Nizhneudinskaya | 36 | 0.0 | 100.0 |
| 19 Okladnikov | 168 | 22.0 | 78.0 |
| 20 Proskuryakova | 47 | 12.8 | 87.2 |
| 21 Razboinich'ya | 607 | 10.9 | 89.1 |
| 22 Sarala | 44 | 6.8 | 93.2 |
| 23 Shestakova | 157 | 32.5 | 67.5 |
| 24 Straschnaya | 527 | 11.4 | 88.6 |
| 25 Ust-Kan | 1654 | 1.9 | 98.1 |
| 26 Ust-Kova | 116 | 15.5 | 84.5 |
| 27 Varvarina Gora | 862 | 9.6 | 90.4 |
| 28 Volchiya Griva | 113 | 46.0 | 54.0 |
| 29 Yelenev | 241 | 0.8 | 99.2 |
| 30 Zhemchuzhnaya | 0 | – | – |

**Table A1.12** End-hollowing (piece count, %)

| Site | No. | Present | Absent | Uncertain |
|---|---|---|---|---|
| 1 Afontova Gora | (5000) | – | – | – |
| 2 Boisman II | 1232 | 3.2 | 96.8 | 0.0 |
| 3 Bolshoi Yakor I | 263 | 0.4 | 99.6 | 0.0 |
| 4 Borabashevskaya | 30 | 0.0 | 100.0 | 0.0 |
| 5 Denisova | 115 | 2.6 | 97.4 | 0.0 |
| 6 Dvuglaska | 205 | 4.4 | 95.6 | 0.0 |
| 7 Gosudarev Log I | 13 | 15.4 | 84.6 | 0.0 |
| 8 Kamenka | 240 | 0.8 | 99.2 | 0.0 |
| 9 Kaminnaya | 628 | 1.9 | 98.1 | 0.0 |
| 10 Kara-Bom | 42 | 4.8 | 95.2 | 0.0 |
| 11 Kirkalinskaya | 4 | 50.0 | 50.0 | 0.0 |
| 12 Krasny Yar | 185 | 30.8 | 69.2 | 0.0 |
| 13 Kurla I | 72 | 0.0 | 100.0 | 0.0 |
| 14 Malaya Seeya | 144 | 0.7 | 99.3 | 0.0 |
| 15 Mal'ta | 186 | 4.3 | 95.7 | 0.0 |
| 16 Maly Yaloman | 116 | 9.5 | 86.2 | 4.3 |
| 17 Mokhovo Mine | 12 | 0.0 | 100.0 | 0.0 |
| 18 Nizhneudinskaya | 36 | 44.4 | 55.6 | 0.0 |
| 19 Okladnikov | 166 | 17.5 | 82.5 | 0.0 |
| 20 Proskuryakova | 47 | 29.8 | 70.2 | 0.0 |
| 21 Razboinich'ya | 543 | 8.1 | 91.9 | 0.0 |
| 22 Sarala | 44 | 15.9 | 84.1 | 0.0 |
| 23 Shestakova | 158 | 1.9 | 98.1 | 0.0 |
| 24 Straschnaya | 527 | 0.9 | 99.1 | 0.0 |
| 25 Ust-Kan | 1654 | 9.5 | 90.5 | 0.0 |
| 26 Ust-Kova | 116 | 3.5 | 96.5 | 0.0 |
| 27 Varvarina Gora | 418 | 1.2 | 98.8 | 0.0 |
| 28 Volchiya Griva | 113 | 0.9 | 99.1 | 0.0 |
| 29 Yelenev | 241 | 0.4 | 99.6 | 0.0 |
| 30 Zhemchuzhnaya | 0 | – | – | – |

**Table A1.13** Notching (piece count, notches per piece, %)

| Site | No. | 0 | 1 | 2 | 3 | 4 | 5 | 6 | 7 | >7 | Total |
|---|---|---|---|---|---|---|---|---|---|---|---|
| 1 Afontova Gora | (5000) | – | – | – | – | – | – | – | – | – | – |
| 2 Boisman II | 541 | 97.4 | 2.0 | 0.2 | 0.2 | 0.2 | 0.0 | 0.0 | 0.0 | 0.0 | 2.6 |
| 3 Bolshoi Yakor I | 263 | 93.5 | 5.3 | 1.1 | 0.0 | 0.0 | 0.0 | 0.0 | 0.0 | 0.0 | 6.5 |
| 4 Borabashevskaya | 30* | – | – | – | – | – | – | – | – | – | – |
| 5 Denisova | 115 | 77.4 | 16.5 | 2.6 | 1.7 | 0.9 | 0.0 | 0.9 | 0.0 | 0.0 | 22.6 |
| 6 Dvuglaska | 205 | 84.9 | 8.8 | 3.4 | 1.5 | 0.5 | 1.0 | 0.0 | 0.0 | 0.0 | 15.1 |
| 7 Gosudarev Log I | 14 | 92.9 | 7.1 | 0.0 | 0.0 | 0.0 | 0.0 | 0.0 | 0.0 | 0.0 | 7.1 |
| 8 Kamenka | 549 | 90.2 | 6.0 | 2.4 | 1.3 | 0.2 | 0.0 | 0.0 | 0.0 | 0.0 | 9.8 |
| 9 Kaminnaya | 768 | 89.2 | 7.2 | 2.7 | 0.7 | 0.0 | 0.1 | 0.0 | 0.0 | 0.1 | 10.8 |
| 10 Kara-Bom | 106 | 95.3 | 4.7 | 0.0 | 0.0 | 0.0 | 0.0 | 0.0 | 0.0 | 0.0 | 4.7 |
| 11 Kirkalinskaya | 4 | 75.0 | 0.0 | 0.0 | 0.0 | 0.0 | 0.0 | 25.0 | 0.0 | 0.0 | 25.0 |
| 12 Krasny Yar | 200 | 91.0 | 2.5 | 1.5 | 3.0 | 0.5 | 0.0 | 1.5 | 0.0 | 0.0 | 0.9 |
| 13 Kurla I | 72 | 97.2 | 0.0 | 2.8 | 0.0 | 0.0 | 0.0 | 0.0 | 0.0 | 0.0 | 2.8 |
| 14 Malaya Seeya | 144 | 88.7 | 8.4 | 2.1 | 0.7 | 0.0 | 0.0 | 0.0 | 0.0 | 0.0 | 11.3 |
| 15 Mal'ta | 186 | 96.8 | 0.5 | 1.1 | 0.5 | 0.0 | 0.5 | 0.5 | 0.0 | 0.0 | 3.2 |
| 16 Maly Yaloman | 178 | 77.0 | 10.1 | 4.5 | 5.1 | 1.7 | 0.6 | 0.6 | 0.0 | 0.6 | 23.0 |
| 17 Mokhovo Mine | 12 | 100.0 | 0.0 | 0.0 | 0.0 | 0.0 | 0.0 | 0.0 | 0.0 | 0.0 | 0.0 |
| 18 Nizhneudinskaya | 36 | 58.3 | 22.2 | 8.3 | 5.6 | 2.8 | 2.8 | 0.0 | 0.0 | 0.0 | 41.7 |
| 19 Okladnikov | 168 | 60.8 | 13.2 | 9.6 | 4.2 | 2.4 | 2.4 | 1.8 | 0.6 | 4.8 | 39.2 |
| 20 Proskuryakova | 47 | 14.9 | 10.6 | 8.5 | 19.1 | 19.1 | 4.3 | 2.1 | 8.5 | 12.8 | 85.1 |
| 21 Razboinich'ya | 618 | 76.4 | 12.9 | 6.3 | 2.7 | 0.8 | 0.6 | 0.2 | 0.0 | 0.0 | 23.6 |
| 22 Sarala | 44 | 75.0 | 15.9 | 4.5 | 0.0 | 4.5 | 0.0 | 0.0 | 0.0 | 0.0 | 25.0 |
| 23 Shestakova | 159 | 98.7 | 1.3 | 0.0 | 0.0 | 0.0 | 0.0 | 0.0 | 0.0 | 0.0 | 1.3 |
| 24 Straschnaya | 527 | 88.6 | 7.8 | 2.5 | 0.6 | 0.2 | 0.2 | 0.0 | 0.2 | 0.0 | 11.4 |
| 25 Ust-Kan | 1653 | 90.7 | 6.8 | 1.5 | 0.6 | 0.1 | 0.0 | 0.1 | 0.0 | 0.1 | 9.2 |
| 26 Ust-Kova | 116 | 97.4 | 2.6 | 0.0 | 0.0 | 0.0 | 0.0 | 0.0 | 0.0 | 0.0 | 2.6 |
| 27 Varvarina Gora | 856 | 94.3 | 4.2 | 1.2 | 0.2 | 0.0 | 0.1 | 0.0 | 0.0 | 0.0 | 5.7 |
| 28 Volchiya Griva | 113 | 100.0 | 0.0 | 0.0 | 0.0 | 0.0 | 0.0 | 0.0 | 0.0 | 0.0 | 0.0 |
| 29 Yelenev | 241 | 94.6 | 5.0 | 0.4 | 0.0 | 0.0 | 0.0 | 0.0 | 0.0 | 0.0 | 5.4 |
| 30 Zhemchuzhnaya | 0 | – | – | – | – | – | – | – | – | – | – |

* Notching not looked for in Borabashevskaya assemblage.

**Table A1.14** Tooth scratches (piece count, scratches per piece, %)

| Site | No. | 0 | 1 | 2 | 3 | 4 | 5 | 6 | 7 | >7 | Total |
|---|---|---|---|---|---|---|---|---|---|---|---|
| 1 Afontova Gora | (5000) | – | – | – | – | – | – | – | – | – | – |
| 2 Boisman II | 1235 | 98.4 | 0.3 | 0.2 | 0.3 | 0.2 | 0.3 | 0.2 | 0.0 | 0.0 | 1.6 |
| 3 Bolshoi Yakor I | 263 | 98.1 | 0.8 | 0.0 | 0.0 | 0.0 | 0.4 | 0.4 | 0.0 | 0.4 | 1.9 |
| 4 Borabashevskaya | 30 | 96.7 | 3.3 | 0.0 | 0.0 | 0.0 | 0.0 | 0.0 | 0.0 | 0.0 | 3.3 |
| 5 Denisova | 115 | 79.1 | 2.6 | 5.2 | 4.3 | 1.7 | 1.7 | 0.9 | 0.0 | 4.3 | 20.9 |
| 6 Dvuglaska | 205 | 61.5 | 10.7 | 7.3 | 5.4 | 4.9 | 2.4 | 0.0 | 0.0 | 8.0 | 38.5 |
| 7 Gosudarev Log I | 14 | 100.0 | 0.0 | 0.0 | 0.0 | 0.0 | 0.0 | 0.0 | 0.0 | 0.0 | 0.0 |
| 8 Kamenka | 541 | 97.2 | 0.2 | 0.7 | 0.0 | 0.0 | 0.4 | 0.0 | 0.0 | 1.5* | 2.8 |
| 9 Kaminnaya | 763 | 84.4 | 4.7 | 2.4 | 2.1 | 1.3 | 0.8 | 0.8 | 1.2 | 2.2 | 15.6 |
| 10 Kara-Bom | 92 | 89.1 | 1.1 | 4.3 | 1.1 | 1.1 | 0.0 | 0.0 | 0.0 | 3.3 | 10.9 |
| 11 Kirkalinskaya | 4 | 25.0 | 25.0 | 0.0 | 0.0 | 0.0 | 25.0 | 0.0 | 0.0 | 25.0 | 75.0 |
| 12 Krasny Yar | 195 | 68.7 | 3.6 | 4.1 | 3.1 | 3.1 | 4.6 | 1.0 | 0.0 | 11.8 | 31.3 |
| 13 Kurla I | 71 | 97.2 | 1.4 | 0.0 | 0.0 | 0.0 | 0.0 | 0.0 | 0.0 | 1.4 | 2.8 |
| 14 Malaya Seeya | 143 | 95.8 | 0.0 | 1.4 | 1.4 | 0.7 | 0.0 | 0.0 | 0.0 | 0.0 | 4.2 |
| 15 Mal'ta | 178 | 97.2 | 0.6 | 0.6 | 0.6 | 0.6 | 0.0 | 0.0 | 0.0 | 0.6 | 2.8 |
| 16 Maly Yaloman | 175 | 71.4 | 5.1 | 4.6 | 2.3 | 3.4 | 4.6 | 0.6 | 0.6 | 7.4 | 28.6 |
| 17 Mokhovo Mine | 12 | 91.7 | 0.0 | 0.0 | 8.3 | 0.0 | 0.0 | 0.0 | 0.0 | 0.0 | 8.3 |
| 18 Nizhneudinskaya | 35 | 51.4 | 11.4 | 5.7 | 8.6 | 5.7 | 8.6 | 0.0 | 2.9 | 5.8 | 48.6 |
| 19 Okladnikov | 168 | 43.0 | 5.5 | 5.5 | 3.6 | 9.1 | 3.6 | 3.0 | 3.6 | 23.0** | 57.0 |
| 20 Proskuryakova | 47 | 2.1 | 2.1 | 0.0 | 6.4 | 0.0 | 4.3 | 4.3 | 0.0 | 80.8 | 97.9 |
| 21 Razboinich'ya | 661 | 61.9 | 11.0 | 7.3 | 6.0 | 4.7 | 1.7 | 1.7 | 1.8 | 3.9*** | 37.1 |
| 22 Sarala | 44 | 52.3 | 20.5 | 11.4 | 6.8 | 4.5 | 0.0 | 2.3 | 0.0 | 2.3 | 47.7 |
| 23 Shestakova | 159 | 100.0 | 0.0 | 0.0 | 0.0 | 0.0 | 0.0 | 0.0 | 0.0 | 0.0 | 0.0 |
| 24 Straschnaya | 526 | 77.1 | 6.3 | 5.5 | 3.0 | 1.5 | 1.0 | 0.6 | 1.1 | 3.8 | 22.9 |
| 25 Ust-Kan | 1654 | 91.0 | 3.0 | 2.5 | 1.4 | 0.8 | 0.2 | 0.3 | 0.1 | 0.8 | 9.0 |
| 26 Ust-Kova | 116 | 95.7 | 0.0 | 1.7 | 1.7 | 0.0 | 0.0 | 0.0 | 0.0 | 0.9 | 4.3 |
| 27 Varvarina Gora | 703 | 98.1 | 0.1 | 0.3 | 0.0 | 0.3 | 0.4 | 0.1 | 0.1 | 0.3 | 1.9 |
| 28 Volchiya Griva | 113 | 100.0 | 0.0 | 0.0 | 0.0 | 0.0 | 0.0 | 0.0 | 0.0 | 0.0 | 0.0 |
| 29 Yelenev | 241 | 98.3 | 0.0 | 0.4 | 0.8 | 0.0 | 0.0 | 0.0 | 0.0 | 0.0 | 1.7 |
| 30 Zhemchuzhnaya | 0 | – | – | – | – | – | – | – | – | – | – |

* One Kamenka piece has 30 scratches.

** Six Okladnikov pieces have 25, 35, 44, 54, 55, and 105 tooth scratches.

*** The maximum number of scratches on a Razboinich'ya piece is 50.

**Table A1.15** Tooth dints (piece count, dints per piece, %)

| Site | No. | 0 | 1 | 2 | 3 | 4 | 5 | 6 | 7 | >7 | Total |
|---|---|---|---|---|---|---|---|---|---|---|---|
| 1 Afontova Gora | (5000) | – | – | – | – | – | – | – | – | – | – |
| 2 Boisman II | 512 | 91.8 | 0.6 | 0.6 | 1.2 | 0.8 | 1.4 | 1.2 | 0.0 | 2.5 | 8.2 |
| 3 Bolshoi Yakor I | 263 | 92.0 | 0.0 | 0.4 | 2.3 | 1.5 | 1.9 | 0.8 | 0.4 | 0.8 | 8.0 |
| 4 Borabashevskaya | 30* | – | – | – | – | – | – | – | – | – | – |
| 5 Denisova | 115 | 77.4 | 7.8 | 0.9 | 4.3 | 0.9 | 3.5 | 0.0 | 0.0 | 5.2 | 22.6 |
| 6 Dvuglaska | 205 | 56.6 | 9.8 | 7.3 | 5.4 | 4.4 | 2.9 | 2.0 | 2.0 | 9.9 | 43.4 |
| 7 Gosudarev Log I | 14 | 100.0 | 0.0 | 0.0 | 0.0 | 0.0 | 0.0 | 0.0 | 0.0 | 0.0 | 0.0 |
| 8 Kamenka | 544 | 96.7 | 0.4 | 0.6 | 0.6 | 0.4 | 0.6 | 0.4 | 0.0 | 0.6 | 3.3 |
| 9 Kaminnaya | 762 | 75.9 | 6.2 | 4.7 | 3.8 | 2.4 | 2.1 | 1.4 | 0.7 | 2.8 | 24.1 |
| 10 Kara-Bom | 89 | 92.1 | 2.2 | 1.1 | 0.0 | 1.1 | 0.0 | 0.0 | 2.2 | 1.1 | 7.9 |
| 11 Kirkalinskaya | 4 | 25.0 | 0.0 | 0.0 | 0.0 | 25.0 | 25.0 | 0.0 | 0.0 | 25.0 | 75.0 |
| 12 Krasny Yar | 194 | 75.3 | 0.5 | 3.1 | 3.6 | 2.6 | 4.1 | 1.5 | 0.5 | 8.8 | 24.7 |
| 13 Kurla I | 72 | 95.8 | 0.0 | 1.4 | 1.4 | 0.0 | 1.4 | 0.0 | 0.0 | 0.0 | 4.2 |
| 14 Malaya Seeya | 143 | 95.1 | 1.4 | 0.0 | 0.7 | 1.4 | 0.7 | 0.0 | 0.0 | 0.7 | 4.9 |
| 15 Mal'ta | 178 | 98.3 | 0.0 | 0.6 | 0.0 | 0.0 | 0.6 | 0.0 | 0.0 | 0.6 | 1.7 |
| 16 Maly Yaloman | 175 | 67.4 | 6.3 | 5.7 | 2.3 | 2.3 | 2.3 | 0.6 | 2.9 | 10.2 | 32.6 |
| 17 Mokhovo Mine | 12 | 91.7 | 0.0 | 0.0 | 0.0 | 0.0 | 0.0 | 0.0 | 0.0 | 8.3 | 8.3 |
| 18 Nizhneudinskaya | 36 | 30.6 | 13.9 | 11.1 | 5.6 | 2.8 | 8.3 | 0.0 | 5.6 | 22.4 | 69.4 |
| 19 Okladnikov | 167 | 44.3 | 7.8 | 9.6 | 5.4 | 6.0 | 4.2 | 1.2 | 1.2 | 20.4 | 55.7 |
| 20 Proskuryakova | 47 | 6.4 | 2.1 | 0.0 | 4.3 | 4.3 | 12.8 | 6.4 | 8.5 | 55.3 | 93.6 |
| 21 Razboinich'ya | 629 | 47.2 | 12.4 | 8.6 | 6.2 | 4.1 | 3.2 | 2.4 | 1.9 | 14.0 | 52.8 |
| 22 Sarala | 43 | 11.6 | 7.0 | 14.0 | 16.3 | 11.6 | 9.3 | 4.7 | 4.7 | 23.3 | 88.4 |
| 23 Shestakova | 159 | 100.0 | 0.0 | 0.0 | 0.0 | 0.0 | 0.0 | 0.0 | 0.0 | 0.0 | 0.0 |
| 24 Straschnaya | 526 | 68.1 | 9.1 | 6.5 | 4.9 | 1.9 | 3.0 | 2.3 | 0.2 | 4.0 | 31.9 |
| 25 Ust-Kan | 1653 | 86.3 | 3.7 | 2.5 | 1.6 | 1.4 | 0.8 | 0.8 | 0.7 | 2.2 | 13.7 |
| 26 Ust-Kova | 116 | 94.8 | 0.0 | 0.0 | 0.9 | 0.9 | 1.7 | 0.0 | 0.9 | 0.9 | 5.2 |
| 27 Varvarina Gora | 682 | 98.1 | 0.0 | 0.3 | 0.6 | 0.4 | 0.0 | 0.4 | 0.0 | 0.1 | 1.9 |
| 28 Volchiya Griva | 113 | 99.1 | 0.0 | 0.9 | 0.0 | 0.0 | 0.0 | 0.0 | 0.0 | 0.0 | 0.9 |
| 29 Yelenev | 241 | 92.1 | 1.2 | 2.1 | 0.8 | 0.8 | 1.2 | 0.4 | 0.0 | 1.2 | 7.9 |
| 30 Zhemchuzhnaya | 0 | – | – | – | – | – | – | – | – | – | – |

* Borabashevskaya definitely has tooth dints, but they were not counted.

**Table A1.16** Pseudo-cuts (piece counts, pseudo-cuts per piece, %)

| Site | No. | 0 | 1 | 2 | 3 | 4 | 5 | 6 | 7 | >7 | Total |
|---|---|---|---|---|---|---|---|---|---|---|---|
| 1 Afontova Gora | (5000) | – | – | – | – | – | – | – | – | – | – |
| 2 Boisman II | 513 | 99.6 | 0.2 | 0.2 | 0.0 | 0.0 | 0.0 | 0.0 | 0.0 | 0.0 | 0.4 |
| 3 Bolshoi Yakor I | 263 | 98.9 | 1.1 | 0.0 | 0.0 | 0.0 | 0.0 | 0.0 | 0.0 | 0.0 | 1.1 |
| 4 Borabashevskaya | 30* | – | – | – | – | – | – | – | – | – | – |
| 5 Denisova | 115 | 70.4 | 12.2 | 10.4 | 3.5 | 1.7 | 0.9 | 0.0 | 0.0 | 0.9 | 28.6 |
| 6 Dvuglaska | 205 | 97.1 | 2.0 | 0.0 | 0.0 | 0.5 | 0.5 | 0.0 | 0.0 | 0.0 | 2.9 |
| 7 Gosudarev Log I | 14 | 100.0 | 0.0 | 0.0 | 0.0 | 0.0 | 0.0 | 0.0 | 0.0 | 0.0 | 0.0 |
| 8 Kamenka | 546 | 99.3 | 0.2 | 0.2 | 0.0 | 0.0 | 0.0 | 0.0 | 0.0 | 0.0 | 0.4 |
| 9 Kaminnaya | 763 | 96.7 | 1.7 | 0.9 | 0.4 | 0.0 | 0.1 | 0.1 | 0.0 | 0.0 | 3.3 |
| 10 Kara-Bom | 90 | 100.0 | 0.0 | 0.0 | 0.0 | 0.0 | 0.0 | 0.0 | 0.0 | 0.0 | 0.0 |
| 11 Kirkalinskaya | 4 | 75.0 | 25.0 | 0.0 | 0.0 | 0.0 | 0.0 | 0.0 | 0.0 | 0.0 | 25.0 |
| 12 Krasny Yar | 200 | 94.5 | 3.5 | 0.5 | 1.0 | 0.5 | 0.0 | 0.0 | 0.0 | 0.0 | 5.5 |
| 13 Kurla I | 72 | 98.6 | 0.0 | 1.4 | 0.0 | 0.0 | 0.0 | 0.0 | 0.0 | 0.0 | 1.4 |
| 14 Malaya Seeya | 144 | 99.3 | 0.0 | 0.0 | 0.0 | 0.0 | 0.0 | 0.7 | 0.0 | 0.0 | 0.7 |
| 15 Mal'ta | 178 | 100.0 | 0.0 | 0.0 | 0.0 | 0.0 | 0.0 | 0.0 | 0.0 | 0.0 | 0.0 |
| 16 Maly Yaloman | 174 | 93.7 | 2.3 | 1.1 | 1.1 | 0.6 | 0.6 | 0.0 | 0.6 | 0.0 | 6.3 |
| 17 Mokhovo Mine | 12 | 100.0 | 0.0 | 0.0 | 0.0 | 0.0 | 0.0 | 0.0 | 0.0 | 0.0 | 0.0 |
| 18 Nizhneudinskaya | 36 | 88.9 | 8.3 | 2.8 | 0.0 | 0.0 | 0.0 | 0.0 | 0.0 | 0.0 | 11.1 |
| 19 Okladnikov | 167 | 92.8 | 3.0 | 3.0 | 0.6 | 0.0 | 0.0 | 0.6 | 0.0 | 0.0 | 7.2 |
| 20 Proskuryakova | 47 | 91.5 | 2.1 | 0.0 | 4.3 | 2.1 | 0.0 | 0.0 | 0.0 | 0.0 | 8.5 |
| 21 Razboinich'ya | 638 | 96.9 | 2.3 | 0.5 | 0.2 | 0.0 | 0.0 | 0.0 | 0.0 | 0.2 | 3.1 |
| 22 Sarala | 41 | 90.2 | 2.4 | 0.0 | 2.4 | 2.4 | 0.0 | 2.4 | 0.0 | 0.0 | 9.8 |
| 23 Shestakova | 159 | 99.4 | 0.0 | 0.0 | 0.0 | 0.0 | 0.0 | 0.0 | 0.0 | 0.6 | 0.6 |
| 24 Straschnaya | 526 | 98.9 | 0.2 | 0.4 | 0.2 | 0.2 | 0.0 | 0.0 | 0.0 | 0.2 | 1.1 |
| 25 Ust-Kan | 1654 | 96.1 | 1.9 | 0.3 | 0.2 | 0.4 | 0.2 | 0.1 | 0.2 | 0.2 | 3.9 |
| 26 Ust-Kova | 116 | 98.3 | 1.7 | 0.0 | 0.0 | 0.0 | 0.0 | 0.0 | 0.0 | 0.0 | 1.7 |
| 27 Varvarina Gora | 689 | 100.0 | 0.0 | 0.0 | 0.0 | 0.0 | 0.0 | 0.0 | 0.0 | 0.0 | 0.0 |
| 28 Volchiya Griva | 113 | 100.0 | 0.0 | 0.0 | 0.0 | 0.0 | 0.0 | 0.0 | 0.0 | 0.0 | 0.0 |
| 29 Yelenev | 241 | 92.1 | 1.2 | 2.1 | 0.8 | 0.8 | 1.2 | 0.4 | 0.0 | 0.4 | 7.9 |
| 30 Zhemchuzhnaya | 0 | – | – | – | – | – | – | – | – | – | – |

* Pseudo-cuts were not looked for in the Borabashevskaya assemblage.

**Table A1.17** Abrasions (piece count, abrasion grooves per piece, %)

| Site | No. | 0 | 1 | 2 | 3 | 4 | 5 | 6 | 7 | >7 | Total |
|---|---|---|---|---|---|---|---|---|---|---|---|
| 1 Afontova Gora | (5000) | – | – | – | – | – | – | – | – | – | – |
| 2 Boisman II | 512 | 98.2 | 0.0 | 0.0 | 0.4 | 0.0 | 0.0 | 0.2 | 0.0 | 1.2 | 1.8 |
| 3 Bolshoi Yakor I | 263 | 98.9 | 0.0 | 0.0 | 0.0 | 0.0 | 0.0 | 0.0 | 0.0 | 1.1 | 1.1 |
| 4 Borabashevskaya | 30 | 100.0 | 0.0 | 0.0 | 0.0 | 0.0 | 0.0 | 0.0 | 0.0 | 0.0 | 0.0 |
| 5 Denisova | 115 | 96.5 | 0.0 | 0.0 | 0.9 | 0.0 | 0.0 | 0.0 | 0.0 | 2.6 | 3.5 |
| 6 Dvuglaska | 205 | 100.0 | 0.0 | 0.0 | 0.0 | 0.0 | 0.0 | 0.0 | 0.0 | 0.0 | 0.0 |
| 7 Gosudarev Log I | 14 | 100.0 | 0.0 | 0.0 | 0.0 | 0.0 | 0.0 | 0.0 | 0.0 | 0.0 | 0.0 |
| 8 Kamenka | 544 | 99.8 | 0.2 | 0.0 | 0.0 | 0.0 | 0.0 | 0.0 | 0.0 | 0.0 | 0.2 |
| 9 Kaminnaya | 763 | 98.4 | 0.1 | 0.0 | 0.0 | 0.1 | 0.1 | 0.0 | 0.3 | 0.8 | 1.6 |
| 10 Kara-Bom | 90 | 98.9 | 0.0 | 0.0 | 0.0 | 0.0 | 0.0 | 1.1 | 0.0 | 0.0 | 1.1 |
| 11 Kirkalinskaya | 4 | 100.0 | 0.0 | 0.0 | 0.0 | 0.0 | 0.0 | 0.0 | 0.0 | 0.0 | 0.0 |
| 12 Krasny Yar | 201 | 98.5 | 0.0 | 0.0 | 0.0 | 0.0 | 0.0 | 0.0 | 0.0 | 1.5 | 1.5 |
| 13 Kurla I | 72 | 98.6 | 0.0 | 0.0 | 0.0 | 0.0 | 0.0 | 0.0 | 0.0 | 1.4 | 1.4 |
| 14 Malaya Seeya | 144 | 100.0 | 0.0 | 0.0 | 0.0 | 0.0 | 0.0 | 0.0 | 0.0 | 0.0 | 0.0 |
| 15 Mal'ta | 177 | 99.4 | 0.0 | 0.0 | 0.0 | 0.0 | 0.0 | 0.0 | 0.0 | 0.6 | 0.6 |
| 16 Maly Yaloman | 173 | 98.8 | 0.0 | 0.0 | 0.0 | 0.0 | 0.6 | 0.0 | 0.0 | 0.6 | 1.2 |
| 17 Mokhovo Mine | 12 | 100.0 | 0.0 | 0.0 | 0.0 | 0.0 | 0.0 | 0.0 | 0.0 | 0.0 | 0.0 |
| 18 Nizhneudinskaya | 36 | 94.4 | 0.0 | 0.0 | 0.0 | 0.0 | 0.0 | 0.0 | 0.0 | 5.6 | 5.6 |
| 19 Okladnikov | 167 | 98.8 | 0.6 | 0.0 | 0.0 | 0.0 | 0.0 | 0.0 | 0.0 | 0.6 | 1.2 |
| 20 Proskuryakova | 47 | 93.6 | 2.1 | 2.1 | 2.1 | 0.0 | 0.0 | 0.0 | 0.0 | 0.0 | 6.4 |
| 21 Razboinich'ya | 623 | 99.5 | 0.2 | 0.0 | 0.0 | 0.2 | 0.0 | 0.0 | 0.0 | 0.2 | 0.5 |
| 22 Sarala | 44 | 100.0 | 0.0 | 0.0 | 0.0 | 0.0 | 0.0 | 0.0 | 0.0 | 0.0 | 0.0 |
| 23 Shestakova | 159 | 100.0 | 0.0 | 0.0 | 0.0 | 0.0 | 0.0 | 0.0 | 0.0 | 0.0 | 0.0 |
| 24 Straschnaya | 526 | 99.8 | 0.0 | 0.0 | 0.0 | 0.2 | 0.0 | 0.0 | 0.0 | 0.0 | 0.2 |
| 25 Ust-Kan | 1653 | 97.3 | 0.1 | 0.1 | 0.1 | 0.1 | 0.1 | 0.4 | 0.4 | 1.7 | 2.7 |
| 26 Ust-Kova | 116 | 99.1 | 0.9 | 0.0 | 0.0 | 0.0 | 0.0 | 0.0 | 0.0 | 0.0 | 0.9 |
| 27 Varvarina Gora | 686 | 99.6 | 0.0 | 0.0 | 0.0 | 0.0 | 0.3 | 0.0 | 0.0 | 0.1 | 0.4 |
| 28 Volchiya Griva | 113 | 100.0 | 0.0 | 0.0 | 0.0 | 0.0 | 0.0 | 0.0 | 0.0 | 0.0 | 0.0 |
| 29 Yelenev | 241 | 96.3 | 0.0 | 0.8 | 0.0 | 0.8 | 0.4 | 0.0 | 0.0 | 1.6 | 3.7 |
| 30 Zhemchuzhnaya | 0 | – | – | – | – | – | – | – | – | – | – |

**Table A1.18** Polishing (piece count, %)

| Sites | No. | None | End | Middle | Both | Uncertain |
|---|---|---|---|---|---|---|
| 1 Afontova Gora | (5000) | – | – | – | – | – |
| 2 Boisman II | 485 | 83.3 | 14.2 | 0.8 | 1.6 | 0.0 |
| 3 Bolshoi Yakor I | 257 | 87.9 | 4.7 | 0.8 | 6.6 | 0.0 |
| 4 Borabashevskaya | 30* | – | – | – | – | – |
| 5 Denisova | 112 | 17.0 | 33.0 | 3.6 | 46.4 | 0.0 |
| 6 Dvuglaska | 203 | 5.4 | 9.4 | 0.5 | 84.7 | 0.0 |
| 7 Gosudarev Log I | 14 | 71.4 | 28.6 | 0.0 | 0.0 | 0.0 |
| 8 Kamenka | 540 | 10.0 | 24.8 | 4.0 | 61.3 | 0.0 |
| 9 Kaminnaya | 770 | 21.7 | 13.4 | 1.2 | 63.8 | 0.0 |
| 10 Kara-Bom | 99 | 24.2 | 7.1 | 2.0 | 66.7 | 0.0 |
| 11 Kirkalinskaya | 4 | 0.0 | 50.0 | 0.0 | 50.0 | 0.0 |
| 12 Krasny Yar | 200 | 7.0 | 5.5 | 3.0 | 84.5 | 0.0 |
| 13 Kurla I | 72 | 48.6 | 18.1 | 1.4 | 31.9 | 0.0 |
| 14 Malaya Seeya | 144 | 18.3 | 13.4 | 2.1 | 66.2 | 0.0 |
| 15 Mal'ta | 181 | 25.4 | 34.8 | 1.7 | 38.1 | 0.0 |
| 16 Maly Yaloman | 178 | 10.7 | 12.9 | 1.1 | 75.3 | 0.0 |
| 17 Mokhovo Mine | 12 | 16.7 | 33.3 | 0.0 | 50.0 | 0.0 |
| 18 Nizhneudinskaya | 36 | 27.8 | 61.1 | 0.0 | 11.1 | 0.0 |
| 19 Okladnikov | 167 | 24.5 | 19.8 | 4.2 | 49.7 | 1.8 |
| 20 Proskuryakova | 47 | 2.1 | 6.4 | 0.0 | 91.4 | 0.0 |
| 21 Razboinich'ya | 598 | 40.6 | 30.6 | 1.7 | 27.1 | 0.0 |
| 22 Sarala | 44 | 20.5 | 15.9 | 2.3 | 61.4 | 0.0 |
| 23 Shestakova | 159 | 95.7 | 1.8 | 1.8 | 0.6 | 0.0 |
| 24 Straschnaya | 511 | 30.3 | 5.7 | 1.0 | 63.0 | 0.0 |
| 25 Ust-Kan | 1651 | 16.8 | 11.6 | 1.5 | 70.0 | 0.0 |
| 26 Ust-Kova | 116 | 66.7 | 24.6 | 4.4 | 4.4 | 0.0 |
| 27 Varvarina Gora | 834 | 4.0 | 24.5 | 0.4 | 71.2 | 0.0 |
| 28 Volchiya Griva | 109 | 100.0 | 0.0 | 0.0 | 0.0 | 0.0 |
| 29 Yelenev | 241 | 29.6 | 30.8 | 2.5 | 37.1 | 0.0 |
| 30 Zhemchuzhnaya | 0 | – | – | – | – | – |

* Polishing present but not counted in the Borabashevskaya assemblage.

**Table A1.19** Embedded fragments (piece count, embedded fragments per piece, %)

| Site | No. | 0 | 1 | 2 | 3 | 4 | 5 | 6 | 7 | >7 | Total |
|---|---|---|---|---|---|---|---|---|---|---|---|
| 1 Afontova Gora | (5000) | – | – | – | – | – | – | – | – | – | – |
| 2 Boisman II | 513 | 95.3 | 1.7 | 1.2 | 1.0 | 0.4 | 0.2 | 0.2 | 0.0 | 0.0 | 4.7 |
| 3 Bolshoi Yakor I | 263 | 97.0 | 1.9 | 0.0 | 0.0 | 0.8 | 0.0 | 0.0 | 0.4 | 0.0 | 0.3 |
| 4 Borabashevskaya | 30* | – | – | – | – | – | – | – | – | – | – |
| 5 Denisova | 83 | 98.8 | 0.0 | 1.2 | 0.0 | 0.0 | 0.0 | 0.0 | 0.0 | 0.0 | 1.2 |
| 6 Dvuglaska | 205 | 99.0 | 1.0 | 0.0 | 0.0 | 0.0 | 0.0 | 0.0 | 0.0 | 0.0 | 1.0 |
| 7 Gosudarev Log I | 14 | 100.0 | 0.0 | 0.0 | 0.0 | 0.0 | 0.0 | 0.0 | 0.0 | 0.0 | 0.0 |
| 8 Kamenka | 547 | 98.5 | 0.9 | 0.2 | 0.4 | 0.0 | 0.0 | 0.0 | 0.0 | 0.0 | 1.5 |
| 9 Kaminnaya | 770 | 97.1 | 0.9 | 1.0 | 0.5 | 0.3 | 0.0 | 0.0 | 0.0 | 0.1 | 2.9 |
| 10 Kara-Bom | 105 | 97.1 | 2.9 | 0.0 | 0.0 | 0.0 | 0.0 | 0.0 | 0.0 | 0.0 | 2.9 |
| 11 Kirkalinskaya | 4 | 100.0 | 0.0 | 0.0 | 0.0 | 0.0 | 0.0 | 0.0 | 0.0 | 0.0 | 0.0 |
| 12 Krasny Yar | 203 | 98.5 | 0.5 | 0.5 | 0.5 | 0.0 | 0.0 | 0.0 | 0.0 | 0.0 | 1.5 |
| 13 Kurla I | 72 | 93.1 | 0.0 | 5.5 | 0.0 | 0.0 | 0.0 | 0.0 | 0.0 | 1.4 | 6.9 |
| 14 Malaya Seeya | 144 | 95.8 | 3.7 | 0.7 | 0.0 | 0.0 | 0.0 | 0.0 | 0.0 | 0.0 | 4.2 |
| 15 Mal'ta | 186 | 97.3 | 0.5 | 0.5 | 0.5 | 0.5 | 0.0 | 0.0 | 0.0 | 0.5 | 2.7 |
| 16 Maly Yaloman | 177 | 97.7 | 1.1 | 0.6 | 0.6 | 0.0 | 0.0 | 0.0 | 0.0 | 0.0 | 2.3 |
| 17 Mokhovo Mine | 12 | 100.0 | 0.0 | 0.0 | 0.0 | 0.0 | 0.0 | 0.0 | 0.0 | 0.0 | 0.0 |
| 18 Nizhneudinskaya | 36 | 66.7 | 11.1 | 5.6 | 11.1 | 0.0 | 0.0 | 0.0 | 0.0 | 5.6 | 33.3 |
| 19 Okladnikov | 168 | 96.4 | 1.2 | 1.2 | 0.6 | 0.6 | 0.0 | 0.0 | 0.0 | 0.0 | 3.6 |
| 20 Proskuryakova | 47 | 93.6 | 2.1 | 2.1 | 2.1 | 0.0 | 0.0 | 0.0 | 0.0 | 0.0 | 6.4 |
| 21 Razboinich'ya | 611 | 91.2 | 5.7 | 1.3 | 0.6 | 0.5 | 0.6 | 0.0 | 0.0 | 0.0 | 8.8 |
| 22 Sarala | 44 | 95.5 | 0.0 | 2.3 | 0.0 | 2.3 | 0.0 | 0.0 | 0.0 | 0.0 | 4.5 |
| 23 Shestakova | 117 | 99.4 | 0.0 | 0.6 | 0.0 | 0.0 | 0.0 | 0.0 | 0.0 | 0.0 | 0.6 |
| 24 Straschnaya | 527 | 99.6 | 0.2 | 0.0 | 0.0 | 0.2 | 0.0 | 0.0 | 0.0 | 0.0 | 0.4 |
| 25 Ust-Kan | 1652 | 98.9 | 0.4 | 0.1 | 0.4 | 0.1 | 0.0 | 0.0 | 0.0 | 0.1 | 1.1 |
| 26 Ust-Kova | 11.6 | 90.5 | 3.4 | 4.3 | 0.9 | 0.9 | 0.0 | 0.0 | 0.0 | 0.0 | 9.5 |
| 27 Varvarina Gora | 858 | 99.1 | 0.1 | 0.3 | 0.2 | 0.2 | 0.0 | 0.0 | 0.0 | 0.0 | 0.9 |
| 28 Volchiya Griva | 113 | 100.0 | 0.0 | 0.0 | 0.0 | 0.0 | 0.0 | 0.0 | 0.0 | 0.0 | 0.0 |
| 29 Yelenev | 241 | 89.6 | 3.7 | 2.9 | 2.5 | 0.8 | 0.0 | 0.0 | 0.0 | 0.4 | 10.4 |
| 30 Zhemchuzhnaya | 0 | – | – | – | – | – | – | – | – | – | – |

* Embedded fragments were not looked for in the Borabashevskaya assemblage.

**Table A1.20** Tooth wear (piece count, wear grade per piece, %, young = 0 and 0–1)

| Site | No. | 0 | 0–1 | 1 | 1–2 | 2 | 2–3 | 3 | 3–4 | 4 | Young |
|---|---|---|---|---|---|---|---|---|---|---|---|
| 1 Afontova Gora | (5000) | – | – | – | – | – | – | – | – | – | – |
| 1 Boisman II | 38 | 10.5 | 65.8 | 15.8 | 2.6 | 2.6 | 2.6 | 0.0 | 0.0 | 0.0 | 76.3 |
| 2 Bolshoi Yakor I | 2 | 0.0 | 0.0 | 0.0 | 100.0 | 0.0 | 0.0 | 0.0 | 0.0 | 0.0 | 0.0 |
| 3 Borabashevskaya | 0 | – | – | – | – | – | – | – | – | – | – |
| 4 Denisova | 2 | 0.0 | 0.0 | 50.0 | 50.0 | 0.0 | 0.0 | 0.0 | 0.0 | 0.0 | 0.0 |
| 5 Dvuglaska | 0 | – | – | – | – | – | – | – | – | – | – |
| 6 Gosudarev Log I | 0 | – | – | – | – | – | – | – | – | – | – |
| 7 Kamenka | 12 | 0.0 | 8.3 | 50.0 | 33.3 | 0.0 | 0.0 | 0.0 | 8.2 | 0.0 | 8.3 |
| 8 Kaminnaya | 23 | 13.0 | 17.4 | 26.1 | 17.4 | 8.7 | 4.3 | 13.0 | 0.0 | 0.0 | 30.4 |
| 9 Kara-Bom | 5 | 0.0 | 40.0 | 20.0 | 20.0 | 20.0 | 0.0 | 0.0 | 0.0 | 0.0 | 40.0 |
| 10 Kirkalinskaya | 0 | – | – | – | – | – | – | – | – | – | – |
| 11 Krasny Yar | 54 | 0.0 | 20.4 | 14.8 | 18.5 | 25.9 | 11.1 | 0.0 | 9.3 | 0.0 | 20.4 |
| 12 Kurla I | 0 | – | – | – | – | – | – | – | – | – | – |
| 13 Malaya Seeya | 0 | – | – | – | – | – | – | – | – | – | – |
| 14 Mal'ta | 1 | 100.0 | 0.0 | 0.0 | 0.0 | 0.0 | 0.0 | 0.0 | 0.0 | 0.0 | 100.0 |
| 5 Maly Yaloman | 34 | 5.9 | 5.9 | 47.1 | 23.5 | 2.9 | 8.8 | 2.9 | 0.0 | 2.9 | 11.8 |
| 16 Mokhovo Mine | 0 | – | – | – | – | – | – | – | – | – | – |
| 17 Nizhneudinskaya | 6 | 16.7 | 83.3 | 0.0 | 0.0 | 0.0 | 0.0 | 0.0 | 0.0 | 0.0 | 100.0 |
| 18 Okladnikov | 18 | 16.7 | 50.0 | 33.3 | 0.0 | 0.0 | 0.0 | 0.0 | 0.0 | 0.0 | 66.7 |
| 19 Proskuryakova | 1 | 100.0 | 0.0 | 0.0 | 0.0 | 0.0 | 0.0 | 0.0 | 0.0 | 0.0 | 100.0 |
| 20 Razboinich'ya | 25 | 0.0 | 36.0 | 44.0 | 8.0 | 8.0 | 4.0 | 0.0 | 0.0 | 0.0 | 36.0 |
| 21 Sarala | 2 | 0.0 | 0.0 | 100.0 | 0.0 | 0.0 | 0.0 | 0.0 | 0.0 | 0.0 | 0.0 |
| 22 Shestakovo | 0 | – | – | – | – | – | – | – | – | – | – |
| 23 Straschnaya | 0 | – | – | – | – | – | – | – | – | – | – |
| 24 Ust-Kan | 13 | 23.1 | 0.0 | 53.9 | 0.0 | 7.7 | 0.0 | 7.7 | 7.7 | 0.0 | 23.1 |
| 25 Ust-Kova | 0 | – | – | – | – | – | – | – | – | – | – |
| 26 Varvarina Gora | 36 | 2.8 | 11.1 | 72.2 | 8.3 | 2.8 | 0.0 | 0.0 | 0.0 | 2.8 | 13.9 |
| 27 Volchiya Griva | 0 | – | – | – | – | – | – | – | – | – | – |
| 28 Yelenev | 21 | 0.0 | 0.0 | 57.1 | 0.0 | 38.1 | 0.0 | 4.8 | 0.0 | 0.0 | 0.0 |
| 29 Zhemchuzhnaya | 0 | – | – | – | – | – | – | – | – | – | – |

**Table A1.21** Acid erosion (piece count, %)

| Sites | No. | Absent | Present | Uncertain |
|---|---|---|---|---|
| 1 Afontova Gora | (5000) | – | – | – |
| 2 Boisman II | 553 | 100.0 | 0.0 | 0.0 |
| 3 Bolshoi Yakor I | 263 | 99.6 | 0.4 | 0.0 |
| 4 Borabashevskaya | 30 | 100.0 | 0.0 | 0.0 |
| 5 Denisova | 115 | 94.8 | 2.6 | 2.6 |
| 6 Dvuglaska | 204 | 77.9 | 22.1 | 0.0 |
| 7 Gosudarev Log I | 14 | 100.0 | 0.0 | 0.0 |
| 8 Kamenka | 550 | 98.7 | 1.3 | 0.0 |
| 9 Kaminnaya | 770 | 82.6 | 17.4 | 0.0 |
| 10 Kara-Bom | 105 | 92.4 | 6.7 | 0.9 |
| 11 Kirkalinskaya | 4 | 100.0 | 0.0 | 0.0 |
| 12 Krasny Yar | 203 | 100.0 | 0.0 | 0.0 |
| 13 Kurla I | 72 | 100.0 | 0.0 | 0.0 |
| 14 Malaya Seeya | 144 | 100.0 | 0.0 | 0.0 |
| 15 Mal'ta | 186 | 98.4 | 1.6 | 0.0 |
| 16 Maly Yaloman | 177 | 52.0 | 48.0 | 0.0 |
| 17 Mokhovo Mine | 12 | 100.0 | 0.0 | 0.0 |
| 18 Nizhneudinskaya | 36 | 100.0 | 0.0 | 0.0 |
| 19 Okladnikov | 168 | 95.2 | 4.8 | 0.0 |
| 20 Proskuryakova | 47 | 100.0 | 0.0 | 0.0 |
| 21 Razboinich'ya | 370 | 92.4 | 7.3 | 0.3 |
| 22 Sarala | 44 | 100.0 | 0.0 | 0.0 |
| 23 Shestakova | 159 | 99.4 | 0.6 | 0.0 |
| 24 Straschnaya | 526 | 71.7 | 28.3 | 0.0 |
| 25 Ust-Kan | 1654 | 77.1 | 22.9 | 0.0 |
| 26 Ust-Kova | 116 | 100.0 | 0.0 | 0.0 |
| 27 Varvarina Gora | 864 | 99.5 | 0.5 | 0.0 |
| 28 Volchiya Griva | 113 | 100.0 | 0.0 | 0.0 |
| 29 Yelenev | 241 | 99.6 | 0.4 | 0.0 |
| 30 Zhemchuzhnaya | 0 | – | – | – |

**Table A1.22** Rodent gnawing (piece count, %)

| Sites | No. | Absent | Present | Uncertain |
|---|---|---|---|---|
| 1 Afontova Gora | (5000) | – | – | – |
| 2 Boisman II | 553 | 94.8 | 5.1 | 0.2 |
| 3 Bolshoi Yakor I | 263 | 100.0 | 0.0 | 0.0 |
| 4 Borabashevskaya | 30 | 100.0 | 0.0 | 0.0 |
| 5 Denisova | 115 | 100.0 | 0.0 | 0.0 |
| 6 Dvuglaska | 205 | 100.0 | 0.0 | 0.0 |
| 7 Gosudarev Log I | 14 | 100.0 | 0.0 | 0.0 |
| 8 Kamenka | 550 | 100.0 | 0.0 | 0.0 |
| 9 Kaminnaya | 771 | 99.4 | 0.6 | 0.0 |
| 10 Kara-Bom | 104 | 99.0 | 1.0 | 0.0 |
| 11 Kirkalinskaya | 4 | 100.0 | 0.0 | 0.0 |
| 12 Krasny Yar | 203 | 95.1 | 4.9 | 0.0 |
| 13 Kurla I | 72 | 100.0 | 0.0 | 0.0 |
| 14 Malaya Seeya | 144 | 100.0 | 0.0 | 0.0 |
| 15 Mal'ta | 186 | 100.0 | 0.0 | 0.0 |
| 16 Maly Yaloman | 178 | 100.0 | 0.0 | 0.0 |
| 17 Mokhovo Mine | 12 | 100.0 | 0.0 | 0/0 |
| 18 Nizhneudinskaya | 36 | 100.0 | 0.0 | 0.0 |
| 19 Okladnikov | 168 | 98.2 | 1.2 | 0.0 |
| 20 Proskuryakova | 47 | 100.0 | 0.0 | 0.0 |
| 21 Razboinich'ya | 611 | 99.2 | 0.8 | 0.0 |
| 22 Sarala | 44 | 100.0 | 0.0 | 0.0 |
| 23 Shestakova | 159 | 100.0 | 0.0 | 0.0 |
| 24 Straschnaya | 527 | 100.0 | 0.0 | 0.0 |
| 25 Ust-Kan | 1654 | 99.5 | 0.5 | 0.0 |
| 26 Ust-Kova | 116 | 99.1 | 0.9 | 0.0 |
| 27 Varvarina Gora | 863 | 100.0 | 0.0 | 0.0 |
| 28 Volchiya Griva | 113 | 100.0 | 0.0 | 0.0 |
| 29 Yelenev | 241 | 100.0 | 0.0 | 0.0 |
| 30 Zhemchuzhnaya | 0 | – | – | – |

**Table A1.23** Insect damage (piece count, %)

| Sites | No. | Absent | Present | Uncertain |
|---|---|---|---|---|
| 1 Afontova Gora | (5000) | – | – | – |
| 2 Boisman II | 513 | 100.0 | 0.0 | 0.0 |
| 3 Bolshoi Yakor I | 263 | 100.0 | 0.0 | 0.0 |
| 4 Borabashevskaya | 30 | 100.0 | 0.0 | 0.0 |
| 5 Denisova | 115 | 100.0 | 0.0 | 0.0 |
| 6 Dvuglaska | 205 | 100.0 | 0.0 | 0.0 |
| 7 Gosudarev Log I | 14 | 100.0 | 0.0 | 0.0 |
| 8 Kamenka | 550 | 100.0 | 0.0 | 0.0 |
| 9 Kaminnaya | 769 | 99.9 | 0.1 | 0.0 |
| 10 Kara-Bom | 104 | 100.0 | 0.0 | 0.0 |
| 11 Kirkalinskaya | 4 | 100.0 | 0.0 | 0.0 |
| 12 Krasny Yar | 203 | 99.5 | 0.5 | 0.0 |
| 13 Kurla I | 72 | 100.0 | 0.0 | 0.0 |
| 14 Malaya Seeya | 144 | 100.0 | 0.0 | 0.0 |
| 15 Mal'ta | 183 | 100.0 | 0.0 | 0.0 |
| 16 Maly Yaloman | 178 | 100.0 | 0.0 | 0.0 |
| 17 Mokhovo Mine | 12 | 100.0 | 0.0 | 0.0 |
| 18 Nizhneudinskaya | 36 | 100.0 | 0.0 | 0.0 |
| 19 Okladnikov | 168 | 100.0 | 0.0 | 0.0 |
| 20 Proskuryakova | 47 | 100.0 | 0.0 | 0.0 |
| 21 Razboinich'ya | 611 | 100.0 | 0.0 | 0.0 |
| 22 Sarala | 44 | 100.0 | 0.0 | 0.0 |
| 23 Shestakova | 159 | 100.0 | 0.0 | 0.0 |
| 24 Straschnaya | 527 | 100.0 | 0.0 | 0.0 |
| 25 Ust-Kan | 1654 | 99.9 | 0.1 | 0.0 |
| 26 Ust-Kova | 116 | 100.0 | 0.0 | 0.0 |
| 27 Varvarina Gora | 864 | 100.0 | 0.0 | 0.0 |
| 28 Volchiya Griva | 113 | 100.0 | 0.0 | 0.0 |
| 29 Yelenev | 241 | 100.0 | 0.0 | 0.0 |
| 30 Zhemchuzhnaya | 0 | – | – | – |

**Table A1.24** Human bone (piece count, perimortem damage type, %)

| Sites | No. | Cut | Break | Abrasion | Polish | Burn | Missing vertebrae |
|---|---|---|---|---|---|---|---|
| 1 Afontova Gora | (5000) | – | – | – | – | – | – |
| 2 Boisman II | 0 | – | – | – | – | – | – |
| 3 Bolshoi Yakor I | 0 | – | – | – | – | – | – |
| 4 Borabashevskaya | 0 | – | – | – | – | – | – |
| 5 Denisova | 0* | – | – | – | – | – | – |
| 6 Dvuglaska | 0 | – | – | – | – | – | – |
| 7 Gosudarev Log I | 0 | – | – | – | – | – | – |
| 8 Kamenka | 0 | – | – | – | – | – | – |
| 9 Kaminnaya | 0 | – | – | – | – | – | – |
| 10 Kara-Bom | 0 | – | – | – | – | – | – |
| 11 Kirkalinskaya | 5 | 0.0 | 100.0 | 0.0 | 80.0 | 0.0 | 100.0 |
| 12 Krasny Yar | 0 | – | – | – | – | – | – |
| 13 Kurla I | 0 | – | – | – | – | – | – |
| 14 Malaya Seeya | 0 | – | – | – | – | – | – |
| 15 Mal'ta | 0 | – | – | - | – | – | – |
| 16 Maly Yaloman | 0 | – | – | – | – | – | – |
| 17 Mokhovo Mine | 0 | – | – | – | – | – | – |
| 18 Nizhneudinskaya | 0 | – | – | – | – | – | – |
| 19 Okladnikov | 0* | – | – | – | – | – | – |
| 20 Proskuryakova | 0 | – | – | – | – | – | – |
| 21 Razboinich'ya | 0 | – | – | – | – | – | – |
| 22 Sarala | 0 | – | – | – | – | – | – |
| 23 Shestakova | 0 | – | – | – | – | – | – |
| 24 Straschnaya | 0 | – | – | – | – | – | – |
| 25 Ust-Kan | 0 | – | – | – | – | – | – |
| 26 Ust-Kova | 0 | – | – | – | – | – | – |
| 27 Varvarina Gora | 0 | – | – | – | – | – | – |
| 28 Volchiya Griva | 0 | – | – | – | – | – | – |
| 29 Yelenev | 65* | 3.1 | 90.8 | 1.5 | 96.9 | 3.1 | 5.8** |
| 30 Zhemchuzhnaya | 0 | – | – | – | – | – | – |

* Human remains were recovered from Denisova, Okladnikov, and Yelenev Caves, but were not stored with their respective faunal collections. See these site sections for details.
** Given that Yelenev has an MNI of five, then seven pieces of vertebrae are only 5.8% of the expected 120.

**Table A1.25** Cut marks (piece count, cuts per piece, %)

| Site | No. | 0 | 1 | 2 | 3 | 4 | 5 | 6 | 7 | >7 | Total |
|---|---|---|---|---|---|---|---|---|---|---|---|
| 1 Afontova Gora | (5000) | – | – | – | – | – | – | – | – | – | – |
| 2 Boisman II | 553 | 78.7 | 6.1 | 3.6 | 2.2 | 3.1 | 1.4 | 1.1 | 0.4 | 3.4 | 21.3 |
| 3 Bolshoi Yakor I | 263 | 87.8 | 1.9 | 2.3 | 1.5 | 1.1 | 0.4 | 1.9 | 0.8 | 2.4 | 12.2 |
| 4 Borabashevskaya | 30 | 100.0 | 0.0 | 0.0 | 0.0 | 0.0 | 0.0 | 0.0 | 0.0 | 0.0 | 0.0 |
| 5 Denisova | 115 | 80.0 | 7.8 | 1.7 | 3.5 | 2.6 | 0.9 | 0.0 | 0.9 | 2.6 | 20.0 |
| 6 Dvuglaska | 205 | 100.0 | 0.0 | 0.0 | 0.0 | 0.0 | 0.0 | 0.0 | 0.0 | 0.0 | 0.0 |
| 7 Gosudarev Log I | 14 | 100.0 | 0.0 | 0.0 | 0.0 | 0.0 | 0.0 | 0.0 | 0.0 | 0.0 | 0.0 |
| 8 Kamenka | 546 | 90.5 | 3.8 | 1.1 | 1.1 | 0.5 | 1.1 | 0.2 | 0.0 | 1.6 | 9.5 |
| 9 Kaminnaya | 769 | 88.6 | 2.2 | 1.4 | 1.0 | 1.2 | 0.5 | 0.7 | 0.7 | 3.6 | 11.4 |
| 10 Kara-Bom | 89 | 95.5 | 1.1 | 0.0 | 1.1 | 1.1 | 0.0 | 0.0 | 0.0 | 1.1 | 4.5 |
| 11 Kirkalinskaya | 4 | 100.0 | 0.0 | 0.0 | 0.0 | 0.0 | 0.0 | 0.0 | 0.0 | 0.0 | 0.0 |
| 12 Krasny Yar | 202 | 99.0 | 0.0 | 0.0 | 0.0 | 0.0 | 0.0 | 0.0 | 0.0 | 1.0 | 1.0 |
| 13 Kurla I | 72 | 65.3 | 4.2 | 0.0 | 2.8 | 4.2 | 5.5 | 2.8 | 4.2 | 11.1 | 34.7 |
| 14 Malaya Seeya | 144 | 86.1 | 1.4 | 0.7 | 2.8 | 1.4 | 0.7 | 1.4 | 0.0 | 5.6 | 13.9 |
| 15 Mal'ta | 182 | 94.0 | 0.5 | 1.1 | 1.6 | 0.5 | 0.0 | 0.0 | 0.0 | 2.2 | 6.0 |
| 16 Maly Yaloman | 174 | 97.7 | 06 | 0.0 | 0.0 | 0.0 | 0.0 | 0.0 | 0.0 | 1.8 | 2.3 |
| 17 Mokhovo Mine | 12 | 91.7 | 0.0 | 0.0 | 0.0 | 0.0 | 8.3 | 0.0 | 0.0 | 0.0 | 8.3 |
| 18 Nizhneudinskaya | 36 | 100.0 | 0.0 | 0.0 | 0.0 | 0.0 | 0.0 | 0.0 | 0.0 | 0.0 | 0.0 |
| 19 Okladnikov | 168 | 94.0 | 0.0 | 1.2 | 0.6 | 1.2 | 0.6 | 0.0 | 0.6 | 1.8 | 6.0 |
| 20 Proskuryakova | 47 | 100.0 | 0.0 | 0.0 | 0.0 | 0.0 | 0.0 | 0.0 | 0.0 | 0.0 | 0.0 |
| 21 Razboinich'ya | 611 | 100.0 | 0.0 | 0.0 | 0.0 | 0.0 | 0.0 | 0.0 | 0.0 | 0.0 | 0.0 |
| 22 Sarala | 44 | 100.0 | 0.0 | 0.0 | 0.0 | 0.0 | 0.0 | 0.0 | 0.0 | 0.0 | 0.0 |
| 23 Shestakova | 159 | 99.3 | 0.6 | 0.0 | 0.0 | 0.0 | 0.0 | 0.0 | 0.0 | 0.0 | 0.7 |
| 24 Straschnaya | 527 | 99.2 | 0.2 | 0.0 | 0.2 | 0.0 | 0.0 | 0.2 | 0.0 | 0.2 | 0.8 |
| 25 Ust-Kan | 1653 | 82.0 | 3.8 | 2.0 | 2.9 | 1.4 | 1.7 | 1.3 | 1.0 | 3.7 | 18.0 |
| 26 Ust-Kova | 116 | 85.3 | 5.2 | 3.4 | 0.9 | 1.7 | 2.6 | 0.0 | 0.9 | 0.0 | 14.7 |
| 27 Varvarina Gora | 714 | 91.2 | 1.5 | 1.7 | 1.5 | 0.1 | 1.1 | 0.4 | 0.1 | 2.0 | 8.8 |
| 28 Volchiya Griva | 113 | 100.0 | 0.0 | 0.0 | 0.0 | 0.0 | 0.0 | 0.0 | 0.0 | 0.0 | 0.0 |
| 29 Yelenev | 241 | 82.2 | 6.6 | 2.5 | 0.8 | 2.1 | 0.8 | 0.8 | 0.0 | 4.1 | 17.8 |
| 30 Zhemchuzhnaya | 0 | – | – | – | – | – | – | – | – | – | – |

**Table A1.26** Chop marks (piece count, chops per piece, %)

| Site | No. | 0 | 1 | 2 | 3 | 4 | 5 | 6 | 7 | >7 | Total |
|---|---|---|---|---|---|---|---|---|---|---|---|
| 1 Afontova Gora | (5000) | – | – | – | – | – | – | – | – | – | – |
| 2 Boisman II | 513 | 91.4 | 2.9 | 1.6 | 1.9 | 0.4 | 0.0 | 0.4 | 0.0 | 1.4 | 8.6 |
| 3 Bolshoi Yakor I | 263 | 97.7 | 0.8 | 0.4 | 0.0 | 0.8 | 0.4 | 0.0 | 0.0 | 0.0 | 2.3 |
| 4 Borabashevskaya | 30 | 100.0 | 0.0 | 0.0 | 0.0 | 0.0 | 0.0 | 0.0 | 0.0 | 0.0 | 0.0 |
| 5 Denisova | 97 | 77.3 | 12.4 | 1.0 | 1.0 | 1.0 | 1.0 | 2.1 | 1.0 | 3.1 | 22.7 |
| 6 Dvuglaska | 205 | 100.0 | 0.0 | 0.0 | 0.0 | 0.0 | 0.0 | 0.0 | 0.0 | 0.0 | 0.0 |
| 7 Gosudarev Log I | 14 | 85.7 | 7.1 | 0.0 | 7.1 | 0.0 | 0.0 | 0.0 | 0.0 | 0.0 | 14.3 |
| 8 Kamenka | 548 | 91.4 | 6.2 | 1.5 | 0.5 | 0.2 | 0.0 | 0.2 | 0.0 | 0.0 | 8.6 |
| 9 Kaminnaya | 770 | 96.6 | 2.2 | 0.5 | 0.1 | 0.0 | 0.1 | 0.0 | 0.1 | 0.2 | 3.4 |
| 10 Kara-Bom | 89 | 93.3 | 1.1 | 1.1 | 1.1 | 0.0 | 0.0 | 2.2 | 0.0 | 1.1 | 6.7 |
| 11 Kirkalinskaya | 4 | 100.0 | 0.0 | 0.0 | 0.0 | 0.0 | 0.0 | 0.0 | 0.0 | 0.0 | 0.0 |
| 12 Krasny Yar | 202 | 100.0 | 0.0 | 0.0 | 0.0 | 0.0 | 0.0 | 0.0 | 0.0 | 0.0 | 0.0 |
| 13 Kurla I | 72 | 69.4 | 13.9 | 9.7 | 0.0 | 0.0 | 1.4 | 0.0 | 2.8 | 2.8 | 30.6 |
| 14 Malaya Seeya | 144 | 99.3 | 0.0 | 0.7 | 0.0 | 0.0 | 0.0 | 0.0 | 0.0 | 0.0 | 0.7 |
| 15 Mal'ta | 183 | 91.3 | 4.4 | 1.6 | 0.5 | 1.6 | 0.5 | 0.0 | 0.0 | 0.0 | 8.7 |
| 16 Maly Yaloman | 174 | 99.4 | 0.0 | 0.0 | 0.6 | 0.0 | 0.0 | 0.0 | 0.0 | 0.0 | 0.6 |
| 17 Mokhovo Mine | 12 | 100.0 | 0.0 | 0.0 | 0.0 | 0.0 | 0.0 | 0.0 | 0.0 | 0.0 | 0.0 |
| 18 Nizhneudinskaya | 36 | 100.0 | 0.0 | 0.0 | 0.0 | 0.0 | 0.0 | 0.0 | 0.0 | 0.0 | 0.0 |
| 19 Okladnikov | 167 | 96.4 | 1.8 | 0.0 | 0.0 | 0.6 | 0.0 | 0.0 | 0.0 | 1.2 | 3.6 |
| 20 Proskuryakova | 47 | 100.0 | 0.0 | 0.0 | 0.0 | 0.0 | 0.0 | 0.0 | 0.0 | 0.0 | 0.0 |
| 21 Razboinich'ya | 611 | 100.0 | 0.0 | 0.0 | 0.0 | 0.0 | 0.0 | 0.0 | 0.0 | 0.0 | 0.0 |
| 22 Sarala | 44 | 97.7 | 0.0 | 0.0 | 0.0 | 0.0 | 0.0 | 0.0 | 0.0 | 2.3 | 2.3 |
| 23 Shestakova | 159 | 100.0 | 0.0 | 0.0 | 0.0 | 0.0 | 0.0 | 0.0 | 0.0 | 0.0 | 0.0 |
| 24 Straschnaya | 527 | 99.8 | 0.2 | 0.0 | 0.0 | 0.0 | 0.0 | 0.0 | 0.0 | 0.0 | 0.2 |
| 25 Ust-Kan | 1654 | 90.5 | 5.8 | 1.6 | 1.0 | 0.4 | 0.4 | 0.1 | 0.1 | 0.2 | 9.5 |
| 26 Ust-Kova | 116 | 94.8 | 2.6 | 0.9 | 0.0 | 0.9 | 0.0 | 0.9 | 0.0 | 0.0 | 5.2 |
| 27 Varvarina Gora | 747 | 86.2 | 8.6 | 2.9 | 1.7 | 0.4 | 0.0 | 0.1 | 0.0 | 0.0 | 13.8 |
| 28 Volchiya Griva | 113 | 100.0 | 0.0 | 0.0 | 0.0 | 0.0 | 0.0 | 0.0 | 0.0 | 0.0 | 0.0 |
| 29 Yelenev | 241 | 90.9 | 7.1 | 1.2 | 0.4 | 0.4 | 0.0 | 0.0 | 0.0 | 0.0 | 9.1 |
| 30 Zhemchuzhnaya | 0 | – | – | – | – | – | – | – | – | – | – |

**Table A1.27** Carnivore perimortem damage (summary percentages, key features; previous tables have sample sizes)

| Site type | End-hollow | Tooth dint | Tooth scratch | Notch | Acid erosion | E-M polish | Mean % |
|---|---|---|---|---|---|---|---|
| **Mainly hyena** | | | | | | | |
| Dvuglaska | 4.4 | 43.4 | 38.5 | 15.1 | 22.1 | 84.7 | 34.7 |
| Maly Yaloman | 9.5 | 32.6 | 28.6 | 23.0 | 48.0 | 75.3 | 36.2 |
| Proskuryakova | 29.8 | 93.6 | 97.9 | 85.1 | 0.0 | 91.4 | 66.3 |
| Razboinich'ya | 8.1 | 52.8 | 37.1 | 23.6 | 7.3 | 27.1 | 26.0 |
| Sarala | 15.9 | 88.4 | 47.7 | 25.0 | 0.0 | 61.4 | 39.7 |
| Straschnaya | 0.9 | 31.9 | 22.9 | 11.4 | 28.3 | 63.0 | 26.4 |
| Unweighted mean | 11.4 | 57.1 | 45.4 | 30.5 | 17.6 | 67.1 | 38.2 |
| Weighted mean | 6.1 | 44.8 | 33.8 | 20.3 | 22.4 | 54.4 | 30.8 |
| **Mixed archaeology and hyena** | | | | | | | |
| Denisova | 2.6 | 22.6 | 20.9 | 22.6 | 2.6 | 46.4 | 19.6 |
| Kaminnaya | 1.9 | 24.1 | 15.6 | 10.8 | 17.4 | 63.8 | 22.3 |
| Kara-Bom | 4.8 | 7.9 | 10.9 | 4.7 | 6.7 | 66.7 | 20.3 |
| Okladnikov | 17.5 | 55.7 | 57.0 | 39.2 | 4.8 | 49.7 | 37.3 |
| Ust-Kan | 9.5 | 13.7 | 9.0 | 9.2 | 22.9 | 70.0 | 22.4 |
| Unweighted mean | 7.3 | 24.8 | 22.7 | 17.3 | 10.9 | 59.3 | 24.4 |
| Weighted mean | 7.8 | 19.2 | 14.3 | 11.8 | 18.9 | 66.0 | 23.2 |
| **Archaeology, no hyena** | | | | | | | |
| Bolshoi Yakor | 0.4 | 8.0 | 1.9 | 6.5 | 0.4 | 6.6 | 4.0 |
| Kamenka | 0.8 | 3.3 | 2.8 | 9.8 | 1.3 | 61.3 | 13.2 |
| Kurla I | 0.0 | 4.2 | 2.8 | 2.8 | 0.0 | 31.9 | 6.9 |
| Malaya Seeya | 0.7 | 4.9 | 4.2 | 11.3 | 0.0 | 66.2 | 14.5 |
| Mal'ta | 4.3 | 1.7 | 2.8 | 3.2 | 1.6 | 38.1 | 8.6 |
| Shestakovo | 1.9 | 0.0 | 0.0 | 1.3 | 0.6 | 0.6 | 0.7 |
| Ust-Kova | 3.5 | 5.2 | 4.3 | 2.6 | 0.0 | 4.4 | 3.3 |
| Varvarina Gora | 1.2 | 1.9 | 1.9 | 5.7 | 0.5 | 71.2 | 3.7 |
| Unweighted mean | 1.6 | 3.6 | 2.6 | 5.4 | 0.5 | 35.0 | 8.1 |
| Weighted mean | 1.5 | 3.3 | 2.3 | 6.3 | 0.7 | 49.3 | 11.2 |

**Table A1.28** Chi-square comparisons of perimortem damage in three groups (1 d.f.; $p < 0.05$ in all cases except those with asterisk)

| Perimortem damage | Hyena and mixed | > / = / < | Hyena and no hyena | Mixed and no hyena |
|---|---|---|---|---|
| End-hollow | 4.2 | < | 45.1 | 76.7 |
| Perimortem breakage | 13.9 | < | 16.5 | 77.8 |
| Tooth dint | 328.0 | < | 958.1 | 287.1 |
| Tooth scratch | 235.9 | < | 695.0 | 210.8 |
| Notch | 58.6 | < | 176.2 | 24.8 |
| Acid erosion | 23.8 | < | 513.7 | 448.4 |
| No polish | 45.9 | > | 0.105* | 62.2 |
| End polish | 9.7 | = | 9.8 | 60.2 |
| End and middle polish | 57.9 | > | 9.8 | 145.8 |
| Completeness[a] | 110.9 | < | 184.8 | 674.4 |
| Pseudo-cut | 7.5 | < | 41.6 | 82.3 |
| Embedding | 21.9 | > | 5.5 | 1.331* |
| $\Sigma?\chi?^2$ | 918.2 | | 2656.2 | 2151.8 |
| Burning | 42.0 | | 9.4 | 24.1 |
| Cutting | 247.7 | | 150.1 | 28.2 |
| Chopping | 124.3 | | 144.6 | 2.3* |
| Abrasions | 18.9 | | 0.162* | 27.9 |
| Perimortem breakage | 13.9 | | 16.5 | 77.8 |
| $\Sigma?\chi?^2$ | 446.8 | | 320.8 | 160.3 |

[a] 2 d.f.

**Table A1.29** Human perimortem damage in three groups (percentages, key features; see Table A1.28)

| Site type | Burn | Chop | Cut |
|---|---|---|---|
| **Mainly hyena** | | | |
| Dvuglaska | 0.5 | 0.0 | 0.0 |
| Maly Yaloman | 0.0 | 0.6 | 2.3 |
| Proskuryakova | 0.0 | 0.0 | 0.0 |
| Razboinich'ya | 0.0 | 0.0 | 0.0 |
| Sarala | 0.0 | 2.3 | 0.0 |
| Straschnaya | 0.0 | 0.2 | 0.8 |
| Unweighted mean | 0.08 | 0.5 | 0.5 |
| Weighted mean | 0.06 | 0.2 | 0.5 |
| **Mixed archaeology and hyena** | | | |
| Denisova | 0.0 | 22.7 | 20.0 |
| Kaminnaya | 3.0 | 3.4 | 11.4 |
| Kara-Bom | 0.0 | 6.7 | 4.5 |
| Okladnikov | 2.4 | 3.6 | 6.0 |
| Ust-Kan | 2.6 | 9.5 | 18.0 |
| Unweighted mean | 1.6 | 9.2 | 12.5 |
| Weighted mean | 2.5 | 7.8 | 15.1 |
| **Archaeology, no hyena** | | | |
| Bolshoi Yakor | 2.3 | 2.3 | 12.2 |
| Kamenka | 0.0 | 8.6 | 9.5 |
| Kurla I | 1.4 | 30.6 | 34.7 |
| Malaya Seeya | 2.8 | 0.7 | 13.9 |
| Mal'ta | 0.5 | 8.7 | 6.0 |
| Shestakovo | 0.0 | 0.0 | 0.7 |
| Ust-Kova | 3.4 | 5.2 | 14.7 |
| Varvarina Gora | 0.1 | 13.8 | 8.8 |
| Unweighted mean | 1.3 | 8.7 | 12.6 |
| Weighted mean | 0.7 | 9.0 | 10.1 |

**Table A1.30** Maximum size sorted by hyena presence (in centimeters)

| Site Type | Mean | S.D. | Range | S.E. | No. |
|---|---|---|---|---|---|
| **Mainly hyena** | | | | | |
| Dvuglaska | 5.2 | 2.8 | 1.7–28.2 | 0.135 | 205 |
| Maly Yaloman | 4.9 | 3.5 | 2.3–28.1 | 0.267 | 177 |
| Proskuryakova | 11.9 | 3.4 | 7.1–18.8 | 0.496 | 47 |
| Razboinich'ya | 6.5 | 3.2 | 2.4–24.4 | 0.13 | 599 |
| Sarala | 8.8 | 4.6 | 3.7–21.2 | 0.693 | 44 |
| Straschnaya | 5.4 | 2.2 | 2.5–16.2 | 0.096 | 527 |
| Unweighted averages | 7.2 | 3.3 | 3.3–22.8 | 0.303 | 1599 |
| **Mixed archaeology and hyena** | | | | | |
| Denisova | 5.2 | 2.2 | 2.3–13.8 | 0.204 | 116 |
| Kaminnaya | 6.2 | 3.0 | 2.4–25.0 | 0.108 | 774 |
| Kara-Bom | 6.6 | 4.2 | 3.0–24.8 | 0.857 | 24 |
| Okladnikov | 7.9 | 4.7 | 2.1–26.9 | 0.350 | 180 |
| Ust-Kan | 4.8 | 1.6 | 2.2–19.9 | 0.056 | 1656 |
| Unweighted averages | 6.1 | 3.1 | 2.4–22.1 | 0.315 | 2750 |
| **Archaeology, no hyena** | | | | | |
| Bolshoi Yakor | 6.0 | 4.0 | 2.5–50.5 | 0.246 | 264 |
| Kamenka | 6.5 | 2.9 | 2.9–20.7 | 0.126 | 550 |
| Kurla I | 4.3 | 1.6 | 2.1–9.5 | 0.189 | 72 |
| Malaya Seeya | 8.3 | 3.4 | 3.2–30.8 | 0.283 | 144 |
| Mal'ta | 16.5 | 12.7 | 3.0–82.5 | 0.931 | 186 |
| Shestakovo | 16.1 | 11.6 | 2.4–68.0 | 0.932 | 155 |
| Ust-Kova | 8.0 | 3.8 | 2.9–28.3 | 0.351 | 117 |
| Varvarina Gora | 9.0 | 5.3 | 2.5–48.5 | 0.184 | 847 |
| Unweighted averages | 9.3 | 5.7 | 2.7–42.3 | 0.405 | 2335 |

**Table A1.31** Hyena elements found in carnivore and Okladnikov Caves (based on identifications by Ovodov and Martynovich 2005)

| Element | Site | | | | | | | |
|---|---|---|---|---|---|---|---|---|
| | Razboinich'ya | | Okladnikov | | Logovo | | Gieny | |
| | No. | % | No. | % | No. | % | No. | % |
| Upper jaw, skull | 42 | 0.6 | 6 | 0.7 | 12 | 4.2 | 9 | 9.5 |
| Mandible | 58 | 0.9 | 27 | 3.5 | 23 | 8.1 | 20 | 20.8 |
| Molar, left lower | 109 | 1.6 | 34 | 4.4 | 15 | 5.3 | 10 | 10.4 |
| Molar, right lower | 137* | 2.0 | 23 | 3.0 | 22 | 7.7 | 11 | 11.5 |
| Teeth | 4541 | 67.8 | 1554 | 72.1 | 157 | 55.1 | 42 | 43.7 |
| Vertebra | 15 | 0.8 | 1 | 0.1 | 0 | 0.0 | 1 | 1.0 |
| Scapula | 7 | 0.1 | 1 | 0.1 | 1 | 0.3 | 0 | 0.0 |
| Radius | 23 | 0.3 | 4 | 0.5 | 4 | 1.4 | 0 | 0.0 |
| Ulna | 17 | 0.2 | 3 | 0.4 | 0 | 0.0 | 0 | 0.0 |
| Humerus | 14 | 0.2 | 1 | 0.1 | 2 | 0.7 | 0 | 0.0 |
| Wrist, small bones | 164 | 2.4 | 3 | 0.4 | 0 | 0.0 | 0 | 0.0 |
| Pelvis | 10 | 0.1 | 0 | 0.0 | 0 | 0.0 | 0 | 0.0 |
| Sacrum | 1 | 0.01 | 0 | 0.0 | 0 | 0.0 | 0 | 0.0 |
| Femur | 10 | 0.1 | 0 | 0.0 | 0 | 0.0 | 0 | 0.0 |
| Patella | 11 | 0.2 | 0 | 0.0 | 0 | 0.0 | 0 | 0.0 |
| Tibia/fibula | 26 | 0.4 | 1 | 0.1 | 1 | 0.3 | 1 | 1.0 |
| Astragalus | 6 | 0.01 | 1 | 0.1 | 0 | 0.0 | 0 | 0.0 |
| Calcaneus | 7 | 0.1 | 0 | 0.0 | 0 | 0.0 | 0 | 0.0 |
| Long bone fragments | 392 | 5.9 | 60 | 7.8 | 28 | 9.8 | 0 | 0.0 |
| Metapodial | 180 | 2.7 | 6 | 0.8 | 3 | 1.0 | 1 | 1.0 |
| Phalanx | 658 | 9.8 | 3 | 0.4 | 2 | 0.7 | 1 | 1.0 |
| Coprolites | 267 | 4.0 | 40 | 5.2 | 15 | 5.3 | 0 | 0.0 |
| Total | 5595 | – | 768 | – | 285 | – | 96 | – |
| Grand total | 7844 | | | | | | | |

* This count is the basis for the MNI of 137 individual hyenas that died in Razboinich'ya Cave.

**Table A1.32** Human teeth from Denisova and Okladnikov Caves

| Cave | Tooth<br>Shpakova | Turner |
|---|---|---|
| Denisova D7 | Upper left I1 | Upper left I1<br>Adult, wear grade 3 |
| Denisova L22 | Lower left m2<br>Disagree | Lower right m1<br>Child, wear 4, chipping |
| Okladnikov B/2, L3 | Lower left P1 | Lower left P1<br>Adult, wear 1–2 |
| Okladnikov, cap E1 L3 | Lower right M3 | Lower right M3<br>Child 12 yrs, unerupted |
| Okladnikov A2, 50, L3 | Lower left M1<br>Disagree | Lower left M3<br>15–18 yrs, wear 0–1 |
| Okladnikov V2, L2 | Lower left M3 | Lower left M3<br>12 yrs, unerupted |
| Okladnikov M6, L7 | Lower right m2 | Lower right m2<br>Child |

Wear scale in Turner: 0, no wear; 0–1, polishing but dentine not exposed; 1, dentine exposed; 2, cusps worn off; 3, pulp exposed; 4, crown mostly or completely worn off. Root stump is functional occlusal surface (Turner *et al.* 1991). Information on wear is also given by Shpakova but she provides no explanation for her wear values.

**Table A1.33** Northern limit of Pleistocene hyenas

| Site | Latitude north | Longitude east | Hyenas (this study) | Hyenas (other studies) | Pleistocene | Mid-mountain relief (forest-steppe) |
|---|---|---|---|---|---|---|
| 1 Afontova Gora | 56°01′ | 96°50′ | No | No | Yes | No |
| 2 Boisman II | 42°47′ | 131°16′ | No | No | No | No |
| 3 Bolshoi Yakor I | 56°22′ | 115°73′ | No | No | Yes | No |
| 4 Borabashevskaya | ? | ? | No | No | Yes | No |
| 5 Denisova | 51°22′ | 84°41′ | Yes | Yes | Yes | Yes |
| 6 Dvuglaska | 54°08′ | 91°04′ | No | Yes | Yes | Yes |
| 7 Gosudarev Log I | 56°04′ | 93°17′ | No | No | Yes | No |
| 8 Kamenka | 51°46′ | 108°20′ | No | No | Yes | No |
| 9 Kaminnaya | 51°17′ | 84°28′ | Yes | Yes | Yes | Yes |
| 10 Kara Bom | 50°10′ | 86°40′ | Yes | Yes | Yes | Yes |
| 11 Kirkalinskaya | ? | ? | No | No | Yes | Yes |
| 12 Krasny Yar | 55°02′ | 82°55 | No | Yes | Yes | No |
| 13 Kurla I | 55°39′ | 109°21′ | No | No | Yes | No |
| 14 Malaya Seeya | 54°50′ | 89°42′ | No | No | Yes | Yes |
| 15 Mal'ta | 52°50′ | 103°32′ | No | No | Yes | No |

**Table A1.33** (cont.)

| Site | Latitude north | Longitude east | Hyenas (this study) | Hyenas (other studies) | Pleistocene | Mid-mountain relief (forest-steppe) |
|---|---|---|---|---|---|---|
| 16 Maly Yaloman | 49°80′ | 86°30′ | Yes | Yes | Yes | Yes |
| 17 Mokhovo Mine | 54°40′ | 86°60′ | No | No | Yes | ? |
| 18 Nizhneudinskaya | 54°28′ | 98°58′ | No | No | Yes | ? |
| 19 Okladnikov | 51°40′ | 84°20′ | Yes | Yes | Yes | Yes |
| 20 Proskuryakova | 54°27′ | 89°28′ | Yes | Yes | Yes | Yes |
| 21 Razboinich'ya | ca. 51°20′ | ca. 84°25′ | Yes | Yes | Yes | Yes |
| 22 Sarala | ca. 54°50′ | ca. 89°42′ | No | No study | Yes | ? |
| 23 Shestakovo | 55°54′ | 87°57′ | No | No | Yes | Yes |
| 24 Straschnaya | 51°75′ | 83°84′ | No | No | Yes | No |
| 25 Ust-Kan | 50°54′ | 84°48′ | Yes | Yes | Yes | Yes |
| 26 Ust-Kova | 58°33′ | 100°33′ | No | No | Yes | ? |
| 27 Varvarina Gora | 51°38′ | 108°10′ | No | No | Yes | No |
| 28 Volchiya Griva | 54°63′ | 80°25′ | No | No | Yes | No |
| 29 Yelenev | 55°58′ | 92°29′ | No | No | No | No |
| 30 Zhemchuzhnaya | 55°28′ | 91°58′ | No | No | Yes | ? |
| Geographic Society | ca. 43° | ca. 133° | No study | Yes | Yes | No |

# Appendix 2

## Scientific names for Siberian Pleistocene species identified in one or more of the 30 faunal assemblages

| Common name | Scientific name |
|---|---|
| Antelope (spiral-horned) | *Spirocerus kiakhtensis* |
| Badger | *Meles meles* |
| Bear (brown) | *Ursus arctos* |
| Beaver | *Caster fiber* |
| Bison | *Bison priscus* |
| Caribou | *Rangifer tarandus* |
| Deer (big-horn) | *Megaloceros giganteus* |
| Deer (red) | *Cervus elaphus* |
| Deer (roe) | *Capreolus capreolus* |
| Elk | *Alces* sp. |
| Fox (Arctic) | *Alopex lagopus* |
| Fox | *Vulpes vulpes* |
| Fox (corsac) | *Vulpes corsac* |
| Gazelle (Mongolian, Zeren) | *Procapra gutturosa* |
| Goat, ibex, markhor | *Capra sibirica* |
| Hare (cape) | *Lepus capensis* |
| Hare | *Lepus timidus* |
| Hyena (cave) | *Crocuta spelaea* |
| Horse | *Equus cf. caballus* |
| Horse | *Equus ex. gr. gallicus* |
| Leopard (snow) | *Uncia uncia* |
| Lion (cave) | *Panthera spelaea* |
| Lynx | *Felis lynx* |
| Mammoth (mammut) | *Mammonteus primigenius* |
| Manul | *Felis manul* |
| Marmot | *Marmota bailbacina* |
| Moose | *Alces alces* |
| Onager (Asiatic wild ass) | *Equus hemionus* |
| Pika | *Ochotona* sp. |
| Rhinoceros | *Coelodonta antiquitatis* |
| Sable | *Martes zibellina* |
| Saiga | *Saiga tatarica* |
| Seal (Baikal) | *Pusa sibirica* |
| Sheep (big-horn) | *Ovis nivicola* |
| Sheep (wild) | *Ovis ammon* |

(*cont.*)

| Common name | Scientific name |
|---|---|
| Wolf (gray) | *Canis lupus* |
| Wolf (red) | *Cuon alpinus* |
| Wolverine (glutton) | *Gulo gulo* |
| Yak | *Poephagus baicalensis* |

For a more complete listing of Pleistocene mammals, see Elaine Anderson 1984. L. I. Galkina and N. D. Ovodov (1975) provide a listing of late Pleistocene fauna found in the Altai Mountains. For the site assemblages herein, the mammalian microfauna are under study by Alexander K. Agadjanian. The birds are being studied by Nicolai V. Martynovich. Species identified with Latin names in the text cited from earlier studies by Ovodov or others are not provided in this appendix.

# Appendix 3

## Listvenka

This open site (Figs. A3.1, A3.2) is below a high railroad bridge crossing a sandy-loess bluff back from the right bank of the Yenisei River. It is about 1 km down-river from the Krasnoyarsk hydroelectric dam. It was tested by Nicolai I. Drozdov and associates. We visited it in August 1998 on our return trip to Krasnoyarsk from a field conference at the Kurtak archaeological station. Listvenka is one of the very few late Pleistocene Siberian sites where human remains have been found. In this case the find was a fragment of a child's lower jaw. The unerupted lower first permanent molars had features more commonly found in Northeast Asian Sinodonts, and less frequently in Eurodonts (Cro-Magnon and modern Europeans; Figs. A3.3, A3.4). The crown morphology is discussed in Chapter 4.

We did not study any non-human animal remains recovered at Listvenka because we had no access to the collection storage building (see discussion on Afontova Gora); hence we do not include Listvenka in our 30 site assemblages.

**Fig. A3.1** Listvenka open site. In the distant background, near the left end of the high railroad bridge, is the Listvenka site. Here, Nicolai Drozdov's archaeological team found a human child's mandible, along with Upper Paleolithic stone artifacts. Listvenka is treated as an appendix herein because we studied only the child's mandibular teeth. The ferry crossing is backed up here on the Yenisei River because of repairs that were taking place on the river-spanning highway bridge, out of sight to the right. This heavily traveled highway, M54, begins in Krasnoyarsk, parallels the Yenisei southward, and ends at the Mongolian border. It was used in our travels to Kurtak and Dvuglaska from Krasnoyarsk (CGT color Listvenka 7-17-00:28).

**Fig. A3.2** The Listvenka site. Standing at the base of the tree-capped loess bluff where he found the Upper Paleolithic remains, Nicolai Drozdov (left) explains to Olga Pavlova and others the archaeological situation. The site is near a large stream, over which the railroad crosses on the high bridge. The stream flows out of the hills in the background and into the nearby Yenisei River. Listvenka was probably a repeatedly used camp site like Afontova Gora, located down-river. Between the two riverside sites is Yelenev Cave, situated high above but nearer to the Yenisei River (CGT color Listvenka 8-6-98:10).

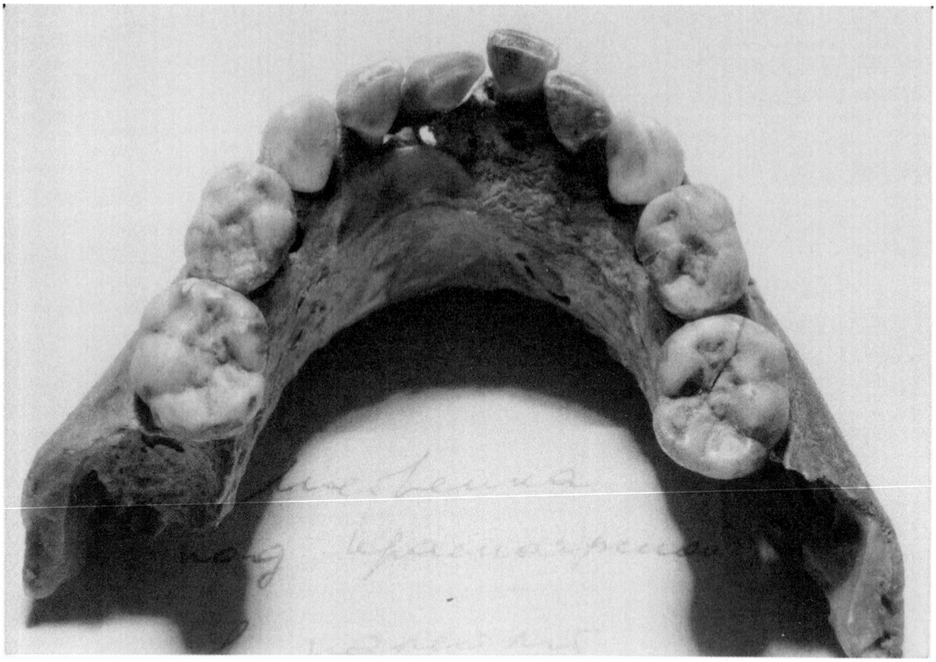

**Fig. A3.3** The Listvenka child's mandible. All of the deciduous teeth had erupted. They show very little wear, no chipping, and no pathology. The first permanent molars had not erupted, so the child was probably 4–6 years of age at the time it died. The morphology of deciduous teeth is much the same in all human groups, so they are rarely used in racial identifications. On the other hand, the morphology of permanent teeth has proven to be of considerable value in epigenetic affinity assessment (CGT neg. IAE 8-8-00:9).

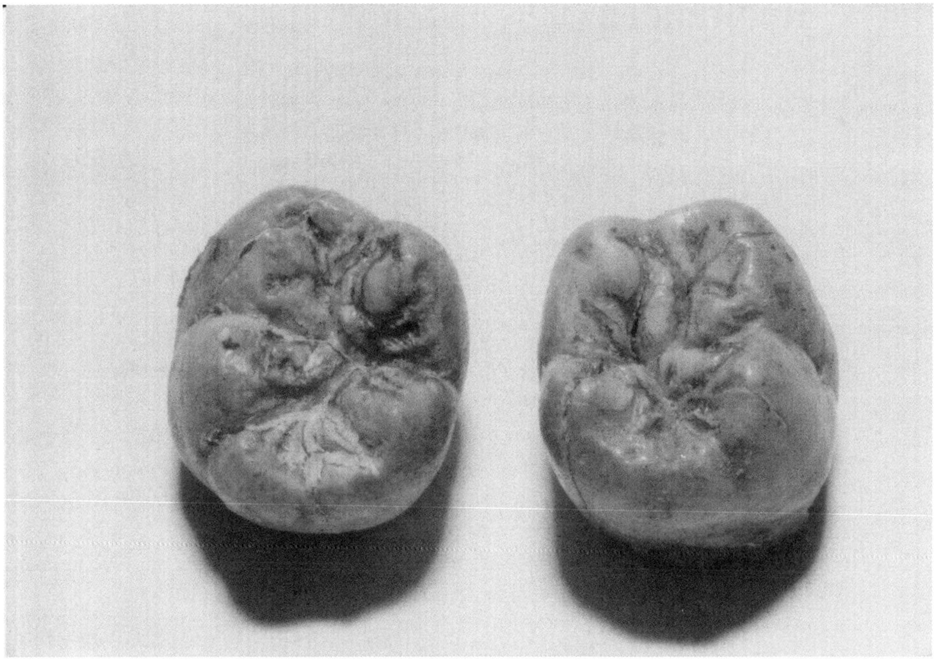

**Fig. A3.4** The Listvenka child's permanent lower first molars, mesial surface at the top. These pristine unerupted teeth exhibit two morphological traits commonly found in modern Asian populations, but rarely in European populations. They are (1) the deflecting wrinkle, and (2) the occurrence of cusp 6. Two other but less diagnostic traits that favor an Asian affiliation are (3) the protostylid pit, and (4) the absence of cusp 7. Hence, the Listvenka child has a higher probability of having had a late Pleistocene Asian, rather than European, affiliation (CGT neg. IAE 8-8-00:12).

# References

Nearly all anthropological publications in Russia are local or regional works under the umbrella of the Russian Academy of Sciences, which could be considered the publisher.

Abdulina, O. I., and V. I. Tashak (1990). A new location of archaeological material on the Braynk River. In *Paleoethnology of Siberia*. Irkutsk, pp. 181–183. (In Russian.)

Abramova, Z. A. (1980). Excavations of Dvuglaska grotto in Khakasia. In *Archaeological Discoveries of 1979*. Moscow, p. 185. (In Russian.)

Abramova, Z. A. (1981). Mousterian grotto Dvuglaska in Khakassia (Preliminary report). *Brief Presentations of the Institute of Archaeology*, **165**:92–98. (In Russian.)

Abramova, Z. A. (1985). Mousterian grotto in Khakasia. *Brief Presentations of the Institute of Archaeology*, **181**:74–78. (In Russian.)

Abramova, Z. A., and N. M. Ermolova (1976). Dvuglaska grotto: Neanderthal dwelling. *Priroda*, **12**:106–107. (In Russian.)

Abramova, Z. A., B. G. Eritsyan, and N. M. Ermolova (1975). New Mousterian site of southern Siberia. In *Archaeological Discoveries of 1975*. Moscow. (In Russian.)

Abramova, Z. A., S. N. Astaknov, S. A. Vasiliev, N. M. Ermolova, and N. F. Litsitsyn (1991). *The Paleolithic of the Yenisei*. Leningrad: Nauka. (In Russian.)

African Wildlife Foundation (2008). Hyena. http://www.awf.org/content/wildlife/detail/1001/

Agadjanyan, A. K. (1999). Small mammals from Holocene sediments of Denisova Cave. *Problems of Archaeology, Ethnography and Anthropology of Siberia and Adjacent Territories*, **5**:226–231. (In Russian.)

Agadjanyan, A. K., and N. V. Serdyuk (2005). The history of mammalian communities and paleogeography of the Altai Mountains in the Paleolithic. *Paleontological Journal*, **6**:S645–S821.

Agadjanyan, A. K., N. V. Serdyuk, and A. V. Postnov (2002). Preliminary chronological interpretation of the Ust-Kan sediments (Gorny Altai) and paleogeographic reconstruction of the conditions of their formation (from data of studying microfauna). In *The Main Regularities of Local and Regional Changes of Climate and Environment in late Cenozoic of Siberia*. Novosibirsk: Institute of Archaeology and Ethnography, pp. 11–22. (In Russian.)

Ageeva, E. A., Y. V. Grichan, and N. D. Ovodov (1978). Pleistocene hyenae in a stone trap. *Priroda*, **12**:14–15. (In Russian.)

Aguilar-Moreno, M. (2006). *Handbook to Life in the Aztec World*. New York: Oxford University Press.

Aigusta, J. (1963). *A Book of Mammals* (trans. M. Schierl). London: P. Hamlyn.

Akimova, E. V., and L. Y. Bleinis (1986). The Paleolithic Ust-Kova site (from materials of 1982). *Proceedings of the Archaeological and Ethnographic Studies in East Siberia (Results and Perspectives)*. Irkutsk, pp. 63–65. (In Russian.)

Alexeev, V. P. (1998). The physical specificities of Paleolithic hominids of Siberia. In *The Paleolithic of Siberia: New Discoveries and Interpretations*, eds. A. Derev'anko, D. B. Shimkin, and W. R. Powers (trans. Inna P. Laricheva). Novosibirsk: Institute of Archaeology and Ethnography, Siberian Division, Russian Academy of Sciences; Urbana and Chicago, IL: University of Illinois Press, pp. 329–335.

Alexeev, V. P. and I. I. Gokhman (1983). *Physical Anthropology of Soviet Asia*. Munich and Vienn: R. Oldenbourg Verlag.

Alexeeva, E. V., and A. M. Maloletko (1984). A new Paleolithic cave monument in the Altai. In *Problems of Studies of the Stone Age of Eurasia*. Krasnoyarsk, pp. 26–28. (In Russian.)

Alexeeva, E. V., and N. K. Vereschagin (1970). Hunters for mammoth in the Baraba steppe. *Priroda*, **1**:71–74. (In Russian.)

Alexeeva, T. I., and N. O. Bader, eds. (2000). *Homo Sungirensis: Upper Paleolithic Man. Ecological and Evolutionary Aspects of the Investigation*. Moscow: Scientific World. (In Russian, with English chapter summaries.)

Anderson, E. (1984). Who's who in the Pleistocene: A mammalian bestiary. In *Quaternary Extinctions: A Prehistoric Revolution*, eds. P. S. Martin and R. G. Klien. Tucson: University of Arizona Press, pp. 40–89.

Anikovich, M. V., A. A. Sinitsyn, J. F. Hoffecker, *et al.* (2007). Early Upper Paleolithic in Eastern Europe and implications for the dispersal of modern humans. *Science*, **315**:223–225.

Anonymous (2006). Mountain lion shot after attacking boy on bike. *Arizona Republic*, April 17, p. A2.

Anonymous (2007). N.J. boy fends off coyote. *Arizona Republic*, April 12, p. A16.

Anonymous (2008). Bears devour 2 Russian miners. *Arizona Republic*, July 24, p. A2.

Anonymous (2009). Coyotes kill musician, 19, on hike in E. Canadian park. *Arizona Republic*, October 29, p. A2.

Anonymous (2010a). Animals, possibly wolves, killed Alaska teacher, autopsy reveals. *Arizona Republic*, March 4, p. A2.

Anonymous (2010b). Bear rampage kills 1, injures 2 at campground. *Arizona Republic*, July 27, p. A2.

Anonymous (2010c). 2 coyote attacks near NYC prompt police response. *Arizona Republic*, July 1, p. A2.

Associated Press (2003). Hyena attack in Malawi. www.CNN.com.

Astakhov, S. N. (1999). *The Yenisey Palaeolithic: The Palaeolithic Sites of Afontova Gora Locality in Krasnoyarsk*. Saint Petersburg: Evropeyskiy Dom. (In Russian.)

Auerbakh, N. K., and G. P. Sosnovsky (1924). The remains of the most ancient culture of man in Siberia. *Life of Siberia*, **5**–6:199–219. (In Russian.)

Bagashev, A. N. ed. (1998). *Studies on the Racial Genesis of West Siberian Peoples*. Tomsk: Tomsk University Press. (In Russian.)

Baikov, N. A. (1925). *The Manchurian Tiger*. Harbin: Society for the Study of Manchuria. (In Russian.)

Bailey, S. E., and C. G. Turner II (1999). A new look at some old teeth: An analysis of non-metric dental traits in Neandertals and Old World modern humans. *American Journal of Physical Anthropology*, **28**:29.

Baryshnikov, G. F. (2003). Bone remains of large mammals from Pleistocene sediments in Denisova Cave. In *Environments and Man in the Paleolithic of Gorny Altai: Conditions of Occupation in the Outskirts of Denisova Cave*, eds. A. P. Derevyanko, M. V. Shunkov, A. K. Agadjanyan, *et al.* Novosibirsk: Institute of Archaeology and Ethnography, pp. 178–232. (In Russian.)

Baryshnikov, G. F., and A. M. Maloletko (1998). Volchiya Cave. *Archaeological Monuments of the Altai Viewed by Geologists*. Barnaul: Altai State University Press, pp. 159–169. (In Russian.)

Baryshnikov, G. F., and N. K. Vereschagin (1996). Brief review of Quaternary hyenas (family Hyaenidae) of Russia and adjacent territories. In *Quarternary and Modern Mammals of Eurasia*, ed. G. F. Baryshinkov. Saint-Petersburg: Reports of the Zoological Institute of RAN (Russian Academy of Sciences).

Baryshnikov, G. F., J. F. Hoffecker, and R. L. Burgess (1996). Palaeontology and zooarchaeology of Mezmaiskaya Cave (northwestern Caucasus, Russia). *Journal of Archaeological Science*, **23**:313–335.

Becker, B., and C. A. Reed (1993). Studies of the bone detritus of the spotted hyena (*Hyaena hyena*) at a site in Egyptian Nubia, and the interpretation of the bone breakage by striped hyenas. In *Skeletons in Her Cupboard: Festschrift for Juliet Clutton-Brock*, eds. A. Clason, S. Payne, and H.-P. Uerpman. Oxford: Oxford Books, pp. 157–182.

Beebe, B. F. (1983). Evidence of carnivore activity in a late Pleistocene/Holocene archaeological site (Blue Fish Cave I), Yukon Territory, Canada. In *Carnivores, Human Scavengers, and Predators: A Question of Bone Technology*, eds. G. M. LeMoine and A. S. MacEachern. Calgary: Archaeological Association of the University of Calgary, pp. 1–14.

Behrensmeyer, A. K., and A. P. Hill, eds. (1980). *Fossils in the Making: Vertebrate Taphonomy and Paleoecology*. Chicago, IL: University of Chicago Press.

Benecke, N., ed. (1999). *The Holocene History of the European Vertebrate Fauna: Modern Aspects of Research*. Randen/Westf: Verlag Marie Leidorf Gmbh.

Billy, G. (1975). Etude anthropologique des restes humains. In *Excavation of the Abri Pataud, Les Eyzies (Dordogne)*, ed. H. L. Movius, Jr. Cambridge, MA: Peabody Museum of Archaeology and Ethnology, Harvard University, pp. 201–261.

Binford, L. R. (1981). *Bones: Ancient Men and Modern Myths*. New York: Academic Press.

Boas, F. (1888). The central Eskimo. *Annual Report of the Bureau of Ethnology, Smithsonian Institution, Washington, D.C.* Reprinted 1964 by the University of Nebraska Press, Lincoln.

Boaz, N. T., and R. L. Ciochon (2004). *Dragon Bone Hill: An Ice Age Saga of* Homo erectus. New York: Oxford University Press.

Boaz, N. T., R. L. Ciochon, Qinqi Xu, and Jinyi Liu (2000). Large mammalian carnivores as a taphonomic factor in the bone accumulation at Zhoukoudian. *Acta Anthropologica Sinica*, **19** (supplement):224–234.

Boddington, A., A. N. Garland, and R. C. Janaway, eds. (1987). *Death, Decay and Reconstruction: Approaches to Archaeology and Forensic Science*. Manchester: Manchester University Press.

Bonch-Osmolovsky, G. A. (1931). About cut marks on Paleolithic bones. *Proceedings of GAMAK*, pp. 25–26. (In Russian.)

Bonnichsen, R., and M. H. Sorg, eds. (1989). *Bone Modification*. Orono, ME: Center for the Study of the First Americans.

Bouchud, J. (1975). Etude de la faune de l'Abri Pataud. In *Excavation of the Abri Pataud, Les Eyzies (Dordogne)*, ed. H. L. Movius, Jr. Cambridge, MA: Peabody Museum of Archaeology and Ethnology, Harvard University, pp. 69–153.

Boulestin, B. (1998). *Approche Taphonomique des Restes Humains: Le Cas des Mesolithiques de la Grotte des Perrats (Agris, Charemte)*. Bordeaux: L'Universite de Bordeaux.

Bower, B. (2010). Recovered DNA suggests a new type of hominid. *Science News*, **177**(9):5–6.

Boyarkina, N. (2000). Tigers. *Arguments and Facts*, **38**:20. (In Russian.)

Brain, C. K. (1981). *The Hunters or the Hunted? An Introduction to African Cave Taphonomy*. Chicago, IL: University of Chicago Press.

Bratlund, B. (1991). A study of hunting lesions containing flint fragments on reindeer bones from Stellmoor, Schleswig-Holstein. In *The Late Glacial in North-West Europe: Human Adaptation and Environmental Change at the End of the Pleistocene*, eds. R. N. E. Barton, A. J. Roberts, and D. A. Roe. London: Council for British Archaeology, pp. 173–207.

Brauer, G. (2007). Origin of modern humans. In *Handbook of Paleoanthropology, Vol. III, Phylogeny of Hominids*, eds. W. Henke and I. Tattersall. Berlin: Springer-Verlag, pp. 1749–1779.

Bunn, H. T. (1991). A taphonomic perspective on the archaeology of human origins. *Annual Review of Anthropology*, **20**:433–467.

Busch, R. H. (2000). *The Grizzly Almanac*. New York: Lyons Press.

Carroll, C. (2005). Misjudged hyenas. *National Geographic*, **207**(6):50–65.

Cavalli-Sforza, L. L., P. Menozzi, and A. Piazza (1994). *The History and Geography of Human Genes*. Princeton, NJ: Princeton University Press.

Chaplin, R. E. (1971). *The Study of Animal Bones from Archaeological Sites*. London: Seminar Press.

Chard, C. S. (1974). *Northeast Asia in Prehistory*. Madison, WI: University of Wisconsin Press.

Chekha, V. P. (1990). Geochemical indicators and some paleogeographic features of the Ust-Kova site. In *Chronostratigraphy of the Paleolithic of North, Central and Eastern Asia and America*. Novosibirsk: Institute of Archaeology and Ethnography, pp. 293–295. (In Russian.)

Chekha, V. P., and N. D. Ovodov (1992). Geology and paleogeography of the Malaya Seeya site. In *Archaeology, Geology, and Paleogeography of Paleolithic Sites in Southern Central Siberia*. Krasnoyarsk: Institute of Archaeology and Ethnography, pp. 109–114. (In Russian.)

Chen, J., R. R. Sokal, and M. T. Ruhlen (1995). Worldwide analysis of genetic and linguistic relationships of human populations. *Human Biology*, **67**(4):595–512.

Chersky, I. D. (1876). Report on the study of Nizhneudinsk Cave. *News of the Siberian Department of the Russian Geographic Society*, **7**(2–3):78–118. (In Russian.)

Clark, G. A., and C. M. Willermet, eds. (1997). *Conceptual Issues in Modern Human Origins Research*. New York: Aldine de Cruyter.

Clark, J. G. D. (1954). *Excavations at Star Carr*. Cambridge: Cambridge University Press.

Clarke, J. (1970). *Man is the Prey*. New York: Pocket Books.

Clary, K. H. (1984). Anasazi diet and subsistence as revealed by coprolites from Chaco Canyon. In *Recent Research on Chaco Prehistory*, eds. W. J. Judge and J. D. Schelberg. Albuquerque, NM: National Park Service, pp. 265–279.

Clottes, J., ed., trans. P. G. Bahn (2003). *Chauvet Cave: The Art of Earliest Times*. Salt Lake City, UT: University of Utah Press.

Conant, E. (2003). Man of 1000 faces. *Archaeology*, July/August:48–53.

Cooper, S. M. (1990). The hunting behaviour of spotted hyaenas (*Crocuta crocuta*) in a region containing both sedentary and migratory populations of herbivores. *African Journal of Ecology*, **28**:131–141.

Crockford, S. J., and Y. V. Kuzmin (2012). Comments on Germorpré *et al.*, *Journal of Archaeological Science* 36, 2009 "Fossil dogs and wolves from Paleolithic sites in Belgium, the Ukraine and Russia. Osteometry, ancient DNA and stable isotopes," and Germopré, Lázkičková-Galetová, and Sabin, *Journal of Archaeological Science* 39, 2012 "Paleolithic dog skulls at the Gravetian Předmostí site, the Czech Republic". *Journal of Archaeological Science*, doi. 10.1016.jas.2012.04.033.

Cruz-Uribe, K. (1991). Distinguishing hyena from hominid bone accumulations. *Journal of Field Archaeology*, **18**:467–486.

Davis, S. J. M. (1987). *The Archaeology of Animals*. New Haven, CT and London: Yale University Press.

Davis, S., and F. R. Valla (1978). Evidence for the domestication of the dog 12,000 years ago in the Natufian of Israel. *Nature*, **276**:608–610.

Debets, G. F. (1946). A fragment of frontal bone of a man from the cultural layer of Afontova Gora II site near Krtasnoyarsk. *Bulleten' Komissii po Izucheniyu Chetvertichnogo Perioda*, **8**:73–77. (In Russian.)

Defleur, A., T. White, P. Valensi, L. Slimak, and E. Cregut-Bonnoure (1999). Neanderthal cannibalism at Moula-Guercy, Ardeche, France. *Science*, **286**:128–131.

de Mortilje, G., and A. de Mortilje (1903). *Prehistoric Life: Origin and Antiquity of Man*. Third Edition, corrected and supplemented with new data (trans. L. Ya. Sternberg). St. Petersburg: "XXth Century" Publishing Co. (In Russian.)

Derevianko, A. P. (2001). The Middle to Upper Paleolithic transition in the Altai (Mongolia and Siberia). *Archaeology, Ethnology & Anthropology of Eurasia*, **3**(7):70–103.

Derevianko, A. P. (2005). The earliest human migrations in Eurasia and the origin of the Upper Paleolithic. In *The Middle to Upper Paleolithic Transition in Eurasia: Hypotheses and Facts*, ed. A. P. Derevianko. Novosibirsk: Russian Academy of Sciences, Siberian Branch, Institute of Archaeology and Ethnography Press, pp. 5–19.

Derevianko, A. P., and M. I. Dergacheva (2002). Holocene sediments of Kaminnaya Cave: Pedogenic features and conditions of formation. *Problems of Archaeology, Ethnography and Anthropology of Siberia and Adjacent Territories*, vol. **8**. Novosibirsk: Institute of Archaeology and Ethnography, pp. 305–308. (In Russian.)

Derevianko, A. P., and Y. V. Grichan (1990). Studies of Kaminnaya Cave: Preliminary excavation results of 1983–1988. *Pleistocene Strata*, **60**:59–80. (In Russian.)

Derevianko, A. P., and S. V. Markin (1992). *Mousterian of the Mountainous Altai*. Novosibirsk: Nauka. (In Russian.)

Derevianko, A. P., and S. V. Markin (1998). Paleolithic of Northwest of Altai-Sayan. *Russian Archaeology*, **4**:17–34. (In Russian.)

Derevianko, A. P., and V. I. Molodin (1994). *Denisova Cave: Part I*. Novosibirsk: Nauka. (In Russian.)

Derevianko, A. P., and V. T. Petrin (1988). Archaeology of the Yaloman Cave. In *Karst of the Altai-Sayany Mountainous Area and Adjacent Mountainous Countries*. Barnaul, pp. 16–19. (In Russian.)

Derevianko, A. P., and E. P. Rybin (2005). The earliest representations of symbolic behavior by Paleolithic humans in the Altai Mountains. In *The Middle to Upper Paleolithic Transition in Eurasia: Hypotheses and Facts*, ed. A. P. Derevianko. Novosibirsk: Russian Academy of Sciences, Siberian Branch, Institute of Archaeology and Ethnography Press, pp. 232–255.

Derevianko, A. P., and M. V. Shunkov (2005). Formation of the Upper Paleolithic traditions in the Altai. In *The Middle to Upper Paleolithic Transition in Eurasia: Hypotheses and Facts*, ed. A. P. Derevianko. Novosibirsk: Russian Academy of Sciences, Siberian Branch, Institute of Archaeology and Ethnography Press, pp. 283–311.

Derevianko, A. P., and A. N. Zenin (1992). Archaeological studies of Srashnaya Cave. *Altaica*, **1**:13–18.

Derevianko, A. P., R. S. Vasilievsky, V. I. Molodin, and S. V. Markin (1985). *Research in Denisova Cave*. Novosibirsk. (In Russian.)

Derevianko, A. P., S. V. Markin, S. V. Nikolaev, and V. T. Petrin (1990). Early Paleolithic complex from Kuzbass. *Chronostratigrahy of Paleolithic of North, Central, and Eastern Asia and America: Reports of the International Symposium*. Novosibirsk, pp. 147–160. (In Russian.)

Derevianko, A. P., V. S. Zykina, S. V. Markin, *et al.* (1992). *The First Early Paleolithic Objects of the Kuznetsk Depression: Stratigraphy, Paleogeography, and Preliminary Archaeological Definitions*. Novosibirsk. (In Russian.)

Derevianko, A. P., S. V. Nikolaev, and V. T. Petrin (1993). Dating Kara-Bom sediments by physical methods. *Altaica*, **3**:3–8. (In Russian.)

Derevianko, A. P., M. I. Dergacheva, and V. T. Petrin (1994). Reconstruction of living conditions of Paleolithic man in the middle-mountainous Altai (based on the example of the Kara-Bom site). *Altaica*, **4**:15–17. (In Russian.)

Derevianko, A. P., S. V. Markin, M. I. Dergacheva, *et al.* (1997a). Complex studies of Pleistocene sediments in Kaminnaya Cave (Gorny Altai) in 1997. *Problems of Archaeology, Ethnography and*

*Anthropology of Siberia and Adjacent Territories*, vol. **3**. Novosibirsk: Institute of Archaeology and Ethnography, pp. 48–51. (In Russian.)

Derevianko, A. P., S. V. Nikolaev, and V. T. Petrin (1997b). Paleoecological aspects of late Pleistocene human existence in the Yelov mountainous area of Gorny Altai (exemplified by the multi-layered site Kara-Bom). *Problems of Archaeology, Ethnography and Anthropology of Siberia and Adjacent Territories*, vol. **3**. Novosibirsk: Institute of Archaeology and Ethnography, pp. 58–63. (In Russian.)

Derevianko, A. P., V. T. Petrin, and L. M. Chevalkov (1997c). Kara-Bom and the origin of the Late Paleolithic in Gorny Altai. *Dulovo Meeting of 1997*. Irkutsk, pp. 11–14. (In Russian.)

Derevianko, A. P., M. V. Shunkov, E. M. Malaeva, V. A. Ulianov, and A. A. Anoikin (1997d). On stratigraphic disjunction of Pleistocene strata in the archaeological cross-section of Denisova Cave. *Problems of Archaeology, Ethnography, and Anthropology of Siberia and Adjacent Territories*, vol. **3**. Novosibirsk: Institute of Archaeology and Ethnography, pp. 80–84. (In Russian.)

Derevianko, A. P., V. N. Zenin, and E. N. Maschenko (1997e). Exploration of the Paleolithic site Shestakovo in 1997. *Problems of Archaeology, Ethnography and Anthropology of Siberia and Adjacent Territories*, vol. **3**. Novosibirsk: Institute of Archaeology and Ethnography, pp. 36–42.

Derevianko, A. P., A. K. Agadjanian, G. F. Baryshnikov, *et al.* (1998a). Okladnikov Cave. In *Archaeology, Geology and Paleogeography of the Pleistocene and Holocene of Gorny Altai*. Novosibirsk: Institute of Archaeology and Ethnography Press, pp. 68–79.

Derevianko, A. P., A. K. Agadjanian, G. F. Baryshnikov, *et al.* (1998b). Denisova Cave: Multi-layered Archaeological Monument of Northwestern Altai. In *Archaeology, geology and paleo-geography of Pleistocene and Holocene of the mountainous Altai*. Novosibirsk: Institute of Archaeology and Ethnography, pp. 24–50. (In Russian.)

Derevianko, A. P., A. K. Agadjanian, G. F. Baryshnikov, *et al.* (1998c). Kaminnaya Cave. In *Archaeology, Geology and Paleogeography of the Pleistocene and Holocene of Gorny Altai*. Novosibirsk: Institute of Archaeology and Ethnography, Siberian Branch, Russian Academy of Sciences, pp. 80–93. (In Russian.)

Derevianko, A. P., *et al.* (1998d). Kaminnaya Cave: Stratigraphy, paleography and culture of ancient man of the Pleistocene epoch (on the basis of work in 1994–1996). *Problems of Paleoecology, Geology, and Archaeology of the Altai Paleolithic*, vol. **4**. Novosibirsk: Institute of Archaeology and Ethnography, pp. 59–80. (In Russian.)

Derevianko, A. P., N. S. Bolikhovskaya, I. V. Foronova, *et al.* (1998e). Stratigraphy and conditions of formation of Pleistocene sediments in Kaminnaya Cave. *Problems of Archaeology, Ethnography and Anthropology of Siberia and Adjacent Territories*, vol. **4**. Novosibirsk: Institute of Archaeology and Ethnography, pp. 50–66. (In Russian.)

Derevianko, A. P., A. K. Agadjanyan, and G. F. Baryshnikov (1998f). The Kara-Bom site. In *Archaeology, Geology, and Paleogeography of Pleistocene and Holocene of Gorny Altai*. Novosibirsk: Institute of Archaeology and Ethnography Press, pp. 106–128. (In Russian.)

Derevianko, A. P., *et al.* (1998g). Geology with elements of paleogeography of the multi-layered site Kara-Bom. In *Problems of Paleoecology, Geology and Archaeology of the Altai Paleolithic*. Novosibirsk: Institute of Archaeology and Ethnography, pp. 133–154. (In Russian.)

Derevianko, A. P., A. K. Agadjanyan, G. F. Baryshnikov, *et al.* (1998h). The Maloyamonskaya Cave. In *Archaeology, Geology, and Paleogeography of Pleistocene and Holocene of Gorny Altai*. Novosibirsk: Institute of Archaeology and Ethnography Press, pp. 157–164. (In Russian.)

Derevianko, A. P., *et al.* (1998i). Stratigraphic elements of the sedimentation process in Straschnaya Cave. In *Problems of Paleoecology, Geology, and Archaeology of the Altai Paleolithic*. Novosibirsk: Institute of Archaeology and Ethnography, Siberian Branch, Russian Academy of Sciences, pp. 96–101. (In Russian.)

Derevianko, A.P., A.K. Agadjanyan, G.F. Baryshnikov, *et al.* (1998j). Ust-Kan Cave. In *Archaeology, Geology, and Paleogeography of Pleistocene and Holocene of Gorny Altai*. Novosibirsk: Institute of Archaeology and Ethnography Press, pp. 100–105. (In Russian.)

Derevianko, A. P., S. V. Glinsky, M. I. Dergacheva, *et al.* (1998k). *Problems of Paleoecology, Geology, and Archaeology of the Altai Paleolithic*. Novosibirsk: Institute of Archaeology and Ethnography, Siberian Branch, Russian Academy of Sciences.

Derev'anko, A., D. B. Shimkin, and W. R. Powers, eds. (trans. Inna P. Laricheva) (1998). *The Paleolithic of Siberia: New Discoveries and Interpretations*. Novosibirsk: Institute of Archaeology and Ethnography, Siberian Division, Russian Academy of Sciences; and Urbana and Chicago, IL: University of Illinois Press.

Derevianko, A. P., M. V. Shunkov, A. K. Agadjanyan, A. A. Anoikin, and V. H. Ulianov (1999a). Study of Holocene layers of Denisova Cave. In *Problems of Archaeology, Ethnography and Anthropology of Siberia and Adjacent Territories*, vol. **5**. Novosibirsk: Institute of Archaeology and Ethnology, pp. 354–359. (In Russian.)

Derevianko, A. P., S. V. Markin, N. S. Bolikhovskaya, *et al.* (1999b). Some results of complex (interdisciplinary) studies of Kaminnaya Cave (northwestern Altai). In *Problems of Archaeology, Ethnography and Anthropology of Siberia and Adjacent Territories*, vol. **5**. Novosibirsk: Institute of Archaeology and Ethnography Press, pp. 98–104. (In Russian.)

Derevianko, A. P., A. K. Agadjanyan, E. G. Vergunov, *et al.* (1999c). Results of complex studies of the Ust-Kan Cave in the field season of 1998. In *Problems of Archaeology, Ethnography, and Anthropology of Siberia and Adjacent Territories*, vol. **5**. Novosibirsk: Institute of Archaeology and Ethnography Press, pp. 195–198. (In Russian.)

Derevianko, A. P., A. V. Postnov, L. M. Chevalkov, *et al.* (1999d). New data on Ust-Kan Cave. In *Problems of Archaeology, Ethnography and Anthropology of Siberia and Adjacent Territories*, vol. **5**. Novosibirsk: Institute of Archaeology and Ethnography Press, pp. 105–110. (In Russian.)

Derevianko, A. P., V. N. Zenin, and M. A. Borisov (1999e). The location of Shestakovo: Main results of studies in 1999. In *Problems of Archaeology, Ethnography, and Anthropology of Eurasia*, vol. **5**. Novosibirsk: Institute of Archaeology and Ethnography, pp. 94–97.

Derevianko, A. P., V. I. Molodin, M. V. Shunkov, and A. A. Anoikin (1999f). Archaeological study of Holocene layers of Denisova Cave. In *Problems of Archaeology, Ethnography and Anthropology of Siberia and Adjacent Territories*, vol. **5**. Novosibirsk: Institute of Archaeology and Ethnography Press, pp. 348–351. (In Russian.)

Derevianko, A. P., A. V. Postnov, L. V. Chevalkov, N. D. Ovodov, and N. H. Kulik (2000a). Preliminary results of excavations of the Ust-Kan Cave in 2000. In *Problems of Archaeology, Ethnography and Anthropology of Siberia and Adjacent Territories*, vol. **6**. Novosibirsk: Institute of Archaeology and Ethnography, pp. 84–89. (In Russian.)

Derevianko, A. P., V. N. Zenin, S. V. Leshchinskiy, and E. N. Maschenko (2000b). Peculiarities of mammoth accumulation at the Shestakovo site in West Siberia. *Archaeology, Ethnology & Anthropology of Eurasia*, **3**(3):42–55.

Derevianko, A. P., V. T. Petrin, and E. P. Rybin (2000c). The Kara-Bom site and the characteristics of the Middle-Upper Paleolithic transition in the Altai (from evidence of the Kara-Bom site). *Archaeology, Ethnology & Anthropology of Eurasia*, **2**(2):33–52.

Derevianko, A. P., A. V. Postnov, and N. D. Ovodov (2001a). Possibilities of paleoecological reconstructions from materials of the Ust-Kan Cave in Gorny Altai. In *Evolution of Life on the Earth*. Tomsk, pp. 483–485. (In Russian.)

Derevianko, A. P., A. V. Postnov, L. V. Chevalkov, N. D. Ovodov, and N. A. Kulik (2001b). Problems of interpretation of geoarchaeological evidences of excavations of the Ust-Kan

Cave in the mountainous Gorny Altai. *Proceedings of the International Symposium Devoted to 130th Anniversary of the Discovery of the Paleolithic in Russia*, editors-in-chief, A. P. Derevianko and G. I. Medvedev. August 1–9, Irkutsk. Novosibirsk: Institute of Archaeology and Ethnography; Irkutsk: Irkutsk Laboratory of Archaeology and Paleoecology, pp. 130–134. (In Russian.)

Derevianko, A. P., P. V. Volkov, and V. T. Petrin (2001c). The Kara-Bom site: Functional analysis of its tool assemblage. In *Modern Problems of Eurasian Paleolithic Science*, eds. A. P. Derevianko and M. I. Medvedev. Novosibirsk: Institute of Archaeology and Ethnography, Siberian Branch, Russian Academy of Sciences, pp. 126–129. (In Russian.)

Derevianko, A. P., S. V. Markin, M. V. Shunkov, *et al.* (2001d). *Paleolithic of the Altai*. Brussels: Richard Liu Foundation, European Institute of Chinese Studies.

Derevianko, A. P., A. K. Agadjanyan, E. G. Vergunov, *et al.* (2002a). Results of complex studies of Ust-Kan Cave in the 2002 field season. *Problems of Archaeology, Ethnography, and Anthropology in Siberia and Adjacent Territories*, vol. **8**. Novosibirsk: Institute of Archaeology and Ethnography, pp. 52–55. (In Russian.)

Derevianko, A. P., M. V. Shunkov, A. K. Agadjanyan, *et al.* (2002b). *Environments and Man in the Paleolithic of Gorny Altai: Conditions of Inhabitants in the Outskits of Denisova Cave*. Novosibirsk: Institute of Archaeology and Ethnography Press. (In Russian.)

Derevianko, A. P., V. N. Zenin, and E. N. Maschenko (2002c). Exploration of the Paleolithic site Shestakovo in 1997. In *Problems of Archaeology, Ethnography and Anthropology of Siberia and Adjacent Territories*, vol. **3**. Novosibirsk: Institute of Archaeology and Ethnography, pp. 36–42. (In Russian.)

Derevianko, A. P. *et al.* (2002d). Complex studies of Pleistocene sediments in Kaminnaya Cave (Gorny Altai) in 1997. *Problems of Archaeology, Ethnography and Anthropology of Siberia and Adjacent Territories*, vol. **3**. Novosibirsk: Institute of Archaeology and Ethnography, pp. 71–75.

Derevianko, A. P., S. V. Markin, and S. A. Yefremov (2002e). Holocene sediments of Kaminnaya Cave: Pedogenic features and conditions of formation. *Problems of Archaeology, Ethnography and Anthropology of Siberia and Adjacent Territories*, vol. **8**. Novosibirsk: Institute of Archaeology and Ethnography, pp. 71–75. (In Russian.)

Derevianko, A. P., M. V. Shunkov, V. A. Ulianov, *et al.* (2003a). New results of studies of the Middle Paleolithic complex of Denisova Cave. In *Problems of Archaeology, Ethnography and Anthropology of Siberia and Adjacent Territories*, vol. **9**, part 1. Novosibirsk: Institute of Archaeology and Ethnography, pp. 112–116. (In Russian.)

Derevianko, A. P., M. V. Shunkov, A. A. Anoikin, and V. A. Ulianov (2003b). New results of studies of the Upper Paleolithic complex of Denisova Cave. In *Problems in Archaeology, Ethnography, and Anthropology of Siberia and Adjacent Territories*, vol. **6**. Novosibirsk: Institute of Archaeology and Ethnography, pp. 93–98. (In Russian.)

Derevianko, A. P., V. I. Molodin, V. N. Zenin, S. V. Lescheninsky, and E. N. Mashchenko (2003c). *The Late Paleolithic Site of Shestakovo*. Novosibirsk: Institute of Archaeology and Ethnography, Siberian Branch, Russian Academy of Sciences. (In Russian.)

Derevianko, A. P., A. V. Postnov, T. A. Gorbunova, *et al.* (2003d). Results of complex studies by the Ust-Kan team in the 2003 field season. In *Problems of Archaeology, Ethnography, and Anthropology of Siberia and Adjacent Territories*, vol. **9**, part 1. Novosibirsk: Institute of Archaeology and Ethnography, pp. 96–100. (In Russian.)

Derevianko, A. P., A. V. Postnov, T. A. Gorbunova, *et al.* (2004). Archaeological studies of the Ust-kan cave and the site Charyshsky Naves in 2004. In: *Problems of Archaeology of Siberia and*

*Adjacent Territories*, vol. **10**, part I, Novosibirsk: Institute of Archaeology and Ethnography, pp. 90–95. (In Russian.)

Dice, M. H. (1993). A Disarticulated Human Bone Assemblage from Leroux Wash, Arizona. Master's thesis, Department of Anthropology, Arizona State University, Tempe.

Diedrich, C. G. (2011). Periodical use of the Balve Cave (NW Germany) as a Late Pleistocene *Crotuta crocuta spelaea* (Goldfuss 1823) den: Hyena occupations and bone accumulations vs. human Middle Palaeolithic activity. *Quaternary International*, **233**:171–184.

Drozdov, N. I. (1981). Stone Age of the Northern Cis-Angara Region. Abstract of Candidate's Dissertation. Novosibirsk. (In Russian.)

Drozdov, N. I., and V. P. Chekha (1990). Paleofrosted phenomena at the Paleolithic site Ust-Kova and problem of preservation of cultural layers. In *Chronostratigraphy of the Paleolithic of North, Central, and East Asia and America (Reports of the International Symposium)*. Novosibirsk: Institute of Archaeology and Ethnography, pp. 174–180. (In Russian.)

Drozdov, N. I., and V. P. Chekha (2003). The geological chronology and periodization of the Late Paleolithic of the Yenisei: Approaches and problems. *Archaeology, Ethnology & Anthropology of Eurasia*, **1**(13):2–10.

Drozdov, N. I., and S. A. Laukhin (1980). Ust-Kova on the Angara River: The most ancient Paleolithic site between the Lena and the Urals dated by the radiocarbon method. *Reports of the U.S.S.R. Academy of Sciences*, **254**:182–185. (In Russian.)

Drozdov, N. I., V. P. Chekha, S. A. Laukhin, *et al.* (1990). The Ust-Kova group of archaeological sites. In *Chronostratigraphy of Palaeolithic Sites of Middle Siberia: Yenisei Basin*. Novosibirsk: IIFiF SO AN SSSR, pp. 147–181. (In Russian.)

Drozdov, N. I., E. V. Artemiev, V. P. Chekha, V. I. Makulov, and A. Tarasov (2000). New pages of investigations in Krasnoyarsk archaeological area. In *Problems of Archaeology, Ethnography and Anthropology of Siberia and Adjacent Territories*, vol. **6**. Novosibirsk: Institute of Archaeology and Ethnography, pp. 118–121. (In Russian.)

Enloe, J. G., F. David, and G. Baryshnikov (2000). Hyenas and hunters: Zooarchaeological investigations at Prolom II Cave, Crimea. *International Journal of Osteoarchaeology*, **10**:310–324.

Ermolova, N. M. (1978). *Terrafauna of the Angara River Valley in the Late Anthropogene*. Novosibirsk: Nauka. (In Russian.)

Ewer, R. F. (1973). *The Carnivores*. Ithaca, NY: Cornell University Press.

Firsov, L. V., and L. A. Orlova (1971). Radiocarbon dating of mammoth bone from the Volchiya Griva site. In *Materials of the Field Studies of the Far Eastern Archaeological Expedition*, **2**:132–134. (In Russian.)

Fischer, G. de Waldheim (1834). Notice sur les ossements de mammiferes trouves dans les cavernes de Tscharych et de Khankara, Gouverment de Tomsk. *Bulletin of the Imperial Society of Nature of Moscow*, **7**:179–186.

Fish, P. R. (1993). *Archaeological Site Recording Manual*. Tucson, AZ: Arizona State Museum.

Fisher, J., N. Simon, and J. Vincent (1969). *Wildlife in Danger*. New York: Viking Press.

Fitzhugh, W. W., and A. Crowell (1988). *Crossroads of Continents: Cultures of Siberia and Alaska*. Washington, DC: Smithsonian Institution Press.

Foronova, I. V. (2000). The paleotheriological characteristic of Kedrovka suite: Estimation of the age of the Paleolithic site Mokhovo-I (Kuznetsk Basin, Southeast of Western Siberia). In *International Scientific Conference "Paleogeography of the Stone Age. Correlation of Natural Events and Archaeological Cultures of Paleolithic Period of Northern Asia and Adjoining Territories"*, ed. N. I. Druzdof. July 22–31. Krasnoyarsk, pp. 137–138. (In Russian.)

Foronova, I. V. (2001a). Paleoterriological characteristics and stratigraphic peculiarities of the Kedrov suite: Possibilities of age estimation of the Mokhovo-1 Paleolithic site. In *Paleoecology of the Pleistocene and Cultures of the Stone Age of North Asia and Adjacent Territories*, vol. **1**. Novosibirsk: Institute of Archaeology and Ethnography, pp. 279–291. (In Russian.)

Foronova, I. V. (2001b). *Quaternary Mammals of Southwestern Siberia: Kuznetsk Depression*. Novosibirsk: Siberian Branch, Russian Academy of Sciences. (In Russian.)

Frank, L. G., S. E. Glickman, and P. Licht (1991). Fatal sibling aggression, precocial development, and androgens in neonatal spotted hyenas. *Science*, **252**(5006):702–704.

Galkina, L. I., and N. D. Ovodov (1975). Anthropogenic terrafauna of caves in the western Altai. In *Systematics, Fauna, and Zoogeography of Mammals and their Parasites: Fauna of Siberia Series*, ed. B. S. Yudin. Novosibirsk: Nauka, pp. 165–180. (In Russian.)

Garutt, W. E. (1964). *Das Mammut:* Mammuthus primigenius *(Blumenbach)*. Leningrad: A. Zaemsan Verlag; Wittenberg. Luther Stadt.

Geist, V. (1989). Did large predators keep humans out of North America? In *The Walking Larder: Patterns of Domestication, Pastoralism and Predation*, ed. J. Clutton-Brock. London: Unwin Hyman, pp. 282–294.

Geist, V., L. Baskin, I. Okhlopkov, and I. Spilenok (2009). Wolves, bears and human anti-predator adaptations. Paper to be presented in Moscow, Russia, August.

Gerasimov, M. M. (1931). *Mal'ta, Paleolithic Site: (Preliminary Data). Results of Studies of 1928–29*. Irkutsk: Historical Museum Press. (In Russian.)

Gerasimov, M. M. (1935). Excavations of a Paleolithic site in the village of Mal'ta: Paleolithic of the U.S.S.R. *Izvestia (News) of the State Academy of Material Culture*, **118**:78–124. (In Russian.)

Gerasimov, M. M. (1940). *Treatment of Bones of the Paleolithic Site of Mal'ta*. Moscow-Leningrad: Moscow Institute of Archaeology. (In Russian.)

Gerasimov, M. M. (1955). *Reconstruction of the Face from the Skull: Modern and Fossil Men*. Moscow: Nauka. (In Russian.)

Gerasimov, M. M. (1958). The Paleolithic site of Mal'ta: Excavations of 1956–57. *Sovetskaya Etnografiya*, **3**:28–52. (In Russian.)

Germonpré, M. (1993). Preliminary results of taphonomy of Denisova Cave. *Altaica*, **2**: 11–16.

Germonpré, M., and L. Lbova (1996). Mammalian remains from the Upper Paleolithic site of Kamenka, Buryatia (Siberia). *Journal of Archaeological Science*, **23**: 35–57.

Gilbert, B. M. (1990). *Mammalian Osteology*. Columbia, MO: Missouri Archaeological Society.

Gittleman, John L., ed. (1996). *Carnivore Behavior, Ecology, and Evolution*. **2** vols. Ithaca, NY: Comstock Publication Associates.

Gittleman, John L., ed. (2001). *Carnivore Conservation*. Cambridge: Cambridge University Press.

Golomshtok, E. A. (1938). *The Old Stone Age in European Russia*. Reprinted from *Transactions of the American Philosophical Society*, Vol. **19**, Part II. Philadelphia and New York: AMS Press.

Goryunova, O. I., A. S. Yendrikhinsky, I. L. Lezhnenko, and P. E. Shmygun (1976). Studies in the North Baikal. In *Archaeological Discoveries of 1975*. Moscow, p. 227. (In Russian.)

Gould, R. A. (2007). *Disaster Archaeology*. Salt Lake City, UT: University of Utah Press.

Grice, G. (2010). *Deadly Kingdom: The Book of Dangerous Animals*. New York: Dial Press.

Grichan, Y. V. (1987). Studies of Kaminnaya cave (the mountainous Altai). In *Problems of Studying Ecology and Protection of Caves*. Kiev, pp. 143–144. (In Russian.)

Gromova, V. I. (1924). The oldest human remains in Siberia and the contemporary fauna. *Life in Siberia*, **5**–6:221–233. (In Russian.)

Gromova, V. I. (1941). Fauna of the Upper Paleolithic site of Mal'ta near Irkutsk. In *Reports of the U.S.S.R. Academy of Sciences*, **33**(1):94–96. (In Russian.)

Gromova, V. I. (1950). Identification of USSR mammals from skeletal elements: Vol. I. Identification from long bones. In *Proceedings of the Commission on the Study of the Quaternary Period*, vol. **9**. Moscow-Leningrad: USSR Academy of Sciences. (In Russian.)

Grupe, G. (2007). Taphonomic and diagenic processes. In *Handbook of Paleoanthropology*, vol. 1, eds. W. Henke and I. Tattersall, Berlin: Springer-Verlag, pp. 241–259.

Gryaznov, M. P. (1932). Human remains from the cultural layer of the Afontova Gora site. In *Proceedings of the Commission for the Study of the Quaternary Period*, vol. **1**. Leningrad: News of the USSR Academy of Sciences, pp. 137–144. (In Russian.)

Guthrie, R. D. (1990). *Frozen Fauna of the Mammoth Steppe: The Story of Blue Babe*. Chicago, IL and London: University of Chicago Press.

Guthrie, R. D. (2005). *The Nature of Paleolithic Art*. Chicago, IL: University of Chicago Press.

Gutieva, N. V., and V. M. Chkhikvadze (1990). Giant Pleistocene frog from Okladnikova Cave (Gorny Altai). In *Chronostratigraphy of the Paleolithic of North, Central and East Asia and America: Reports of the International Symposium*. Novosibirsk: Institute of Archaeology and Ethnography, pp. 132–134. (In Russian.)

Haeussler, A. M. F. (1996). Dental anthropology of Russia, Ukraine, Georgia, Central Asia: Evaluation of five hypotheses for Paleo-Indian origins. PhD dissertation, Department of Anthropology, Arizona State University, Tempe.

Haeussler, A. M. F., and C. G. Turner II (2000). The Upper Paleolithic children of Russia: Comparative dental anthropological analysis of the permanent teeth. *American Journal of Physical Anthropology*, **30** (supplement):169–170.

Haglund, W. D., and M. H. Sorg, eds. (1997). *Forensic Taphonomy: The Postmortem Fate of Human Remains*. Boston, MA: CRC Press.

Haynes, G. (1982). Utilization and skeletal disturbances of North American prey carcasses. *Arctic*, **35**(2):266–281.

Haynes, G. (1988). Longitudinal studies of African elephant death and bone deposits. *Journal of Archaeological Science*, **15**:131–157.

Haynes, G. (1991). *Mammoths, Mastodons, and Elephants*. Cambridge: Cambridge University Press.

Haynes, G. (2000). Mammoths, measured time, and mistaken identifies. *Radiocarbon*, **42**(2):257–269.

Haynes, G. (2002a). Archeological methods for reconstructing human predation on terrestrial vertebrates. *The Paleontological Society Papers: The Fossil Record of Predation*, **8**:51–67.

Haynes, G. (2002b). *The Early Settlement of North America: The Clovis Era*. Cambridge: Cambridge University Press.

Heizer, R. F. (1967). Analysis of human coprolites from a dry Nevada cave. *University of California Archeological Reports*, **70**(1):1–20.

Herzog, H. (2010). *Some We Love, Some We Hate, Some We Eat: Why It's so Hard to Think Straight About Animals*. New York: HarperCollins.

Herzog, W. (2005). *Grizzly Man*. Documentary. California, Image Entertainment.

Hoffecker, J. F. (2005). *A Prehistory of the North: Human Settlement of the Higher Latitudes*. Piscaway, NJ: Rutger University Press.

Hornocker, M. (1997). Siberian tigers. *National Geographic*, **191**(2):100–109.

Horwitz, L. K., and J. Kerbis (1991). Hyenas at home. *Israel Land & Nature*, **16**(4):152–165.

Horwitz, L. K., and P. Smith (1988). The effects of striped hyaena activity on human remains. *Journal of Archaeological Science*, **15**:471–481.

India News (2005). Ranchi: A village in Jharkhand has credited a dog with saving the lives of a woman and her child during an attack by hyenas. www.indianews.com

Ineshin, E. M., and A. B. Tetenkin (1995). New data for the model of human activity on the Pleistocene–Holocene border in the Vitim basin within Baikal-Patomsk highlands. In *The Latest Archaeological and Ethnographic Discoveries in Siberia*. Novosibirsk, pp. 100–102. (In Russian.)

Ineshin, E. M., and A. V. Tetenkin (2000). On problems in geoarchaeological studies of the Baikal-Patomsk highland. *Problems of Archaeology, Ethnography and Anthropology*, vol. **6**. Novosibirsk: Institute of Archaeology and Ethnography, p. 141. (In Russian.)

Ineshin, E. M., A. G. Revenko, and A. P. Sekerin (1998). Exotic types of raw materials for late Pleistocene artifacts of the Vitim River basin (Baikalian Siberia) and the means of their transporting. In *Problems of Archaeology, Ethnography and Anthropology of Siberia and Adjacent Territories*, vol. **4**. Novosibirsk: Institute of Archaeology and Ethnography, pp. 108–113. (In Russian.)

Ineshin, E. M., L. D. Sulerzhitsky, and A. V. Tetenkin (2003). Radiocarbon dating of archaeological and paleontological localities of the Baikal-Patomsk mountain region. In *Current Problems of Eurasian Paleolithic Science*. Novosibirsk: Institute of Archaeology and Ethnography Press, pp. 160–170. (In Russian.)

Ineshin, E. M., A. B. Klementiev, N. V. Martynovich, *et al.* (2004). Problems and issues of methodology of zooarchaeological studies of Bolshoi Yakor 1. *News of the Archaeological Laboratory*. Irkutsk: Irkutsk State Technical University, pp. 49–80. (In Russian.)

Ineshin, E. M., A. B. Klementiev, and A. B. Tetenkin (2005). Planigraphic analysis of cultural horizons of multilayered Bolshoi Yakor 1 site on the lower Vitim. *News of the Archaeological Laboratory*. Irkutsk: Irkutsk State Technical University, pp. 74–108. (In Russian.)

Ivaniev, L. N., A. M. Sizikov, and A. A. Khamzina (1981). Geology of locations of large mammals of the northern Cis-Baikal area. In *Relief and Quaternary Sediments of the Stanovoi Ridge*. Moscow: Nauka, pp. 100–110. (In Russian.)

Ivleva, N. G. (1990). Microteriological materials from Okladnikova and Denisova caves in the Altai. In *Complex Studies of Paleolithic Objects of the Anuj River Basin*, eds. R. S. Vasilievsky and Yu. P. Kholyushkin. Novosibirsk: Institute of Archaeology and Ethnography, pp. 82–104. (In Russian.)

Jacobson, E. (2005). Aral Tolgoi. In *Petroglyphs of Aral Tolgoi (Mongolia)*, ed. N. Batbold. Ulan-Bataar: Mongolia Academy of Sciences, Institute of Archaeology, pp. 103–132.

Jones, K. T. (1983). Forager archaeology: Ache' of Eastern Paraguay. In *Carnivores, Human Scavengers & Predators: A Question of Bone Technology*, eds. G. M. LeMoine and A. S. MacEacjerm. Alberta: Archaeological Association of the University of Calgary, pp. 171–191.

Kalmykov, N. P. (2000). Evolution of biota of the Lake Baikal Anthropogene. In *Problems of Reconstruction of Climate and Environments of Siberian Holocene and Pleistocene*, vol. **2**, ed. S. Markin. Novosibirsk: Russian Academy of Sciences, Siberian Branch, Institute of Archaeology and Ethnography, pp. 261–268. (In Russian.)

Kanowski, M. (1987). *Old Bones: Unlocking Archaeological Secrets*. Melbourne: Longman.

Kaschenko, N. F. (1901). Mammoth skeleton with signs of using some parts of the body of this animal as food by contemporary humans. *Memoirs, Imperial Academy of Sciences of St. Petersburg*, 11(7). (In Russian.)

Keates, S. G. (2003). Biostratigraphy, taphonomy, paleoenvironment and hominid diet in the Middle and Late Pleistocene of China. In *Current Research in Chinese Pleistocene Archaeology*, eds. Chen Shen and S. G. Keates. Oxford: Archaeopress, pp. 37–56.

Kehoe, T. F. (1983). Carnivores or humans? Gull Lake Plains hunters vs. Schrecrensee German Neolithic hunter-farmers. In *Carnivores, Human Scavengers & Predators: A Question of Bone*

*Technology*, eds. G. M. LeMoine and A. S. MacEacjerm. Alberta: Archaeological Association of the University of Calgary, pp. 157–170.

Khenzykhenova, F. I., and G. G. Shushpanova (2001). Terrafauna of the Mal'ta site. In *Modern Scientific Problems of the Eurasian Paleolithic: Proceedings of the International Symposium Devoted to the 130th Anniversary of Paleolithic Discoveries in Russia*, eds. A. P. Derevianko and G. I. Medvedev. Novosibirsk: Institute of Archaeology and Ethnography, Siberian Branch, Russian Academy of Sciences, pp. 374–378. (In Russian.)

Kim, A. (2005). Man survives grizzly bear attack, describes his experience. *Arizona Republic*, September 11, p. B10.

Kimura, H. (1997). *The Study of Paleolithic Culture in Siberia*. Sapporo: Hokkaido University Press. (In Japanese.)

Kimura, H. (2003). Blade industry of Mal'ta site. *Archaeology, Ethnography & Anthropology of Eurasia*, **1**(13):11–33.

Klein, R. G. (1975). Palaeoanthropological implications of the nonarchaeological bone assemblage from Swartklip I, South-Western Cape Province, South Africa. *Quaternary Research*, **5**:275–288.

Konopatsky, A. K. (2001). *The Great Explorer of the Past (Academician A. P. Okladnikov: Pages of Biography)*. Novosibirsk: Siberian Chronograph. (In Russian.)

Konopatsky, A. K. (2009). *The Great Explorer of the Past: Academician A. P. Okladnikov (Pages of Biography Part II, 1950s to the mid-1970s)*. Novosibirsk: Siberian Chronograph. (In Russian.)

Koop, B. F., M. Burbridge, A. Byun, *et al.* (2000). Ancient DNA evidence of a separate origin for North American indigenous dogs. In *Dogs Through Time: An Archaeological Perspective*, ed. S. J. Crockford. Oxford: Archaeopress.

Kovalskiy, V. V. (1974). *Geochemical Ecology*. Moscow: Nauka. (In Russian.)

Kozintsev, A. G. (2003). The Middle to Upper Paleolithic transition: Adaptation and assimilation (the Sungir problem revisited). *Archaeology, Ethnology & Anthropology of Eurasia*, **1**(13):58–64.

Krakhmalnaya, T. V. (1999). Carnivore remains from late Pleistocene and Holocene deposits in the Ukraine. In *The Holocene History of the European Vertebrate Fauna: Modern Aspects of Research*, ed. N. Benecke. Rahden/Westf: Verlag Marie Leidorf GmbH, pp. 223–235.

Krause, J., L. Orlando, D. Serre, *et al.* (2007). Neanderthals in central Asia and Siberia. *Nature*, **449**:902–904.

Kruuk, H. (1972). *The Spotted Hyena: A Study of Predation and Social Behavior*. Chicago, IL and London: University of Chicago Press.

Kruuk, H. (2002). *Hunter and Hunted: Relationships Between Carnivores and People*. Cambridge: Cambridge University Press.

Kulik, N. A., and S. V. Markin (2003). Petrography of the industry of Okladnikova Cave (northwestern Gorny Altai). In *Problems of Archaeology, Ethnography and Anthropology of Siberia and Adjacent Territories*, vol. **9**, part 1. Novosibirsk: Institute of Archaeology and Ethnography, pp. 148–153. (In Russian.)

Kurtén, B. (1976). Trans-Beringian carnivore populations in the Pleistocene. In *Beringia in the Cenozoic Era*, ed. V. L. Kontrimovichus. Vladivostok: Academy of Sciences of the USSR Far Eastern Scientific Center, pp. 339. Translated by R. Chackravarty, US Department of the Interior and the National Science Foundation, Amerind Publishing Co.

Kurtén, B., and E. Anderson (1980). *Pleistocene Mammals of North America*. New York: Columbia University Press.

Kuzmin, Y. V. (1994). Prehistoric colonization of northeast Siberia and migration to America: Radiocarbon evidence. *Radiocarbon*, **36**(3):367–376.

Kuzmin, Y. V., and L. A. Orlova (1998). Radiocarbon chronology of the Siberian Paleolithic. *Journal of World Prehistory*, **12**:1–53.

Kuzmin, Y. V., and K. B. Tankersley (1996). The colonization of eastern Siberia: An evaluation of the Paleolithic Age radiocarbon dates. *Journal of Archaeological Science*, **23**:577–585.

Larichev, V. E. (1978) The art of the Upper Paleolithic settlement Malaya Syya: Dating, types and images, the artistic style, and the problem of interpretation. A preliminary report. *Series Observations in Science*, **3**(2):104–119.

Larichev, V., U. Khol'ushkin, and I. Laricheva (1992). The Upper Paleolithic of northern Asia: Achievements, problems, and perspectives. III. Northeastern Siberia and the Russian Far East. *Journal of World Prehistory*, **6**(4):441–476.

Laukhin, S. A., N. I. Drozdov, A. V. Starikov, *et al.* (1980). Radiocarbon dating of Pleistocene–Holocene sediments in the valley of the latitudinal course of the Angara. In *Geochronology of the Quaternary Period*. Moscow: Nauka, pp. 213–223. (In Russian.)

Law, J. (2007). *Crocuta crocuta*: Spotted hyena. University of Michigan Museum of Zoology Animal Diversity Web. http://animaldiversity.unmz.umich.edu/site/accounts/information/Crocuta_crocuta.html.

Lbova, L. V. (1991a). The Palaeolithic site of Kamenka-1: Preliminary results of investigation. In *The Archaeological and Ethnographic Problems of Siberia and the Far East*, ed. N. Drozdov. Krasnoyarsk: Pedagogical Institute of Krasnoyarsk, pp. 90–91. (In Russian.)

Lbova, L. V. (1991b). The Paleolithic settlement of Kamenka-1: Preliminary results of research in 1990. In *Problems of Archaeology and Ethnography of Siberia and the Far East: Conference Abstracts*, vol. **1**. Krasnoyarsk, pp. 54–56. (In Russian.)

Lbova L. V. (1996). The Kamenka Paleolithic settlement. *New Paleolithic Sites of the Trans-Baikal*. Chita, pp. 24–47. (In Russian.)

Lbova, L. V. (2000). *The Paleolithic of the Northern Zone of Western Trans-Baikal*. Ulan Ude: Buryatia National Center of the Siberian Branch, Russian Academy of Sciences. (In Russian.)

Lbova, L. V. (2002). The transition from the Middle to Upper Paleolithic in the Western Trans-Baikal. *Archaeology, Ethnology & Anthropology of Eurasia*, **1**(9):59–75.

Lbova, L. V., and M. Germonprè (1994). Problems of seasonal inhabitance at the Kamenka Paleolithic site, Trans-Baikal. In *Cultural Traditions of Peoples of Siberia and America: Continuity and Ecology. Abstracts of International Symposium*, Chita, pp. 86–87. (In Russian.)

Lbova, L., and E. A. Khamzina (1999). The first academic expeditions (1724, 1733–43, 1768–1772). D. G. Messerschmidt (Diaries of 1742). In *The Map of Archaeological Monuments*, ed. A. P. Derevianko. Ulan-Ude: Publishing House of the Buryat Scientific Center, pp. 1–221. (In Russian.)

Lbova, L. V., and A. A. Shteinokova (1996). Specific features of planigraphic structure of Complex A of the Kamenka locality. In *Archaeology, Paleo-ecology, and Ethnology of Siberia and Far East*. Irkutsk, pp. 115–118. (In Russian.)

Lbova, L. V., and P. V. Volkov (1993). The tool kit of the Paleolithic Kamenka-1 locality: Typology and functions. In *Cultures and Monuments of the Early Metal Epoch*. Novosibirsk: Nauka, pp. 3–20. (In Russian.)

Lbova, L. V., I. N. Rezanov, N. P. Kalmykov, *et al.* (2003). *Environments and Man in Neopleistocene (Western Trans-Baikal and South-eastern Cis-Baikal)*. Ulan-Ude: Publishing House of Buryat Scientific Center of the Siberian Branch of Russian Academy of Sciences.

Leontiev, V. P., and N. I. Drozdov (2003). Some materials of the Stone Age from the Ust-Kova site in northern Cis-Angara. In *Problems of Archaeology, Ethnography, and Anthropology of Siberia and Adjacent Territories*, vol. **9**, part 1. Novosibirsk: Institute of Archaeology and Ethnography, pp. 408–410. (In Russian.)

Leontiev, V. P., and S. G. Stepanov (1986). Archaeological studies at the multi-layered site Ust-Kova in the Kezhemsky district, Krasnoyarsk region. In *Monuments of Ancient Cultures of Siberia and the Far East*. Novosibirsk: Institute of History, Philosophy, and Philology, pp. 125–128. (In Russian.)

Leontiev, V. P., N. I. Drozdov, V. I. Makulov, and A. L. Zaika (2000). Archaeological studies at the site Ust-Kova in the middle near-Angara area in 2000. In *Problems of Archaeology, Ethnography, and Anthropology of Siberia and Adjoining Territories*, vol. **6**. Novosibirsk: Institute of Archaeology and Ethnography, pp. 332–333. (In Russian.)

Leroi-Gourhan, A. (trans. Norbert Guterman) (1967). *Treasures of Prehistoric Art*. New York: Harry N. Abrams, Inc.

Leshchinskiy, S. V. (1998). Geology and paleo-geography of the Late Paleolithic Shestakovo encampment. In *Pleistocene Paleoecology and Cultures of the Stone Age in North Asia and Contiguous Territories*, vol. **1**, ed. A. P. Derevianko. Novosibirsk: Institute of Archaeology and Ethnography, pp. 209–220. (In Russian.)

Leshchinskiy, S. V. (1999). Influence of regional geological structures on the distribution of Paleolithic man (southeast of the West Siberian plain). In *Geochemistry of Landscapes, Paleoecology and Ethnogenesis*. Ulan-Ude: Buryat Siberian Branch, Russian Academy of Sciences, pp. 472–473. (In Russian.)

Leshchinskiy, S. V. (2000). Stratigraphy and paleogeography of the Pleistocene of the southeast of the west Siberian plain. Candidate for Doctor of Science (Geology-Mineralogy) Dissertation. Tomsk State University, Tomsk. (In Russian.)

Leshchinskiy, S. V. (2001a). The late Pleistocene *beast solonetz* of West Siberia: "Mineral oases" in mammoth migration paths, foci of Palaeolithic man's activities. In *The World of Elephants*, eds. G. Cavarretta, P. Gioia, M. Mussi, and M. R. Palobo. Rome: Consiglio Nazionale delle Ricerche, pp. 293–298.

Leshchinskiy, S. V. (2001b). New data on geology and genesis of the site of Volchiya Griva. In *Modern Problems on Eurasian Palaeolithic Science*, eds. A. P. Derevianko and G. I. Medvedev. Novosibirsk: Institute of Archaeology and Ethnography; Irkutsk: Irkutsk Laboratory of Archaeology and Paleoecology, pp. 244–251. (In Russian.)

Leshchinskiy, S. V. (2012). Paleoecological investigation of mammoth remains from the Kraków Spadzista Street (B) site. *Quaternary International* (in press).

Leshchinskiy, S. V., E. M. Burkanova, and L. A. Orlova (2003). Kochegir: A new mammoth site. Animal paleosalt place in the southeastern West Siberian lowland. *Bulletin of the Tomsk State University Series "Sciences on Earth (Geology, Geography, Meterology, Geodesy)"*, 3(11):n.p. (In Russian.)

Lisitsyn, N. F. (1997) Some results of the study of the Stone Age on the high terraces of the Yenisei. In *Paleolithic and Neolithic of the USSR*. Novosibirsk: Nauka, pp. 33–38.

Lipnina, E. A., and G. I. Medvedev (1993). New data on the Mal'ta Paleolithic location. In *Historical Experience of Colonization of Eastern Regions of Russia*. Abstracts of Reports and Presentations of International Scientific Conference. Vladivostok, pp. 70–73.

Lipnina, E. A., G. I. Medvedev, N. H. Saveliev, and N. G. Torshina (1995). Malta archaeological locality: Historiographic narration. In *Archaeology, Paleoecology, and Ethnology of Siberia and the Far East*. Abstracts of Reports of the 36th Regional Archaeological Conference of Students. Part 1. Irkutsk, pp. 54–56. (In Russian.)

Lyman, R. L. (1994). *Vertebrate Taphonomy*. Cambridge: Cambridge University Press.

Lyman, R. L. (2002). Taphonomic agents and taphonomic signatures. *American Antiquity*, **67**(2):361–365.

Lynch, T., ed. (2005). *El Lobo: Readings on the Mexican Gray Wolf.* Salt Lake City, UT: University of Utah Press,

Malaeva, E. M. (1999). Paleonology of Pleistocene sediments of the near entry area of Denisova Cave. In *Problems of Archaeology, Ethnography and Anthropology of Siberia and Adjacent Territory*, vol. **5**. Novosibirsk: Institute of Archaeology and Ethnography, pp. 163–168. (In Russian.)

Maloletko, A. M., and V. A. Panychev (1989). Late Pleistocene deposits of the Denisova Cave (Altai): Conditions of accumulation and age. In *Karst of the Altai-Sayan Mountainous Area and Adjacent Mountainous Countries*. Barnaul, pp. 35–38. (In Russian.)

Marean, C. W. (1991). Measuring the postdepositional destruction of bone in archaeological assemblages. *Journal of Archaeological Science*, **18**:677–694.

Marean, C. W., and L. Bertino (1994). Intrasite spatial analysis of bone: Subtracting the effect of secondary carnivore consumers. *American Antiquity*, **59**(4):748–768.

Marinina, A. M. (1966). Karsk processes in the basin of rivers Kaysla and Sarala. In *Problems of Geology and Geomorphology of West Siberia*. Barnaul, p. 24.

Markin, S. V. (2000). The Neolithic burial of northwestern Altai. *Archaeology, Ethnology & Anthropology of Eurasia*, **2**(2):53–64.

Markin, S. V., E. J. T. Jull, L. A. Orlova, and Y. V. Kuzmin (2001). Interpretation of new radiocarbon dating for the Kaminnaya site (northwestern Altai). In *Modern Problems of Eurasian Paleolithic Science*, eds. A. P. Derevyanko and G. I. Medvedev. Novosibirsk: Institute of Archaeology and Ethnography, pp. 262–266. (In Russian.)

Martin, P. S. (1984). Prehistoric overkill: The global model. In *Quaternary Extinctions: A Prehistoric Revolution*, eds. P. S. Martin and R. G. Klein. Tucson, AZ: University of Arizona Press, pp. 354–403.

Martynovich, N. V. (1998). Birds of the Late Pleistocene from Okladnikova Cave as a source for paleolandscape reconstructions. In *Complex Studies of Paleolithic Objects of the Anuj River Basin*, eds. R. S. Vasilievsky and Yu. P. Kholyushkin. Novosibirsk: Institute of Archaeology and Ethnography, pp. 66–81. (In Russian.)

Maschenko, E. N., and S. V. Leshchinskiy (2001). Composition and morphology of mammoth remains of the Volchiya Griva locality. In *Proceedings of the 2nd International Symposium, Evolution of Life on the Earth*, November 12–15, ed. V. M. Podobina, pp. 507–511. (In Russian.)

Maschenko, E. N., and S. V. Leshchinskiy (2002). The first find of a lower jaw of a mammoth fetus (Proboscidea, *Mammuthus*) in the Upper Pleistocene of the Tomsk region, Russia. *Paleontological Journal*, **36**(2):195–203.

Medvedev, G. I. (1998). Upper Paleolithic sites in south-central Siberia. In *The Paleolithic of Siberia: New Discoveries and Interpretation*, eds. A. Derev'anko, D. B. Shimkin, and W. R. Powers (trans. I. P. Laricheva). Novosibirsk: Institute of Archaeology and Ethnography, Siberian Division, Russian Academy of Sciences; Urbana and Chicago, IL: University of Illinois Press, pp. 122–132.

Medvedev, G. I., N. Kov, G. Vorobieva, *et al.* (1996). *Malta Paleolithic Locality.* Irkutsk: Arkom.

Medvedev, G. I., N. Kov, G. Vorobieva, *et al.* (1998a). *The Mal'ta Paleolithic Location*. Brussels: Korolevski Museum of Art and History of Brussels; Irkutsk: Department of Archaeology, Irkutsk State University.

Medvedev, G. I., G. A. Vorobieva, V. Sitlivy, *et al.* (1998b). New look on the dynamics of the formation of cultural layers at the Mal'ta Palaeolithic site. In *Pleistocene Paleoecology and Stone Age Cultures of Northern Asia and Adjacent Territories*, ed. A. P. Derevyanko. Novosibirsk: Russian Academy of Sciences, Siberian Branch, Institute of Archaeology and Ethnography, p. 160.

Melkheyev, M. N. (1965). Nizneudinskaya Cave. *News of the East-Siberian Department of the Geographic Society of the USSR*, **63**:177–189. (In Russian.)

Mellers, P. (1996). *The Neanderthal Legacy: An Archaeological Perspective from Western Europe*. Princeton, NJ: Princeton University Press.

Mills, M. G. L. (1990). *Kalahara Hyaenas: Comparative Behavioural Ecology of Two Species*. London: Unwin Hyman.

Mochanov, Y. A. (2007). *Dyuktai Bifacial Tradition of North Asia Paleolith (Its Definition and Research History)*. Yakutsk: Academy of Sciences, Sakha Republic (Yakutia), Center of Arctic Archaeology and Human Paleoecology. (In Russian.)

Mochanov, Y. A. (2010). *Fifty Years in the Stone Age of Siberia (Archaeological Studies in the Asian Part of Russia)*, vols. **1** and **3**. Yakutsk: Academy of Sciences, Sakha Republic (Yakutia), Center of Arctic Archaeology and Human Paleoecology. (In Russian.)

Mochanov, Y. A., and S. A. Fedoseeva (trans, R. L. Bland) (2010). *Archaeology, the Paleolithic of Northeast Asia, a Non-Tropical Origin for Humanity and the Earliest Stages of the Settlement of America*. Burnaby, BC: Archaeology Press.

Morey, D. F. (2010). *Dogs: Domestication and the Development of a Social Bond*. Cambridge: Cambridge University Press.

Mott, M., and J. K. Kozlowsky (2005). The transition from the Middle to the Upper Paleolithic in north Eurasia. In *The Middle to Upper Paleolithic Transition in Eurasia: Hypotheses and Facts*, ed. A. P. Derevianko. Novosibirsk: Institute of Archaeology and Ethnography Press, pp. 20–31.

Movius, H. L., Jr., ed. (1975). *Excavation of the Abri Pataud, Les Eyzies (Dordogne)*. Cambridge, MA: Peabody Museum of Archaeology and Ethnology, Harvard University.

Nikolaev, S. V., and S. V. Markin (1990). First materials of the Lower Paleolithic in southeastern West Siberia. In *Chronostratigraphy of Paleolithic of North, Central, and East Asia and America*. Novosibirsk: Reports of International Symposium, pp. 242–245.

O'Connor, T. (2000). *The Archaeology of Animal Bones*. College Station, TX: Texas A&M University Press.

O'Hanlon, L. (2010). Ancient humans may have dined on hyenas. http://news.discovery.com

Okladnikov, A. P. (1983). The Paleolithic site of Kara-Bom in Gorny Altai (based on excavations in 1980). In *The Paleolithic of Siberia*. Novosibirsk, pp. 5–20. (In Russian.)

Okladnikov, A. P., and I. I. Kirilov (1980). *The Southeastern Trans-Baikal Area in the Stone and Bronze Epoch*. Novosibirsk: Nauka. (In Russian.)

Okladnikov, A. P., and N. D. Ovodov (1972). A new Paleolithic monument in the Altai with Levallois technique of stone flaking. In *Archaeology and History of the Altai*. Barnaul, pp. 3–6. (In Russian.)

Okladnikov, A. P., and N. D. Ovodov (1977). The Paleolithic site of Denisova Cave in the Altai. In *Archaeological Discoveries*. Moscow: Nauka, pp. 266–268. (In Russian.)

Okladnikov, A. P., V. M. Muratov, N. D. Ovodov, and E. O. Fridenberg (1973). Straschnaya Cave: A new monument of the Altai Paleolithic. In *Materials on Archaeology of Siberia and the Far East*, vol. **2**. Novosibirsk, pp. 3–54. (In Russian.)

Okladnikov, A. P., N. D. Ovodov, and S. A. Rybakov (1975). Proskuryakov grotto: A new Paleolithic site in Khakasia. *Bulletin of the Commission on Studying the Quaternary Period*, **44**:111–117. (In Russian.)

Olson, S. J. (1985). *Origins of the Domestic Dog: The Fossil Record*. Tucson, AZ: University of Arizona Press.

Olson, S. J., and J. W. Olson (1977). The Chinese wolf, ancestor of New World dogs. *Science*, **197**:533–535.

Orlova, L. A., Y. V. Kuzmin, V. S. Volkova, and I. D. Zolnikov (2000a). Mammoth (*Mammuthus primiogenius* Blum.) and primitive map [sp. correct] in Siberia: Joint analysis of habitats of populations on the basis of C-14 dates. In *Problems of Reconstruction of Climate and Environment of Siberian Holocene and Pleistocene*, vol. **2**, ed. V. S. Zykin. Novosibirsk: Archaeology and Ethnography Press, pp. 1–11. (Text in English, tables and figures in Russian.)

Orlova, L. A., S. V. Leshchinskiy, V. N. Zenin, and M. A. Borisov (2000b). Radiocarbon and stratigraphic studies of the Volchiya Griva site in 2000. In *Problems of Archaeology, Ethnography, and Anthropology of Siberia and Adjacent Territories*, vol. **6**. Novosibirsk: Institute of Archaeology and Ethnography, pp. 188–191. (In Russian.)

Otte, M., and J. K. Kozlowski (2001). The transition from the Middle to Upper Paleolithic in north Eurasia. *Archaeology, Ethnology & Anthropology of Eurasia*, **3**(7):51–62.

Ovodov, N. D. (1970). Remains of brown bears (*Ursus arctos L.*) in the caves of Siberia and the Far East. *Bulletin of the Moscow Society of Nature Explorers: Biological Section*, **75**(4):116–126. (In Russian.)

Ovodov, N. D. (1972). Late Anthropogene mammalian fauna from caves of West Altai. In *Archaeology and History of the Altai*. Barnaul, pp. 11–15. (In Russian.)

Ovodov, N. D. (1973a). Late Pleistocene mammalian fauna in Straschnaya and Logovo Gieny caves of the Altai Mountains and the problem of paleogeography. *Sixth International Congress of Speleology, Abstracts*. Olomouc, pp. 158–159.

Ovodov, N. D. (1973b). Original peopling of Siberia based on materials of speleological study. In *Ethnogenesis of the Peoples of North Asia*. Novosibirsk, pp. 21–23. (In Russian.)

Ovodov, N. D. (1973c). Comparative taphonomic features of Siberian caves based on mammalian remains. In *Materials on Archaeological Research in Siberia and the Far East*, vol. **2**. Novosibirsk, pp. 328–338. (In Russian.)

Ovodov, N. D. (1973d). Taphonomy of caves of Siberia and the Far East based on mammalian remains, and the problem of peopling of caves by Paleolithic man. *International Congress of Speleology: Abstracts of Papers*. Olomouc, pp. 158–159.

Ovodov, N. D. (1974). Subfossil bat remains from caves in Siberia and the Far East. In *Proceedings of the First All-Union Conference on Bats*. Leningrad, pp. 84–91. (In Russian.)

Ovodov, N. D. (1975). Fauna of Paleolithic sites of Siberia and problems of chronological and paleoenvironmental interpretations. In *Correlation of Ancient Cultures of Siberia and Adjacent Territories of the Pacific Ocean Basin*. Novosibirsk, pp. 35–51. (In Russian.)

Ovodov, N. D. (1977a). Cave behavior of brown bears. *Natura [Priroda]*, **9**:126–128. (In Russian.)

Ovodov, N. D. (1977b). Late Anthropogene mammalian fauna in the southern Ussurijsk region. In *Fauna and Systematics of Vertebrates of Siberia*, ed. B. S. Yudin. Novosibirsk: Nauka, pp. 157–176. (In Russian.)

Ovodov, N. D. (1977c). Late Pleistocene mammalian fauna from Straschnaya and Logovo Gieny caves in the Altai and problems of paleogeography. *Proceedings of the Sixth International Congress of Speleology*, vol. **5**, ed. V. Panos. Praha: Academia-Praha, pp. 293–298.

Ovodov, N. D. (1979). Paleofaunal study of caves. In *General Methods of Studying the History of Modern Ecosystems*. Moscow: Nauka, pp. 102–128. (In Russian.)

Ovodov, N. D. (1987a). Fauna of the Paleolithic settlements Tolbaga and Varvarina Gora in West Trans-Baikal. In *Environments and Ancient Man in the Late Anthropogene*, ed. I. N. Rezanov. Ulan-Ude, pp. 122–140. (In Russian.)

Ovodov, N. D. (1987b). A wrong trend in the estimation (identification) of osteological material from Paleolithic caves. *1987 Conference Proceedings*, ed. I. A. Pecherkin. Kiev: Institute of Geological Sciences, pp. 144–145. (In Russian.)

Ovodov, N. D. (1991). Razboinich'ya Cave. In *Trip Guide for the Altai: Archaeology and Paleoecology of the Paleolithic of Gorny Altai. Excursion No. 1. Guidebook of the International Symposium "Chronostratigraphy of the Paleolithic of Northern, Central, Eastern Asia and America (Paleoecological Aspect)."* eds. A. P. Derevianko, Yu. V. Grichan, M. I. Dergacheva, and A. N. Zenin. Novosibirsk: Institute of History, Philosophy and Philology, Siberian Branch, U.S.S.R. Academy of Science, pp. 98–104. (In Russian.)

Ovodov, N. D. (1997a). Cave hyena and Paleolithic man. In *First International Symposium "Evolution of Life on the Earth."* Tomsk, pp. 125–126. (In Russian.)

Ovodov, N. D. (1997b). Taphonomy of caves of Siberia and the Far East from remains of mammals, and problems of peopling caves by Paleolithic man. *Proceedings of the Sixth International Congress of Speleology*, vol. **6**, ed. V. Panos, Praha: Academia-Praha, pp. 47–51.

Ovodov, N. D. (1999). Hunting with dogs for mammoth. In *Ecology of Ancient and Modern Societies*, ed. N. P. Matveyeva. Tyumen: Institute for Problems of Colonization of the North, pp. 79–82. (In Russian.)

Ovodov, N. D., and N. G. Ivleva (1985). Pleistocene fauna of Denisova Cave (Altai) from data of 1982–1984 excavations. In *Quaternary Geology and Primitive Archaeology of Southern Siberia*, vol. **1**. Ulan-Ude, pp. 88–90. (In Russian.)

Ovodov, N. D., and Y. V. Kuzmin (2006). Did Neanderthals have dogs? In *Almanac Eniseiskaya Provintsiya*, vol. 2. Krasnoyarsk, pp. 171–177. (In Russian.)

Ovodov, N. D., and N. V. Martynovich (1992). New data on mammals and birds of the Dvuglaska grotto in Khakasia. In *Problems of Archaeology, Ethnography and History of Near Yenisei Region*. Krasnoyarsk, pp. 78–83. (In Russian.)

Ovodov, N. D., and N. V. Martynovich (1994). Yeleneva Cave as a unique Paleofaunistic indicator of the Yenisei water level. In *All-Russian Conference on the Study of the Quaternary Period*. Moscow, p. 182. (In Russian.)

Ovodov, N. D., and N. V. Martynovich (1999). Activity of early Holocene people in the Middle Yenisei, with respect to useful vertebrate animals. In *The Ecology of Ancient and Modern Societies*, ed. N. P. Matveyeva. Tyumen: Institute for Problems of Colonization of the North, pp. 107–109. (In Russian.)

Ovodov, N. D., and N. V. Martynovich (2000). Vertebrates from caves of the Birysa Karst area. In *Problems of the Reconstruction of Climate and Environments in Holocene and Pleistocene Siberia*, vol. **2**, ed. S. V. Markin. Novosibirsk: Institute of Archaeology and Ethnography, pp. 373–382. (In Russian.)

Ovodov, N. D., and N. V. Martynovich (2004). Okladnikov Cave in the Altai: Preliminary taphonomic evaluation. In *Problems of Archaeology, Ethnography, and Anthropology of Siberia and Adjacent Territories*, vol. **10**, part 2. Novosibirsk: Institute of Archaeology and Ethnography, pp. 175–184.

Ovodov, N. D., and N. V. Martynovich (2005). Some oddities in the behavior of cave hyenas (*Crocuta spelaea Goldf.*). In *Problems of Archaeology, Ethnography, and Anthropology of Siberia and Adjacent Territories*, vol. **11**, part 1. Novosibirsk: Institute of Archaeology and Ethnography, pp. 181–183.

Ovodov, N. D., and N. V. Martynovich (2009). To find human tooth in the hyena's coprolite. In *Caves: Protection, History, Present Conditions and Perspectives of Scientific Research in the Caves on the Territory of the Former USSR*, ed. I. Burmak, Krasnoyarsk, pp. 66–70. (In Russian.)

Ovodov, N., and N. Orlova (1977). Fauna of mammals of the eastern slopes of Kuznetsky Ala-Tau in the late Holocene–Pleistocene. In *Volume of Papers: Ecology, Methods of Studies and Organization for Protection of Mammals in Mountainous Areas*. Sverdlovsk, pp. 1–2.

Ovodov, N. D., V. M. Muratov, V. A. Panichev, and L. A. Orlova (1992a). New data on geology and teriofauna of Proskuryakov grotto (Khakasia). In *Peotr Alekseyevich Kropotkin: Humanist, Scientist, Revolutionary*, ed. M. V. Konstantinov. Chita: Chita Pedagogical Institute, pp. 43–45. (In Russian.)

Ovodov, N. D., N. V. Martynovich, A. A. Pozdnyakov, and L. A. Orlova (1992b). Mammals and birds in the surroundings of Krasnoyarsk at the Pleistocene–Holocene boundary. In *Paleoecology and Peopling of Ancient Man in North Asia and America*, ed. A. P. Derevianko. Krasnoyarsk, pp. 197–200. (In Russian.)

Ovodov, N. D., N. V. Martynovich, and A. Nadakhovsky (1999). Taphonomic features of the Razboinich'ya Cave (Altai). In *Papers of the Sixth Regional Scientific-Practical and Methodical Conference*, ed. A. Krasnoyarsk, pp. 65–66. (In Russian.)

Ovodov, N. D., N. V. Martynovich, and L. A. Orlova (2003). Fauna of the Malayamanskaya Cave (Central Altai). In *Problems of Archaeology, Ethnography, and Anthropology of Siberia and Adjacent Territories*, vol. **1**. Novosibirsk: Institute of Archaeology and Ethnography, pp. 192–196. (In Russian.)

Ovodov, N. D., Y. V. Kuzmin, and R. J. Cruz (2010). AMS 14C dating of "exotic" mammals from south Siberian sites. *Current Research in the Pleistocene*, **27**:181–183.

Ovodov, N. D., S. J. Crockford, Y. V. Kuzmin, *et al.* (2011). A 33,000-year-old incipient dog from the Altai Mountains of Siberia: Evidence of the earliest domestication disrupted by the Last Glacial Maximum. *PLoS One*, **6**(7):1–7.

Owens, M., and D. Owens (1984). *Cry of the Kalahari*. Boston, MA: Houghton Mifflin Co.

Panichev, A. M. (1990). *Lithophagy in the Life of Animals and Human Beings*. Moscow: Nauka. (In Russian.)

Panteleyev, A. V. (2002). Bone remains of birds from Pleistocene sediments of Denisova Cave. In *Problems of Archaeology, Ethnography, and Anthropology of Siberia and Adjacent Territories*, vol. **8**, eds. A. P. Derevianko and V. I. Molodin. Novosibirsk: Institute of Archaeology and Ethnography, pp. 173–178. (In Russian.)

Pavlova, O. (2004). *Green Miles*. Novosibirsk: Ed. S. Popkov. (In Russian).

Pavlova, O. (2008). *Daria's Garden*. Novosibirsk: Svinyn and Sons. (In Russian).

Pavlova, O. (2011). *Colors of My Life*. Novosibirsk: Svinyn and Sons. (In Russian).

Payne, S., and P. J. Munson (1985). Ruby and how many squirrels? Animal bones from Asiki Huyuk. *Anatolian Studies*, **35**:109–122.

Pemberton, M. (2010). Man mauled twice by bear. *Arizona Republic*, June 25, p. A10.

Perkins, S. (2001). Ancient whodunit: Scientists indict wee suspects in ancient deaths. *Science News*, **161**:275–276.

Pickering, T. R. (2002). Reconsideration of criteria for differentiating faunal assemblages accumulated by hyenas and hominids. *International Journal of Osteoarchaeology*, **12**:127–141.

Pickering, T. R., and J. Wallis (1997). Bone modifications resulting from captive chimpanzee mastication: Implications for the interpretation of Pliocene archaeological faunas. *Journal of Archaeological Science*, **24**:1115–1127.

Pickering, T. R., R. J. Clarke, and J. Moggi-Cecchi (2004). Role of carnivores in the accumulation of the Sterkfontein Member 4 Hominid assemblage: A taphonomic reassessment of the complete hominid fossil sample (1936–1999). *American Journal of Physical Anthropology*, **125**(1):1–15.

Pickrell, J. (2002). Rebranding the hyena. *Science News*, **161**:267–269.

Pijoan, C. M. A. (1997). Evidencias de Sacrificio Humano y Cannibalismo en Restos Oseos: El Caso del Entierro Numero 14 de Tlatelolco, D.F. Doctoral dissertation, Mexico, Universidad Nacional Autynoma de Mexico.

Plummer, T. (2004). Flaked stones and old bones: Biological and cultural evolution at the dawn of technology. *Yearbook of Physical Anthropology*, **47**(39):118–164.

Popov, A. N., and D. R. Yesner (2006). Early maritime adaptation on the southern coast of the Far East of Russia in ancient times. In *Beyond the Steppe and the Sown: Proceedings of the 2002 University of Chicago Conference on Eurasian Archaeology (Colloquia Pontica 13)*, eds. D. L. Peterson, L. M. Popova, and A. T. Smith. Boston, MA: Brill, pp. 469–476.

Popov, A. N., T. A. Chikisheva, and E. G. Shpakova (1997). *The Boisman Archaeological Culture of Southern Primary: The Multilayered Site of Boisman-2*. Novosibirsk: Institute of Archaeology and Ethnography, Russian Academy of Sciences, Siberian Branch. (In Russian.)

Postnov, A. V., and Y. B. Karavaeva (2005). A comparative analysis (description) of the finds from the multilayered site, Ust-Kan Cave. In *Roots, Formation, and Development of Eurasian Polcultures: Cultures and Societies of North Asia in Historical Past and at Modern Times*. Irkutsk, pp. 149–150.

Postnov, A. V., and N. A. Kulik (1998). On the problem of the origin of Ust-Kan Cave. In *Problems of Archaeology, Ethnography, and Anthropology of Siberia and Adjacent Territories*, vol. **4**. Novosibirsk: Institute of Archaeology and Ethnography, pp. 67–70. (In Russian.)

Postnov, A. V., and N. A. Kulik (2003). Reconstruction of water regime in the Ust-Kan Cave from results of petrographic and mineralogic studies of plates surrounding the hearth (layer no. 9). In *Problems of Archaeology, Ethnography, and Anthropology of Siberia and Adjacent Territories*, vol. **9**, part 1. Novosibirsk: Institute of Archaeology and Ethnography, pp. 197–201. (In Russian.)

Prashkevich, G. M. (2000). Ivan Antonovich Yefremov. In *The Most Famous Scientists of Russia*, ed. G. M. Prashkevich. Moscow: Veche, pp. 537–550. (In Russian.)

Prashkevich, G. M. (2002). *Numerous Perfect Things: Part II. People. Ivan Antonovich Yefremov*. Moscow: ESLI(IF). (In Russian.)

Proskuryakov, P. S. (1889). The caves of Iuys (the letter by Proskuryakov of 18.11.1888 from Krasnoyarsk). *News of Eastern-Siberian Department of the Russian Geographic Society*, **20**(2):32–34. (In Russian.)

Proskuryakov, P. S. (1890). Report on the preliminary study of the Iuys caves. *News of the Eastern-Siberian Department of the Russian Geographic Society*, **21**(4):20–25. (In Russian.)

Proskuryakov, P. S. (1893). On evidences of the post-Pliocene epoch in the vicinity of Krasnoyarsk: Torgashinskaya cave. In *Reports of the Society of Doctors of the Yenisei Guberniya (region) for the Period of 1892–1893*. Krasnoyarsk. (In Russian.)

Quammen, D. (2003). *Monster of God: The Man-Eating Predator in the Jungles of History and the Mind*. New York: W. W. Norton.

Rouse, I. (1986). *Migrations in Prehistory: Inferring Population Movement from Cultural Remains*. New Haven, CT: Yale University Press.

Rudenko, S. I. (1960). Ust-Kan: A Paleolithic cave site. In *Materials and Studies on the Archaeology of the USSR*. Moscow and Leningrad: U.S.S.R. Academy of Sciences, pp. 104–125. (In Russian.)

Rudenko, S. I. (trans. C. S. Chard and H. Marie Wormington) (1961). Ust-Kanskaia Paleolithic Cave site, Siberia. *American Antiquity*, **27**(2):203–215.

Russell, M. D. (1987). Bone breakage in the Krapina hominid collection. *American Journal of Physical Anthropology*, **72**(3):373–379.

Saks, V. N., ed., I. A. Volkov, and S. A. Arkhipov, compilers (1978). *The Quaternary Deposits in the Novosibirsk Area (Informational Material)*. Novosibirsk: U.S.S.R. Academy of Sciences, Siberian Branch, Institute of Geology and Geophysics. (In Russian and English.)

Savolainen, P., Y.-P. Zhang, K. Luo, *et al.* (2002). Genetic evidence of an East Asian origin of domestic dogs. *Science*, **298**:1610–1636.

Scott, G. R., and C. G. Turner II (1997). *The Anthropology of Modern Human Teeth: Dental Morphology and Its Variation in Recent Human Populations*. Cambridge: Cambridge University Press.

Selvaggio, M. M. (1994). Carnivore tooth marks and stone tool butchery marks on scavenged bones: Archaeological implications. *Journal of Human Evolution*, **27**:215–228.

Shipman, P. (1981). *Life History of a Fossil*. Cambridge, MA: Harvard University Press.

Shmygun, P. E. (1978). Works on the northern Baikal. In *Archaeological Discoveries of 1977*. Moscow, p. 285. (In Russian.)

Shmygun, P. E. (1981). Pre-ceramic complexes from Quaternary sediments of northern Baikal. In *Relief and Quaternary Sediments of the Stanovoi Ridge*. Moscow: Nauka, pp. 121–128. (In Russian.)

Shmygun, P. E., and A. K. Filippov (1982). The lower complex of the Kurla sites. In *Material Culture of the Ancient Population of Eastern Siberia*. Irkutsk: Irkutsk University Press, pp. 15–24. (In Russian.)

Shmygun, P. E., and A. M. Sizikov (1977). Work on the northern Baikal. In *Archaeological Discoveries of 1976*. Moscow, pp. 254–255. (In Russian.)

Shmygun, P. E., and A. S. Yendrikhinsky (1978). The Kurla non-ceramic complex in northern Baikal (preliminary report). In *Ancient History of the Peoples of the Southern Part of Eastern Siberia*, **4**:46–69. (In Russian.)

Shpakova, E. G. (2001). Paleolithic human dental remains from Siberia. *Archaeology, Ethnology & Anthropology of Eurasia*, **4**(8):64–76.

Shpakova, E. G. (2005). Paleolithic human dental remains from Siberia. In *The Middle to Upper Paleolithic Transition in Eurasia: Hypotheses and Facts*, ed. A. P. Derevianko. Novosibirsk: Institute of Archaeology and Ethnography Press, pp. 414–426.

Shpakova, E. G., and A. P. Derevianko (2000). The interpretation of odontological features of Pleistocene human remains from the Altai. *Archaeology, Ethnology & Anthropology of Eurasia*, **1**(1):125–138.

Shunkov, M. V., and A. K. Agadjanyan (2000). Paleogeography of the Paleolithic of Denisova Cave. *Archaeology, Ethnography & Anthropology of Eurasia*, **2**(2):2–19.

Skinner, J. D., M. A. Haupt, M. Hoffmann, and H. M. Dott (1998). Bone collecting by brown hyaenas *Hyaena brunnea* in the Namib Desert: Rate of accumulation. *Journal of Archaeological Science*, **25**:69–71.

Slodkevich, V. S. (1936). Results of excavations of Quaternary mammals in Nizhneudinsk Cave (East Siberia). *Proceedings of the Paleozoological Institute of the U.S.S.R. Academy of Sciences* **5**:235–241. (In Russian.)

Soffer, O. (1985). *The Upper Paleolithic of the Central Russian Plain*. Orlando, FL: Academic Press.

Soffer, O. (1987). *The Pleistocene Old World: Regional Perspectives*. New York: Plenum Press.

Sokolova, Z. P. (2000). The bear cult. *Archaeology, Ethnology & Anthropology*, **2**(2):121–130.

Stringer, C., and C. Gamble (1995). *In Search of the Neanderthals: Solving the Puzzle of Human Origins*. New York: Thames and Hudson.

Tattersall, I., and J. H. Schwartz (2000). *Extinct Humans*. New York: Westview Press.

Thiessen, M. (2008). Grizzly bear mauls woman returning to Alaskan Lodge. *Arizona Republic*, July 25, p. A8.

Thornton, M., and J. Fee (2001). Rodent gnawing as a taphonomic agent: Implications for archaeology. In *People and Wildlife in Northern North America: Essays in Honor of R. Dale Guthrie*, eds. S. C. Gerlach and M. S. Murray. Oxford: Archaeopress, n.p.

Travis-Henikoff, C. A. (2008). *Dinner with a Cannibal: The Complete History of Mankind's Oldest Taboo*. Santa Monica, CA: Santa Monica Press.

Trinkaus, E., and J. Svoboda (2005). *Early Modern Human Evolution in Central Europe: The People of Dolni Vestonice and Pavlov.* Oxford: Oxford University Press.

Tseitlin, S. M. (1974). Mal'ta. In *Geology of the Paleolithic of North Asia*, ed. Yu. M. Vasiliev. Moscow: Nauka, pp. 177–182. (In Russian.)

Turner, C. G. II (forthcoming). The Sopka dentition. *A Comparative Study of West Siberian Dental Characteristics*, ed. Y. Molodin. Novosibirsk: Institute of Archaeology and Ethnography.

Turner, C. G. II (1983a). Sinodonty and Sundadonty: A dental anthropological view of Mongoloid microevolution, origin, and dispersal into the Pacific basin, Siberia, and the Americas. In *Late Pleistocene and Early Holocene Cultural Connections of Asia and America*, ed. R. S. Vasilievsky. Novosibirsk: USSR Academy of Sciences, Siberian Branch, pp. 72–76. (In Russian.)

Turner, C. G. II (1983b). Taphonomic reconstruction of human violence and cannibalism based on mass burials in the American Southwest. In *Carnivores, Human Scavengers, and Predators: A Question of Bone Technology*, eds. G. M. LeMoine and A. S. MacEachern. Calgary: Archaeological Association of the University of Calgary, pp. 219–240.

Turner, C. G. II (1983c). Dental evidence for the peopling of the Americas. In *Early Man in the New World*, ed. R. Shutler, Jr. Beverly Hills, CA: Sage Publications, pp. 147–157.

Turner, C. G. II (1984). Black Sea to Bering Strait: Review of physical anthropology of Soviet Asia. *Quarterly Review of Archaeology*, **5**(3):5.

Turner, C. G. II (1990a). Upper Paleolithic Mal'ta child (Siberia). *News of Siberian Branch, U.S.S.R. Academy of Sciences*, **2**:70–71. (In Russian.)

Turner, C. G. II (1990b). Paleolithic Siberian dentition from Denisova and Okladnikov caves, Altayskiy Kray, U.S.S.R. *Current Research in the Pleistocene*, **7**:65–66.

Turner, C. G. II (1990c). Paleolithic teeth of the Central Siberian Altai Mountains. In *International Conference on Chronostratigraphy of the Paleolithic in North, Central, East Asia and America.* Novosibirsk: Institute of History, Philology and Philosophy, Siberian Branch of the Academy of Sciences, pp. 239–243.

Turner, C. G. II (1991). *The Dentition of Arctic Peoples.* New York: Garland Publishing, Inc.

Turner, C. G. II (1992a). New World origins: New research from the Americas and the Soviet Union. In *Ice Age Hunters of the Rockies*, eds. D. J. Stanford, and J. S. Day. Niwot, CO: Denver Museum of Natural History and University Press of Colorado, pp. 7–50.

Turner, C. G. II (1992b). Microevolution of East Asian and European populations: A dental perspective. In *The Evolution and Dispersal of Modern Humans in Asia*, eds. T. Akazawa, K. Aoki, and T. Kimura. Tokyo: Honkusen-Sha Pub. Co., pp. 415–443.

Turner, C. G. II (1995). Shifting continuity: Modern human origin. In *The Origin and Past of Modern Humans as viewed from DNA*, eds. S. Brenner and K. Hanihara. Singapore: World Scientific, pp. 216–243.

Turner, C. G. II (1998). An update on dental anthropological interpretation of the people of Siberia and the Americas. In *Pleistocene Paleoecology and Stone Age Cultures of Northern Asia and Adjacent Territories*, ed. A. P. Derevianko. Novosibirsk: Institute of Archaeology and Ethography, Siberian Branch of the Russian Academy of Sciences, pp. 180–188.

Turner, C. G. II (2002). Teeth, needles, dogs, and Siberia: Bioarchaeological evidence of the colonization of the New World. In *The First Americans*, ed. N. Jablonski. San Francisco, CA: University of California Press, pp. 123–158.

Turner, C. G. II (2003). Mystery of Yelenev Cave. *Day and Night*, **39**(1–2):317–319. (In Russian.)

Turner, C. G. II (2007). *Bioarchaeology.* Novosibirsk: Institute of Archaeology and Ethnography, Siberian Branch, Russian Academy of Sciences. (In Russian.)

Turner, C. G. II (2010). The Russian–American perimortem taphonomy project in Siberia: A tribute to Nicolai Dimitrevich Ovodov, pioneering Siberian vertebrate paleontologist-cave archaeologist. *Alaska Journal of Anthropology*, **8**:87–105.

Turner, C. G. II and J. A. Turner (1974) Progress report on evolutionary anthropological study of the Akun Strait district, eastern Aleutionas, Alaska, 1970–1971. *Anthropological Papers of the University of Alaska*, **16**(1):27–57.

Turner, C. G. II, and J. A. Turner (1999). *Man Corn: Cannibalism and Violence in the Prehistoric American Southwest*. Salt Lake City, UT: University of Utah Press.

Turner, C. G. II, C. R. Nichol, and G. R. Scott (1991). Scoring procedures for key morphological traits of the permanent dentition: The Arizona State University dental anthropology system. In *Advances in Dental Anthropology*, ed. M. A. Kelley and C. S. Larsen. New York: Wiley-Liss, pp. 13–31.

Turner, C. G. II, N. D. Ovodov, N. V. Martynovich, *et al.* (2000). Perimortem taphonomy of Late Pleistocene Siberia faunal assemblages from natural and cultural settings. In *Program of International Scientific Conference "Paleogeography of the Stone Age: Correlation of Natural Events and Archaeological Cultures of Paleolithic Period of Northern Asia and Adjoining Territories"*, ed. N. I. Drozdov. Krasnoyarsk: Krasnoyarsk Pedagogical University, pp. 70–71.

Turner, C. G. II, N. D. Ovodov, N. V. Martynovich, *et al.* (2001a). Perimortem bone taphonomy of late Pleistocene human and hyena refuse deposits in Siberia. *Current Research in the Pleistocene*, **18**:88–90.

Turner, C. G. II, N. D. Ovodov, and O. V. Pavlova (2001b). Animal remains from open (Ust-Kova) and cave (Ust-Kan) sites. In *Program of Work of the International Scientific Symposium: Current Scientific Problems of the Eurasian Paleolithic Dedicated to the 130th Anniversary of the Discovery of the First Paleolithic Site in Russia, "The Military Hospital," in 1871*. August 1–9. Irkutsk-Usolie Sibirskoye, pp. 421–423.

Turner, C. G. II, N. D. Ovodov, and O. V. Pavlova (2003). Perimortem taphonomy and dental affinity assessment of the human skeletal remains from Yelenev Cave, Krasnoyarsk Territory, Siberia. *Archaeology, Ethnology & Anthropology of Eurasia*, **1**(3):49–57.

Ubelaker, D. H. (1989). *Human Skeletal Remains: Excavation, Analysis, Interpretation*. Second Edition. Washington: Taraxacum..

Ukralutseva, V. V., L. Agenbroad, and J. I. Mead (1996). A paleoenvironmental reconstruction of the "Mammoth Epoch" of Siberia. In *American Beginnings: The Prehistory and Paleoecology of Beringia*, ed. F. H. West. Chicago, IL: University of Chicago Press, pp. 129–146.

van der Post, L. (1961). *The Heart of the Hunter*. New York: William Morrow and Co.

van Lawick, H., and J. van Lawick-Goodall (1970). *Innocent Killers*. New York: Ballantine Books.

Vasili'ev, S. A. (1996). *Late Paleolithic of the Upper Yenisei (as Represented by Multicomponent Sites Near the Maina Village)*. St. Petersburg: Russian Academy of Sciences, Institute for the Material Culture History, and Romisch-Germanisches Zentral Museum Forschunginstitut fur Vor- und Frugeschichte. (In Russian.)

Vasili'ev, S. A. (2000). Hunted fauna and economic activity of Paleolithic inhabitants of Siberia. In *Paleogeography of the Stone Age: Correlation of Natural Events and Archaeological Cultures of the Paleolithic of North Asia and Adjacent Territories*. Krasnoyarsk, pp. 24–25. (In Russian.)

Vasili'ev, S. A., Y. V. Kuzmin, L. A. Orlova, and V. N. Dementiev (2002). Radiocarbon-based chronology of the Paleolithic in Siberia and its relevance to the peopling of the New World. *Radiocarbon*, **44**(2):503–530.

Vasiliev, S. K. (1995). Mammalian fauna of the last interglacial of Novosibirsk near the Ob area. In *Abstracts of the First International Mammoth Conference*, October 16–22. Saint Petersburg, pp. 604–605. (In Russian.)

Vasiliev, S. K. (2002). Large mammal fauna of Kazantsev and Kargin times of Novosibirsk near the Ob area (from materials of the Krasny Yar site). In *Problems of Archaeology, Ethnography and Anthropology of Siberia and Adjacent Territories*, vol. **8**. Novosibirsk: Institute of Archaeology and Ethnography, pp. 38–43. (In Russian.)

Vasilievsky, R. S. (trans. S. Bundev) (1999). Alexei Pavlovich Okladnikov. In *The Great Archaeologists*, vol. **2**, ed. T. Murray. Santa Barbara, CA: ABC-CLIO, pp. 567–580.

Vasilievsky, R. S., V. V. Burilov, and N. I. Drozdov (1988). *Archaeological Sites of the Northern Cis-Angara Area*. Novosibirsk: Nauka. (In Russian.)

Vereschagin, N. K. (1971). *Why Did Mammoths Go Extinct*? Lenigrad: Nauka. (In Russian.)

Vereschagin, N. K. (1981). *Notes of a Paleontologist: Prints of Our Ancestors*. Leningrad: Nauka. (In Russian.)

Vereschagin, N. K., and G. F. Baryshinkov (1982). Origin and composition of the mammoth fauna. In *Paleoecology of Beringia*, eds. D. M. Hopkins, J. V. Matthews, Jr., C. E. Schweger, and S. Young. New York: Academic Press, pp. 268–279.

Vereschagin, N. K., and G. F. Baryshinkov (1984). Quaternary mammalian extinctions in northern Eurasia. In *Quaternary Extinctions: A Prehistoric Revolution*, eds. P. S. Martin and R. G. Klein. Tucson, AZ: University of Arizona Press, pp. 483–516.

Volkov, P. V. (2006). *Experimental Archaeology*. Novosibirsk: Federal Agency for Science and Innovations. (In Russian.)

Werdelin, L., and N. Solounias (1991). *The Hyaenidae: Taxonomy, Systematics and Evolution*. Oslo: Universitetsforlaget.

West, D. (1990). Horse hunting, processing, and transport in the Middle Danube. In *Paleolithic in the Middle Danube Region: Anniversary Volume to Bohueslav Klima*, ed. J. S. Svoboda. Brno: Institute of Archaeology, pp. 209–246.

West, F. H. (ed.) (1996). *American Beginnings: The Prehistory and Paleoecology of Beringia*. Chicago, IL: University of Chicago Press.

White, T. (1992). *Prehistoric Cannibalism at Mancos 5MTUMR-2346*. Princeton, NJ: Princeton University Press.

Wilson, M. C. (1983). Canid scavengers and butchering patterns: Evidence from a 3000-year-old bison bone bed in Alberta. In *Carnivores, Human Scavengers, and Predators: A Question of Bone Technology*, eds. G. M. LeMoine and A. S. MacEachern. Calgary: Archaeological Association of the University of Calgary, pp. 95–139.

Wolfe, T. (1942). *You Can't Go Home Again*. Garden City, NY: Sun Dial Press.

Wrinn, P. J. (n.d.). Hominin foragers on the northern periphery: Subsistence and land use in the Upper Pleistocene of the Altai Mountains, Siberia. Uncompleted PhD Dissertation, University of Arizona, Tucson.

Wrinn, P. J. (2010). Middle Paleolithic settlement and land use in the Altai Mountains, Siberia. In *Settlement Dynamics of the Middle Paleolithic and the Middle Stone Age*, Vol. **3**, eds. N. J. Conard and A. Delagnes. Tubingen: Kerns Verlag, pp. 163–194.

www.abc.net (1999). Hyena's last laugh. abc.net.au/quantum/scripts99/9904/hyenascpt.htm

www.Herald-Sun.com (2005). Killer hyena hunted down in Malawi.

www.igorilla.com (1999). Three Ethiopians eaten by hyenas.

Yefremov, I. A. (1940). Taphonomy: new branch of paleontology. *Pan-American Geologist*, **74**(2):81–93.

Yendrikhinsky, A. S., V. A. Banchev, L. V. Plysnina, A. M. Sizikov, and P. E. Shmygun (1978). Geological structure and age of northern Baikalian paleontological-archaeological monument. *Reports of the U.S.S.R. Academy of Sciences*, **243**(3):719–722. (In Russian.)

Yesner, D. R. (2001). Studying northern bones: Perspectives on paleobiology and zooarchaeology in high latitude environments. In *People and Wildlife in Northern North America: Essays in Honor of R. Dale Guthrie*, eds. S. C. Gerlach and M. S. Murray. Oxford: Archaeopress, pp. 1–12.

Zenin, V. N., and S. I. Konovalenko (2001). Paleolithic cache from Shestakovo. In *Proceedings of the International Symposium Devoted to the 130th Anniversary of the Discovery of the Paleolithic in Russia*, eds. A. P. Derevianko and G. I. Medvedev, Russia 1–9 August, Irkutsk. Novosibirsk and Irkutsk: Institute of Archaeology and Ethnography, and the Irkutsk Laboratory of Archaeology and Paleoecology, pp. 154–159. (In Russian.)

Zenin, V. N. and E. N. Maschenko (2000). The geological features and faunal composition of large mammals from the Shestakovo site (Kemerovo region). In *Conference Devoted to 100th Anniversary of the Birthday of Professor V. A. Khakhlov*, eds. V. M. Podobina and S. A. Rodygin. Tomsk: Tomsk State University, pp. 130–132. (In Russian.)

Zenin, V. N., L. A. Orlova, and E. van der Pleicht (2000a). Radiocarbon dating of the Shestakovo site. *International Scientific Conference "Paleogeography of the Stone Age: Correlation of Natural Events and Archaeological Cultures of the Paleolithic Period of Northern Asia and Adjoining Territories"*, ed. N. I. Drozdov, July 22–31, Kurtak, pp. 61–62. (In Russian.)

Zenin, V. N., S. V. Leshchinskiy, I. N. Fedeneva, L. A. Orlova, and M. A. Borisov (2000b). About accumulation of mammoth bones in the district of the Shestakovo site. In *International Scientific Conference "Paleogeography of the Stone Age: Correlation of Natural Events and Archaeological Cultures of the Paleolithic Period of Northern Asia and Adjoining Territories"*, ed. N. I. Drozdov, July 22–31. Kurtak, pp. 59–60. (In Russian.)

Zoubov, A. A. (1977). Odontoglyphics: The laws of variation of the human molar crown microrelief. In *Orofacial Growth and Development*, eds. A. A. Dahlberg and T. M. Graber. The Hague: Mouton, pp. 269–282.

Zoubov, A. A. (2000). Morphological study of children's teeth from the Sunghir burial (Sunghir 2, Sunghir 3). In *Homo Sungirensis: Upper Paleolithic Man – Ecological and Evolutionary Aspects of the Investigation*, eds. T. I. Alexeeva and N. O. Bader. Moscow: Scientific World, pp. 256–268 (English summary pp. 268–270). (In Russian, with English summaries.)

Zudin, A. N., S. V. Nikolaev, L. I. Galkina, *et al.* (1983). Making grounds for a stratigraphic scheme of Neogenic and Quaternary sediments of the Kuznetsk Depression. In *Problems of Stratigraphy and Paleogeography of the Siberian Pleistocene*. Novosibirsk: Nauka, pp. 133–149. (In Russian.)

# Index

**Page numbers in bold denote figures**